FOOD HOARDING IN ANIMALS

STEPHEN B. VANDER WALL

FOOD HOARDING IN ANIMALS

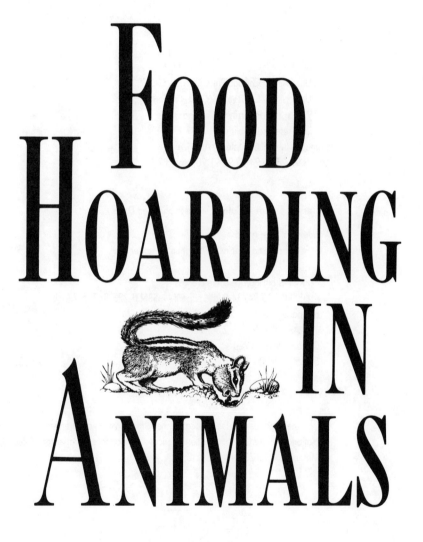

THE UNIVERSITY OF CHICAGO PRESS

Chicago and London

Stephen B. Vander Wall is assistant professor of biology at the University of Nevada, Reno.

The University of Chicago Press, Chicago 60637
The University of Chicago Press, Ltd., London
© 1990 by the University of Chicago
All rights reserved. Published 1990
Printed in the United States of America
99 98 97 96 95 94 93 92 91 90   5 4 3 2 1

Library of Congress Cataloging in Publication Data

Vander Wall, Stephen B.
    Food hoarding in animals / Stephen B. Vander Wall.
       p.   cm.
    Includes bibliographical references.
    ISBN 0-226-84734-9 (alk. paper). — ISBN 0-226-84735-7 (pbk. :
alk. paper)
    1. Animals—Food. 2. Animal behavior. I. Title.
QL756.5.V36 1990
591.53—dc20   89-20535
                                                      CIP

For Kathie and Erin

CONTENTS

8. *Food-Hoarding Mammals* 217

9. *Food-Hoarding Birds* 283

PREFACE

Food hoarding is an important component of the adaptive strategies of many animals. The benefits derived from hoarding food include improving an individual's chances of surviving a period of food scarcity, allowing animals to optimize foraging and feeding with regard to different criteria, improving an animal's competitive status when foraging for limited resources, and ensuring a continuous, even flow of food to young in the reproductive season. Awareness of the importance of hoarding in the survival and reproductive strategies of animals has grown in recent years, judging from the recent flush of research papers and major reviews, but for many species, the natural history and ecological importance of hoarding are still poorly understood. The literature related to food storing is large and widely scattered in journals in fields such as mammalogy, ornithology, entomology, ecology, behavior, psychology, physiology, botany, and forestry. The distribution of the hoarding literature itself reflects the broad taxonomic range of food-storing animals and suggests that the behavior is complex and that its effects may be far-reaching.

Certain aspects of the study of food hoarding have developed in nearly total isolation. For example, experimental analyses of the environmental and physiological factors that influence hoarding have proceeded almost completely independently of ecological studies of hoarding, although the two are tightly interdependent. Further, studies of hoarding and mass provisioning by spiders, bees, wasps, ants, and beetles have been largely ignored by those studying the hoarding behavior of vertebrates. In general, many of those who study food-hoarding behavior have not profited as much as they might have from findings elsewhere in the field. The reason for this is that, despite several fine recent reviews, a detailed synthesis that covers the diverse aspects of hoarding and that draws broadly from the wealth of scattered literature is not available. This book provides such a synthesis. It is meant to serve as a reference for those actively working on problems concerning food-hoarding animals, as a source for those who wish to initiate such studies, and, I hope, as interesting reading for those more generally interested in ecology, evolution, or behavior.

The organization of this book reflects two important goals. First, I address the major ecological, evolutionary, and behavioral questions and issues central to food hoarding. These questions and issues are treated in chapters 1 through 7 and chapter 11. In these eight chapters, I ignore, as much as possible, taxonomic boundaries and adopt a comparative, integrative approach. In three other chapters (chapters 8 through 10), I consider hoarding within taxonomic groups (mammals, birds, and arthropods), focusing on the behavior and natural history of food-hoarding animals. The treatment of different taxonomic groups (usually at the family level) is unbalanced, but this simply reflects the relative importance of hoarding in different groups and our depth of understanding. The combining of these two approaches in a single volume, I feel, emphasizes the importance and consequences of food hoarding without losing track of the animals and how they behave.

This book would still be in preparation if it were not for the help and encouragement of numerous colleagues and friends. Ron Lanner gave a great deal of support and encouragement early in the project when finding and organizing the diverse literature seemed like an impossible task. The Department of Biology, Utah State University—especially James MacMahon, Ivan Palmblad, and Keith Dixon—gave me both moral and direct support during my tenure there as a Research Assistant Professor. Vince Tepedino and Phil Torchio of the U.S. Department of Agriculture, Bee Biology and Systematics Laboratory, at Utah State University gave a great deal of assistance with the literature and behavior of bees and wasps. Any misstatements or overgeneralizations about bees and wasps that still exist in the pages that follow probably remain because I didn't always follow their advice. Carl Cheney, Barry Gilbert, Ron Lanner, Vince Tepedino, and Phil Torchio read several chapters, and Steve Jenkins, Chris Smith, Kimberly Smith, and an anonymous reviewer read the entire manuscript and made numerous helpful suggestions that have greatly improved its organization and content. Stephen Weiss, Susan McBride, and Mary Walton of the Merrill Library, Utah State University, and Judy Sokol of the Getchell Library, the University of Nevada, Reno, provided invaluable assistance in locating and obtaining obscure publications and government documents. Marilyn Hoff Stewart skillfully transformed my crude instructions into beautiful artwork that illustrates the hoarding behavior of various animals. Numerous others have allowed their photographs, figures, or tables to be reproduced here. They are credited in the appropriate captions, and I thank them all. And last I thank my wife, Kathie, and daughter, Erin, who showed an incredible amount of patience with me as I labored, at their expense, to complete the manuscript.

June 1989
Reno, Nevada

1

What Is Food Hoarding?

Food-hoarding animals have the capacity to control the availability of food in space and time, which gives them a marked advantage over non-hoarders. When unstored food in the environment is in short supply, which may happen daily, seasonally, or unpredictably, food hoarders can turn to their food reserves; nonhoarders may be forced to migrate, enter torpor, hibernate, or suffer losses in body mass. When food is abundant but environmental contingencies such as inclement weather or risk of predation prevent foraging, hoarders can retire to the safety of their burrow or territory and feed. When it is time to reproduce, an accumulation of food in a sealed nest chamber provides a secure environment in which an insect larva can develop. And when it is rewarding for an animal to invest time in activities other than foraging (e.g., courtship, territorial display), the presence of stored reserves that can be quickly exploited allows the animal to reallocate foraging time to those activities. By permitting animals a measure of control over their food supply, food hoarding has become an important element in adaptive strategies for circumventing problems of food limitation. Animal ecologists and behaviorists are just beginning to unravel how food hoarding fits into and shapes these adaptive strategies, and their activities have been the subject of several recent reviews (Roberts 1979; Shettleworth 1983, 1985; Smith and Reichman 1984; Sherry 1984b, 1985, 1987; Covich 1987; Balda et al. 1987; Vander Wall and Smith 1987; Källander and Smith 1990).

Food hoarding can be defined as the handling of food to conserve it for future use. (I consider the terms *caching* and *storing* to be synonymous with hoarding.) Two points are essential to this definition. First, consumption of food items is deferred. The period of deferment varies greatly among food-hoarding species, ranging from a few minutes, as in the Barbados green monkey (*Cercopithecus aethiops*), which hid an apple until dominant group members moved away from its foraging site (Baulu, Rossi, and Horrocks 1980), to as much as 2 years, as has been documented in red squirrels (*Tamiasciurus hudsonicus*) storing white spruce (*Picea glauca*) cones in Alaska (M. C. Smith 1968). Food may be stored for virtually any length of time between these two extremes. This definition excludes the accumulation of food in a crop or other diverticula of the digestive tract by various grouse, finches, hummingbirds, and the like, which consume food items as they encounter them but defer digestion.

Second, food items must be handled in some way that deters other organisms from consuming them. Handling encompasses a diverse array of behaviors including preparation, transportation, placement, and concealment of food items, but not all of these are exhibited in all situations.

Preparation is the special handling of food immediately after capture or collection. For example, short-tailed shrews (*Blarina brevicauda*) and many wasps inject venom into prey to immobilize them. Gray squirrels (*Sciurus carolinensis*) remove the husks from walnuts (*Juglans nigra*), and red-headed woodpeckers (*Melanerpes erythrocephalus*) husk and fragment acorns before transporting them to storage sites. And dung beetles may form dung into spherical masses.

Transport of prey is often a conspicuous and important component of food-hoarding behavior (Smith and Reichman 1984). Transport occurs over distances ranging from a few centimeters (e.g., carabid beetle larvae; Alcock 1976) to many kilometers (e.g., nutcrackers; Vander Wall and Balda 1981). But prey need not be transported. Kruuk (1964), for example, described how a red fox (*Vulpes vulpes*) buried an oystercatcher under the oystercatcher's own nest. The spider wasp *Anoplius marginalis* (Pompilidae) attacks burrowing wolf spiders in their retreats and stores them there as food for their larvae (Gwynne 1979). The araneid spider *Argiope* wraps prey at the capture site in its web (Robinson, Mirick, and Turner 1969; Vollrath 1979). Although transport of prey is not an essential component of food hoarding, handling the food to reduce the probability that another organism will consume it is essential. A scavenger that drags bones away from a carcass and returns to feed on them again has not necessarily stored the bones; however, a predator that leaves a kill in place but rakes dirt and grass over the carcass, an act that camouflages the prey and that may reduce the rate at which the meat spoils, has stored that prey.

Placement in its simplest form includes the selection of a cache site and deposition of the food item. A nuthatch (*Sitta* spp.) wedging a seed in a bark crevice is a familiar example. Many other food hoarders first prepare the cache site by digging a small hole to receive the food. Some food-storing animals place items in hollow trees and in underground chambers, and many bees and wasps expend considerable time and energy constructing special receptacles in which they deposit food.

Concealment of prey is a common component of hoarding behavior. If an animal buries food in the ground, it may draw soil, plant litter, or other debris over the cache to camouflage it. Animals that place food in bark crevices often tuck bits of lichen, bark, or moss around the item. But not all food hoarders conceal their prey. Shrikes (*Lanius* spp.) impale prey on exposed barbs and thorns (e.g., Smith 1972); MacGregor's bowerbirds (*Amblyornis macgregoriae*) simply wedge fruit into branch forks (Pruett-Jones and Pruett-Jones 1985). Pikas (*Ochotona* spp.) construct hay piles on exposed rocks (e.g., Millar and Zwickel 1972a). At all stages in the caching process, the hoarder handles the food so as to increase the probability that the food will be present and in good condition when it returns.

1.1 CACHE DISPERSION

Food-hoarding animals distribute stored food in a variety of ways, ranging from highly clumped to highly dispersed (fig. 1.1). The end points of this spectrum of cache-dispersion patterns have been termed *larder hoarding* and *scatter hoarding*. Larder hoarding, in the strictest sense, is the concentration of all food at one site. A good example of this type of hoarder is the honey bee (*Apis mellifera*), which places all of the colony's stored food supply in a set of combs in close proximity (Seeley, Seeley, and Akratanakul 1982). This highly concentrated type of food storage is relatively rare. Most larder hoarders store food in several discrete, closely spaced sites. Heteromyid rodents, for example, store seeds in several chambers within their burrow system (e.g., Vorhies and Taylor 1922). Red squirrels usually store all their cones within one midden (an accumulation of cone debris), but cones are not distributed uniformly throughout the midden, nor are they stored all together. Rather, they are buried in groups of twenty to forty in numerous chambers under the litter (e.g., Kendall 1980). Western harvester ants (*Pogonomyrmex occidentalis*) store seeds in dozens of subterranean chambers ranging from a few centimeters to 3 m or more below ground (Lavigne 1969).

The term *scatter hoarding* was coined by Morris (1962) to describe the tendency of captive green acouchis (*Myoprocta acouchy*) to bury food items widely spaced. Scatter hoarding, in its most extreme form, means one food item stored at each cache site. This form of food storage is practiced by several species of nut-storing rodents. A familiar example is the caching of single black walnuts (*Juglans nigra*) in shallow surface caches by fox squirrels (*Sciurus niger*) (Stapanian and Smith 1978). Another is the storing of single seeds in bark crevices, moss, and soil by marsh tits (*Parus palustris*) (Cowie, Krebs, and Sherry 1981). Some scatter

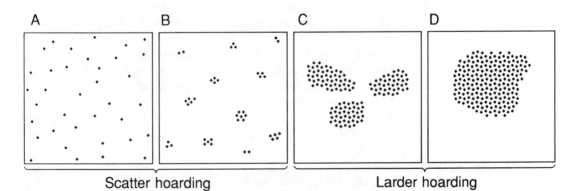

Scatter hoarding   Larder hoarding

Figure 1.1   The spacing of food items stored by animals ranges from highly dispersed to highly clumped. Symbols represent food items; clumps of symbols represent clusters of food items. A, Black walnuts buried in soil by fox squirrels (Stapanian and Smith 1978). B, Piñon pine seeds cached in soil by Clark's nutcracker (Vander Wall and Balda 1977). C, Seeds of forbs and grasses stored in underground chambers by bannertail kangaroo rats (Vorhies and Taylor 1922). D, Honey stored in combs by honey bees (Seeley and Morse 1976).

hoarders, particularly those that store relatively small nuts and seeds, often store several items at each site. Clark's nutcracker (*Nucifraga columbiana*), for example, store one to fourteen pine seeds in shallow caches in soil (Vander Wall and Balda 1981), and white-footed mice (*Peromyscus leucopus*) store about twenty-five to thirty eastern white pine (*Pinus strobus*) seeds in subterranean caches (Abbott and Quink 1970).

Larder hoarders and scatter hoarders converge on similar dispersion patterns near the middle of the spectrum. The harvester ant (*Pogonomyrmex barbatus*), for example, falls near the middle of the spectrum, storing a few grams of seeds in each of 100 or more chambers (McCook 1879). But the two types of hoarding can be rigorously distinguished by a single behavioral criterion. Scatter hoarders typically make but a single visit to each cache site when making a cache. Larder hoarders, on the other hand, make numerous visits to their larder, adding more food with each visit. By such a definition, harvester ants are clearly larder hoarders.

Highly clumped and widely dispersed arrangements of stored food are correlated with a number of other characteristics of the storage site, most of which concern the protection of stored food. Larders, because of the concentration of food, are highly attractive resources to other foragers. Consequently, they are usually placed in protected sites within tree cavities or deep underground, where they can be easily defended by the hoarder, which is usually vigilant and well equipped to protect the larder from most animals that attempt to raid it. Scattered caches, on the other hand, are much less attractive. Their dispersed nature make them difficult to defend effectively, and the hoarder often seems inattentive. What protection the food caches do receive is conferred largely by their inconspicuousness.

Despite the sharp distinction between larder hoarding and scatter hoarding, species cannot always be neatly categorized. A number of animals store food in larders and in scattered surface caches. These species include some kangaroo rats (*Dipodomys*), chipmunks (*Tamias*), woodmice (*Apodemus*), flying squirrels (*Glaucomys*), the red fox, white-footed mouse (*Peromyscus leucopus*), and red-headed woodpecker. Acorn woodpeckers (*Melanerpes formicivorus*) store acorns singly in holes they chisel into tree bark, but the stored acorns are so closely spaced and so conspicuous that these granaries are best considered larders. Because so many species store food both in concentrated larders and in small, scattered caches, it seems prudent to use the terms larder hoarder and scatter hoarder to describe behavior and not as a means of classifying species.

1.2   MASS PROVISIONING: STORING FOOD FOR THE NEXT GENERATION

Food-hoarding animals store food for themselves, for their social group, or to feed their young. In some arthropods, food is stored by adults specifically to nourish the next generation, a behavior called mass provisioning. These insects construct nest chambers, stock them with sufficient food to allow larval development, lay an egg on one of the food items, and seal the nest chamber. Subsequently, the adults usually take no part in feeding the larvae, although they may guard the nest site and periodically open the chambers to monitor the larvae's progress. Mass provisioning is performed by dung beetles, burying beetles, stingless bees, some bumblebees, and numerous species of solitary wasps and bees.

Some related forms of bees and wasps provision offspring progressively (Stephen, Bohart, and Torchio 1969). In progressive provisioning, an adult actively feeds developing larvae, either by regurgitating small quantities of partially digested food or by placing larvae on recently delivered food items. Fresh food is continually supplied at a rate roughly corresponding to larval needs. Food storage is not necessarily involved, because adults can collect food and feed it directly to their larvae. Many social Hymenoptera such as paper wasps, honey bees, ants, and certain bumblebees provision progressively. Some of these insects, most notably honey bees, are progressive provisioners that also store food in other contexts.

The distinction between progressive and mass provisioning is not as sharp as the previous discussion implies (Evans 1966b; Parker, Tepedino, and Vincent 1980). Some progressive provisioners supply food somewhat faster than the larvae can consume it, so a small surplus accumulates. If mass provisioning is drawn out by a paucity of prey or inclement weather, it then resembles progressive provisioning. The line that divides these two types of provisioning is in some cases somewhat arbitrary. This book is concerned primarily with mass provisioning as a form of food storage, that is, any situation in which larvae are provided with food far in excess of their immediate needs.

1.3 PREVIEW

The ten remaining chapters of this book can be divided into two groups: chapters 2 through 7 and chapter 11 address a series of ecological, evolutionary, and behavioral issues central to food hoarding, whereas chapters 8 through 10 are taxonomically arranged overviews of the hoarding behavior of all groups of animals known to store food.

Chapter 2 examines when and how animals use the food they store and the benefits they gain by storing food. For most animals, the duration of storage fits one of two patterns: short-term, ranging from hours to days, and long-term, ranging from months to over a year. Among the most important and most frequently cited benefits derived from these two types of food storage are that the hoarder avoids periods of food scarcity, but this is not the only reason animals store food. Other advantages include gaining a competitive advantage over other individuals when foraging for limited resources, increasing reproductive output, and permitting an individual a greater range of behavioral options.

Chapter 3 addresses the ultimate driving forces likely responsible for the evolution of hoarding and presents some ideas on the proximate conditions that may have promoted evolution of the behavior. A number of species that rely on stored food have evolved specializations for collecting, transporting, and storing food, and some animals have evolved behavior patterns that allow them to efficiently exploit stored food. These specializations are described, as are some coevolutionary interactions between certain food hoarders and the propagules that they store.

Chapter 4 describes the types of losses to which stored foods are subjected and the ways that food-hoarding animals protect stored food. For stored food to be of any use to a hoarder, it must be present and in good condition when the hoarder needs it. Food hoarders display rich behavioral repertoires that function to safeguard stored food, and in many cases they appear to actively manage their food reserves throughout the period of storage. An earlier version of this chapter has appeared elsewhere (Vander Wall and Smith 1987).

Chapter 5 considers how various internal and environmental factors influence hoarding behavior. This begins by examining evidence for the genetic basis of hoarding and how experience during development influences the expression of hoarding in adults. Experimental studies and theoretical arguments concerning the physiological regulation (e.g., hypothalamic and hormonal control) of hoarding are considered next, followed by a survey of the environmental variables (e.g., photoperiod, temperature, light intensity, social context, types of food) that further modify hoarding behavior.

Chapter 6 reviews experiments and observational studies of how animals find the food they have hidden several hours to many months previously. This is a fascinating problem, which has received much attention in the last decade. The manner in which scatter hoarders successfully return to inconspicuous cache sites after long periods is not only of intense interest to those concerned with food-hoarding behavior but is of great interest to animal behaviorists in general. Food-hoarding species can tell us much about how animals perceive and remember details of their environment.

Chapter 7 examines how animals that store plant propagules influence plant dispersal and establishment. A diverse array of plants, including shrubs, forbs, grasses, and dominant trees of temperate deciduous forests, montane coniferous forests, and tropical forests, are dispersed principally by animals that bury their seeds. Many of these plants produce nuts, seeds, and supporting structures that attract animals and facilitate their collecting activities, suggesting that the plants have made an evolutionary commitment to these mutualisms. This is an important aspect of food hoarding from the point of view of community ecology, for it is here that certain food-storing animals have had far-reaching effects on community processes.

Chapters 8 through 10 summarize our knowledge of the hoarding behavior and related natural history of food-storing mammals, birds, and arthropods. Coverage of literature pertaining to the vertebrate groups is more thorough than that for arthropods, largely because the literature on provisioning in arthropods is very large and has been reviewed elsewhere (e.g., Evans 1966a; Krombein 1967; Stephen, Bohart, and Torchio 1969;

Spradbery 1973; Michener 1974; Iwata 1976; Halffter and Edmonds 1982). These chapters are organized taxonomically, with food-hoarding behavior usually described at the family level. For large families that exhibit diverse means of storing food (e.g., Sciuridae, Cricetidae, Corvidae, Formicidae), information is presented for subgroups that store similarly. A table summarizing the ways each taxon distributes stored food items, the types of food stored, the substrates or locations where they stored it, and the relative lengths of storage periods can be found near the beginning of each chapter. These three chapters present pertinent information on each taxon so that workers in the field or persons interested in initiating studies on food-storing taxa will have a handy reference to information on hoarding and related behavior. I have attempted to include all major works so that the chapters will provide easy entry to the literature.

Chapter 11 examines hoarding from a broader perspective than that used in the first ten chapters. Food hoarding has increased species diversity in varying environments and altered the outcome of competitive interactions between hoarders and nonhoarders. Further, hoarder-plant mutualisms have benefited many nonhoarding species. These factors appear to have had major impacts on the structures of some biotic communities and guilds.

The remainder of the book consists of two appendixes that summarize material from the text and indexes designed to increase the usefulness of this book as a reference. Appendix I lists all species of food-hoarding birds and mammals mentioned in the text, and appendix II lists all species of plants mentioned in the text that have been found in animal caches. These animals and plants are also listed in the comprehensive subject index. The list of works cited also serves as an author index. Each entry is followed by a list of page numbers in brackets indicating where it has been cited. This should be especially useful to those wishing to find more information concerning a particular paper or author.

2

How Do Animals Use Stored Food?

The ability of animals to control the availability of food in space and time by storing it has far-reaching consequences, the extent of which depends on at least three things. First, the amount of food stored, including its nutritional as well as caloric content, strongly influences the impact the food will have when the hoarder retrieves it. The quantity of food an animal could potentially store is constrained by the amount of food available, the time available to store food, and foraging specializations of the animal. Second, the exact point or points in time when a hoarder chooses to retrieve hidden food is important. The optimal time for recovering an item is a complex interaction between maximizing the energetic or nutritional benefits derived from the food and minimizing the probability that some other forager or microbe finds the food. Recovery of hidden reserves should occur to satisfy needs that cannot be met by foraging for other foods or when recovery of stored food conserves time that can be better used for activities other than foraging. Third, the use a hoarder makes of stored items is of great consequence. The hoarder may consume the food itself, feed it to its mate, feed it to young, or restore it for later use. The options available depend on when the food is recovered.

The question how do animals use stored food does not have a simple answer. Thousands of species of birds, mammals, and arthropods store food. For these species, food hoarding benefits the hoarder in one or more ways. First, and perhaps most important among vertebrate food hoarders, animals use stored food to survive periods of food scarcity. Many ecologists living in temperate climates view food scarcity as a seasonal phenomenon and use vertebrate hoarding intended to provide calories for a temperate zone winter as a subconscious model. This bias is evident, for example, in Roberts' (1979) review of the evolution of food hoarding in birds. But food scarcity, as detailed in the next section, can also occur on a daily or tidal cycle or occur stochastically. These short-term fluctuations in the availability of resources can significantly affect an animal's nutritional status. Second, some species store food not because of an imminent shortage but because stored food permits an individual to minimize foraging time so that it can invest time in other important activities such as courtship, incubating eggs, nursing young, or interacting with conspecifics at a roost site. In these situations, a small quantity of stored food permits, in a sense, an animal to keep its options open. Third, stored food may contribute to reproductive success and in particu-

lar serve as an energy source for the hoarder's offspring. This is especially true of nest-provisioning bees, wasps, and beetles (e. g., Krombein
1967; Halffter and Edmonds 1982) but is also the case for some birds and
mammals. These benefits are not mutually exclusive.

2.1　FOOD SCARCITY AND NUTRITIONAL DEFICIT

Food hoarding often is associated with periods of food scarcity but is certainly not restricted to these conditions. More specifically, hoarding often
appears to be associated with situations in which food is available in
excess (and thus some food is not consumed and potentially could be
stored) followed by a period where energy demand exceeds available resources and an animal requires additional food to sustain itself. An animal's energy demand fluctuates over time in a complex way that depends
on behavioral states, circadian and reproductive cycles, seasonal changes
in the environment, and many other variables. Food availability is equally
variable; it rises with pulses of production and decreases through the
action of foragers and microbes. Interacting patterns of food (energy)
availability and energy demand can be very complex, and any attempt to
generalize on these patterns can be criticized as oversimplified. Nevertheless, figure 2.1 illustrates three general ways that food availability and
energy demand can interact to produce conditions conducive to food storage. One such situation is produced when energy demand is relatively
constant over time while availability fluctuates (fig. 2.1A). This situation
applies to food hoarders such as the western harvester ant (*Pogonomyrmex occidentalis*) during the summer months. Conditions in the un-

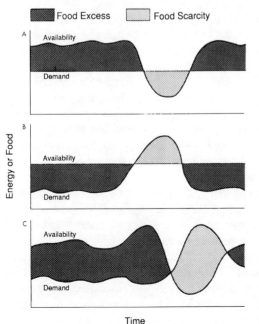

Figure 2.1
Three types of food (or energy)
scarcity resulting from fluctuation
in availability (A), demand (B), or
both (C). See text for examples.

derground nest are nearly constant over time so that the energy demand of a mature colony may stay relatively uniform. Above ground, however, delivery of food to the nest fluctuates significantly, depending on the weather and presence of predators (Willard and Crowell 1965; MacKay and MacKay 1984). On cool, cloudy days, foraging may be precluded. Seeds collected during periods of accessibility are used to sustain the colony during periods when foraging is impossible or unprofitable.

Other food hoarders are confronted with conditions that are the converse of those just described: relatively uniform availability of food but fluctuations in demand (fig. 2.1B). Such a situation occurs with snowy owls (*Nyctea scandiaca*) breeding on the arctic tundra. The abundance of lemmings, their primary prey, is relatively constant over short periods, and in most years more than adequate to sustain adult owls in the early spring. But when young hatch and the adults have to feed six to ten hungry mouths, the demand for food quickly outstrips the adults' ability to provide it. Consequently, snowy owls stockpile lemmings at the nest during the incubation period in preparation for the nestling period. Pitelka, Tomich, and Treichel (1955), for example, counted eighty-three lemmings encircling a snowy owl nest near Barrow, Alaska, in June, just after the eggs had hatched.

A third, more common situation occurs when both food availability and energy demand fluctuate markedly over time. For a species like the gray squirrel (*Sciurus carolinensis*), food availability peaks in fall with the production of nuts and acorns but quickly declines as foragers deplete the supplies (fig. 2.1C). On the other hand, energy demand, especially that expended in thermoregulation, is relatively low during summer but increases markedly with the onset of winter. The interplay of these two variables results in a period of great resource abundance in the fall quickly followed by a long period of intense resource scarcity. By storing nuts in the fall, the gray squirrel in effect extends the period of food availability (Thompson and Thompson 1980).

The time period over which food is used in the previous examples varies. Stored seeds affect the daily lives of harvester ant colonies, stockpiled lemmings supplement the diets of nestling snowy owls over a period of weeks, and cached nuts influence the lives of gray squirrels months later. It is convenient to divide food hoarding into two types based on duration of storage: short-term hoarding and long-term hoarding (e.g., Dorka 1980). Short-term hoarding, as defined here, refers to storage of food for periods of less than about 10 days but in a few cases slightly longer. Long-term hoarding, on the other hand, involves storage periods of several months to more than a year. The delineation of these two forms of storage is somewhat arbitrary—some species store food for intermediate periods, other species store certain foods for short intervals and other foods for long intervals, and long-term hoarders often retrieve and consume some food items within a short time of caching them. Nevertheless, this distinction is useful because, as detailed in the following discussion, members of these two groups often use stored food in different ways.

Unlike the previous examples, some short-term food hoarders store food when it is readily available and consume the stored food when other food in the environment is abundant. The absolute abundance of food in the environment appears to have a negligible influence, at least in some circumstances, on the adaptive value of hoarding in these animals. Two examples illustrate this point. Red tree voles (*Arborimus longicaudus*), one of the few rodents that stores food only for short periods, collect the foliage of Douglas-fir (*Pseudotsuga menziesii*) trees during the night and stack it on top of or near their arboreal nests (Howell 1926). During the day, they pull the foliage into their nests and feed on the needles and bark. Because they eat only fresh, unwilted foliage, they restock their larder each day. Douglas-fir foliage is an abundant and readily available food resource throughout the year. Male MacGregor's bowerbirds (*Amblyornis macgregoriae*) store fruit near their bowers, where they spend much of their time waiting for females, which they court and attempt to mate (Pruett-Jones and Pruett-Jones 1985). In the tropical New Guinea forests where these bowerbirds live, fruit is abundant during the bird's reproductive season, but the male bowerbirds must leave their bowers to forage and may thus miss mating opportunities. Males use the stored fruit to increase their bower attendance. Other examples of short-term food storage when food is abundant can be found in the next section.

The adaptive significance of hoarding in these animals appears to be that food hoarding at one point in time permits the animal to engage in activities other than foraging at some later time. Red tree voles apparently avoid increased risk of predation by having food available near the seclusion of their nests, and male MacGregor's bowerbirds increase their bower attendance and consequently their chances of mating with females by storing food near their bowers. If these animals did not store food, one or two consequences would result: they would have to forage, potentially increasing their risk of predation or sacrificing opportunities to mate, or they would incur a nutritional deficit. Food hoarding allows these animals to avoid nutritional deficits while keeping their options open regarding nonforaging activities, which may yield greater returns in fitness than does foraging. This advantage of hoarding food will be discussed further in the next section, but these examples indicate that even though the effects on food hoarders of storing food to avoid food scarcity and to avoid nutritional deficits are similar, food scarcity, either real or potential, is not the only reason animals store food.

2.2 SHORT-TERM FOOD STORAGE

One usually associates food hoarding with the accumulation of large quantities of food in the fall for use during an energy-demanding winter. Nut burial by squirrels and honey storage by honey bees are familiar examples. But well over half the animals that store food do so for brief periods of usually less than 10 days, and such storage is as likely to occur during the spring or breeding season as during fall or winter. The reason for such short storage periods is in many cases obvious. Many animals store food items that are perishable (carcasses of animals or ripening fruit), which limits the maximum time over which they can be maintained. As we will see in chapter 4, some species exhibit behaviors that permit them to extend the time food can be stored, but for most of these, the time gained is only a few days.

Animals that store food for short periods include predators and insectivores. Large carnivores (e.g., tigers, bobcats, wolves, foxes, bears) that kill prey larger than they can consume in one meal often store the remains and continue to feed on them for several days (e.g., Murie 1936; Schaller 1967; Kurt and Jayasuriya 1968; Nellis and Keith 1968; Mech 1970; Elgmork 1982). Most hawks that store prey do so for only a few hours to 2 or 3 days (e.g., Schnell 1958; Oliphant and Thompson 1976; Ashmole 1987), but American kestrels (*Falco sparverius*) have been observed to store food for 5–7 days (Stendell and Waian 1968; Balgooyen 1976). New Zealand falcons (*Falco novaeseelandiae*) have been observed retrieving prey after 10 days (Fox 1979). Owls may store large numbers of prey for 2–3 weeks under the cool conditions of early spring (Wallace 1948; Pitelka, Tomich, and Treichel 1955), but the maximum duration of prey storage is usually much shorter during summer (e.g., Korpimäki 1987). Northwestern crows (*Corvus caurinus*) usually recover clams they have stored within one day of caching them (James and Verbeek 1984). During the breeding season and summer, shrikes (*Lanius* sp.) store prey for periods of several hours to several days (Owen 1948; Durango 1956; Beven and England 1969; Applegate 1977), but prey may be stored for up to 2 months in the winter (Karasawa 1976). The araneid spiders *Nephila* and *Argiope* consume within several hours the prey they have ensnared and wrapped in silk in their webs (Robinson, Mirick, and Turner 1969).

Not all short-term food hoarders store highly perishable flesh. Chickadees and tits inhabiting regions with moderate winters store seeds for short periods. Marsh tits (*Parus palustris*) that participated in a field experiment in Wytham Woods, England, retrieved stored sunflower seeds in a mean of 7.7 hours and recovered nearly all seeds within about 20 daylight hours of storing them (Cowie, Krebs, and Sherry 1981; see also Stevens and Krebs 1986). The larvae of an unidentified species of ground beetle (Carabidae) store seeds in burrows and consume them within 1–3 days (Alcock 1976). And MacGregor's bowerbirds replace fruit they have stored near their bowers on average once every 6–7 days (Pruett-Jones and Pruett-Jones 1985).

Some food hoarders not only store food for short periods but exhibit storage and retrieval as part of a daily cycle of foraging and feeding. South Island robins (*Petroica australis*) store and retrieve food throughout the day, but the relative frequencies of these two activities change in a predictable fashion. On winter mornings and early afternoons, South Island robins store more invertebrates than they retrieve (fig. 2.2). In contrast, the robins retrieve more than they store on late afternoons and evenings. Powlesland (1980) estimated that 58% of the items stored by South Island robins were recovered the same day, suggesting that items stored early in the day are retrieved in the late afternoon. Similar daily cycles are exhibited by European kestrels (*Falco tinnunculus*) (Rijnsdorp, Daan, and Dijkstra 1981), American kestrels (Mueller 1974; Collopy 1977), merlins (*Falco columbarius*) (Oliphant and Thompson 1976), and white-breasted nuthatches (*Sitta carolinensis*) (Waite and Grubb 1988). In American Kestrels, caching occurred frequently at midday, whereas most instances of cache retrieval occurred in the couple of hours before the falcons went to roost (fig. 2.3). Northwestern crows forage in the intertidal zone and, in this case, caching and recovery are timed to the tidal cycle (fig. 2.4); caching occurs as the tide is falling and recovery of the cached items occurs as the tide is rising (James and Verbeek 1984). Daily cycles of hoarding and recovery probably exist in many other short-term hoarders.

What is the adaptive value of storing food for such brief periods? More specifically, what benefits are derived from caching and retrieving food on a daily cycle? As yet, we do not know enough about cyclic patterns of caching and recovery and how these patterns correlate with the timing of energy demands and the pattern of other behaviors through the day to answer these questions definitively, but several hypotheses (not mutually exclusive) have been proposed. A small reserve of food may dampen the effects of fluctuations in food availability. Short-term storage, for example, may be important as a hedge against inclement weather, when foraging might be impaired. A small cache would lessen the impact of sudden inclement weather or rapid changes in accessibility of prey (Owen 1948; Tordoff 1955; Durango 1956; Collopy 1977a; Fox 1979; Powlesland 1980; Carlson 1985; Warkentin and Oliphant 1985). Henry (1986) stated that red foxes (*Vulpes vulpes*) require about 1 pound (0.45 kg) of meat per day and that he had followed foxes on days when hunting success was poor and nearly all of their food came from caches. The digger wasp *Aphilanthops frigidus* temporarily stores queen ants of the genus *Formica* to use later in provisioning cells. Nuptial flights of *Formica* ants are distinctly periodic, and thus the supply of ants is episodic (Evans 1962a). Storage of queen ants allows the wasp to provision nests at a more regular pace, independent of the periodicity of ant nuptial flights.

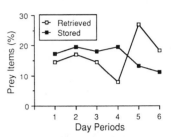

Figure 2.2
Diurnal pattern of storage and retrieval of prey by South Island robins in New Zealand. Days are divided into six periods of equal length. Redrawn from Powlesland 1980.

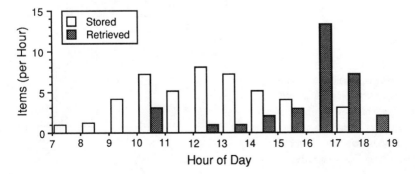

Figure 2.3
Storage and retrieval of cached prey by female American kestrels during the day in the Arcata Bottoms, Humboldt County, California, in the winters of 1972–73 and 1973–74. After Collopy 1977.

Cyclic patterns of caching and recovery (figs. 2.2–2.4) may have more specific functions. Recovery and ingestion of prey near the end of daylight may act to ensure adequate food for maintenance during nocturnal inactivity. Powlesland (1980), for example, suggested that prey stored by South Island robins early in the day serve as a readily available source of energy during the evening, when South Island robins require more energy to sustain themselves through long, cool winter nights. Also, robins consume some stored prey in early morning before invertebrates become active. The result of this pattern of hoarding and recovery is that food is readily accessible just before the robins go to roost. This explanation also may account for cycles of caching and recovery in raptors (e.g., Collopy 1977) and white-breasted nuthatches (Waite and Grubb 1988). An additional advantage to deferring prey consumption until just before going to roost in the evening is to maintain low body mass during the day, thus avoiding the detrimental influences of a full gut on flight dynamics (Rijnsdorp, Daan, and Dijkstra 1981) or minimizing the risk of predation (Lima 1986; Waite and Grubb 1988). In this view, short-term food hoarding is a component of optimal body mass management.

In general, opportunities to capture prey may not coincide with the foragers' need to ingest prey (Daan 1981). Storing prey thus allows animals to time hunting with regard to prey and environmental variability while timing ingestion according to different criteria (Rijnsdorp, Daan, and Dijkstra 1981). Animals that forage for mobile, secretive, uncommon prey must be opportunistic (Solheim 1984). The unpredictability of a predator finding prey when it is actively foraging has promoted a situation in which predators kill suitable prey whenever opportunities present themselves and store the prey until it is needed. Hawks, for example, may kill and store a series of prey far in excess of their short-term needs (e.g., Nunn et al. 1976; Fox 1979).

A number of rodent species generally considered to be long-term hoarders are also known to cache and recover food from their larders on a

Figure 2.4
Frequency distribution of clams and other intertidal invertebrates stored (A) and retrieved (B) by northwestern crows in relation to the time before and after low tide. Redrawn from James and Verbeek 1984.

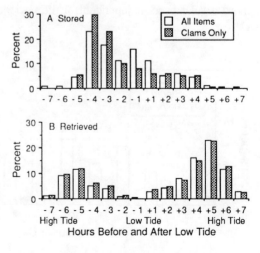

daily cycle. Syrian gold hamsters (*Mesocricetus auratus*) make a net contribution to their larders during the night and a net withdrawal during the day (Toates 1978). Silky pocket mice (*Perognathus flavus*) feed from their larders near the end of the day, just before nocturnal foraging (Wolff and Bateman 1978). Wolff and Bateman suggested that this pattern of feeding increases the efficiency of food gathering and storing in the early part of the night, the most favorable period for foraging. And Jaeger (1982) suggested that the deer mouse (*Peromyscus maniculatus*) forages and stores in the early part of the night to allow for late night feeding before the light phase. In each of these species, the temporal pattern of consumption of food already stored in the larder not only maintains an individual's nutritional status but increases its efficiency in gathering unstored food and thus contributes to the accumulation of more food. In other words, short-term consumption of food in the larder facilitates the long-term establishment of the larder.

A better understanding of the adaptive value of short-term hoarding awaits detailed comparison of the nutritional and energetic contents of food reserves with the requirements of the hoarder. Only a few such studies have been made. One is that of Pitelka, Tomich, and Treichel (1955) cited previously. Another is the field analysis of hoarding energetics of northwestern crows conducted by James and Verbeek (1984), who estimated that male northwestern crows needed a mean of 2.7 average size clams per hour to satisfy their energetic demands in spring. Thus, male crows have to store 16.4 average size clams to meet their energetic demands for the approximately 6-hour period when the intertidal zone is covered by water. Color-banded males actually stored a mean of 6.5 clams and 2.6 other items during low tides. The clams could account for about 40% of the crow's energy requirements during high tide if all were recovered, and the other items contributed an unknown amount of energy. Given that crows do not recover all the food they store and that some of the recovered food is delivered to incubating females, it seems certain that male crows store at low tide only a portion of the energy necessary to sustain themselves at high tide. This means that crows must either meet their food requirements by feeding in excess of immediate needs at low tide (and thus have internal energy reserves to sustain them at high tide) or feed elsewhere for unstored prey during high tide. Even so, the stored food comprises a significant proportion of the crow's diet and decreases the amount of time that would otherwise have to be allocated to foraging at high tide.

2.3 LONG-TERM FOOD STORAGE

Long-term hoarding is often associated with marked seasonal cycles in food availability. In north-temperate and arctic regions, there is usually a late summer peak in seeds, nuts, or green vegetation followed by a prolonged winter season during which these and other foods are scarce. In deserts, food may be abundant following pulses of plant production after rains but much less abundant at other times of year. In tropical regions, similar changes in food abundance occur between the wet and dry seasons. Many food hoarders store large quantities of food and draw on this resource nearly continuously over long periods for sustenance. As part of an adaptive strategy that maximizes the utility of stored food reserves, some of these species exhibit physiological (e.g., adaptive hypothermia) and behavioral (e.g., social huddling) traits that conserve stored food. In yet other species of long-term food hoarders, storage of food is unrelated or only weakly related to seasonal cycles in resource abundance. These species store food for relatively long periods and retrieve food during short, intense periods of especially unfavorable conditions when foraging for other foods is impossible or unprofitable. Thus, similar hoarding habits may be components of several divergent adaptive strategies.

Surviving Prolonged Food Scarcity

Several groups of food hoarders remain active throughout long periods of food scarcity, and during this time derive much of their energy from stored food. For some of these groups, the quantity of stored food is sufficient to account for the complete diet of the food hoarders, whereas in others it only accounts for a small yet extremely important portion. Perhaps the best example of total dependence on stored food is the honey bee (*Apis mellifera*). Production of honey by wild honey bee colonies in temperate areas of North America is about 60 kg (Seeley 1985). Some of this honey is used in summer, but about 25 kg is stored as an energy source for the colony during winter. Honey reserves vary considerably with climate and colony size; in tropical areas, where bees experience only moderate seasonal food shortage, honey bees usually store small quantities of honey (Seeley 1985; Danka et al. 1987; Rinderer et al. 1985, 1986). Honey bees are the only insects that remain active throughout the winter in temperate regions (Seeley and Visscher 1985). The queen and several thousand workers slowly consume the honey during winter, maintaining the hive temperature well above ambient temperature (Owens 1971).

For stingless bees of the tropics, stored honey and pollen are important food reserves for the queen and workers during the wet season, when there is in many areas a dearth of flowers. Roubik (1982b) found that honey and pollen stores of *Melipona fulva* and *M. favosa* colonies are greatest at the end of the dry season and gradually decline during the wet season (fig. 2.5). The decline in colony food reserves is associated with a conservative shift in worker foraging behavior and brood production. Mean worker longevity during periods of resource dearth was nearly double that during periods of resource abundance, suggesting that workers actually forage less to conserve energy rather than increase their foraging effort to find more resources during periods of resource scarcity. At the same time, brood production drops to nearly zero. Workers consume larval provisions and, in some cases, the young larvae themselves, but workers never eat pupae. Thus, colonies curb reproduction in favor of maximizing the number of workers that survive until floral nectar and pollen again become abundant.

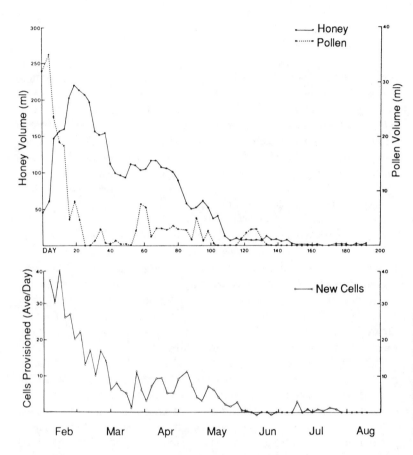

Figure 2.5 Stores of honey and pollen and average number of new cells constructed per day, recorded at 3-day intervals in the nest of a *Melipona fulva* colony. Negative brood production indicates the consumption of provisioned cells by adults. From Roubik 1982b.

For tits and chickadees of the northern boreal forests, food stored in fall forms a significant share of the winter diet. Individual tits in Norwegian spruce forests have been estimated to store from 50,000–80,000 spruce seeds in the fall (Haftorn 1959). Willow tits (*Parus montanus*) and crested tits (*P. cristatus*) form small, highly stable winter flocks occupying territories where group members stored food the previous autumn (Ekman 1979; Nakamura and Wako 1988). More than 50% of the winter diet of crested tits in Norway comes from food stores (Haftorn 1954), and 60–95% of the stomach contents of willow tits in January and February consists of seeds, many of them apparently taken from caches (Haftorn 1956b). Access to supplemental food during the long winters in Sweden significantly increases the survivorship of these two species of tits (Jansson, Ekman, and von Brömssen 1981), and Haftorn concluded that the food-storing habits of crested and willow tits allow them to exist in northern latitudes in winter. Varied tits (*P. varius*) and willow tits in Japan start to recover stored seeds in late fall, and these stored foods are the main source of nourishment for adults in winter (Higuchi 1977; Nakamura and Wako 1988). This pattern of caching and usage contrasts sharply with the short-term hoarding patterns of marsh tits (*P. palustris*) and black-capped chickadees (*P. atricapillus*) of more southerly regions

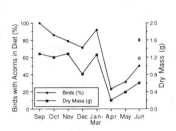

Figure 2.6
Acorn consumption from September to June in a population of jays in The Netherlands. Dry mass refers to mean dry mass of acorns per bird that had acorns in its stomach. The open symbols in June represent data for fledglings. Sample size ranges from nine to twenty-five adults for each sampling period and thirty-five fledglings. Redrawn from Bossema 1968.

(Cowie, Krebs, and Sherry 1981; Sherry 1984a; Stevens and Krebs 1986).

Nut storage by jays and nutcrackers has received a great deal of attention. Under favorable conditions, these birds store thousands of nuts and seeds in scattered surface caches during fall, and these items are a major component of their winter diet. Eurasian jays (*Garrulus glandarius*) in the Netherlands, for example, relied heavily on stored acorns as a food source in the winter of 1963–64 (fig. 2.6), and piñon pine seeds comprised a major component of the pinyon jay (*Gymnorhinus cyanocephalus*) diet following a large seed crop in the fall of 1969 (Ligon 1978). Over 60% of the food items eaten by Florida scrub jays (*Aphelocoma coerulescens coerulescens*) in fall and winter were acorns, and about half of these came from caches (DeGange et al. 1989; see fig. 4.6A). Dow (1965) and Rutter (1972) have argued convincingly that gray jays (*Perisoreus canadensis*) could not persist in boreal forests in winter without their stored food reserves. Seventy percent to 100% of Clark's nutcracker (*Nucifraga columbiana*) and pinyon jay diets between November and February is comprised of stored seeds following good seed years (Giuntoli and Mewaldt 1978; Ligon 1978). The seed stores of Clark's nutcrackers have been estimated to contain from 1.8–3.3 times an individual's energetic requirements during the subalpine winter months (mid-October to mid-April) (Vander Wall and Balda 1977; Vander Wall 1988), and 3–5 times its requirement for spring and early summer (April to July) (Tomback 1982). Nutcrackers apparently store more food than they require to survive the winter because some seeds may spoil, rodents may find some caches, some caches may not be found, and a portion of the stored seeds may be needed after mid-April to assist reproduction (Vander Wall and Balda 1977, 1981; Tomback 1978; Vander Wall and Hutchins 1983).

A large cache of conifer cones is necessary if red squirrels (*Tamiasciurus hudsonicus*) are to pass the winter in good health (C. C. Smith 1968; M. C. Smith 1968; Kemp and Keith 1970; Rusch and Reeder 1978). Territory size varies in proportion to habitat productivity (i.e., type of coniferous forest) so that there is sufficient food in a territory to sustain one squirrel for the winter in most years (C. C. Smith 1968; Rusch and Reeder 1978; Gurnell 1984). Two to 15 bushels of cones have been recovered from red squirrel middens, but these amounts underestimate the actual quantities of cones stored because cone collectors seldom take all of the cones from a midden. Finley (1969) estimated that one-half of an adult red squirrel's annual energetic requirements (21,350 kcal), approximately that needed to pass the winter, can be met with 8.6 bushels of blue spruce (*Picea pungens*) cones, 13 bushels of Douglas-fir cones, or 24 bushels of lodgepole pine (*Pinus contorta*) cones. Red squirrels derive some energy from fungi, shoots, lichens, cambium, and other sources, so it is not necessary to obtain all energetic requirements from stored seeds. But red squirrels residing in deciduous forests, where they cannot accumulate cones, have a higher probability of winter mortality and lower spring body weights than squirrels from prime coniferous forests in years of cone abundance (Rusch and Reeder 1978). The estimate of Yeager (1937) that red squirrels require only ½ bushel of cones to survive the winter is erroneous. Red squirrels in the Colorado Front Range stored in middens only enough lodgepole pine cones to meet their energetic needs for 3–4 weeks (Gurnell 1984). Since, however, most lodgepole pine cones in this region are serotinous, they are available on trees throughout the winter. Red squirrels in other habitats may often store an excess of cones. M. C. Smith (1968) found that some red squirrels in Alaska had sufficient stored cone reserves to last through two winters following a white spruce (*Picea glauca*) cone crop failure in the fall of 1964. Fifteen of eighteen middens remained active during the winter following the first cone crop failure, but when the second cone crop failed in the fall of 1965 and no new cones were available, only six of eighteen middens remained active. The squirrels consumed the last cones in their middens during April and May 1965, about 21 months after the most recent white spruce cone crop. M. C. Smith (1968) estimated that red squirrels in Alaska typically cached 12,000–16,000 white spruce cones each fall and that unused cones accumulated in the midden. After a series of years with normal cone production, enough excess cones may accumulate in a midden to get a squirrel through a year of low cone production.

Tree squirrels (*Sciurus* spp.) also rely heavily on stored nuts to survive the winter. Gray squirrels in Toronto in winter 1974–75 retrieved most cached nuts from December through February (fig. 2.7). By March and April, squirrels' use of cached nuts had diminished greatly, but they still included some cached nuts in their diet. Monitoring of marked squirrel caches has provided an indication of the proportion of cached nuts recovered by squirrels. Cahalane (1942) checked 251 nuts in April that squirrels had cached the previous fall and found that all but 6 (97.6%) were missing, presumably recovered by squirrels. Thompson and Thompson (1980) found that at least 27.1% of the nuts eaten by a population of gray squirrels in Toronto during winter came from caches made the previous fall. The percentage may have been higher because Thompson and Thompson could not determine the source of many of the nuts they saw squirrels eat. Nevertheless, this indicates that cached nuts comprised a substantial portion of the winter diet. Of 500 horse chestnuts experimentally buried on 21 November to simulate squirrel caches, 84.6%

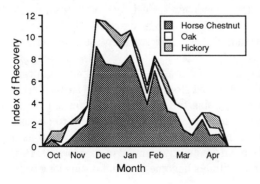

Figure 2.7
Seasonal index of cache recovery behavior (the percentage of observation periods in which a recovery of that species occurred) of gray squirrels in Mount Pleasant Cemetery, Toronto, Canada. Redrawn from Thompson and Thompson 1980.

were removed by May (see fig. 4.4). Thompson and Thompson (1980) found that the rate of removal of experimentally cached nuts was correlated with amount of cache recovery activity occurring in the population, indicating that experimentally cached nuts were being recovered at about the same rate as squirrel-cached nuts. Survival of adult (but not subadult) gray squirrels in Ohio was significantly correlated with size of the hickory nut crop (Nixon, McClain, and Donohoe 1975). Although caching was not a part of this study, it seems certain that the link between large nut crops and high winter survivorship is the hiding of large quantities of nuts for use during winter.

Extensive winter use of stored food has been documented in many groups of small mammals, including voles (Couch 1925; Murie 1961; Formozov 1966; Litvinov and Vasil'ev 1973; Le Louarn 1974; Gates and Gates 1980; Wolff 1984); gerbils (Naumov and Lobachev 1975); woodrats (*Neotoma*) (Horton and Wright 1944); field mice (*Apodemus*) (Imaizumi 1979); dormice (*Myoxus*) (Koenig 1960); and moles (*Talpa*) (MacDougall 1942; Evans 1948; Skoczen 1961; Funmilayo 1979).

Surviving Intermittent Food Scarcity

Animals that store food for long periods derive important benefits by having a reserve food supply to supplement their daily diets, but this is not the only benefit. The food supply also may serve as emergency rations during periods when foraging is precluded for various reasons. In fact, this appears to be the main reason for food storage in some species.

Fossorial small mammals usually maintain small stores of underground plant parts in chambers near their sleeping nest. The adaptive significance of these caches does not seem to be related to passing a long season of food scarcity, as the food caches are small and these mammals usually remain active year-round foraging underground for roots, tubers, corms, and bulbs, which are available throughout the year. Rather, food caches seem to serve primarily as a reserve when burrowing is precluded by waterlogged or frozen soils (Nevo 1961; Genelly 1965; Stuebe and Andersen 1985). Stuebe and Andersen (1985) determined that a subnival cache of the northern pocket gopher (*Thomomys talpoides*) in Utah contained enough energy to meet the requirements of a male for 5.7 days and a female for 6.3 days. The presence of two or more caches per burrow, a common occurrence, would obviously extend the period over which stored food could sustain an individual. Such food reserves are sufficient to enable northern pocket gophers to rely exclusively on cached food during commonly encountered adverse weather (Stuebe and Andersen 1985).

For some species, such as harvester ants, the interruption of foraging is much more predictable, and food storage seems to be an adaptation to frequent, short periods unfavorable to foraging. Harvester ant workers collect seeds on warm summer days and transport them to storage chambers in their subterranean nests. Stored seeds are not used as a winter food supply—as has been frequently claimed (e.g., Cole 1934)—because harvester ants of midlatitudes become dormant during long periods of cold weather (Willard and Crowell 1965; Lavigne 1969; MacKay and MacKay 1984). Rather, stored seeds seem to function primarily to sustain the colony when above-ground foraging is impossible or unprofitable. The times that harvester ants can forage are limited by the harsh physical environment they inhabit. *Pogonomyrmex owyheei* of central Oregon, for example, can only forage when the soil surface is between 20° and 54°C (Willard and Crowell 1965). On clear summer days, soil temperatures do not permit foraging until after 8:00 am. By 11:30 or 12:30, the soil is often too hot, and foraging is suspended for 1–5 hours. At 15:30–16:30, foraging is again possible but stops for the day at about sunset. On cool spring and fall days, or on cloudy or rainy days during midsummer, foraging time is even more restricted. But the growth and well-being of the colony requires a continuous and uninterrupted supply of food, especially during periods of brood production. Seed storage permits harvester ant colonies to remain active underground independent of surface conditions. See MacKay and MacKay (1984) for some alternative hypotheses.

Acorn woodpeckers (*Melanerpes formicivorus*) store impressive quantities of acorns in granary trees, which serve as communal larders for groups of two to fifteen woodpeckers. Koenig (1980) estimated that over a 3-year period the population of woodpeckers at Hastings Natural History Reservation, California, stored 43,600 acorns per year, equivalent to 344 acorns per bird each year. Koenig (1980) estimated this to be 6–7% of the total annual metabolic requirement for an individual, and 10–11% of the metabolic requirement for the 8 months of the year when acorns were not available directly from trees. Koenig and Mumme (1987), working with data for a 6-year period, calculated that granaries at their peak contained energy to maintain the woodpecker population for 6.6–24 days (mean = 14 days). By June these granaries still contained a 4-day supply of food for each woodpecker. It seems clear that, at least at Hastings, the stored food contributes only a fraction of the needed energy to pass the winter. However, these statistics obscure the importance of the granary to the acorn woodpeckers. The value of the stored acorns may be to get the colony through brief periods of food shortage caused by adverse conditions and to supplement the diet of adults and sometimes the young during the breeding season (Koenig and Mumme 1987). Adult acorn woodpeckers in New Mexico belonging to groups with large acorn storage facilities survive longer than individuals from groups with small or medium acorn storage facilities (fig. 2.8A) (Stacey and Ligon 1987). Furthermore, at Hastings, the benefits of having large acorn stores extends beyond individual survivorship. The persistence of lineages in particular territories is significantly related to territory quality as measured by the number of acorn storage holes (fig. 2.8B) (Koenig and Mumme 1987). Lacking a sizeable quantity of stored acorns, acorn woodpeckers abandon their breeding territories in late fall (Hannon et al. 1987; Koenig and Mumme 1987).

The North American Pika (*Ochotona princeps*), Pallas' pika (*O. pallasi*), and probably all other pikas that make hay piles use stored hay to

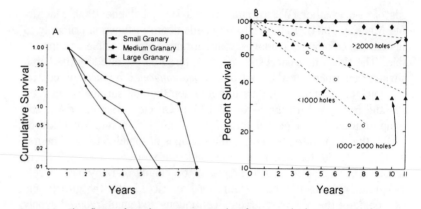

Figure 2.8   The influence of stored acorns on survivorship of acorn woodpeckers. A, Survivorship curves for acorn woodpeckers first banded as adults and assigned an age of 1 year when captured. Median lifetimes for birds living in territories with small granaries (<1,000 storage holes), medium granaries (1,000–3,000 storage holes), and large granaries (>3,000 storage holes) were 1.65 (N = 46), 1.73 (N = 45), and 2.31 (N = 46) years, respectively. B, Survivorship curves for lineages as a function of territory quality as indicated by the number of storage holes. Solid diamonds are territories with >2,000 holes, solid triangles territories with 1,000–2,000 holes, and open circles territories with <1,000 holes. Values plotted are the mean percent surviving; lines were drawn by eye. Differences among groups in A and B are significant (P < 0.01). A, from Stacey and Ligon 1987; B, from Koenig and Mumme 1987.

supplement their winter diets of lichens, bark, roots, and other available plant materials (Johnson and Maxell 1966; Okunev and Zonov 1980; Conner 1983). Pikas do not hibernate to reduce their metabolic demands but stay active, using tunnels in the snow that lead from talus slopes to exposed trees and bushes. Conner (1983) concluded that most hay piles do not contain enough material to allow pikas to feed exclusively on them throughout the winter. Captive pikas consume 26–36 g of dry hay per day. Millar and Zwickel (1972a) calculated that each pika needs about 6 kg of hay to get through the 6-month long winter. Only 3 of 105 hay piles they weighed contained this much dried vegetation. Adult male and female mortality rates were similar, even though males had significantly larger hay piles (Millar and Zwickel 1972a, 1972b). Also, mortality rates of pikas in different habitats did not differ, even though the amount of hay stored varied with productivity of the habitat (Millar and Zwickel 1972a). Conner (1983) suggested that hay storage may be important in the unpredictable habitats that pikas inhabit, primarily as a safeguard against extreme environmental conditions that preclude foraging.

Conservation of Stored Food

Laying in a large store of food is not the only means available to food hoarders for surviving prolonged food scarcity. Behavioral and physiological adaptations that reduce energy demand are also important, because they reduce the rate at which animals consume stored food and consequently increase the length of time over which the food store can sustain the hoarder. Some larder-conserving mechanisms are adaptive hypothermia, fat accumulation, nest insulation, opportunistic foraging for unstored food items, and communal huddling.

The beaver (*Castor canadensis*) exhibits a well-integrated strategy that incorporates many of these components. The caches of branches and logs submerged by beavers provide the primary food source for the colony during the icebound period, which at northern latitudes may last about 150 days (Novakowski 1967). During periods of activity, beavers leave the lodge to feed on the food cache, often dragging stems back to the lodge to be eaten. The useable weights of five caches examined by Novakowski (1967) in Wood Buffalo National Park, Canada, ranged from 40 to 65 kg per beaver—equivalent to 0.26–0.43 kg of edible material per beaver per day. The complete caches contained 31,000–70,000 kcal (mean = 49,600 kcal) of digestible energy per beaver. Nevertheless, these caches were not sufficient to supply the complete energetic needs of active beavers for the icebound period. Three of the five colonies lacked sufficient digestible energy to maintain basal metabolism. Adult beavers lost weight during the winter, suggesting that fat catabolism provided the needed energy for maintenance, but young beavers (less than 2 years old) actually gained weight, possibly because they foraged more extensively from the food cache. Beavers appear to conserve the energy in their larder by living communally (which allows mutual warming of their living space), constructing a well-insulated lodge, periodically entering dormancy, drawing on fat reserves, and occasionally foraging for unstored food under the ice to supplement the food cache (Novakowski 1967; Aleksiuk 1969).

For many food hoarders that hibernate or regularly enter torpor, food stored in or near the hibernaculum, sometimes augmented with varying amounts of body fat, is essential to successful wintering. The animal periodically arouses, feeds from its food stores, and then becomes hypothermic again. For these species, food storage is a component of a complex physiological and behavioral strategy for surviving the winter. Such a strategy is found in some paper wasps (Vespidae) and in numerous small mammals.

Female paper wasps (*Polistes annularis*) in south Texas store a viscous honey in their nests, which colony members use as a food source to survive the winter (Strassman 1979). Unlike honey bees, these wasps must abandon their nests when the weather becomes cold (usually November) in favor of protected hibernacula. On sunny winter days when ambient temperatures exceeded 21°C for at least 110 minutes (or cloudy days above 26°C), wasps emerged from hibernacula and fed on honey supplies at the nest. Such days occur about fifty times during the winter and, in the year of Strassman's study, at least five times each 15-day period. Honey was probably converted to abdominal fat, which was then used to maintain the wasps in their hibernacula during cold periods. Strassman (1979) demonstrated the importance of the stored honey by simply removing some nests while the wasps were in their hibernacula. Survival rates for wasps that had their nests and stores of honey experimentally removed were significantly lower than those of wasps with intact nests.

Chipmunks (*Tamias* spp.) show considerable inter- and intraspecific variation in levels of winter dormancy, ranging from shallow to deep hibernation, and they deposit relatively little fat as a winter energy source (Panuska 1959; Cade 1963; Jameson 1964; Forbes 1966; Jaeger 1969; Brenner and Lyle 1975; Wrazen and Wrazen 1982). During arousal, which occurs about every sixth day (Panuska 1959; Jameson 1964), chipmunks feed from their larders. The seeds, nuts, and bulbs stored in the nesting burrow are critical for winter survival, as they are the sole source of food for chipmunks during the winter (Broadbooks 1958; States 1976; Elliott 1978; Kawamichi 1980). But the role of the food cache seems to vary somewhat in the overwintering strategy of individual chipmunks. Wrazen and Wrazen (1982), for example, found that cache size was inversely related to the amount of food consumed just before winter, whereas body mass in winter was directly related to consumption of food from caches. In other words, individuals seem to pursue one of several strategies. Some individuals consumed relatively little food during fall, gained little or no mass, but made large caches and consumed more during frequent periods of winter activity. Other individuals ate more during fall, gained more mass, accumulated small caches, and were more likely to become torpid. Some chipmunks adopted intermediate strategies. Cade (1963) calculated that a 50 g yellow pine chipmunk (*Tamias amoenus*) requires 720 kcal during 120 days of hibernation. This energy can be found in 294 g of seeds (assuming 70% assimilation efficiency). Three seed caches of yellow pine chipmunks found by Broadbooks (1958) weighed 70, 170, and 190 g (some of these chipmunks were still adding food to their larders). Cade (1963) concluded that if the chipmunk stored 10–20 g of body fat, the two larger caches were probably sufficient to sustain the chipmunks. Other chipmunks store more relative to their winter needs. Two least chipmunk (*T. minimus;* adult mass about 35 g) caches contained 529 and 465 g (Criddle 1943). Elliott (1978) found that eastern chipmunks (*T. striatus*) have 75–921 g of food still available in burrows in spring, enough to last the chipmunks for 30–330 days at resting summer metabolic rates. At least at northern latitudes, winter survival is probably impossible without a sizeable food reserve. Elliott (1978), however, contends that the food hoard is at times also critical during spring and summer and that eastern chipmunks sometimes rely on food stored over a year previously.

Some species of ground squirrels and even sexes within a species (see Sexual Differences in Hoarding, chapter 5) show strikingly different seasonal patterns of cache use, and these patterns appear to be coordinated with timing and extent of hibernation. In arctic ground squirrels (*Spermophilus parryii*), males lose less weight during winter than do females, suggesting that males probably use stored food during periods of arousal from hibernation (McLean and Towns 1981). But for most species of ground squirrels, stored food is a food reserve for a brief period at emergence from hibernation (Shaw 1925; Krog 1954; Mayer and Roche

1954). Hibernating ground squirrels are subject to variable and sometimes extreme conditions in early spring (Morton and Sherman 1978), and fresh vegetation or seeds are probably seldom available at emergence. The presence of a small food cache may permit ground squirrels to emerge earlier than they might otherwise, and thus the trait may yield important reproductive advantages. Rock squirrels (*S. variegatus*), on the other hand, are active throughout the winter in much of their range, and food reserves are probably used at that time (Howell 1938).

Heteromyid rodents with body masses less than 40 g often enter torpor to conserve energy reserves (Brown and Bartholomew 1969; Brower 1970; Kenagy 1973; Wolff and Bateman 1978; MacMillen 1983). Wolff and Bateman (1978) found that the silky pocket mouse could subsist on less than 20% of its normal daily food ration if it entered torpor each day, and Kenagy (1973) determined that little pocket mice (*Perognathus longimembris*) that regularly became torpid in winter survived on 26–30% of normothermic energy requirements. Captive pale kangaroo mice (*Microdipodops pallidus*) adjusted time in torpor to ambient temperature and food supply so that they maintained body mass and still accumulated a food store. The size of larders maintained by pocket mice in the wild has not been well documented. Reed (1987) described a mid-July cache of the plains pocket mouse (*P. flavescens*) that weighed 70.3 g and contained 356 kcal. Reed, using metabolic rate data for the similar-sized silky pocket mouse (Wolff and Bateman 1978), estimated that this cache would maintain an active plains pocket mouse for about 35 days. Winter caches of this and other species of *Perognathus* are probably larger than midsummer caches (e.g., Bailey 1929; Lawhon and Hafner 1981). Whether heteromyids undergo torpor as an energy-conserving response to brief periods when foraging is precluded, as in silky pocket mice (Wolff and Bateman 1978), or during long periods of cold nocturnal temperatures, as in pale kangaroo mice (Brown and Bartholomew 1969), little pocket mice (Kenagy 1973), and Great Basin pocket mice (*P. parvus*) (MacMillen 1983), the period of time over which food caches can sustain an individual is greatly increased. Most kangaroo rats (*Dipodomys* spp.), however, are too large to enter torpor to conserve energy. The quantity of food stored varies greatly both inter- and intraspecifically. Bannertailed kangaroo rats (*Dipodomys spectabilis*) are avid seed hoarders, having stored food reserves during winter and early spring (October–March) of 2.4 kg (range 0.3–5.8 kg, $N = 11$) (Vorhies and Taylor 1922). Great Basin kangaroo rats (*D. microps*), on the other hand, establish small larders containing an average of only 50 g (Kenagy 1973). These food stores provide an important food source during short periods of inclement weather as well as during the winter when low night-time temperatures often preclude foraging. Heteromyid rodents also might obtain significant quantities of preformed water by storing hygroscopic seeds in humid portions of their burrow, where the seeds absorb moisture from the air (Morton and MacMillen 1982; Frank 1988b).

Food storage is associated with adaptive hypothermia in several other groups of rodents. Stored food and daily torpor may be used to varying degrees by members of the genus *Peromyscus* as components of overwintering strategies (Tannenbaum and Pivorun 1984, 1987). Hamsters do not deposit body fat but store very large quantities of food before hibernation. The caching and recovery of stored food have never been studied in the wild (Murphy 1971), but in captive Syrian golden hamsters, the possession of a food store facilitates initiation of hibernation under cold temperatures (Lyman 1954). During periodic arousals, they feed on stored food, which is their only form of sustenance. Hamsters probably cannot survive the winter without stored food.

Overwintering in a social group sharing a communal nest or den where individuals conserve body heat through huddling is practiced by several species of rodents and by honey bees (see preceding discussion). The edible dormouse (*Myoxus glis*), beaver, southern flying squirrel (*Glaucomys volans*), mid-day gerbil (*Meriones meridianus*), and several species of voles are known to occupy winter nests in groups of from five to fifteen or more individuals (Formozov 1966; Novakowski 1967; Muul 1970; Wolff and Lidicker 1981; Nowak and Paradiso 1983; Madison 1984; Madison, FitzGerald, and McShea 1984). Members of the group contribute to a communal larder from which they all feed during the winter. Through huddling, these rodents reduce surface area across which heat can be lost and consequently reduce their energetic expenditure. Taiga voles (*Microtus xanthognathus*) living in communal groups maintain nest temperatures significantly higher than voles nesting singly (Wolff and Lidicker 1981). Further, nest temperatures remain nearly constant because at least some members of the group are always in the nest. Ponugaeva (1953, cited in Formozov 1966) estimated energetic savings of 38% in winter aggregations of social voles (*Microtus socialis*). The individual gains in several ways from its social involvement. First, the energetic savings it experiences acts to extend the time over which a given quantity of stored food can sustain the individual. Second, any mortality in the group benefits a surviving individual because it becomes the heir to a portion of the deceased individual's food stores. But there also may be costs to a food-hoarding individual if other members of the group store little or no food. The evolutionary consequences of this interesting situation are considered in the next chapter.

Many animals pursue a mixed strategy of accumulating both body fat and food, which leads one to ask, what are the relative advantages and disadvantages of these two forms of energy storage? Maximum fat deposition increases with body mass$^{1.0}$ (Morrison 1960; French 1988) whereas maximum food storage is not constrained by body size. This means that animals, especially small animals, can accumulate much greater energy reserves in the form of stored food than they can in the form of body fat. Further, stored food is more economical than body fat because fat contributes to body mass, and metabolic rate increases with body mass. In other words, there is a metabolic expense to maintaining fat. Excessive fat accumulations may also have a detrimental effect on an animal's ability to avoid predators (Lima 1986; Rogers 1987; Waite and Grubb 1988). And, if maintaining a high body temperature is advantageous (e.g., French

1976), animals might be expected to accumulate more energy in the form
of a food store than as body fat. On the other hand, stored food may deteriorate over time, may be removed by cache robbers, or may simply be
lost (see chapter 4). Many animals must expend energy managing and
protecting their food stores. Eating food and converting it to fat avoids
these types of losses and the energetic costs of managing stored food. A
large accumulation of body fat adds to an animal's fasting capacity, especially large animals, permitting some animals to enter prolonged dormancy
in the relative security of a hibernaculum. Thus, both fat accumulation and
food storage have some decided advantages. The ecological and evolutionary forces that cause animals to allocate resources to either fat reserves or food reserves as they do are not well understood and this area
is a profitable one for research.

2.4 REPRODUCTIVE BENEFITS

It is difficult to draw a sharp distinction between the benefits gained by
using stored food to pass the winter or some other period of food scarcity
and the impact of stored food on a subsequent breeding effort. An individual must endure periods of resource dearth in good condition if it is to
breed successfully when conditions are favorable. Stored food could significantly affect reproductive success even if the supply is completely depleted before the breeding season begins; however, such an impact may
be very difficult to detect. Some benefits of food hoarding for reproductive success are more direct and thus much easier to discern. Potential
benefits arising from use of stored food during the reproductive season
include (1) promoting early breeding and increasing the number of individuals in breeding condition, (2) facilitating courtship, (3) increasing litter or
clutch sizes, (4) increasing nest attentiveness, and (5) supplementing the
diet of young. For many solitary bees, wasps, and beetles, food is gathered and stored away solely in pursuit of reproductive success. Because
their methods of food storage are strikingly different from those of birds
and mammals, these reproductive efforts are considered separately.

Early Breeding and Number of Individuals in Breeding Condition

The timing of reproductive seasons is cued by certain environmental
variables that change predictably (Immelmann 1971). In mid and high latitudes, photoperiod is usually the cue, or zeitgeber, that triggers hormonal changes initiating sexual activity in birds (e.g., Wingfield et al.
1987). But the ultimate causes for the timing of breeding seasons often
are availability of food and increased offspring survivorship resulting from
breeding as early in spring as possible (Perrins 1970; Yom-Tov and Hilborn 1981). Animals that store sufficiently large quantities of food are to
some extent released from the constraint imposed by seasonal cycles in
food availability. As a consequence, some food-hoarding animals breed
early in the spring, long before most nonhoarding species.

The pinyon jay is one of the earliest nesting passerines in North
America, initiating breeding in February with eggs being laid as early as
late February (Balda and Bateman 1973; Ligon 1978). Under certain con-

ditions, pinyon jays may even breed in fall (Ligon 1978). Balda and Bateman (1973) and Ligon (1978) demonstrated that the timing of pinyon jay
breeding did not depend on weather conditions just before nesting but on
the size of the cone crop the previous fall. Large cone crops allow the
pinyon jays to store large quantities of pine seeds, which presumably provide the energy for gonadal development and courtship, both of which
occur under winter conditions, as well as a portion of the energy for nest
building, egg laying, and incubation. Piñon pine seeds and cones are sufficient stimuli under the appropriate photoperiod to cause gonadal development in captive pinyon jays during late winter (Ligon 1974, 1978).
Clark's nutcrackers breed at about the same time as pinyon jays and at a
higher elevation, where energetic demands of early breeding are even
greater (e.g., Mewaldt 1956; Vander Wall and Balda 1981). Stored conifer seeds comprise between 70% and 98% of the nutcracker's diet, with a
frequency of 100% ($N = 117$ stomachs) during the months of March and
April following years of seed production (Giuntoli and Mewaldt 1978).
The boreal owl (*Aegolius funereus*) is one of the earliest breeding raptors
in the Holarctic region (Korpimäki 1989), and stored prey plays an important role in buffering the attending female and her young from food shortage caused by late winter snow storms (Korpimäki 1987). Among acorn
woodpeckers, breeding seems to depend strongly on the presence of
stored acorns, and groups with stored acorns breed about 3 weeks earlier than groups that lack food stores (Koenig and Mumme 1987). Honey
bees begin brood rearing in late winter or early spring, a month or more
before the first worker bees initiate foraging. The protein to finance this
reproductive effort is stored in the form of pollen the previous fall (Seeley
and Visscher 1985).

Mice, voles, squirrels, and various other rodents breed earlier following autumns with large seed and nut crops (Smyth 1966; Ashby 1967;
Gashwiler 1979; Jensen 1982; King 1983). Breeding may even occur in
the fall or winter following large crops. For example, 41.6% of adult female deer mice were in breeding condition and 17.8% were pregnant during fall and winter (September to March) following large Douglas-fir and
western hemlock seed crops in western Oregon (Gashwiler 1979). Following poor seed crops only 4.7% of females were in breeding condition
and 2.3% were pregnant. No doubt many seeds are stored following large
seed crops, but most of the studies of rodent reproductive response to
seed crops have not documented the extent of seed caching. Consequently, the importance of stored seeds in winter breeding has yet to be
elucidated.

Courtship and Mating Behavior

MacGregor's bowerbird is the only vertebrate known to use stored
food solely to facilitate courtship. Pruett-Jones and Pruett-Jones (1985)
suggested that male bowerbirds store fruit to extend the time they spend
at the bower. A well-maintained bower is essential to the reproductive
success of male MacGregor's bowerbirds (Pruett-Jones and Pruett-Jones
1982). Stored fruit supplements males' diets so that they do not have to

leave their bowers as often or for as long to forage. Spending more time at the bower means that males are more likely to be present to drive off marauding male bowerbirds, which try to destroy the bower, and to court females if they should visit the bower. Consistent with this hypothesis, consumption of cached fruit was greatest in the morning, when both bower attendance by males and visitation by females was greatest.

Elsewhere among food-hoarding vertebrates, the presence or size of a food cache has seldom been shown to influence mating success. Male northern shrikes (*Lanius excubitor*) in the Negev Desert, Israel, that were deprived of their food cache failed to attract a mate and did not breed (Yosef and Pinshow 1989). In contrast, male shrikes whose caches had been artifically augmented not only mated but attracted a female significantly earlier in the reproductive season than did control (caches unaltered) males. In this case, the large, conspicuous caches appeared to have stimulated unmated females to settle and form pair bonds with territorial males. This effect of stored food on mating behavior may be more widespread than currently realized and should be looked for in species that construct larders where females can relatively easily assess the extent of stored resources.

Litter and Clutch Size

For those animals that store considerable food before the reproductive season, one might expect a relationship between the amount of food stored and litter or clutch size. But the data on this point are equivocal. Data collected by Mohana Rao (1980) show that seasonal changes in litter size of lesser bandicoot rats (*Bandicota bengalensis*) in southern India are proportional to changes in grain hoarded in burrows with a 1-month lag (fig. 2.9). This suggests that breeding is cued to food availability, but it is uncertain whether more stored food causes larger litter sizes or that changes in both variables are caused by some unmeasured variable. C. C. Smith (1968, 1981) and Kemp and Keith (1970) found that litter size of red squirrels varied in response to conifer seed production, but in both

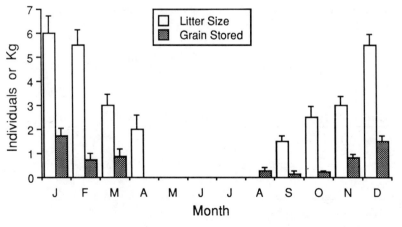

Figure 2.9
Mean litter size and mean quantity of grain hoarded (kilograms per burrow) in the lesser bandicoot rat in southern India rice fields during 1975 and 1976. Based on data from Mohana Rao 1980.

cases large litters occurred just before large seed crops, which suggests that change in litter size was a response to developing cones (or some other factor) and not stored cones. Groups of acorn woodpeckers with acorn stores still present in spring had significantly larger clutches (4.6 eggs) than groups that lacked food stores during the breeding season (3.6 eggs) (Koenig and Mumme 1987). Eurasian nutcrackers (Swanberg 1981) laid larger clutches following years of medium to heavy hazel nut production (table 2.1), but the amount of food actually stored by birds was not determined. Supplementing the larders of boreal owls in early spring results in significantly larger clutches (Korpimäki 1989). Cache size of northern shrikes did not significantly affect clutch size; male northern shrikes with augmented caches, however, fathered three broods, whereas control males (those with unaltered caches) produced only one or two broods, resulting in a significant increase in total eggs laid and total young fledged in the territories of augmented males (Yosef and Pinshow 1989). A number of other studies have failed to find a link between quantity of stored food and litter or clutch size (e.g., Balda and Bateman 1973; Gashwiler 1979; Jensen 1982).

Nest Attentiveness

Short-term food caching during the breeding season may play an integral role in the reproductive behavior of some species. Male shrikes impale prey a short distance from their nest for consumption by females, which spend much of their time incubating eggs and brooding young (Beven and England 1969; Applegate 1977). Similarly, male South Island robins feed stored prey to their mates, a behavior that allows females to be more attentive to their nests. Male crows store food in trees near their nests (Kilham 1984b; McNair 1985), and incubating females retrieve cached prey as need arises. A detailed study of the effects of

Table 2.1 Clutch size of Eurasian nutcrackers (*Nucifraga caryocatactes caryocatactes*) relative to size of hazelnut crop the previous autumn (a relative measure of parent birds' winter food stores) and to ad libitum feeding of nutcrackers with hazelnuts during the winter before breeding

| | Number of Clutches | | | |
| | Size of Hazelnut Crop | | | Provisioned with |
Clutch Size	Below average	Average	Heavy	Hazelnuts
3	8	4	6	2
4	0	0	12	33
5	0	0	3	1

Source: After Swanberg 1981.

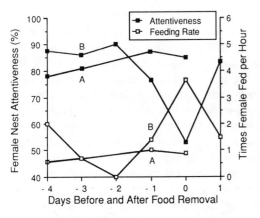

Figure 2.10
Nest attentiveness of two female northwestern crows (A and B) and the rates of feeding of those females by their mates before and after removal of male B's stored food. Redrawn from James and Verbeek 1984.

cached food on nest attentiveness of female northwestern crows was conducted by James and Verbeek (1984). During spring, female northwestern crows incubate and attend nests while males forage in the intertidal zone and store clams and other prey on hillsides well above the high tide line. Males feed females periodically from these food caches. James and Verbeek monitored attentiveness of females at two nests for 4 consecutive days and found that on average they spent 84.7% of their time at the nest (fig. 2.10). On the fifth day, the male of one nest was continually observed, and all food caches were removed by the experimenters immediately after he made them. During the next high tide, this male had no cached food to feed to the female, and she had to leave the nest to forage for herself. The female spent only 53.3% of her time at the nest compared to 88.9% attentiveness at the control nest. On the following day, the male's caches were undisturbed, and attentiveness of the female returned to its predisturbance level. James and Verbeek (1984) concluded that stored food maintains high nest attentiveness in northwestern crows. A high level of nest attentiveness probably reduces nestling predation and, for very young nestlings, thermal stress.

Some long-term food hoarders also use stored food to increase nest attentiveness. For example, stored acorns comprised 29% of the identified food items that breeding female Florida scrub jays received from their mates during the time of egg laying and incubation (DeGange et al. 1989).

Diet of Offspring

Several species of hawks and owls are known to store prey near their nests during the incubation and nestling periods (Pitelka, Tomich, and Treichel 1955; Van Camp and Henny 1975). Incubating female American kestrels may cache prey received from the male. The female may eat these prey (Balgooyen 1976; Oliphant and Thompson 1976) or use them to maintain an even delivery of food to nestlings (Schnell 1958; Lyons and Mosher 1982). The male also may cache prey away from the nest, which he may later consume himself or present to the female (Balgooyen 1976)—options that may dampen fluctuations in the delivery of prey to the female. Male red-backed shrikes (*Lanius collurio*) cache prey near the nest to maintain an even delivery of food to the young (Carlson 1985). Pitelka, Tomich, and Treichel (1955) described a snowy owl nest encircled by eighty-three lemmings, nearly all of which disappeared during a 3-week period when the young were developing rapidly. When the young fledged, only two or three quite old lemmings remained. Pitelka, Tomich, and Treichel (1955) did not observe whether adults or young consumed the lemmings, but it is safe to conclude that the stored prey benefited the reproductive effort. The eighty-three lemmings weighed an estimated 6.0 kg. This impressive quantity of stored food apparently buffered fluctuations in delivery of fresh prey to the nest. Adult snowy owls consume about four lemmings each day. A cache of eighty-three lemmings could supply a family of eight young and two adults for several days if fresh prey could not be caught because of inclement weather (Pitelka, Tomich, and Treichel 1955).

Cached nuts do not comprise a major portion of nestling diets for most nut-storing bird species. A granivorous diet is rare among nestling birds (Lack 1968; Morton 1973). Stored *Castanopsis* nuts comprise 3–5% of the food items that varied tits feed to nestlings during late April and May (Higuchi 1977). Pinyon jays feed nestlings about 10% piñon pine (*Pinus edulis*) seeds following large pine seed crops (Balda and Bateman 1973); however, stored nuts may be used extensively during brief periods when unfavorable environmental conditions preclude capture of insects (Ligon 1978). Acorn woodpeckers feed their young some acorn meat when arthropods are temporarily unavailable, but during fair weather acorn meat is an insignificant component of the nestling diet (MacRoberts and MacRoberts 1976; Koenig and Mumme 1987). Nevertheless, groups of acorn woodpeckers with acorn stores remaining during the breeding season have a higher proportion of eggs hatch and fledge more young than groups without acorn stores (Koenig and Mumme 1987). Apparently, acorn consumption by adults permits the adults to deliver more insects (which the adults would eat if acorns were not available) to the young.

In marked contrast to these birds, Clark's and Eurasian nutcrackers feed their young stored pine seeds almost exclusively (Mewaldt 1956; Reimers 1959a; Kishchinskii 1968). Very young nestlings receive finely mashed seeds mixed with a few insects; older nestlings are fed whole shelled seeds. This diet is possible at least in part because of the high proportion of protein and lipids found in pine seeds (Botkin and Shires 1948). When nestling nutcrackers fledge, they follow their parents to caching areas, where they continue to be fed pine seeds almost exclusively until midsummer (Volker and Rudat 1978; Vander Wall and Hutchins 1983). By the time cached seeds are depleted, the current year's crop is nearly ripe on the trees. Thus, under good conditions, young nutcrackers eat pine seeds throughout development (Mezhennyi 1964; Vander Wall and Hutchins 1983).

The contribution of stored food to the nutrition of young mammals is much more difficult to estimate. The caloric cost of lactation for a number of small rodents with litters of five ranges from 66–147% above the energetic costs of nonbreeding adult females (Millar 1979 and references therein). For those rodents that maintain a larder during the reproductive season, a portion of this additional energy presumably comes from stored food. No one has attempted to estimate this energetic contribution; nevertheless, circumstantial evidence suggests that it may be significant. For example, female hamsters hoard more than males (Bruce and Hindle 1934; Smith and Ross 1950a; Etienne et al. 1983); following mating, females hoard all the available food, including food previously hoarded by the male (Lisk, Ciaccio, and Catanzaro 1983). Miceli and Malsbury (1982) found that permitting captive lactating and virgin hamsters to accumulate food in their nests significantly increased their likelihood of showing maternal behavior toward foster pups. Female Norway rats (*Rattus norvegicus*) and wood mice (*Apodemus sylvaticus*) store more than males, and high levels of female storage often are associated with lactation (Calhoun 1963; Schenk 1979). These findings suggest that the larder may act as a necessary food reserve for lactating females in that a lactating female with a larder could reduce the amount of time she spent foraging while maintaining a continuous supply of energy to her litter and increasing her nest attentiveness.

Larval bees and ants are major benefactors of stored food. Honey bee larvae are fed directly by workers. Stored pollen is used as a protein source for the brood, but it is not fed to them directly. Workers eat the pollen and secrete a proteinaceous substance from their hypopharyngeal glands, which then is mixed with honey and fed to the larvae. Stingless bees, on the other hand, are mass provisioners (see fig. 10.19C) and provision cells with a glandular secretion mixed with pollen and honey (Kerr and Laidlaw 1956). Harvester ants feed seed fragments to larvae; some species secrete digestive enzymes on the fragments to partially predigest carbohydrates for the larvae (Wheeler 1910; Delage 1962). Honey production also is important in the process of colony reproduction or swarming. Honey bee swarms usually occur in spring and early summer, when stored food reserves in hives are increasing rapidly (fig. 2.11). Unlike honey bees, stingless bees transport large amounts of honey, pollen, and cerumen from the mother hive to future nest sites before swarming (Kerr and Laidlaw 1956). Nogueira-Neto (1954, cited in Michener 1974) demonstrated this transfer of food stores by coloring honey in the mother nest and observing the appearance of the dye in the honey pots of the daughter colony. Relocation of these resources apparently increases the probability that daughter colonies will succeed.

There seems to be little doubt that stored food has significant reproductive benefits for some species, but the causal links between the food cache and number and health of offspring have rarely been clearly identified. Acorn woodpecker colonies with large storage facilities produce significantly more fledglings (Stacey and Ligon 1987; table 2.2), but we are just beginning to understand how stored acorns are converted into young woodpeckers (Koenig and Mumme 1987). Peak density of bank voles (*Clethrionomys glareolus*) in Denmark follows 1 year after peak beech mast production (Jensen 1982), and Kawamichi (1980) found

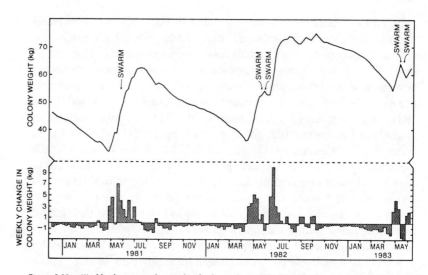

Figure 2.11   Weekly changes in the weight of a honey bee colony (hive plus bees and stored food) in Connecticut over 2.5 years. Swarming (colony reproduction) is associated with periods of rapid weight change (honey storage and brood production) of the colony. From Seeley and Visscher 1985.

that when the acorn crop failed, resulting in a small fall seed harvest, Siberian chipmunk (*Tamias sibiricus*) spring reproduction (measured as number of juveniles trapped) decreased. What role does stored food play in these population changes? Experimental approaches are needed to help establish these causal links. An example of a fruitful approach is provided by Strassman (1979). Female paper wasps that were experimentally deprived of their winter honey stores not only were less likely to survive the winter than control wasps but those deprived wasps that did survive produced significantly smaller spring nests (Strassman 1979).

Table 2.2 Effect of size of mast-storage facilities on reproductive success of acorn woodpecker groups

Mast-storage Category	Number of Storage Holes	N[a]	Probability That Stores Are Present During Breeding	Mean Number ($\pm$ SE) of Young Fledged
Small	<1,000	62	0.09	0.96 ± 0.18
Medium	1,000–3,000	47	0.23	1.76 ± 0.28
Large	>3,000	55	0.58	2.33 ± 0.23
Statistic			$G = 34.7$	$X^2 = 16.5$[b]
P			<0.001	<0.001

Source: Stacey and Ligon 1987.
[a] Number of territory years.
[b] Kruskal-Wallis test.

These nests were not observed through the summer to determine ultimate reproductive success.

Mass Provisioning

Mass provisioning is an integral component of the reproductive behavior of various groups of bees, wasps, and beetles. Many solitary wasps and bees, unlike the birds and mammals discussed previously, provide opportunities to precisely determine energetic and nutritional budget components under natural conditions. The reason is that these insects place all of the food to which a larva will ever have access in a cell, lay an egg on or near these provisions, and then seal the cell (see chapter 10). The mass of all the provisions can be easily determined, as can the mass of uneaten provisions, larvae, prepupae, pupae, and emerging adults (e.g., Cross et al. 1978). Furthermore, in healthy cells, no energy is lost through degradation of provisions by microbes, and cast skins, excreta, and secretory products (e.g., silk) are available within the cell for analysis. Due largely to these advantages, our understanding of the role of stored food in the reproductive behavior of certain Hymenoptera, although limited, is still well ahead of that for other food hoarders.

The nature and quantity of provisions vary greatly in different groups of nest-provisioning Hymenoptera. Nearly all wasps provision with spiders or insect prey. Female wasps may provision cells with one large prey or numerous small prey. Bees and a few wasps stock cells with loaves of pollen moistened with nectar as larval food. The larva typically consumes all of the food placed in a cell. How much food do female wasps and bees place in a cell and how does this amount compare to that required for normal development? Unfortunately, most descriptions of provisions consist of counts or linear dimensions of prey that cannot be objectively related to the requirements of the larvae. The mass of provisions has been reported in relatively few studies, and I have summarized some of these in table 2.3. The mass of provisions varies enormously among wasp and bees species. Much of this variability can be accounted for by expressing the amount of food stored in a cell as a provisioning ratio, that is, the mass of the provisions divided by the mass of the female wasp or bee that provisioned the cell. For the few sphecid wasps (Sphecidae) for which there are data, the provisioning ratio ranges from 1.7–13.0; for eumenid wasps (Eumenidae) it ranges from 6.4–12.7; and for spider wasps (Pompilidae) it ranges from 2.1–4.2. Female solitary wasps and bees often are much larger than males, so one would expect provisioning ratios to be greater for female-producing cells than for male-producing cells. Table 2.3 indicates that this prediction holds true for the few species for which there are sufficient data. Provisioning ratios also should vary with the nutritional qualities of the provisions. Bees, which provision with rich mixtures of pollen and nectar, should be expected to have lower provisioning ratios than wasps that use invertebrate prey. Dung beetles, on the other hand, should have very high provisioning ratios. Unfortunately, there appear to be no data on bees to test this prediction, but in two species of dung beetles (*Heliocopris japetus* and *H. hamadryas*), pro-

Table 2.3 Relationship between mass of provisions and mass of provisioning female wasp

Family Species	Prey per Cell	Provision Mass (mg)			Adult Female Mass (mg)			Sex of Off- spring	Provi- sion- ing Ratio*	Reference
		Mean	Range	N cells	Mean	Range	N			
Pompilidae										
Anoplius semirufus	1	61.4	37–92	6	25.4	11–46	6	—	2.4	Kurczewski et al. 1987
Anoplius marginatus	1	45.0	24–88	12	21.8	13–31	12	—	2.1	Kurczewski et al. 1987
Arachnospila scelestus	1	131.3	49–233	3	54.0	35–83	3	—	2.4	Kurczewski et al. 1987
Sphecidae										
Ammophila procera	1	1,080		1	180		1	M	6.0	Evans 1959
Ammophila nigricans	1	1,300		1	100		1	—	13.0	Evans 1959
Cerceris flavofasciata	12–30	252	168–322	9	46	36–61	3	—	5.5	Kurczewski and Miller 1984
Cerceris fumipennis	2–15	189	128–233	7	85	74–95	2	M	2.2	Kurczewski and Miller 1984
Cerceris fumipennis	2–15	345	308–430	6	85	74–95	2	F	4.1	Kurczewski and Miller 1984
Cerceris fumipennis	2–5	206	98–259	4	122	85–149	6	M	1.7	Kurczewski and Miller 1984
Cerceris fumipennis	2–5	424	409–441	4	122	85–149	6	F	3.5	Kurczewski and Miller 1984
Cerceris rufopicta	4–9	135	82–177	5	36	33–39	2	—	3.8	Kurczewski and Miller 1984
Cerceris rufopicta	3–20	123	60–217	14	40	34–44	5	—	3.1	Kurczewski and Miller 1984
Plenoculus davisi	2–24	19.8	10–35	41	5	4–6	6	—	4.0	Kurczewski 1968
Eumenidae										
Ancistrocerus adiabatus		51		251	8		693	M	6.4	Cowan 1981
Ancistrocerus adiabatus		82		712	8		693	F	10.3	Cowan 1981
Euodynerus foraminatus		130		416	16		285	M	8.1	Cowan 1981
Euodynerus foraminatus		204		323	16		285	F	12.7	Cowan 1981

*Mean mass of provisions divided by mass of provisioning female wasp.

visioning ratios ranged from approximately 10–30 (estimated from data in Klemperer and Boulton 1976).

The ability of larvae to convert provisions into larval biomass has been termed the food conversion factor (i.e., mass of provisions in a cell divided by the mass of the adult that emerges from the cell) by Freeman

(1981a). This important ratio has been virtually ignored by those studying the nesting behavior of wasps and bees. For mud-dauber wasps
(*Sceliphron assimile*) studied by Freeman (1981a), the food conversion
factor was 3.76 ± 0.12 (means ± SD) (the factors for sons and daughters
were not significantly different). A food conversion factor of 3.98 for
cicada killer wasps (*Sphecius speciosus*) can be calculated from data in
Dow (1942). For the megachilid bee, *Osmia lignaria,* food conversion
factors are 4.05 for male offspring and 3.59 for females (calculated from
data in Phillips and Klostermeyer 1978). And for the anthophorid carpenter bee, *Ceratina calcarata,* Johnson (1988) calculated food conversion factors of 2.87 ± 0.75 (mean ± SD) for females and 2.88 ± 0.71 for
males. Obviously, much more data are needed before any generalizations
can be drawn, but in these cases, conversion of provisions to offspring in
wasps and bees varies relatively little. Wasps and bees exhibit considerable intraspecific variation in adult body mass, and much of this variability
is attributable to the quantity of provisions placed in cells by the female
parent (Klostermeyer et al. 1973; Freeman 1981a, 1981b; Cowan 1981;
Johnson 1988). Phillips and Klostermeyer (1978) determined that the
mass of adult *O. lignaria* was positively correlated with the estimated
mass of provisions (fig. 2.12). Freeman (1981a) concluded that size of
mud-dauber wasps was significantly affected by mass of stocked prey and
the efficiency of food conversion. Thus the question of how much provision is required for normal development is not a simple one. Also, some
of the provisioning ratios listed in table 2.3 are much less than the conversion factors of 2.9–4.0, which suggests that the mass of the offspring will
be considerably less than the mass of the maternal wasp. In most cases
these offspring are males.

How are provisions used by a developing larva? Detailed analyses of
energy transfer from provisions to developing larvae have been conducted for only a few species (e.g., Cross et al. 1978; Wightman and
Rogers 1978). Cross et al. (1978) found that the mean caloric content of
spider prey of the organ-pipe mud-dauber (*Trypoxylon politum*) was 4,999
cal/g unashed dry mass, and Mitchell and Hunt (1984) found the spider
provisions of the black and yellow mud-dauber (*Sceliphron caementarium*)
contained 5,576 cal/g ash-free dry mass. These results are similar if one
takes into account that Cross et al. (1978) did not adjust their caloric
value for percent ash content. Mud-daubers provision cells with a mean
of about 209 mg (dry mass) of spiders, equivalent to 1,046 calories (un-

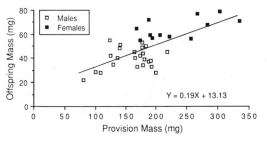

Figure 2.12
Relation between provision mass
and mass and sex of adult offspring of the megachilid bee *Osmia lignaria.* Redrawn from
Phillips and Klostermeyer 1978.

corrected for ash content). Only 77% of these calories are actually as-similated by the larva, 23% of the calories in the provisions being egested in feces (fig. 2.13). This is a high assimilation efficiency considering that the larva consumes all the food over a very short time (about 4.5 days at 24°C) and is apparently due to the larva retaining all the ingesta in its gut until after it spins its cocoon (4–8.5 days, depending on time of ingestion) (Cross et al. 1978). Of the calories assimilated, 474 are used to produce the fully formed prepupa, 99 go into the constituents of the cocoon, and 229 are used in respiration. Additional calories were expended in respiration during the transition of the prepupa into a pupa (43 calories) and from a pupa into an adult (104 calories) (fig. 2.13). Comparison of the nutritional content of provisions and feces of the black and yellow mud-dauber indicates the selective assimilation of proteins and lipids by developing larvae (Mitchell and Hunt 1984). An analysis of energy and nitrogen dynamics between larval leaf-cutter bees, *Megachile pacifica,* and their provisions of pollen and nectar has been conducted by Wightman and Rogers (1978).

As mentioned earlier, females of most wasp species are considerably larger than males (e.g., Dow 1942; Evans 1971; Cross et al. 1978; Cowan 1981; Jayasingh and Taffe 1982; Hastings 1986). Male cicada killer wasps (*Sphecius grandis*), for example, weigh 95 mg and females weigh 256 mg (Hastings 1986). The sexual size dimorphism of wasps is controlled by the nest building and provisioning behavior of the maternal wasp (fig. 2.14). Maternal wasps construct larger cells and pack more provisions into these cells when preparing to lay female-producing eggs than when preparing to lay male-producing eggs (e.g., White 1962; Klostermeyer et al. 1973; Cowan 1981; Freeman 1981a; Jayasingh and Taffe 1982). Female-producing cells of the bee *Osmia lignaria* (Megachilidae) averaged 14.3 mm long and contained pollen masses 8.2 mm long, whereas male-producing cells averaged 10.8 mm long and contained 5.2-mm pollen masses (Krombein 1967). Maternal *Euodynerus foraminatus* (Eumenidae) provision male-producing cells with a mean of 130 mg

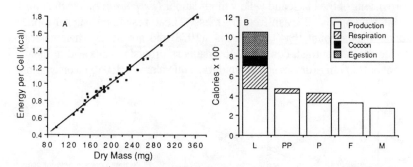

Figure 2.13   A, Regression of calories per cell on dry mass of spider provisions in that cell in nests of organ-pipe mud-dauber wasps ($r^2 = 0.978$). Cell biomass represents total consumption of the wasp larva. B, Use of the energy in provisions by different life stages of the organ-pipe mud-dauber wasp. L, larva; PP, prepupa; P, pupa; F, adult female; M, adult male. Redrawn from Cross et al. 1978.

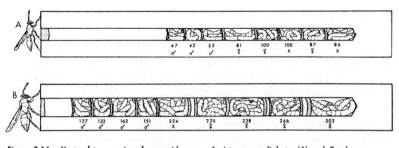

Figure 2.14   Nests of two species of eumenid wasps: *Ancistrocerus adiabatus* (A) and *Euodynerus foraminatus* (B). Numbers below cells indicate mass in milligrams of cell contents. Symbols indicate the sex of the wasp reared from each cell. X indicates wasp died prematurely. Female-producing cells contain more provisions and are located deeper in the nest. From Cowan 1981.

of caterpillars and female-producing cells with 204 mg of caterpillars, 57% more food per cell for females (Cowan 1981). Males emerging from these cells weighed 9.4 mg, whereas emerging females weighed 16.1 mg; the females weighed 72% more than the males. The disparity between females and males is even greater than the disparity between the amount of provisions provided them, suggesting that female *E. foraminatus* larvae use provisions more efficiently. There is, however, no difference in efficiency of provision use by male and female carpenter bees (*C. calcarata*) (Johnson 1988). If one experimentally reverses the quantity of provisions in male- and female-producing cells, as White (1962) did with mud-dauber wasps (*Sceliphron spirifex*), males become larger than females by approximately the same magnitude. Thus, sexual size dimorphism in at least some species is a consequence of how maternal wasps allocate resources to cells.

Many wasps and bees make their nest in hollow stems or empty tunnels bored in wood or soil by insects that partition the cylindrical cavity into a linear series of cells. These wasps and bees are sometimes referred to as trap-nesters because they readily nest in cylindrical cavities furnished by researchers (e.g., Krombein 1967). The positions of female-producing and male-producing cells within a nest often occur in a predictable pattern (fig. 2.14). Female-producing cells are usually provisioned first and consequently are the innermost cells in the nest (Krombein 1967; Gerber and Klostermeyer 1972; Cowan 1981; Jayasingh and Taffe 1982; Johnson 1988). Again, in these nests the female-producing cells are larger and consistently have more provisions than male-producing cells (Klostermeyer, Mech, and Rasmussen 1973; Phillips and Klostermeyer 1978). After finishing a nest, the maternal wasp or bee may begin provisioning another nest, again starting with females and ending with male cells. One can conclude from this pattern of provisioning and egg laying that many maternal wasps and bees have the ability to coordinate the sex of the egg with length of cell and amount of provisions. Sex determination in Hymenoptera is haplodiploid; male-producing eggs are unfertilized (haploid), and female-producing eggs are fertilized (diploid). Thus, females can determine the sex of offspring by controlling the release of sperm from the spermatheca.

How do females control the quantity of provisions provided to male-producing and female-producing eggs? Jayasingh (1980) suggested that since female-producing eggs are larger and take longer to develop than male-producing eggs, the greater quantity of provisions in female-producing cells may simply be a function of the greater time available for collecting prey. A second hypothesis states that female wasps have "an inherent ability to relate the amount of food placed within a cell to the sex of the egg" (Cowan 1983). For eumenid wasps, which lay eggs in empty cells before provisioning, these two hypotheses yield mutually exclusive predictions. The first hypothesis maintains that the amount of provisions placed in a cell is determined by the sex (and, consequently, development time) of the egg for the *next* cell, whereas the second hypothesis holds that the amount of provisions is determined by the sex of the egg for the *current* cell. The interesting point in the provisioning process with regard to these hypotheses is when the maternal wasp switches from provisioning female-producing to male-producing cells. The first hypothesis predicts that the last female-producing cell in a sequence should contain the same quantity of provisions normally placed in a male-producing cell because a male egg (which develops relatively quickly) is developing in the ovary. The second hypothesis predicts that a quantity of provisions usually found in a female-producing cells should be found in the last female-producing cell because provisioning time is independent of the sex of egg that is currently developing. Cowan (1983) tested these hypotheses by comparing the mass of provisions in the last female-producing cell in a sequence with the quantity of provisions found in other female-producing cells in two species of eumenid wasps, *E. foraminatus* and *Ancistrocerus adiabatus*. There was no significant difference between the provisions in these two types of cells, allowing Cowan (1983) to reject the first hypothesis; egg development time does not influence the quantity of provisions placed in cells.

The differential allocation of provisions to female- and male-producing cells and the ability of wasps and bees to control sex of eggs have many interesting implications for the evolution of parental investment and sex ratios in nest-provisioning Hymenoptera. Many of these implications lie beyond the scope of this book. Those wishing to pursue these issues should consult Trivers and Hare (1976), Freeman (1981a, 1981b), Frohlich and Tepedino (1986), and references therein. Some of these consequences, however, directly influence the efficiency of food hoarding. Provisioning ability in many Hymenoptera is positively related to female size (Freeman 1981a, 1981b,; Willmer 1985; Hastings 1986; but see Tepedino and Torchio 1982). Large females can carry larger prey and often make more provisioning trips per day than small females. This relationship is nonlinear in female mud-dauber wasps (*S. assimile*); females weighing 120 mg, for example, carry mean loads 4.3 times heavier than 60-mg females. Consequently, large females provision cells with a few large prey while small females provision with numerous smaller prey.

Very small females often cannot successfully construct or provision cells. The consequence of these foraging and provisioning differences is that large females construct and provision cells much more quickly than small females. Large females often place more provisions in both male- and female-producing cells than do smaller females, resulting in body size of sons and daughters being correlated with body size of the maternal wasp (e.g., Freeman 1981a).

2.5 CONCLUSIONS

Food-hoarding animals store food for a variety of reasons. Most hoarders store food to meet energetic and nutritional demands during acute or chronic food scarcity. Others store food so that activities that compete with foraging can be performed without incurring a nutritional deficit. Yet others store food to finance a reproductive effort. The impacts of stored food seem to be great. Indeed, many food-hoarding species of temperate, boreal, and subarctic regions probably could not survive in these areas without stored food.

For many long-term food hoarders, food storing is simply one element of a complex behavioral and physiological strategy to survive periods of food scarcity. These species establish a reserve of energy in the form of a food cache and then employ a number of energy-saving mechanisms to conserve this energy. The effect of these adaptations is that stored food becomes a reliable food source for considerably longer than it might otherwise have been. However, the energetic impacts of these adaptations on food hoarders, in most cases, have not been examined quantitatively.

The nutritional implications of food storage have been largely ignored. Although the nutritional characteristics of stored food may be relatively unimportant to short-term food hoarders, long-term hoarders may be severely affected if the quality of food deteriorates in the cache or if cached food is of lower quality than foods that are consumed when foragers encounter them. Stuebe and Andersen (1985), for example, found that pocket gophers preferentially consumed food items that were high in nitrogen (i.e., high protein) and stored items that were low in nitrogen. What is the consequence of this foraging decision for the nutritional health of the hoarder during periods of food scarcity? Do animals that rely on stored food to pass long periods of food scarcity select items that are sufficient to meet their nutritional as well as energetic requirements? Having selected these food items, do animals consume them in a way that maximizes long-term nutritional benefits? Only a few studies have addressed these questions (e.g., Reichman and Fay 1983; Stuebe and Andersen 1985).

The extent of cache use and implications for other aspects of life history are poorly known. For example, the amount of food stored by most food hoarders is not well documented. A great deal of descriptive work is needed on the quantities of food stored, the energetic and nutritional qualities of this food, and the position of stores relative to the hibernaculum or area of activity. The timing of recovery and use also is poorly known. Progress in our understanding of the use of stored food will come in three areas. First, we need more precise information on the temporal pattern of consumption of stored food by free-ranging animals. This has been done precisely on arthropods that provision cells, as illustrated by Cross et al. (1978), but there is great potential for this type of study in some birds and mammals. Second, modeling of food consumption derived from laboratory measurements of energetic rates and field data on quantity of food stored will also contribute to our understanding of the importance of stored food to animals. This technique has been used by Novakowski (1967), Brown and Bartholomew (1969), Wolff and Bateman (1978), and Wrazen and Wrazen (1982). And third, direct experimental manipulations of animal food stores in the field and laboratory are needed. The consequences of accelerated depletion and augmentation of food stores will be very enlightening. The studies of Strassman (1979) and James and Verbeek (1984) illustrate this approach.

3
The Evolution of Food Hoarding

The term *food hoarding* covers a variety of behaviors that are united by two common criteria: postponement of food consumption and food conservation through special handling. A diverse array of activities is included. Some animals hide small amounts of food to be eaten later the same day, whereas others store large quantities of food during seasons of plenty to be recovered months later when food is scarce. Certain species store food in small parcels specifically to nourish their offspring; others accumulate food for their own consumption. Some animals concentrate food in a large larder, and others scatter food items in numerous inconspicuous sites throughout their home range. What selective forces have caused this diversity of hoarding behaviors? The habit of storing food has arisen independently in numerous evolutionary lines in taxa as diverse as spiders, beetles, bees, woodpeckers, jays, bowerbirds, moles, carnivores, rodents, and pikas. In some groups (e.g., bees, carnivores, rodents), the behavior has apparently evolved more than once. The frequency with which the habit has been acquired and the taxonomic diversity of the animals involved suggest that the environmental conditions that promote hoarding are widespread and the behavioral precursors to hoarding are not uncommon.

In this chapter, the origin and elaboration of food hoarding are considered. I ask four major questions: (1) Why has food hoarding evolved? (2) What proximate conditions have led to the initial establishment of the behavior? (3) How have the morphology and behavior of species that have become highly dependent on stored food for survival or reproduction been affected? and (4) What evolutionary interactions do food hoarders have with the items they store?

3.1 WHY HAS FOOD HOARDING EVOLVED?

Food hoarding (exclusive of provisioning) is thought to have evolved in response to temporal food scarcity or nutritional deficit caused by the interaction of two variables: food availability and energy demand. When periods of food surplus precede food shortages or nutritional deficits, food hoarding is an evolutionary option to avoid the energetic "crunch." Food hoarding can potentially evolve whenever two conditions are met: (1) the fitness of a hoarder exceeds that of a nonhoarder, that is, whenever the presence of stored food increases the probability that a hoarder will survive, mate, or successfully rear young above that probability experienced by a nonhoarder, and (2) an animal exhibits some genetically based, food-handling behavior on which natural selection can act to produce incipient hoarding. The second condition is the subject of the next section of this chapter.

Andersson and Krebs (1978) have expressed the first condition, the adaptive value of hoarding, more precisely in a mathematical model:

G = fitness gain from eating one stored item during a future shortage period

C = fitness cost of hoarding one item

p = probability that the hoarder recovers any one of the items it has hoarded

p_r = probability that an item not hoarded remains available until the shortage season and is then found by the forager

m = multiplication factor for food items not hoarded, between hoarding and shortage seasons (for seeds, $m \leq 1$ normally, whereas $m > 1$ is possible for growing food items; $m < 1$ refers to decay of nonstored seeds)

Andersson and Krebs (1978) then reasoned that a hoarder (H) has an expected fitness gain of

$$F_H = Gp - C$$

for each item stored. In comparison, a nonhoarder (N) that does not store an item but returns to the site where it was found at a later date expects a fitness gain for the item of

$$F_N = p_r mG.$$

For food hoarding to be adaptive, it is necessary for the fitness gain of hoarders to be greater than that of nonhoarders or $F_H > F_N$. Hence,

$$Gp - C > p_r mG.$$

Obviously, the probability that an animal retrieves an item it has stored (p) has an important influence on the likelihood that hoarding will evolve. Specifically, it is necessary that

$$p > \frac{C}{G} + p_r m.$$

If animals do not recover a sufficiently high proportion of the food they store, the fitness costs of hoarding may outweigh the fitness gains and food hoarders will eventually disappear from the population. Moreno, Lundberg, and Carlson (1981) have argued that p in Andersson and Krebs' (1978) formulation actually consists of two different probabilities. First, the probability that a hidden food item remains available until needed, and second, the probability that the individual that hid the item actually finds it. This distinction is very important (although it in no way invalidates Andersson and Krebs' treatment) because it focuses on the idea that food hoarders have two distinct ways in which they can manipulate p: cache-protecting behavior and cache-recovery behavior. These two important areas of food-storing behavior are treated at length in chapters 4 and 6.

This model examines fitness gain from eating stored food items during a future shortage period but is not concerned with the nature of the food shortage. The model applies equally well to short-term hoarders, such as those that experience food shortages on a daily or tidal cycle, and long-term hoarders, such as those that experience food shortages on a seasonal cycle. Further, the model applies to situations where animals store food so that they can opt to refrain from foraging at some future time, allowing them to allocate time in other activities (e.g., incubating eggs, nursing young) that contribute more to fitness than does foraging (see Food Scarcity and Nutritional Deficit, chapter 2). In these situations, the "shortage period" is simply a consequence of the animal not taking time to forage, but the nutritional deficit the animal would have experienced had it not stored food is, nevertheless, real.

What effect might "cheaters"—nonhoarding individuals that exploit food stored by conspecifics—have on the evolution of hoarding? Cheating could be an important constraint on the evolution of hoarding in social animals. Nonhoarding individuals have foraging behaviors and perceptive abilities similar to those of hoarding conspecifics and consequently might find hidden food effectively. If cheating is sufficiently common in a group, the probability that the hoarder will recover the food may be insufficient to sustain hoarding. There appears to be two ways that hoarders have adapted to this problem. One way is to hide food in a manner that minimizes the possibility that other group members can find it. This is achieved in the same way that hoarders protect food from "robbers" (any animals, including hoarders and nonhoarders, that exploit food stored by other individuals), such as carefully selecting cache sites, camouflaging stored food, or aggressively defending caches. African wild dogs (*Lycaon pictus*), for example, use all three of these techniques to protect cached food from group members (Malcolm 1980). If the hoarder is sufficiently more likely to find food it has stored than are other animals, hoarding may be adaptive even in the presence of nonhoarding conspecifics (Andersson and Krebs 1978). Second, if cheaters are related to the hoarder and if consumption of some stored food by cheaters increases the hoarder's inclusive fitness, then cheating might not constrain or prevent the evolution of hoarding. Communal hoarding is known for numerous social species in which groups are formed of close kin (e.g., honey bees, stingless bees, acorn woodpeckers).

Andersson and Krebs (1978) did not intend their mathematical treatment of the evolution of hoarding to apply to mass provisioning by insects and, indeed, it does not apply. In determining the adaptive value of hoarding, the pertinent comparison is the potential fitness gain at some future time resulting from stored versus unstored items. It is the conserving of food through time that confers adaptedness to hoarders. Movement of food in space also may be important but is secondary in importance to the temporal aspect. In mass provisioning, the relevant comparison is whether fitness gain to the provisioner (in the form of offspring survivorship) is greater if a food item with egg attached is placed in some protected nest or left exposed at the site of capture. In this case, the spatial displacement of food to a favorable microsite confers adaptedness. An appropriate model would be a formulation similar to that of Andersson and Krebs (1978) that compares the fitness gain and cost of moving food to a protected site versus fitness gain and cost of ovipositing on food where it is found.

3.2   HOW DID HOARDING EVOLVE?

Hoarding has evolved many times in taxa with disparate food-handling behavior, which suggests that the proximate steps leading to hoarding behavior may take a diversity of forms. Nevertheless, by comparing hoarding behavior of various groups it is possible to identify several likely paths that have led to hoarding behavior. Here I describe the hypotheses that have been proposed and, when none seems adequate, formulate new hypotheses to explain how hoarding might have come about. These hypotheses are not intended to be mutually exclusive and, in many cases, two or more have apparently acted in concert to foster the development of storing behavior.

Feeding Site Hypothesis

Some animals store food at sites that physically resemble feeding stations. It seems highly probable that the initial step in the evolution of hoarding for these species was leaving food, originally found or captured at some remote site, at a feeding perch (Richards 1958; Turcek and Kelso 1968). Ancestors of these animals may have left food at feeding perches because of satiation, the need to engage in some other behavior (e.g., territorial defense), or to avoid predators. Whatever the reason for leaving the food, the same individual could return and consume the food at some later time. If this residual food proved to be of value to the animal that left it, then any elaboration of the behavior that increased the amount of food left behind or the animal's chance of retrieving the food, such as wedging it securely in a crevice or covering it, would be selected for. Eventually the behavior might acquire independent motivation and cease to be a component of feeding behavior (Ewer 1968).

This situation may have been a critical step in the development of hoarding behavior for most birds. Nuthatches (*Sitta* spp.), for example, wedge food items too large to be ingested into bark crevices and then peck them into small pieces (Richards 1958). They feed in this manner because their feet, which are used to cling to vertical surfaces, cannot be used to hold food. Food items secured in cracks for feeding differ from stored food only in that they are not concealed (fig. 3.1). Woodpeckers often feed on large items by inserting them into cracks and fragmenting them with the bill. Several species of melanerpine woodpeckers, including Lewis' woodpecker (*Melanerpes lewis*), red-headed woodpeckers (*M. erythrocephalus*), and red-bellied woodpeckers (*M. carolinus*), also store insects, acorn fragments, and other food in cracks (Kilham 1958c, 1963; Bock 1970). Hoarding by tits and chickadees (*Parus* spp.) may have originated by the birds leaving food remnants at anvils where they break seeds and large prey apart. Numerous hawks and owls store prey by simply draping them over tree branches (Mumford and Zusi 1958; Stendell and Waian 1968; Catling 1972; Oliphant and Thompson 1976), similar to the way they hold prey at feeding perches. Shrikes (*Lanius*) impale prey or wedge them in branch forks to feed on them. This means of prey handling may have originated when shrikes carried prey to thorny branches where they could more easily dismember them (Smith 1972). Shrikes initially handle prey they store and prey they immediately consume similarly. Australian butcherbirds (*Cracticus* spp.) feed and store prey in a manner similar to shrikes (Amadon 1951). South Island robins (*Petroica australis*) of New Zealand forage for invertebrates on the ground but carry prey to tree branches to consume them (Powlesland 1980). Feeding perches and storage sites share many important characteristics.

For corvids, most of which store food in the soil, obvious parallels between use of feeding sites and hoarding sites are hard to find. Most corvids forage on the ground and fly to elevated perches to feed. At least two scenarios for the initiation of hoarding seem possible. First, hoarding may have arisen in ground foragers that hammered food items apart on the ground surface and may have inadvertently driven food fragments into the soil. Second, food hoarding may have had its beginnings at feeding sites in tree branches or other elevated sites in a manner similar to that described for tits and chickadees. Subsequently, these hoarding sites may have been modified, having shifted over evolutionary time from tree branches to the soil or, in the case of the genus *Perisoreus*, to foliage. The adaptive value of this presumed shift may be that soil and foliage provide more secure sites in which to hide food.

Figure 3.1
Eurasian nuthatch placing a piece of lichen over a seed it has cached in a bark crevice. Original drawing by Marilyn Hoff Stewart.

Security Hypothesis

A requirement for secure feeding sites may have set conditions favorable to the evolution of hoarding. Bindra (1948b), who investigated the influence of a secure feeding site on the hoarding behavior of white rats in a behavioral (not evolutionary) context, termed the motivation to carry food to a refuge the *security hypothesis*, maintaining that areas where animals must forage often are less secure than refuges (e.g., burrows, nests, tree cavities) where they sleep or reproduce. To minimize risk (or maximize security), animals should return to a refuge to eat food items they have gathered. Food left in runways or chambers could be returned to and eaten during inclement weather or during that portion of the light-dark cycle when the animal cannot forage. If food discarded in the nest is sufficient to increase the fitness of the resident, then behaviors that increase the amount of food left in the burrow and its management to prevent spoiling would be selected for. This hypothesis is similar to the feeding site hypothesis in that the incipient food hoarding occurs at a feeding station, but here we have the important constraint that the feeding sites are selected to reduce the probability of predation.

This hypothesis adequately explains hoarding when it is conducted in or near a burrow or cavity in which the hoarder also lives, a practice of moles, shrews, pikas, and numerous rodents. Virtually all small mammals take food to a refuge at the least sign of danger. Mountain beavers (*Aplodontia rufa*), for example, consume food in a special feeding chamber, which also has been called a temporary food storage chamber (Martin 1971) because of the considerable edible refuse that accumulates. Red tree voles (*Arborimus longicaudus*) store foliage in entrances to arboreal nests at night and feed on the material in the security of the nest during the day (Howell 1926). And the larvae of ground beetles store seeds in protected burrows (Alcock 1976).

Hay making has evolved independently in pikas, certain voles, and at least one species of gerbil (*Rhombomys opimus*) and thus is likely to have its origin in some simple behavior pattern (Ewer 1968). All these species construct hay piles near the security of burrows or refuges (see fig. 8.31). A plausible scenario for the evolution of hay making is that ancestors to these species carried vegetation to secure feeding stations near burrow entrances where they consumed the most palatable portion, discarding less palatable parts. Later in the winter, when other sources of food are depleted or inaccessible, the animal may emerge to feed on food remnants at the feeding site (Formozov 1966). Ewer (1968) pointed out that from such a simple beginning, the evolution of hay making requires only two things: an increase in the tendency to cut more than is immediately required and postponement of hay cutting until late in the season, when vegetation is most available. Caches of twigs at burrow entrances (e.g., Nagorsen 1987) may have a similar origin.

Food Envy Hypothesis

Competition for food may have been the driving force behind evolution of hoarding in some species. Food envy, a nearly universal trait of higher vertebrates, often results in an animal protecting food even when it is not hungry to prevent competitors from eating it (Ewer 1968). Envy, in this context, is not meant to imply desire but only that animals in possession of food often behave so as to retain possession. To maintain food as an exclusive resource, an animal carries the food or deposits it in some inconspicuous site. This latter act may readily evolve into hoarding. Any behavior that decreases the conspicuousness of food, such as covering it with debris or placing it in dense vegetation, will increase the probability that the animal that hid it will eventually consume it.

This hypothesis may account for hoarding in situations where a relatively rich resource, which cannot be continually guarded, is likely to attract numerous competitors. Large carnivores that risk losing unguarded prey to scavengers may have evolved storing in this way. Tigers, mountain lions, and bears drag prey into dense vegetation and cover them with grass, soil, or plant litter (Wright 1934; Schaller 1967; Elgmork 1982); hyenas submerge chunks of carrion in lakes (Kruuk 1972); and leopards drag prey up trees to avoid scavengers (Eltringham 1979). Oksanen (1983) suggested that small mustelids (e.g., weasels) hoard as a means of protecting prey from larger members of the carnivore guild, but this hypothesis was rejected by Korpimäki (1987). Barbados green monkeys (*Cercopithecus aethiops*) hide food to prevent stealing by dominant members of the social group (Baulu, Rossi, and Horrocks 1980). Raptors have already been mentioned in the context of the feeding site hypothesis, but it also is possible that hoarding arose as a consequence of hiding prey from competitors.

Dung beetles and burying beetles also appear to have acquired hoarding in response to competition for food. Piles of dung and carrion are rich resources that are sought by numerous scavengers (Heinrich and Bartholomew 1979). A large assemblage of flies and beetles lay eggs on or consume dung and carrion. Animals that use these substances to nourish themselves and their young benefit by appropriating the resources so that other individuals do not interfere. Burying beetles have accomplished this by digging soil out from under a carcass until it is interred, and most dung beetles transport loads of dung to underground burrows. In both cases, any activity that acted to relocate the food may have reduced competition. Dung beetles, for example, that formed brood masses at the dung-soil interface under a mass of dung may have benefited by provisioning the brood masses slightly deeper under a layer of soil. Intensification of this behavior may have resulted in more extensive burrowing and relocation of large quantities of dung. Burying of feces by dung beetles may have served an additional function by preventing the dung from drying out (Halffter and Edmonds 1982).

Food Delivery Hypothesis

Delivery of food to a nest site faster than food can be consumed by offspring or nest mates also may result in incipient food storing (Smith and Reichman 1984). A forager may simply deposit excess food in or near the nest and then leave to continue foraging. Initially, nestlings or nest mates may have consumed this accumulated food within minutes or hours of delivery, but if the food was nonperishable or if the forager handled perishable food to preserve it, the duration of storage may have gradually lengthened.

Owls that accumulate prey at nest sites (Wallace 1948; Pitelka, Tomich, and Treichel 1955; Ligon 1968; Van Camp and Henny 1975; Ritchie 1980; Rich and Trentlage 1983) apparently developed the habit from bringing food to the nest faster than the nestlings could consume it. Similar behavior is exhibited in several species of hawks (e.g., Schnell 1958; Walter 1979).

The development of large larders by honey bees, bumblebees, and stingless bees is related to the evolution of sociality; numerous foragers can, at times, deliver nectar and pollen to the colony faster than it can be used to provision cells or be consumed by larvae or adults. To maintain a continuous flow of these resources into the nest, foraging bees ancestral to modern apids probably deposited their loads in any convenient receptacle: empty brood cells in the case of honey bees and recently vacated cocoons in the case of bumblebees. This method of storage is still prevalent among honey bees, but bumblebees also store some food in specially constructed honey pots and pollen pots, and stingless bees store all pollen and honey in special pots (Michener 1964).

The three types of food storage exhibited by ants (harvester ants, honey ants, and certain carnivorous ants) have parallel but independent origins. Foragers of many ant species deliver food to the nest faster than it can be consumed by adults or larvae, which are fed progressively as they develop. This food is dropped in chambers or passed to other workers in the nest. Only in the honey ants is there a serious storage problem to overcome: where do they put the nectar? It seems likely that this problem was first "solved" by transferring nectar to receiving ants that carried the nectar in their slightly distended gasters. In certain extant ants that store only small quantities of nectar, represented by *Prenolepis imparis* (fig. 3.2), this task is conducted by certain callow workers (repletes), which can hold more nectar because the sclerites of the abdomen have not yet hardened. Early in the evolution of repletes, they probably were able to move about the nest and care for larvae (Wheeler 1908), but eventually, as capacity increased, repletes became helpless receptacles (see fig. 10.11).

Nest-Building Hypothesis

Some have proposed that food hoarding might have evolved from nest-building behavior. Miller and Viek (1944), for example, suggested this as a possibility for Norway rats but eventually rejected the hypothesis as inadequate. The nest-building hypothesis might at first seem plausible because some rodents (e.g., chipmunks) store food under or in the lining of their nests (Broadbooks 1958), and some rodents (e.g., muskrats, voles) occasionally eat portions of their nest (e.g., Stamatopoulos and Ondrias 1987). However, in many ways the hypothesis seems less

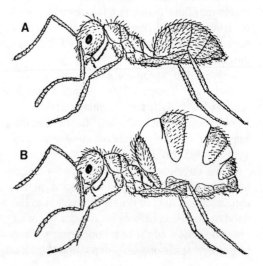

Figure 3.2
The honey ant *Prenolepis imparis.*
A, Worker in ordinary condition.
B, Replete. From Wheeler 1908.

compelling than the security hypothesis. For example, it cannot explain the storage of animal matter in burrows by shrews, moles, and mustelids nor storage of nuts and seeds in burrows by various rodents. The nest-building hypothesis seems a possible evolutionary explanation only for the storage of vegetation, but most vegetation is stored outside the burrow in the form of haystacks. It seems clear that nest building has not been an important force in the origin of storage behavior. The nest-building hypothesis may, however, explain cache construction by beavers. Beavers (*Castor canadensis*) construct food caches by submerging branches and logs so that they remain accessible under the ice. There are similarities in the ways beavers construct dams, food caches, and lodges, and it seems probable that dam and cache construction may have been derived from lodge construction (Richard 1964). A second example of hoarding developing from building activities, although not nest building, occurs in MacGregor's bowerbird (*Amblyornis macgregoriae*). Fruit storage in this bowerbird occurs near the bower and may have developed from bower decorating (Pruett-Jones and Pruett-Jones 1985).

Protected Site Hypothesis

As pointed out in a previous section, solitary wasps and bees that mass provision larvae gain fitness by the spatial displacement of food to protected microsites where their larvae can develop undisturbed. This behavior almost certainly evolved from a more generalized parasitoid ancestor that paralyzed prey and then laid a single egg on it but made no attempt to conceal or protect its developing offspring (Evans 1958; Malyshev 1968). In living Hymenoptera that exhibit this parasitoid behavior (e.g., members of the superfamily Scolioidea), the single prey has to be as large or larger than the parasitoid to allow the complete development of the larva. These prey are left exposed to physical (e.g., sun, wind, rain) and biotic (e.g., scavengers and other parasitoids) hazards. Any activity of ancestoral parasitoids that moved or concealed both the developing larva and its food supply may have increased the larva's chance of surviving. A number of wasps, including several bethylids (small parasitoid wasps) and some primitive pompilids (spider wasps), move paralyzed prey to natural cavities or crevices—behaviors that may represent the initial stage in the evolution of provisioning. The bethylid wasp *Epyris extraneus*, for example, attacks and paralyzes the larvae of ground beetles (Tenebrionidae). It drags a larva a short distance to a crack in the soil, buries it 2–3 cm deep (without constructing a cell), and finally lays an egg on its abdomen (Williams 1919). In some situations, the struggle between the prey and the parasitoid may bury, partially or completely, the intended provisions (Malyshev 1968). These evolutionary stages, through a gradual series of refinements, apparently have led to the provisioning behavior exhibited by some modern spider wasps and tiphidid wasps (Tiphiidae), which attack large prey that live in burrows and then hide the victims in their own retreats. Further evolutionary refinements of provisioning behavior in spider wasps include dragging prey to suitable storage sites where the wasp digs a special chamber for the single item.

In the most advanced form of provisioning in this group, some spider wasps prepare the nest before searching for a spider.

The evolutionarily more advanced sphecid wasps exhibit more complex provisioning behavior than that practiced by spider wasps. First, many sphecid wasps provision nests with numerous small prey rather than one large prey. This change allowed these wasps to exploit a wider range of prey and transport each item more quickly, usually in flight, to the nest (Evans 1959). Second, nest architecture has become more diverse and complex. Sphecid wasps nest in rotten wood, construct chambers from mud, or burrow in the soil. Most species construct nests with multiple chambers for rearing numerous larvae.

It seems ironic that some of these later improvements in efficiency of provisioning behavior should eventually lead to the loss of mass provisioning in some groups of sphecid wasps and in vespid wasps (Vespidae), but this nevertheless seems to be the case. The use of numerous small prey to provision each cell set the stage for a shift from mass provisioning, where all prey are taken to a cell before the egg hatches, to progressive provisioning. The step was simple. Some wasps acquired the habit of laying eggs on the provisions before all the prey were brought to the nest (Evans 1966a; Malyshev 1968). Interruption of provisioning by inclement weather or darkness may have resulted in the egg hatching before provisioning was complete. This simple event has had profound consequences. It permitted the larva and the adult to come in contact for the first time. Adults could monitor larval development, bring more prey as needed and, in certain vespid wasps, partially prepare food by masticating it and then feeding the young directly. Furthermore, the maternal wasp could protect the larvae from parasites and predators. Mortality rates of eggs and young larvae are much lower in progressive provisioners than in mass provisioners (Evans 1966a). This transition is illustrated especially well by digger wasps of the genus *Ammophila* (Evans 1959; Powell 1964; Parker, Tepedino, and Vincent 1980). This sort of contact between adults and young is termed subsocial behavior and is an important step in the evolution of eusociality (Wilson 1971; Michener 1974).

A major distinction between bees and wasps is that most bees provision nests with a mixture of pollen and nectar, whereas most wasps feed their young or provision nests with animal matter. Bees, which except for some advanced forms mass provision nests or are parasitic on species that mass provision, are thought to have evolved from sphecid wasps that provisioned nests progressively. Many sphecid wasps feed as adults on pollen and nectar, so it is not difficult to envision how some wasps may have begun feeding young with a mixture of these. Once a diet of nectar and protein-rich pollen mixed to a thick consistency by the maternal wasp had been adopted, mass provisioning may have eventually occurred by providing these in quantities that exceeded immediate needs. Primitive bees behave similarly to wasps but with the innovation that they often line cells with salivary secretions. This lining apparently prevents drying or flooding of the provisions in the cell (Michener 1964). Some of the more advanced bees (i.e., honey bees and bumblebees—Apidae), have returned to progressive provisioning of larvae. Honey bee workers, for example, feed on honey and pollen and then feed larvae secretions from their pharyngeal glands. But in both of these groups, and in stingless bees as well, nectar and pollen are stored elsewhere in the colony in specially constructed pots or in cells identical to brood cells.

Provisioning and hoarding have been associated with and, in some cases, have had important evolutionary impacts on nesting behavior, parental care, and the development of sociality. A full account of the role of food storage on evolution of behavior in the Hymenoptera would take us far beyond the objectives of this book. Readers seeking more background on provisioning and the evolution of sociality should consult Evans (1958, 1966a), Malyshev (1968), Wilson (1971), Michener (1974), and Iwata (1976). Details of arthropod provisioning and hoarding behavior are discussed in chapter 10.

3.3 EVOLUTION OF SPECIALIZED FOOD HOARDING

Animals that face periods of food scarcity often require more than just a few food items accumulated in a burrow or a handful of nut fragments hidden away near feeding perches to overcome the energetic demands of a lean period. Environmental situations that place animals in energetic or nutritional stress, whether of short or long duration, favor individuals that accumulate large quantities of food. The benefits derived from a large food store have acted as strong selection pressures for the evolution of behavioral and morphological adaptations that increase the efficiency of harvesting and storing. These include the ways in which food items are collected, transported to storage sites, cached, protected, and eventually retrieved. Behaviors not directly related to hoarding also have been influenced in some evolutionary lines, for example, where animals spend the lean season, when they reproduce, and what they feed their young. Here I consider how the behavior and morphology of food hoarders have been modified in those evolutionary lines that have incorporated hoarding as a major component of their adaptive strategy. I emphasize food transport, cache dispersion patterns, hoarding intensity, eruptive migrations in response to shortages of storable foods, and evolution of food storage in social species. Two important types of adaptive specializations, cache-protecting behavior and cache-recovery behavior, will be treated in subsequent chapters.

Specializations for Food Transport

Animals that harvest ephemeral, widely scattered resources often benefit if they can transport more than one food item at a time. This advantage is especially important to animals that store food because the quantity of stored food may be important to the hoarder's well-being during future food scarcity. Larders and areas where scatter hoarders concentrate hoarding activities are central places to which foragers frequently return (Orians and Pearson 1979; Kramer and Nowell 1980; Smith et al. 1979; Carlson 1985; Covich 1987; Elliott 1988). An efficient forager must, among other things, carry larger loads as it forages at greater distances from its central place to compensate for the greater time and energy expended traveling. In response to the selection pressure to maintain efficient food transport, various morphological structures and associated behaviors have evolved to increase the number of items carried (i.e., multiple-prey loading) or to carry a single item more quickly or less encumbered. Obviously, food-transporting structures and efficient food-carrying behavior are not unique to food-hoarding animals. Many non-hoarding birds such as alcids, grouse, hummingbirds, and certain finches have crops or pouches in which they carry food (e.g., Portenko 1948; Fisher and Dater 1961). Furthermore, food-transporting structures of hoarders may be used in contexts other than hoarding (e.g., carrying food to nestlings). Nevertheless, a comparison of these structures in food hoarders can tell us much concerning how food storing has evolved.

Cheek pouches of rodents are paired, distensible cavities opening at the angle of the lips (Chiasson 1954; Long 1976; Hardy et al. 1986; Ryan 1986; Brylski and Hall 1988a, 1988b). In some species of the super-families Sciuroidea (e.g., chipmunks and ground squirrels) and Muroidea (e.g., hamsters, deer mice, pouched mice [*Saccostomus*], and African giant rats), cheek pouches are evaginations of the oral cavity opening just inside the mouth (fig. 3.3). The pouches of kangaroo rats, pocket mice (Heteromyidae), and pocket gophers (Geomyidae), in contrast, are fur-lined structures located just outside and lateral to the mouth. Pouch volume of heteromyid rodent species is proportional to body mass (fig. 3.4), although energy density of preferred food also may be important in predicting pouch capacity (e.g., herbivorous species have larger cheek pouches than granivorous species of similar body mass) (Morton, Hinds, and MacMillen 1980). In general, the quantity of food that heteromyids can carry in their cheek pouches is approximately equivalent to their daily energetic requirements (Morton, Hinds, and MacMillen 1980). Cheek pouch volumes of most other groups of rodents appear not to have been

Figure 3.3
The eastern chipmunk transports seeds, nuts, and other foods in its internal cheek pouches. Photograph courtesy Lang Elliott.

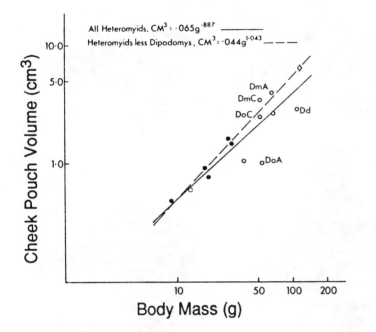

All Heteromyids, $CM^3 = \cdot065g^{\cdot887}$ ⎯⎯⎯
Heteromyids less Dipodomys, $CM^3 = \cdot044g^{1\cdot043}$ ⎯ ⎯ ⎯

Figure 3.4
Double logarithmic relationship between cheek pouch volume and body mass in heteromyid rodents. Species symbols: Filled circle, *Peromyscus* spp.; open square, *Microdipodops megacephalus*; open circle, *Dipodomys* spp.; open diamond, *Thomomys bottae*; Dd, *Dipodomys deserti*; DmA and DmC, *Dipodomys microps* from Arizona and California; DoA and DoC, *Dipodomys ordii* from Arizona and California. The regression lines are derived from the absolute value for each specimen, and their equations are indicated in the figure. From Morton et al. 1980.

measured, but pouch contents give an indication of relative capacity. For example, the cheek pouches of deer mice (*Peromyscus maniculatus*) are able to accommodate a "small pea" (Hamilton 1942), and hamsters (*Cricetulus*) can transport forty-two soybeans (Nowak and Paradiso 1983). Studies that compare pouch capacity of related, similar-sized species that differ ecologically (e.g., diet, seasonality of habitat, patchiness of food resources, tendency to deposit fat rather than store food as an energy reserve) would help to clarify the adaptive significance of cheek pouches, as would studies that explore the relationships between maximum foraging range and cheek pouch capacity.

Corvids are the only food-hoarding birds that possess special structures in which to transport food to cache sites. (Some hawks have a crop, but they apparently never carry prey in it that they intend to cache.) Eurasian jays (*Garrulus glandarius*), blue jays (*Cyanocitta cristata*), Steller's jays (*Cyanocitta stelleri*), and pinyon jays (*Gymnorhinus cyanocephalus*) have distensible esophagi in which they carry acorns and nuts (Novikov 1948; Eigelis and Nekrasov 1967; Bock, Balda, and Vander Wall 1973; Vander Wall and Balda 1981). Magpies, crows, and ravens carry food in an antelingual pouch, a distensible sac between the forks of the lower mandible (Eigelis and Nekrasov 1967), and nutcrackers have a sublingual pouch (Bock, Balda, and Vander Wall 1973). The sublingual pouch and antelingual pouch are similar except that the sublingual pouch is positioned under the tongue apparatus instead of in front of it (Bock, Balda, and Vander Wall 1973). The capacity of these structures varies considerably (fig. 3.5, table 3.1)

Dung beetles exhibit a variety of behavioral and morphological adaptations for moving dung. Burrowers, such as *Heliocopris* and *Onthopha-*

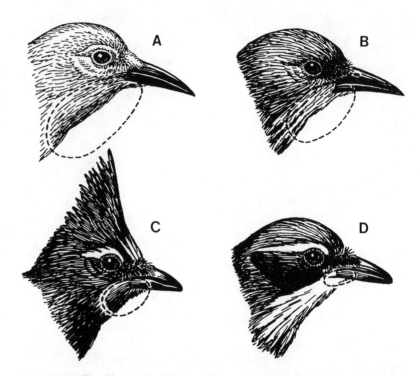

Figure 3.5   Profiles of four conifer-seed-storing corvids. A, Clark's nutcracker. B, Pinyon jay. C, Steller's jay. D, Scrub jay. The broken lines indicate the relative capacity of food transporting structures. Original drawing by Marilyn Hoff Stewart.

gus, are robust with a pronotum shaped like the blade of a bull-dozer, powerful legs, shovellike front tibiae, and spines on the middle and hind legs that endow good traction as they plow through the soil (see fig. 10.2; Heinrich and Bartholomew 1979; Halffter and Edmonds 1982). But the most intriguing dung beetles that transport food are the ball-rollers. Ball-rolling dung beetles have evolved a complex repertoire of behaviors for cutting, forming, and rolling balls of dung away from a fecal deposit. They have relatively long, thin legs suited for cursorial locomotion and rakelike front tibiae that they use to form the ball. A single beetle or a mated pair working as a team fashion a nearly spherical ball from the margin of a dung mass. Position of the sexes during ball rolling varies among genera. For example, male *Sisyphus spinipes* often push the ball with their hind legs as they walk backward while females pull the ball with their front legs (fig. 3.6).

Burying beetles can move carcasses many times their own weight to suitable burial sites. Burying beetles, usually working in groups comprised of one or more mated pairs, move a carcass by burrowing under it on the side nearest the future burial site. Just under the carcass they roll over on their backs, lift the carcass off the ground slightly, and crawl upside down under the carcass. As they move under the carcass, they push it in the opposite direction a few millimeters closer to the burial site (fig. 3.7). Upon emerging on the opposite side of the carcass, the beetles

Table 3.1 Capacity of food-transporting structures of jays, crows, nutcrackers and related corvids

Species	Transport Structure	Capacity[a]		Percent of Body Mass[a]	Reference
		Vol. (ml)	Mass (g)		
Scrub jay	Mouth	1.6	1.5	1.9	Vander Wall and Balda 1981
Steller's jay	Distensible esophagus	5.8	5.4	5.1	Vander Wall and Balda 1981
Pinyon jay	Distensible esophagus	17.9	16.8	15.5	Vander Wall and Balda 1981
Jackdaw	Antelingual pouch	4	6	3	Eigelis and Nekrasov 1967
Black-billed magpie	Antelingual pouch	8	11	5	Eigelis and Nekrasov 1967
Carrion crow	Antelingual pouch	18	19	4	Eigelis and Nekrasov 1967
Rook	Antelingual pouch	18	22	5	Eigelis and Nekrasov 1967
Clark's nutcracker	Sublingual pouch	28.5	30.6	21.6	Vander Wall and Balda 1981

[a] Data from Vander Wall and Balda (1981) are estimates of maximum capacity; data from Eigelis and Nekrasov (1967) are apparently estimates of mean capacity.

run half-way around the carcass and crawl under it again to repeat the process. Burying beetles exhibit surprising plasticity in behavior when moving carcasses, maneuvering around obstructions and cutting roots and vegetation that get in the way (Milne and Milne 1976). Burying beetles can move a full-grown mouse 1 m/hr, and they can maintain this pace for as much as 2.5 hours (Milne and Milne 1944).

Wasps have evolved prey-carrying mechanisms that increase rate of transport and that free the mandibles and legs, allowing the wasp to engage in other activities while transporting prey (Evans 1962b). Wasps that exhibit the most primitive means of carrying prey (tiphidids, bethylids, and some pompilids) paralyze and then grasp the prey with the mandibles and drag it while walking backward toward a hole or nest site. Because these prey are often very large, progress is slow and laborious. Some more advanced pompilids and some sphecids straddle prey and carry it in their mandibles forward under the body. Most wasps that carry prey this way have long legs and high stance so that the prey does not impede walking. Some pompilid wasps amputate the legs of spiders, permitting less encumbered transport (e.g., Kurczewski and Spofford 1986). Those wasps that provision with several small prey carry prey in flight. Among these wasps, there has been an evolutionary trend to carry

Figure 3.6   A pair of dung beetles (*Sisyphus spinipes*) rolling a dung ball away from a dung pat. Courtesy of CSIRO, Australia.

the prey more posteriorly. In the most primitive condition, many sphecids and all vespids carry prey in their mandibles alone. Some of the more advanced sphecid wasps carry prey in the mandibles assisted by the legs or in the legs alone (fig. 3.8), but the most advanced forms carry prey near the tip of the abdomen, either on the barbed sting (e.g., *Oxybelus* spp.) or on a highly modified apical segment (e.g., *Aphilanthops haigi*) (Evans 1962b; Steiner 1978). The more advanced forms of transport free the mandibles and forelegs, enabling wasps to open nests without dropping prey and leaving them unattended, thereby reducing the probability of kleptoparasitism.

Bees have evolved special structures for transporting nectar and pollen. Bees carry nectar in a honey stomach located in the anterior part of the abdomen. Honey bees can transport on average 61 μl of nectar, equivalent to 71 mg, or 82% of body mass (Wells and Giacchino 1968). Bees carry pollen in pollen baskets or corbiculae, formed from a ring of specialized setae on the outer surface of the hind tibiae (fig. 3.9) or adhered to specialized scopal setae located on the abdomen or hind legs. Honey bees carry pollen loads ranging from 12–29 mg (Park 1922). The sizes of pollen loads that bees carry are proportional to the sizes of their corbiculae (Milne and Pries 1986), and nectar loads are proportional to the capacity of their honey stomachs (Wells and Giacchino 1968). Provisioning loads carried by the solitary bee *Osmia lignaria* can be as much

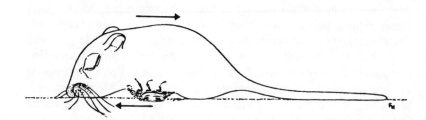

Figure 3.7   Movement of a mouse by a burying beetle (*Nicrophorus* spp.). From Halffter et al. 1983.

Figure 3.8
Female sand wasp (Sphecidae) in flight with a paralyzed horsefly. Photograph courtesy William Sheehan.

as 20.4% of body mass (mean = 7.1%) (Phillips and Klostermeyer 1978). Certain solitary bees of the families Melittidae and Anthophoridae transport floral oils used in mass provisioning on specialized setae located on the ventral surface of the abdomen and on the tarsi (Roberts and Vallespir 1978; Neff and Simpson 1981; Cane et al. 1983).

As mentioned previously, these various food-transporting structures and related behaviors allow food-hoarding animals to carry larger loads or to carry food more quickly. But another equally important benefit is that spacious cheek pouches, sublingual pouches, corbiculae, and the like allow food-hoarding animals to travel greater distances to exploit scattered resources. Corvids illustrate this point well. The maximum distances these birds transport food is related to the capacity of their food-transporting structures (fig. 3.10). Species such as the scrub jay, with relatively unspecialized food-transporting structures, carry only a few pine seeds or small acorns in their mouth and bill and rarely fly more than 1 km (Vander Wall and Balda 1981; DeGange et al. 1989). Clark's nutcrackers, on the other hand, have a voluminous pouch (fig. 3.11), in which they can carry a hundred or more pine seeds (depending on pine species) as far as 22 km (Vander Wall and Balda 1977). The capacities of other corvids are between these two extremes, as are the distances that they carry seeds. Further, Vander Wall and Balda (1981) have pointed out that corvids that transport large loads of seeds long distances generally are much faster and bolder fliers. This difference in flight speed is partially because those species that carry large loads are larger, but the wing shape and flight behavior of species like nutcrackers make them much stronger fliers than the jays and magpies. The combination of mor-

Figure 3.9
A, The corbiculae (c) on the hind leg of a stingless bee, *Trigona pallida.* B, The corbiculae with a large pollen mass. From Michener et al. 1978.

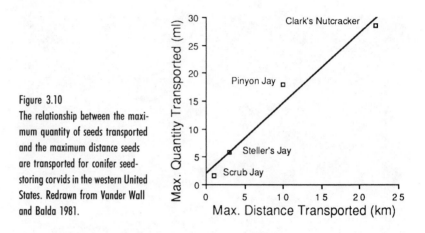

Figure 3.10
The relationship between the maximum quantity of seeds transported and the maximum distance seeds are transported for conifer seed-storing corvids in the western United States. Redrawn from Vander Wall and Balda 1981.

phological and behavioral adaptations of nutcrackers results in a much more efficient exploitation of patchily distributed nut crops than is possible by scrub jays.

The efficiency of food transport has scarcely been examined. Källander (1978) estimated that walnuts (*Juglans regia*) stored by rooks (*Corvus frugilegus*) contained about 28 kcal and that the energetic cost of flight was 0.65 kcal/km. For an average flight of 3 km (1.5 km each way) with one walnut and an assimilation efficiency of 80%, the rook achieves a net gain of 20.5 kcal each trip. This, of course, assumes that the rook finds all the nuts it stores. The efficiency of transport was 1150%, i.e., the energetic gain was 11.5 times greater than the energetic cost. It would be interesting to know more about how energetic efficiency of food transport varies among taxa with different modes of locomotion and between related taxa with food-transporting structures of different capacities. But time expenditure may be more important than energy expenditure in understanding the efficiency of food transport when the harvest season is short. As distance traveled increases, the time per transport trip necessarily increases, resulting in fewer trips and a smaller food store. The distance that honey bees must travel to collect nectar is inversely related to honey production (Ribbands 1953), and pikas near feeding meadows construct larger haypiles than those located at some distance from forage (fig. 3.12; Millar and Zwickle 1972a). These smaller larders are not so much the result of the greater amounts of "fuel" consumed in long transport trips but of the greater time expended in longer trips.

Modification of Cache Dispersion Patterns

How animals distribute stored food—whether in a concentrated larder, widely dispersed caches, or some intermediate pattern (see fig. 1.1)—is a consequence of the conditions under which hoarding evolved, modified through evolutionary time by agents that cause the loss of stored food and by selection for hoarders that efficiently cache, manage, and recover items. Evolutionary changes in hoarding patterns are for the most part obscure because an adequate record is lacking, although there

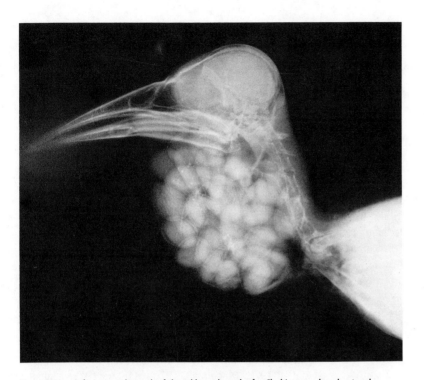

Figure 3.11   Soft-tissue radiograph of the sublingual pouch of a Clark's nutcracker showing the position and size of a large load of singleleaf piñon pine (*Pinus monophylla*) seeds: 28 seeds, 30.6 g, and 28.5 ml. Photograph by author.

is some very fragmentary fossil and subfossil information (Miller 1950; Voorhies 1974). By comparing hoarding behavior of related taxa, however, it is possible to identify situations in which the dispersion pattern appears to have been markedly modified.

Secondarily modified cache-dispersion patterns are especially evident in rodents that exhibit two modes of food storage. Chipmunks, ground squirrels, flying squirrels, and red squirrels (Sciuridae), kangaroo rats and pocket mice (Heteromyidae), deer mice and African giant rats

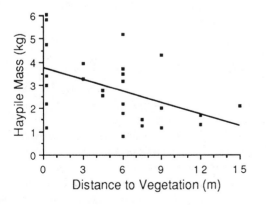

Figure 3.12
Regression of size of hay piles (kg) of mature male pikas in late September against distance (m) from hay piles to feeding meadows in a population of pikas in the Front Range of Colorado. Redrawn from Millar and Zwickel 1972a.

(Cricetidae), and wood mice (Muridae) store food in larders and in scattered surface caches (Scheffer 1938; Hawbecker 1940; Ewer 1965, 1967; Abbott and Quink 1970; Muul 1970; Elliott 1978; Imaizumi 1979; Hurly and Robertson 1987). Larder hoarding seems to be the predominant form of food storage in all these groups and, although it is impossible to know with certainty, it is likely that ancestors of these rodents initially evolved larder hoarding. Scatter hoarding may have been, in some way, derived from larder hoarding or, as Ewer (1968) suggested for the African ground squirrel (*Xerus erythropus*), it may have a separate evolutionary origin. The function of scatter hoarding in these rodents is not always clear. Kangaroo rats may make shallow surface caches to cure seeds before they transport them to burrows for permanent storage (Shaw 1934). Chipmunks may store food in surface caches as a means of reducing risk of cache loss to robbers (Yahner 1975). Whatever the functional significance, the simultaneous presence of scattered caches and a larder suggests that one form of hoarding can lead to and possibly even replace another over time. For example, rodents that store food exclusively or almost exclusively in scattered caches, such as tree squirrels (*Sciurus* spp.) (Stapanian and Smith 1978), may be derived from ancestors that stored food in both larders and scattered caches.

Another example of shifts in cache dispersion patterns can be found in canids. Most canids are scatter hoarders (e.g., van Lawick-Goodall and van Lawick-Goodall 1970; Macdonald 1976; Magoun 1979; Malcolm 1980), although red foxes occasionally store prey in small larders (Fisher 1951). Occasionally, however, red foxes construct large larders that contain dozens of prey (Maccarone and Montevecchi 1981), and arctic foxes regularly construct larders (Gibson 1922; Degerbøl and Freuchen 1935; Pedersen 1966). Scatter hoarding, because it is more common, is likely to be the ancestral pattern, larder hoarding having been derived from scatter hoarding.

Shifts in cache dispersion pattern also are evident in some birds. As pointed out earlier in this chapter, most woodpeckers store food in natural crevices. This behavior probably evolved at scattered feeding sites where only one or a few items may have been left. This pattern is still exhibited, only slightly modified, by Lewis', red-bellied, and red-headed woodpeckers (Kilham 1958c, 1963; Bock 1970). The acorn woodpeckers' habit of excavating numerous, closely spaced holes to receive whole acorns and other nuts has probably evolved from a primitive crack-storing behavior. When more acorns are available than storage holes, acorn woodpeckers hammer acorns into desiccation cracks. Sometimes they widen these cracks to receive acorns (MacRoberts and MacRoberts 1976), a habit occasionally exhibited by Lewis' and red-headed woodpeckers (Bock 1970; Moskovits 1978). Acorn woodpeckers store whole and fragmented acorns in cracks in some tropical regions and in parts of southeastern Arizona (Skutch 1969; Stacey and Bock 1978). The evolution of hole construction for food storage had two important benefits. First, these holes accommodated whole acorns and permitted the woodpeckers to store acorns over longer periods of time with less deterioration of nutritional value (Bock 1970). Second, these holes allowed acorn woodpeckers to store a large amount of food in a small area where they could defend it more economically. The energy required to drill holes was more than compensated for by a large store that could be more easily protected from robbers and microbes and consequently serve the wood-peckers for a longer period of time (Bock 1970).

A shift toward larger caches also can be seen in some of the more specialized nut-storing corvids. Unspecialized jays, such as scrub jays, store single items at each cache site, whereas more specialized food hoarders (e.g., pinyon jays and nutcrackers) store one to eighteen items at each site (Swanberg 1951; Vander Wall and Balda 1981; Balda and Kamil 1989). The larger caches of pinyon jays and nutcrackers may be a result of selection to increase the efficiency of caching and foraging for stored food. Nutcrackers, for example, may store more than 100,000 pine seeds (Hutchins and Lanner 1982; Mattes 1982). By storing seeds in clusters rather than singly, nutcrackers greatly reduce the number of caches they have to make, manage, and retrieve. Why don't nutcrackers make even larger caches and further increase efficiency? The answer, very likely, is that cache size is a dynamic balance between two opposing forces: benefits of increased handling efficiency and costs of greater cache loss to rodents as cache size and olfactory signals increase. Larger caches might be preyed upon by rodents to the extent that nutcrackers ultimately retrieve fewer seeds.

Evolution of Hoarding Intensity

The most fundamental consequence of hoarding, which affects all food-hoarding animals to some degree, is that food stores can supplement the diet during short periods of food scarcity. For species that store only small amounts of food, the impacts of hoarding may not extend very far beyond this simple yet extremely important role. But as temporal food scarcity exerts a greater limiting effect on populations, the adaptive advantage of hoarding increases. Consequently, as severity and length of the lean period increases, the quantity of food stored and the length of time over which stored food contributes significantly to the diet should increase. For warm-blooded species, body size plays a role in this interaction between organism and environment. Metabolic rate increases with body mass$^{0.75}$, whereas maximum internal energy storage increases with body mass$^{1.0}$. Consequently, fasting duration decreases with decreasing body mass (Morrison 1960; French 1988), making small-bodied organisms more susceptible to food scarcity. This is especially true under cold (energetically demanding) conditions at high latitudes. Thus, under identical conditions, small species should hoard more food (as a proportion of total energetic needs) than large species (Smith and Reichman 1984; Korpimäki 1987).

An excellent illustration of differences in hoarding intensity as a function of environmental variability occurs in European (temperate adapted) and Africanized (tropical adapted) races of honey bees. Wild honey bees of temperate regions store more honey than conspecifics adapted to tropical environments (Rinderer et al. 1982, 1985, 1986; Seeley 1985; Danka et al. 1987). Tropical honey bees invest more foraging time accumulating pollen to be used in brood production, which occurs throughout much of the year (Danka et al. 1987). Availability of pollen may limit tropical colonies more than nectar production. In north-temperate zones, on the other hand, colonies must store large quantities of honey to survive the winter successfully (Danka et al. 1987).

Some data suggest that the quantity of food animals store changes along gradients (e.g., moisture, latitude, elevation) of environmental seasonality. Southern flying squirrels (*Glaucomys volans*) in eastern North America, for example, store many more nuts in Michigan than in Louisiana and Florida where more moderate winters are the rule (Avenoso 1968; Muul 1970; Goertz, Dawson, and Mowbray 1975). For beavers, severity of winter food shortage is determined largely by how long ice on ponds prevents terrestrial foraging. In areas where extensive winter ice does not form, construction of food caches often does not occur. Hill (1982), for example, stated that at low elevations south of 38° N latitude, beaver food caches are seldom observed. Horton and Wright (1944) found that dusky-footed woodrats (*Neotoma fuscipes*) below 1,370 m elevation in southern California chaparral usually stored less than 100 g of food for winter consumption, but in similar habitat above 1,370 m, mean winter cache size exceeded 1,250 g. Finley (1958) noted that woodrats at low elevations, such as white-throated woodrats (*N. albigula*), stored less food than bushy-tailed (*N. cinerea*) and Mexican (*N. mexicana*) woodrats, which live at higher elevations. Populations of deer mice and white-footed mice (*Peromyscus*) from high latitudes with severe winters store more than those from low latitudes with mild winters (Barry 1976) but, countering this example, some low-altitude populations store more than high-altitude populations (e.g., Tannenbaum and Pivorum 1987). Tengmalm's owl (*Aegolius funereus*) stores more food in its nest boxes in Finland (2.4 items per nest) than in Germany (1.6 items per nest) (Korpimäki 1987).

Tropical representatives of food-storing groups often store much less food than their temperate or arctic zone counterparts. Eighteen species of owls, all in north-temperate or arctic regions, are known to store food. This pattern is probably caused by the combined effects of reduced importance and accelerated decomposition of stored prey in the tropics. Numerous tropical jays have been observed storing food in captivity (Goodwin 1976), but food caching by tropical jays in the field has rarely been reported, which suggests that tropical jays store little food compared to temperate zone jays. Except for acorn woodpeckers, tropical woodpeckers do not store food, and even acorn woodpeckers in the tropics store much less than do northern races (Skutch 1969; MacRoberts and MacRoberts 1976; Kattan 1988). Not all groups, however, seem to follow this pattern. Tree squirrels (*Sciurus*), for example, store nuts and fruit extensively in the tropics (e.g., Heaney and Thorington 1978; Emmons 1980). Some tropical heteromyid rodents hoard seeds intensively and, furthermore, Fleming and Brown (1975) found little difference in hoarding intensity of *Liomys salvini*, a species that inhabits seasonal dry forests, and *Heteromys desmarestianus*, which lives in wet tropical forests that lack strong seasonality.

Tits and chickadees (*Parus*) exhibit an interesting array of food-hoarding behaviors that seem to be correlated with severity of winter food shortages. Marsh tits (*P. palustris*) and black-capped chickadees (*P. atricapillus*) store a small amount of food each day and retrieve it within 2–3 days (Cowie, Krebs, and Sherry 1981; Sherry 1984a). Tits and chickadees of more northern sites store for longer periods of time. Crested tits (*P. cristatus*), coal tits (*P. ater*), and willow tits (*P. montanus*) in Norway and boreal chickadees (*P. hudsonicus*) in northern Canada store extensively in the fall (Haftorn 1956c, 1959, 1974), and stored food is recovered throughout the winter. The total amount of food stored by the short-term and long-term hoarding tits has never been quantitatively compared, but tits of more northern regions seem to store more intensely during a shorter harvest season.

Eruptive Migrations

Most animals that depend on stored food for sustenance during winter are permanent residents within a territory or home range. Insufficient storable food, however, may occasionally force some of these animals to leave their usual abode and wander in search of food (Formozov 1933; Schorger 1949; Davis and Williams 1957, 1964; Svärdson 1957; Stacey and Bock 1978). These eruptive migrations are especially common in animals that store nuts and conifer seeds, which are characterized by great between-year fluctuations in production (e.g., Smith and Balda 1979). The tendency for food-hoarding animals to emigrate when storable food is scant or absent is important in the discussion of evolution of specialized food hoarding because the severity of eruptive migration is an indirect measure of their dependence on stored food as a winter food.

Several species of food-storing birds (e.g., red-breasted nuthatches, *Sitta canadensis*) move southward or to low elevations during the fall when food is scarce (Bock and Lepthien 1976). Populations of acorn woodpeckers in southern Arizona migrate when storable food is scant, but they store food and spend the winter when acorns are abundant (Stacey and Bock 1978). Nutcrackers are noted for their eruptive migrations. Clark's nutcrackers irregularly show up in coastal areas, deserts, plains, and low mountains, well outside their normal breeding range and preferred habitats (Davis and Williams 1957, 1964; Fisher and Myers 1979). Clark's nutcrackers initiate eruptions during mid-August shortly after they normally begin foraging on new cones (Vander Wall, Hoffman, and Potts 1981). The timing of these movements suggests that the absence of green cones is the stimulus that causes Clark's nutcrackers to emigrate. Eruptions of Eurasian nutcrackers from Siberia often extend to western Europe (Formozov 1933; Andersen-Harild et al. 1966; Hollyer 1970; Conrads and Balda 1979). Notable eruptions of Eurasian nutcrackers have occurred fifty times in a period of about 180 years (1753–1933; Formozov 1933). Phillips, Marshall, and Monson (1964) reported numerous extralimital occurrences of Clark's nutcrackers in Arizona during 14 years between 1864 and 1964.

Among corvids that store pine seeds in western North America, the strengths of eruptive migrations vary and are correlated with the degree to which species are specialized to harvest and store seeds (Vander Wall and Balda 1981). Clark's nutcrackers, as already mentioned, engage in extensive eruptions and are highly specialized both behaviorally and morphologically to harvest pine seeds. Pinyon jays, which share many but not all of the nutcracker's specializations for harvesting seeds, make less extensive eruptive migrations than those taken by nutcrackers (Henderson 1920; Westcott 1964). Steller's and scrub jays rarely emigrate (Mailliard 1927; Westcott 1969), and these species are relatively unspecialized for harvesting pine seeds. Vander Wall and Balda (1981) interpreted this pattern to mean that although all four species of corvids consumed stored pine seeds during winter, nutcrackers and pinyon jays were more dependent on stored food reserves than Steller's and scrub jays.

Eruptive migrations seem to be much less common among mammals, partially due to their more limited mobility, but nevertheless occur in some species. Squirrels (*Sciurus* spp.) migrate from their normal home ranges in response to severe food shortages (Formozov 1933; Schorger 1949; Kiris 1956; Sharp 1959). Extensive eruptions, in which thousands of individuals have been reported swimming rivers or crossing roads (Schorger 1949; Kiris 1956), were reported much more frequently in the past than they are today. Eastern gray squirrels begin deserting their home range by early August during years when little or no mast is produced, and virtually all squirrels may be gone by early September (Sharp 1959). When acorns and nuts are produced, they are conspicuous on trees by late July. The coincidence of emigration and nut ripening suggests that the lack of ripening nuts may trigger migrations long before a food shortage actually occurs.

The frequency of widespread occurrence of animals outside their normal geographic ranges probably underestimates the frequency of eruptive movements and distorts our perception of the role of these movements. Vander Wall, Hoffman, and Potts (1981) observed large numbers of Clark's nutcrackers migrating within their normal geographic range during three consecutive falls; in none of these years were extralimital occurrences of nutcrackers reported. The most likely explanation for these observations is that regional cone crop failures occurred somewhere in the nutcracker's range and that these migrations resulted in birds finding cone crops elsewhere within their range. Widespread cone crop failure occurs infrequently, and only in these years do numerous extralimital sightings occur. Kiris (1956) stated that little attention has been paid to the more frequent but less dramatic migrations of Eurasian red squirrels (*Sciurus vulgaris*), which take place virtually unnoticed within extensive forests. These movements are apparently part of an adaptive strategy to find and exploit food resources early in the fall at a time when food can still be stored for winter. Extralimital occurrences of species like nutcrackers and sightings of squirrels in unsuitable habitats are simply an indication that this adaptive strategy sometimes fails.

Communal Food Storage

Andersson and Krebs (1978) have argued on theoretical grounds that the conditions promoting the evolution of food hoarding are more restrictive for social or group-living species than for solitary animals because some members of a social group could "cheat" by consuming food stored by others without storing food themselves, thus avoiding the associated costs of food storing. The presence of cheaters reduces the potential fitness gains to hoarders and thereby reduces the selective advantage of hoarding. Nevertheless, numerous social species store food. This behavior has taken two general forms. First, members of some social groups hide food in inconspicuous sites, often away from the group's center of activity. The individual that stored the food item is much more likely to find it than are other members of the group. African wild dogs, wolves (*Canis lupus*), Florida scrub jays, and pinyon jays store food this way (Balda and Bateman 1971; Magoun 1979; Malcolm 1980; DeGange et al. 1989). Hoarders might share the retrieved food with other group members, especially young, and some stored items may be "accidentally" discovered by group mates, but the behavior of the storing individual suggests that stored food is private property. Second, group members construct a communal cache and then share the food supply freely among themselves. The food store is often a focal point for social activity. Acorn woodpeckers, several mammals (e.g., Brandt's vole, taiga vole, Mongolian gerbil, and beaver), and a variety of social insects (i.e., all food-storing ants, some paper wasps (Vespidae), and social bees of the family Apidae) store food in this manner (Howard and Evans 1961; Formozov 1966; Novakowski 1967; MacRoberts and MacRoberts 1976; Strassman 1979; Wolff and Lidicker 1981; Seeley, Seeley, and Akratanakul 1982; MacKay and MacKay 1984; Ågren, Zhou, and Zhong 1989). All group members have the same probability of retrieving a particular stored item.

The second type of social food storing is truly communal. In addition to those taxa already mentioned, several species have been described as communal food hoarders. Haftorn (1954), for example, argued that food stored by crested tits in the spruce forests of Norway was a collective resource used by all members of the local population. Haftorn (1956b) made a similar assertion for the coal tit. These and similar claims for scatter-hoarding species have not been verified and it seems prudent, in light of the findings of Sherry, Avery, and Stevens (1982), Shettleworth and Krebs (1982), and Sherry (1984a)—which showed that memory of specific cache sites strongly influences the probability of cache recovery—to assume that these tits do not use stored food collectively until it is demonstrated otherwise. Further, species such as Clark's nutcracker that share a common or communal cache area but that do not share the cached food (Vander Wall and Balda 1977) are not communal food storers, as Roberts (1979) has suggested. Numerous species of halictid bees and social sphecid wasps store food communally, but these stores are used as provisions for developing larvae (e.g., Evans 1958; Batra 1964; Salbert and Elliott 1979; Alcock 1980; Evans and Hook 1982), not as a food reserve for the adults.

Two conditions seem to be common to all or nearly all species that freely share a communal food cache. First, most groups that store food communally are composed of closely related individuals that maintain highly integrated social units persisting over long periods. These conditions have apparently fostered the evolution of communal hoarding through the actions of kin selection or reciprocity (e.g., Mumme and de Queiroz 1985). Under these conditions, the evolutionary constraints imposed by cheaters are lessened because the inclusive fitness gain to the hoarder increases if the cheater and hoarder are closely related. Voles appear to be an important exception to this generalization. Groups of five to ten taiga voles (*Microtus xanthognathus*) that construct communal winter food caches appear to be random aggregations of unrelated individuals (Wolff and Lidicker 1981). Further, the groups do not persist for longer than one winter season (8–9 months). Formozov (1966) described the communal winter groups of Brandt's vole (*M. brandti*) as "families" but gives no information on relatedness of group members. Communal construction of a large winter food cache and nest endows participants with several advantages, including thermoregulatory savings, reduction of risk to predators, and accessibility to mates in late winter. The mechanism that promotes cooperative food storing by unrelated individuals is, however, more difficult to envision. Wolff and Lidicker (1981) offer several possible explanations, including interdemic selection (groups with few cheaters would survive better than groups with many cheaters) and the possibility that isolated populations of taiga voles are inbred and thus even though groups are not families, individuals are still closely related.

The second condition common to all communal food hoarders is that food is always stored in a larder where it is conspicuous and available to all members of the group. Equal accessibility to the food store is an essential element of the communal system, and this requirement can only be met in the larder form of hoarding. The communal food store is a central feature in the social organization of many social food-hoarding species. Members of the group are constantly moving to and from the food store to deliver, consume, or manage items. Consequently, the communal larder is a hub of activity for the group, a physical entity that acts to promote interactions and reinforce bonds among group members (fig. 3.13). The food store also is the centerpiece of the social group's adaptive strategy for surviving periods of resource scarcity and is crucial to maintaining the structure of the society, as has been demonstrated very persuasively for acorn woodpeckers (Stacey and Bock 1978; Koenig and Mumme 1987; Stacey and Ligon 1987). Some acorn woodpeckers of southeastern Arizona and southwestern New Mexico exhibit a social system markedly different from that exhibited elsewhere in the western United States, and this difference appears to be directly attributable to availability of storage facilities and a consistent supply of acorns to store

(Stacey and Ligon 1987). Most acorn woodpeckers are communal, permanent residents and form strong social bonds that persist for many years. Only one pair breeds in each social group, but the young are protected, fed, and cared for by all group members (MacRoberts and MacRoberts 1976; Koenig and Mumme 1987). In southeastern Arizona, where acorn production is highly variable, the situation is different. Although some groups of acorn woodpeckers excavate numerous storage holes in which they hammer acorns, most groups store only small quantities of acorns and place these in natural crevices. These birds usually deplete their food stores in early fall and then leave the area individually (Stacey and Bock 1978). These migratory birds return in the spring to breed but do not form communal groups. Furthermore, migratory birds show no mate fidelity, and there is no communal feeding or defense of nestlings. The few groups that maintain granaries, however, behave nearly identically to communal breeding populations studied in central California. This illustrates an amazing behavioral plasticity in these woodpeckers, due largely to the woodpeckers' opportunity to form a communal food store. Stacey and Ligon (1987) argued that the underlying basis for delayed dispersal and cooperative breeding in these woodpeckers is the benefit of philopatry to juveniles, specifically, a high-quality territory with large acorn storage facilities.

Figure 3.13
Acorn woodpeckers perched in a granary tree. The granary is the focal point of acorn woodpecker social groups. Original drawing by Marilyn Hoff Stewart.

3.4 COEVOLUTION OF PLANT PROPAGULES AND FOOD HOARDERS

Plant propagules, including seeds, nuts, bulbs, corms, tubers, and roots, are commonly stored in the ground by food hoarders. A plant propagule that has been stored will experience one of several fates: (1) the propagule may be relocated or discovered by an animal and eaten; (2) if not found, the propagule may nevertheless be in a site unsuitable for establishment, in which case it will eventually decompose; and (3) if hidden in a suitable site, the propagule may establish a new plant (see chapter 7). Thus, some food-hoarding animals may act as dispersal agents as well as predators of plant propagules. If the impact of hoarders is consistently beneficial and affects a significant proportion of the propagules, both the hoarder and the plant may evolve adaptations that facilitate the harvesting of those plant products. Depending on a number of variables, including length of time during which the hoarder and plant interact and the effectiveness of other means of dispersal, the evolutionary response of the mutualists may range from dramatic to negligible (see following examples). Alternatively, if the hoarder acts solely as a predator of propagules, the plant may evolve adaptations that discourage the harvesting of those plant parts by animals. A range of interactions has been documented between food-hoarding animals and their food plants, and some of these interactions have led to strikingly different sets of coevolved (sensu Janzen 1980) traits.

A third group of organisms that plays a role in these plant-animal interactions is ectomycorrhizal fungi. These fungi, which are often overlooked in discussions of plant dispersal by food hoarders, may arrive at a cache site through the actions of the food hoarder or by other means; by facilitating water and nutrient uptake from the soil, they often greatly increase the probability that propagules will establish and survive. Although the significance of the ectomycorrhizal fungi in these plant-animal interactions is not yet fully understood, it seems likely that dispersal of ectomycorrhizal spores to cache sites by food hoarders may have facilitated the mutualistic interactions between food hoarders and plant propagules (Pirozynski and Malloch 1988).

The effect of the red squirrel (*Tamiasciurus hudsonicus*) on lodgepole pine (*Pinus contortus*) is an excellent example of a food-hoarding animal acting as an important selective agent on plant morphology. The result is a set of traits that discourages foraging. Red squirrels store the cones of lodgepole pine and other conifers under a thick layer of cone debris called a midden. Seeds from these cones have virtually no chance of establishing seedlings. When, as in this situation, the food hoarder acts exclusively as a seed predator, the evolutionary response of the plant is to produce seeds that are unattractive and difficult for that predator to harvest. The cones of the serotinous race of lodgepole pine, which have been shaped by the discriminatory feeding of red squirrels (Smith 1970; Elliott 1974) (fig. 3.14A) are small, sessile, spiny and very hard, especially near the base, causing red squirrels to invest much energy when

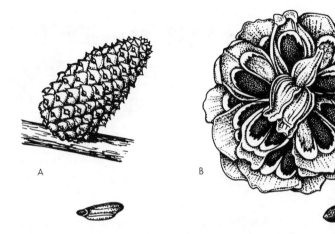

Figure 3.14
The structure of some pine cones are in part a result of selection by food-hoarding animals. Lodgepole pine cones (A) are hard, spiny, and animals have difficulty extracting the small, winged seeds. In contrast, singleleaf piñon pine cones (B) display their large, dark brown, wingless seeds to foragers. See text for further explanation. Original drawing by Marilyn Hoff Stewart.

removing them from the branch and gnawing off bracts to get to the seeds. Cones contain relatively few seeds, and these are small, so the energetic reward for harvesting cones is small (Smith 1970). The seeds are winged and thus disperse by wind after intense heat from a forest fire causes the cones to open. The cones and seeds of lodgepole pine contrast with the wingless-seeded pines dispersed by corvids (see below) in nearly every trait relevant to a vertebrate consumer. The red squirrel, for its part, has evolved stronger jaw musculature to open the sturdy cones (Smith 1970) and a territorial system that permits it to harvest, store, and defend seasonal food supplies (C. C. Smith 1968).

A number of species of soft pines (subgenus *Strobus*) have cone and seed traits that contrast strikingly with those of lodgepole pine. The seeds of these pines are widely collected and stored in the soil by several species of jays and nutcrackers. Besides being wingless (an obvious shift away from wind dispersal, which is the predominant means of seed dispersal in the Pinaceae), seeds are large, nutritious, thin hulled, and borne in cones that lack spines and other protective structures (Vander Wall and Balda 1977; Balda 1980a; Lanner 1982a; Mattes 1982; Saito 1983b). Further, in many species, cones retain their seeds where birds can easily harvest them (fig. 3.14B). This constellation of traits results in a highly attractive and easily harvested food resource. The broad taxonomic and geographic distribution of the wingless-seeded soft pines suggest a polyphyletic origin of winglessness within the subgenus (Lanner 1982b). The group includes all of the approximately nine species of piñon pine (section Cembroides) from southwestern United States and northern Mexico, all five species of stone pines (section Cembrae)—one from the western United States, one from Europe, and three from northern and eastern Asia, two species of pines (section Gerardianae) from southern Asia, and five species of white pine (section Strobi)—three from western United States and northern Mexico, one from central China, and one from Japan (see table 7.3). The composite geographic range of the five corvids known to harvest and store these seeds overlaps broadly the composite range of the wingless-seeded soft pines (Lanner 1982b).

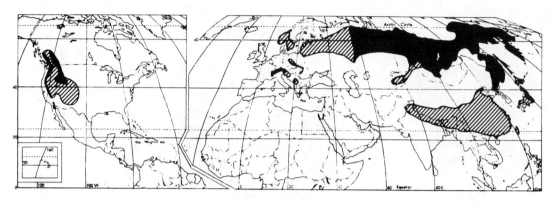

Figure 3.15
The geographic distribution of five species of stone pines (subsection *Cembrae*) (dark areas) and two species of nutcrackers (*Nucifraga*) (dark areas and hatched areas). Nutcrackers are found wherever stone pines occur. In areas where stone pines are absent, nutcrackers forage for and store other wingless pine seeds and hazelnuts.

　　　　　The foraging specializations of food hoarders (described in the previous section) complement the cone and seed adaptations of these pines, suggesting that some of these mutualistic interactions are coevolved (Vander Wall and Balda 1977, 1981; Lanner 1982a; Mattes 1982; Saito 1983b; Tomback 1983). Examination of the probable biogeographic histories of the mutualists strengthens this notion. The nutcracker–stone pine mutualism is perhaps the best example. Stone pines and nutcrackers have broadly overlapping geographic ranges (fig. 3.15). Both groups are thought to have an Old World origin based on their greater taxonomic diversity there (Amadon 1944; Lanner 1982b). Eurasian nutcrackers effectively disperse the four Old World species of stone pine (e.g., Reimers 1953; Mattes 1982; Saito 1983a). Clark's nutcracker is apparently derived from an ancestral nutcracker that crossed the Bering land bridge sometime during the Pleistocene. Similarly, whitebark pine is apparently the descendant of stone pines that colonized the Nearctic region about the same time. The dependence of stone pines on seed dispersal by nutcrackers observed today (e.g., Hutchins and Lanner 1982; Mattes 1982; Saito 1983a, 1983b) strongly supports the notion that these two species colonized the Nearctic region simultaneously, the nutcracker, in effect, bringing the pine with it (Lanner 1982b; Tomback 1983).

　　　　　The biogeographic history of the pinyon jay—a jay of uncertain taxonomic affinity (Amadon 1944)—is poorly known, but the close association of the jays with piñon pines (Balda and Bateman 1971; Ligon 1978) suggests a long evolutionary history. The piñon pines evolved and differentiated in the Mexican highlands and later moved into what is now the southwestern United States (Lanner 1981). Pinyon jays are found wherever piñon pine occurs in the United States, but these jays do not occupy much of the southern half of the composite piñon pine distributional range. Until the mechanics of piñon pine seed dispersal in Mexico are understood, it is premature to speculate on how pinyon jays may have affected the evolution of piñon pines.

The limber pine–southwestern white pine–Mexican white pine complex of western North America also seems to have originated in the Mexican highlands. Mexican white pine (*P. ayacahuite*) is of special interest because its three named varieties differ in seed size and wing length (Lanner 1982b). Typical varieties of southern Mexico have relatively small seeds with large wings (fig. 3.16). In the northern mountains, variety *brachyptera* has larger seeds with only a remnant of a wing. This variety of Mexican white pine intergrades with southwestern white pine (*Pinus strobiformis*) (which has large wingless seeds) in northern Mexico, which in turn intergrades with the wingless-seeded limber pine in northern Arizona (fig. 3.16). Thus, members of the species complex appear to be wind dispersed in southern Mexico and dispersed by seed-caching corvids in the western United States. Lanner (1982b) described the morphological changes in cones, cone orientation, and tree growth form that accompanies these changes in seeds. Is this an example of the evolutionary transition to large, wingless pine seeds? The selective force that initiated this evolutionary change may have been that large, winged seeds established better in the arid southwestern mountains. Seed-caching corvids were attracted to these slightly larger seeds and harvested them preferentially. Being buried deeply in substrates where there was more soil moisture further benefited the pines so that trees from cached seeds predominated in future generations and passed on the genes for larger seeds. Ancestral jays, perhaps antecedents of Steller's jays, may have contributed by dispersing the seeds, as they do those of limber and southwestern white pine today.

Nutcrackers are known to disperse the seeds of both limber and southwestern white pines (Benkman, Balda, and Smith 1984) and piñon pines (Vander Wall and Balda 1977; Ligon 1978), but because of their relatively recent arrival in western North America, nutcrackers could not have been important in the early evolutionary steps toward winglessness in these pines. Today, the range of Clark's nutcracker extends far beyond that of whitebark pine, and nutcrackers opportunistically forage for seeds of all wingless-seeded pines within their range. Although nutcrackers are unlikely to have been important selective agents on these pines in the past, their acquired role in establishing these wingless-seeded pines may be an important evolutionary force today. Clark's nutcrackers appear to be coevolved with whitebark pine and coadapted with limber and piñon pines. A parallel but more obvious distinction between these two types of interactions occurs between Eurasian nutcrackers and the seeds and nuts they disperse. Throughout much of eastern Europe and Asia, Eurasian nutcrackers disperse the seeds of stone pines (e.g., *P. sibirica* and *P. cembra*), with which they are thought to have coevolved (Formozov 1933; Mattes 1982). But in southern Sweden, where neither of these pines occurs, Eurasian nutcrackers store and are an important disperser of hazel nuts (*Corylus avellana*) (Swanberg 1951). Clearly, these nutcrackers and hazels are coadapted but not coevolved.

Another major interaction between food hoarders and plant propagules involves nut-bearing trees such as walnuts, hickories, beeches,

Figure 3.16
The distributions of limber pine
(A), southwestern white pine (B),
and Mexican white pine (C1, C2,
and C3) in western North America.
The morphological differences in
cones and seeds are illustrated to
the right (all cones × 1/4; all
seeds × 2/3). Note the changes in
cone length, cone orientation on
the branch, seed size, and length
of seed wing as one proceeds north
from southern Mexico. See text for
further explanation. Cones drawn
by Marilyn Hoff Stewart.

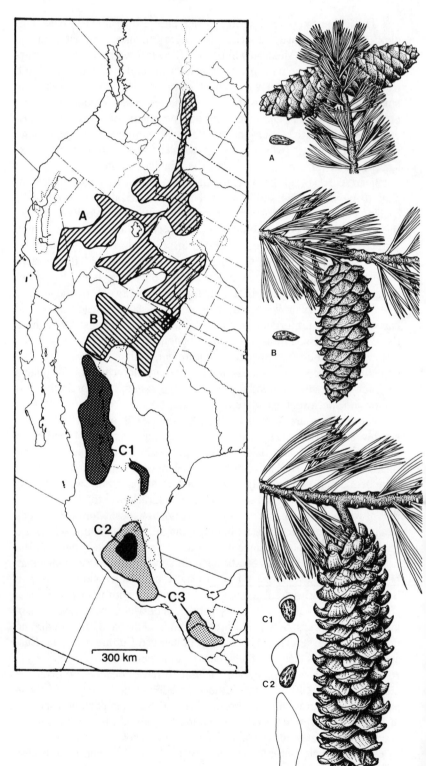

horse chestnuts, and oaks and some tropical trees, which are dispersed by rodents and whose fruits are thought to have evolved in response to selection by rodents. Such nuts are large, nutritious, often thin-hulled, and enclosed in a fibrous, semiwoody husk that, in many species, dehisces at maturity, permitting the nut to fall to the ground. The woody husk of nuts is derived from the floral involucre and, in some cases, is homologous with the fleshy pericarp of fruits. This is especially evident for the almond (*Prunus dulcis*)—a close relative of peach (*P. persica*) and apricot (*P. armeniaca*)—which has a pericarp that is astringent and tough. The practice of food hoarders eating the fleshy pericarp of some fruits and then caching the woody nut within (e.g., Smythe 1970; 1978; Bradford and Smith 1977; Vandermeer 1979) is likely to be the evolutionary antecedent of these nuts, where the reward was gradually transformed from the fleshy pericarp to the embryo and endosperm within the hull. The husk acts as a protective barrier to animals and microbes before nut ripening.

Hull thickness is a malleable character, subject to selective pressures exerted by predators and dispersers of nuts. For example, English walnut (*Juglans regia*), which is native to the eastern Mediterranean region and southern Asia, has much thinner shells than North American walnuts. In England, the thin-shelled species of walnut is stored by rooks (*Corvus frugilegus*) (Källander 1978). One variant of this species called "titmouse walnut" (*J. r.* var. *fragilis*) has shells so thin that even small birds can break the hull (Bean 1981). The Japanese walnut (*Juglans ailantifolia*) also has a thin-shelled variety (Bean 1981).

The evolution of nut characteristics is due largely to the combined selective effects of nut predators, nut dispersers, and germination requirements. Unlike the corvid–pine seed dispersal interactions, it is difficult to identify particular interactions that are likely to result from species-species coevolution. Instead, coevolution between nuts and hoarders seems diffuse, with an array of hoarders having a potential evolutionary impact on a suite of nut-bearing trees and vice versa. Some interactions seem, at first glance, "tighter" than others, such as that between *Juglans* and *Sciurus* in North America, but even these break down when viewed on a larger geographic scale. Other interactions, such as that between agoutis and guapinol, are clearly not coevolved but fortuitous consequences of fruit characteristics and rodent behavior (Hallwachs 1986). This does not mean that the dependence of nut-bearing plants on animals is weaker than that between corvids and pines, only that it is less specific. A more detailed analysis of coevolutionary relationships between hoarders and nut-bearing trees must await better understanding of nut dispersal and the evolutionary history of these plants.

The interaction between propagule and disperser goes beyond the act of dispersal in some oaks. White oaks (subgenus *Lepidobalanus*) and live oaks have evolved a pattern of germination and growth that is apparently an adaptation to reduce acorn predation by squirrels and jays (Barnett 1977; Fox 1982). Unlike acorns of most black oaks (subgenus *Erythrobalanus*), most white oak acorns germinate in fall within a few

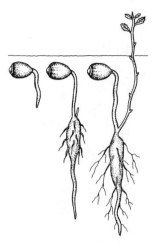

Figure 3.17
The acorns of white oaks (e.g., *Quercus virginianus*) germinate in the fall (left) and translocate energy reserves to a thick and fleshy primary root before winter (middle). The future above-ground stem remains short and inconspicuous until spring (right). Original drawing by Marilyn Hoff Stewart; after Lewis 1911.

weeks of falling from trees. Squirrels and jays store many nondormant acorns before they begin to germinate (e.g., Barnett 1977; Fox 1982). Early stages of seedling growth in white oaks and live oaks are markedly different from those of dormant black oaks. In *Quercus virginiana,* for example, the primary root becomes very thick (about 1 cm diameter) and fleshy soon after elongation (Lewis 1911), the engorged root functioning as an overwintering food-storage organ (fig. 3.17). The future above-ground portion of seedlings is slow growing and inconspicuous. The result of the white oak and live oak pattern of growth is that acorns are abandoned by seedlings as quickly as possible, a pattern that Fox (1982) interpreted as escape from seed predation. If an acorn is recovered by an animal during winter or spring, the connection between seedling and acorn breaks, leaving the seedling relatively unharmed.

This evolutionary parry by the oak is not always successful, however. Gray squirrels have found a way to sidestep the unusual antipredator adaptation of white oaks. Before storing white oak acorns, gray squirrels notched the apices of acorns to excise the embryos (Wood 1938; Fox 1982). Fifty-two percent of *Q. alba* acorns stored in acorn-handling experiments were notched by gray squirrels (Fox 1982). Furthermore, excision of embryos from white oaks has been found to be common in populations of gray squirrels from Michigan, Illinois, New Jersey, and Florida. Squirrels did not excise embryos from acorns of the black oak group, which do not germinate until spring (Fox 1982). By removing embryos from *Q. alba* acorns, gray squirrels effectively prevent fall germination of acorns, avoiding loss of that portion of acorns that would normally be translocated to seedling taproots (Fox 1982). The complexity of interactions between squirrels and white oaks underscores the fact that food hoarders may be both dispersers and predators of propagules.

The acorns of the white and black oak subgenera also differ markedly in chemical composition. Acorns of the black oak subgenus contain relatively high levels of lipids (typically 18–25%) and tannins (6–10%), whereas acorns of the white oak group contain only 5–10% lipids and 0.5–2.5% tannins (Smallwood and Peters 1986 and references therein). The lipids are nutritious, high-energy foodstuffs, and the tannins are thought to interfere with the efficiency of protein digestion. These compositional differences influence choice of acorns by squirrels in experimental trials (Smith and Follmer 1972; Smallwood and Peters 1986) and also in nature where representatives of the two oak subgenera grow in mixed stands (e.g., Lewis 1982). Smith and Follmer (1972) and Lewis (1982) found that squirrels preferred acorns with higher lipid content, whereas Smallwood and Peters (1986) found that squirrels avoided artificial "acorns" with higher tannin content.

The causes and consequences of the compositional differences of acorns are many and, in most cases, are not well understood. Among other effects, acorn composition may have some important influences on the behavior of food hoarders. Smallwood and Peters (1986) have hypothesized that gray squirrels use tannin content as a cue to recognize winter-dormant and, hence, less perishable black oak acorns and that they preferentially store these acorns. On the other hand, they hypothesized, gray squirrels are more likely to eat acorns of the white oak group. Experimental support for these hypotheses is lacking, but if correct, they would have several interesting implications. First, gray squirrels may be more effective disseminators of black oak acorns than of white oak acorns because they are less likely to eat and more likely to store black oak acorns when they first encounter them. This does not mean, however, that gray squirrels would necessarily be more effective in establishing black oak seedlings because the white oak acorns that they bury may have a greater chance of escaping predation due to their lack of winter dormancy. Second, the gray squirrel foraging and caching strategy may not maximize daily energy gain, because they cache the more calorific black oak acorns. If the squirrels recover a higher proportion of the less perishable black oak acorns over the fall and winter, however, their caching of these acorns may be a means of maximizing long-term energy gain. The complex interrelations of acorn composition, packaging, and dormancy, and the foraging and caching behavior of gray squirrels and other predators and dispersal agents comprise a fascinating arena in which to study plant-animal mutualism. This complex knot is only just beginning to be unraveled.

3.5 CONCLUSIONS

Despite the great diversity of food-hoarding species, the conditions that promote the behavior are relatively simple. Food hoarding (exclusive of mass provisioning) is an adaptive response to food shortages or nutritional deficits caused by fluctuations in food availability and energy demand (see fig. 2.1). Food storage conserves food through time. The time scale over which food shortages are experienced may vary from hours to months, and food shortages may be due to environmental changes in food availability or to suspension of foraging by animals so that they can engage in other activities that provide greater returns in fitness at that time. Mass provisioning has evolved in response to different ultimate causes: offspring survivorship is greater when food intended for larval development is moved to a protected microsite. Although mass provisioning is similar to hoarding in many respects, they differ in that mass provisioning is primarily an adaptive response to a spatial problem, whereas hoarding is primarily an adaptive response to a temporal problem. Food-handling behaviors are important proximal steps in the evolution of food hoarding. Nearly all hypotheses proposed to explain the initial steps in the evolution of hoarding concern peculiarities in the ways animals handle prey preparatory to consumption.

Specialized food-hoarding species often exhibit similar adaptive strategies, presumably because the underlying causative factors that shape the behavior are similar for all food-hoarding species. A plausible evolutionary path taken by specialized food-storing species is illustrated in fig. 3.18, which I have adapted from Vander Wall and Balda (1981) as a

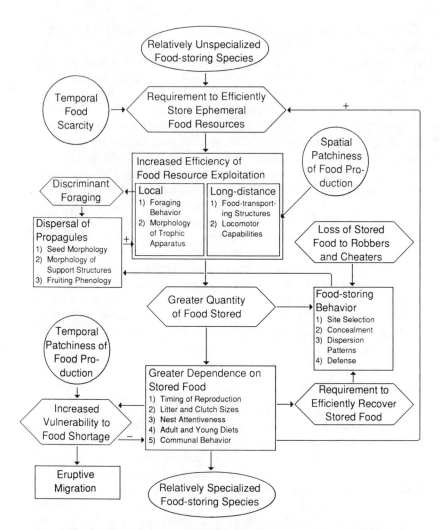

Figure 3.18   A model describing a possible evolutionary path taken by relatively specialized food-storing species. Compartment shapes designate the following: ovals, end points of the path from relatively unspecialized to specialized food-storing species; circles, external driving variables (ultimate causes); hexagons, selection pressures impinging on (or resulting from the activities of) food-storing species; rectangles, adaptive responses of food-hoarding species (or their food resources) to selection pressures. Numbered items within rectangles are major attributes of the adaptive responses. Arrows show the direction of the effects; signs refer to positive or negative feedbacks. See text for further explanation.

convenient means of summarizing the ideas presented in this chapter. This model is general; no one species is likely to exhibit all of the traits described. Moreover, the evolution of provisioning departs from this descriptive model in many important ways. The evolutionary journey starts with a relatively unspecialized food-storing species, represented by an oval at the top of figure 3.18. The ultimate driving variable responsible for the development of food-storing behavior is temporal food scarcity (Andersson and Krebs 1978; Roberts 1979), which for simplicity in the model includes nutritional deficits. This results in selection for individuals that efficiently store ephemeral food resources. The adaptive response of these food hoarders is an increase in efficiency of food resource exploitation. Attributes of this adaptive response can be partitioned into two components: local and long-distance. The local component of resource exploitation includes those behavioral and morphological attributes that influence the harvesting and handling of food items, i.e., preferences of squirrels for the most nutritious nuts (Smith and Follmer 1972), bill morphology of New World jays (Zusi 1987), cone-handling behavior of nutcrackers (Vander Wall and Balda 1981), sorting of edible from inedible seeds by pinyon jays without breaking the hull (Ligon and Martin 1974), strong jaw musculature of red squirrels (Smith 1970), stinging behavior of wasps (Evans 1966a), and pollen-collecting techniques of bees (Michener, Winston, and Jander 1978). The long-distance component of resource exploitation refers to a different set of adaptations that have presumably evolved in response to spatial patchiness in food production. These adaptations include food-transporting structures and locomotor specializations that permit animals to move efficiently between food-harvesting areas and food-storage areas. These adaptations not only allow animals to carry larger loads but permit them to range over larger areas to procure food (Vander Wall and Balda 1981).

The combined result of the local and long-distance components of food resource exploitation is that animals can store greater quantities of food. As the quantity of stored food increased over evolutionary time, the size of the food store would exceed previous needs, the "excess" food then permitting food hoarders to evolve greater levels of dependence. Attributes of this greater dependence include, for example, early spring breeding by pinyon jays (Balda and Bateman 1971), increased prevalence of stored food in the diet of adult, nestling, and fledgling Clark's nutcrackers (Mewaldt 1956; Giuntoli and Mewaldt 1978; Vander Wall and Hutchins 1983), and the fostering of communal behavior in taiga voles, honey bees, and acorn woodpeckers (MacRoberts and MacRoberts 1976; Wolff and Lidicker 1981; Seeley 1985). A high level of dependence on stored food can only evolve and be maintained if food-hoarding animals acquire efficient means of recovering stored items. This and two other selective pressures—the need to manage efficiently the great quantity of food items stored and the threat of loss of stored items to cache robbers, microbes, and cheaters—have molded the ways in which food is stored. Important attributes of this adaptive response are the types of sites selected for storage, concealment of food items, dispersion patterns of stored items, and aggressive defense of larders. Further, the behavior of food-storing animals, especially their propensity to cache in soil and to scatter food caches, has resulted in the dispersal of plant propagules, primarily seeds and nuts, but also roots, corms, and tubers. Plants have entered into a mutualistic relationship with some of these food hoarders by evolving seeds, supporting structures, and phenology of seed production that attract potential dispersal agents and that increase the seeds' chances of establishing from cache sites. These plant characteristics act as a positive feedback, increasing the efficiency of the local component of food resource exploitation.

Due to temporal patchiness of food production (e.g., the fact that the level of food production is not constant from year to year), greater dependence on stored food has resulted in increased vulnerability to food shortages. Specialized food-hoarding species have adapted to this component of food scarcity by eruptive migrations during which they search for productive areas beyond their home range where they can establish residence and store food if such areas exist or, alternatively, by wandering eventually outside their normal habitat or geographic range. This behavior, although it may increase the probability that the emigrants will survive, probably often results in low reproductive success the next breeding season compared to periods of food abundance. Vulnerability to temporal food shortages is depicted as a negative feedback loop in figure 3.18, in that it tends to oppose the evolution of greater dependence. In other words, too great a level of dependence so increases the level of vulnerability to widespread food shortages that such highly dependent individuals are selected against. Greater dependence on stored food, on the other hand, exerts positive feedback, which increases the requirement of hoarders to efficiently store ephemeral food resources. It is this effect that drives species toward greater levels of specialization and dependence. The point at which these two opposing feedback systems balance each other influences the level of dependence on stored food attainable in a particular environment.

Given that food-hoarding animals reap important benefits that make them better able to cope with varying environments, one must ask: Why don't all animals store food? After all, all environments vary to a greater or lesser degree, so it seems that all animals could gain by setting aside a small amount of food to sustain themselves when they are unable to forage or when conditions make it unprofitable. I have placed the many reasons why some animals do not store food into three groups: First, it is important to remember that just because a particular behavior (or structure or metabolic process) would be advantageous does not mean it will evolve. A species or population must exhibit some genetically controlled trait to serve as a raw material from which natural selection can produce the advantageous characteristic. As pointed out earlier in this chapter, hoarding behavior appears to have evolved most frequently from different sorts of food-handling behavior. It seems likely that in many cases where hoarding seems advantageous but has not yet evolved, it may be because the animals do not exhibit a behavior that can act as a precursor to hoarding. A seed-eating sparrow may have as much to gain from storing seeds as do corvids, tits, and nuthatches, but if the sparrow does not exhibit a genetically controlled behavior that can qualify as incipient hoarding, such as hammering and breaking seeds at a feeding perch where some seed fragments could be left behind and returned to later (which appear to have been the initial steps of the evolution of hoarding in several groups of seed-eating birds), then hoarding, regardless of how advantageous an activity it may be, will never evolve.

Alternatively, the necessary genetically based food-handling behavior may be present, but the probability of recovering a stored food item may not be sufficiently high to cause the behavior to persist or spread in a population. If an animal does not retrieve a sufficiently high proportion of the items it stores, the fitness costs of storing will exceed the fitness gains (Andersson and Krebs 1978). Under such conditions, hoarding can neither evolve nor persist. A wide variety of factors can influence the probability that a individual will retrieve the food it has stored, including perishability of the food (i.e., food is lost to microbes); lack of exclusive use of the area where food is stored (i.e., food is lost to robbers); and lack of a sufficiently keen spatial memory to relocate the food (i.e., food is simply lost). Thus food hoarding is much less likely to evolve in species that forage for highly perishable foods and those that lack individual feeding areas.

The third reason that some species have not evolved food-hoarding behavior is that even if a species exhibits an appropriate genetically based, food-handling behavior to act as a precursor for hoarding and even if individuals can effectively retrieve food items they have stored, food hoarding will not evolve or persist in a population if no advantage is gained by storing food. If a nonhoarder can find food in its environment with the same probability than a hoarder can return to a cache site and find a food item which it has stored, there is no advantage to storing food (Andersson and Krebs 1978). This situation may exist in some tropical regions where daily and seasonal patterns of food scarcity are weak or absent. It is, of course, possible (even likely) that two or more of these factors interact to explain why a particular species does not store food.

4

Cache Loss and Cache-Protecting Behavior

Loss of stored food is detrimental to food-hoarding animals. Whether animals store food to bridge long periods of food scarcity, survive short periods when foraging is precluded, or raise offspring, the disappearance or deterioration of stored food may have dire consequences for the hoarder. To avoid these consequences, food-hoarding animals have evolved handling and managing behaviors that increase the probability that stored food will persist at the storage site until it is needed (Vander Wall and Smith 1987). This chapter describes how food-hoarding animals handle and manage food before and after storing it to minimize cache loss. Cache loss can be divided into five types. First, fungi, bacteria, and other microbes may cause cached food to decompose. Second, relatively large vertebrate and invertebrate cache robbers and parasites may find stored items before the original storer can retrieve or use them. The effects of the organisms in these first two types of loss on the food hoarder are similar, but I treat them separately because food-hoarding animals use different handling techniques to protect stored food from each type of loss. Third, cached food may escape. Seeds commonly escape by germinating, but animal prey also may escape if stored alive and relatively unharmed. Fourth, cached prey may be lost when environmental factors such as snow, ice, and flooding prevent a food-hoarding animal from gaining access to stores. And fifth, the caching animal may lose stored items by forgetting the location or otherwise not being able to find cache sites. This last type of loss is discussed in chapter 6.

4.1 LOSS TO DECOMPOSERS

Many food-hoarding animals store food that is susceptible to attack by various decomposers. Further, storage sites are frequently cool and moist, conditions that favor microbial growth and exacerbate the threat of cache loss to decomposers. Decomposers may rapidly make cached prey unpalatable or toxic (Janzen 1977), but since food-hoarding animals have had a long history of association with microbial competitors, they have acquired behaviors that reduce or prevent microbial growth in food caches.

Seed, Fruit, Vegetation, and Fungus Storage

Seed-hoarding species have largely circumvented the problem of cache spoilage because some seeds are adapted to persist for long periods in a dormant state and are generally resistant to microbial attack

(Smith and Reichman 1984). Many seed-hoarding birds and mammals further reduce the possibility of seed spoilage by sorting seeds as they collect them, discarding unhealthy seeds. Jays and nutcrackers use visual, tactile, and auditory cues to sort nuts with a very high degree of accuracy (Ligon and Martin 1974; Vander Wall and Balda 1977; Bossema 1979). Squirrels are equally adept at rejecting wormy nuts even when they appear healthy (Dennis 1930; Mailliard 1931). Such keen abilities to discriminate the quality of food is not, however, universal among food-hoarding animals. Horton and Wright (1944), for example, found that 32% of the acorns stored by woodrats (*Neotoma*) were wormy or of poor quality.

Seeds and nuts are much less susceptible to microbial attack if dried before storage. Shaw (1934) observed that giant kangaroo rats (*Dipodomys ingens*) cured seed pods of peppergrass (*Lepidium nitidum*) in hundreds of shallow surface pits before final storage in underground larders. Caches containing unripe seed pods were closely spaced near burrow entrances and covered by 1 or 2 cm of soil. Using dyed seeds, Shaw found that surface caches were retrieved in summer, after the seeds were thoroughly dried, and moved to deeper storage chambers in the burrow. Woodpeckers that store pieces of acorn also are faced with problems of fungal growth. Their frequent manipulations of stored acorns expose new surfaces for drying, which may retard fungal growth (Kilham 1958c; Bock 1970; Moskovits 1978). During heavy rains, the storage chambers of ants may be flooded, following which the ants carry seeds to the ground surface where the seeds dry in the sun (Costello 1947; Hutchins 1966).

Storage of fleshy fruits is much less common than seed storage, primarily because most fruits spoil rapidly (Janzen 1977). Male MacGregor's bowerbirds (*Amblyornis macgregoriae*) store green or fresh ripe fruits near their bower (Pruett-Jones and Pruett-Jones 1985). Bowerbirds recover cached fruits within a few hours to several days but they abandon or discard fruits that have rotted.

Pikas, some voles, and the great gerbil (*Rhombomys opimus*) harvest fresh green vegetation during late summer and fall and transport this material to hay piles (see fig. 8.31) near their burrows, where it slowly cures (Murie 1961; Formozov 1966; Millar and Zwickel 1972a; Naumov and Lobachev 1975; Conner 1983). Some species turn the vegetation on top of the pile to assist thorough drying. Nevertheless, the hay sometimes becomes moldy and inedible during the winter (Hayward 1952; Millar and Zwickel 1972a). Hay piles may be placed under rock overhangs or transferred to underground chambers to protect the hay from rain, snow, and other herbivores (Loukashkin 1940; Broadbooks 1965; Formozov 1966; Millar and Zwickel 1972a).

Red squirrels (*Tamiasciurus hudsonicus*) and Eurasian red squirrels (*Sciurus vulgaris*) collect, dry, and store mushrooms for winter use (Buller 1920; C. C. Smith 1968; Sulkava and Nyholm 1987). These squirrels carry mushrooms into trees and place them in hollow cavities, on top of old bird nests, or in foliage, where they dry. Later, red squirrels may transport many of these mushrooms to protected sites such as the atrium of a hollow tree. As long as the mushrooms stay dry, they are resistant to decay, but if they become moist, they quickly decompose (e.g., Hatt 1929). Drying of fungi also may change the physical and chemical properties of fungi in ways that benefit squirrels (Moller 1983). In contrast, red squirrels do not dry conifer cones before storage because these cones open or disintegrate when dried. Instead, red squirrels cache cones in cool, moist piles of cone debris called middens (Shaw 1936; Finley 1969).

As the previous examples attest, preventing or controlling microbial degradation of stored items is a major problem confronting food hoarders. But Rebar and Reichman (1983) and Reichman and Rebar (1985) have suggested that some heteromyid rodents may actually prefer eating seeds that are slightly moldy. Rock pocket mice (*Perognathus intermedius*), for example, consume more uninfected and slightly moldy panicgrass (*Panicum virgatum*) seeds than very moldy seeds (Rebar and Reichman 1983), and banner-tailed kangaroo rats (*Dipodomys spectabilis*) eat more mold-free or slightly moldy barley seeds that had imbibed moisture than seeds that were dry or had imbibed moisture and were very moldy (fig. 4.1) (Reichman amd Rebar 1985). This fungal activity may increase the nutritional value of seeds by partially digesting them or by producing metabolic by-products (Rebar and Reichman 1983; Reichman, Fattaey, and Fattaey 1986). Metabolic by-products can be beneficial or detrimental, depending on their concentration and the conditions under which they are produced. The management of seeds to achieve intermediate levels of moldiness may be an attempt on the rodent's part to

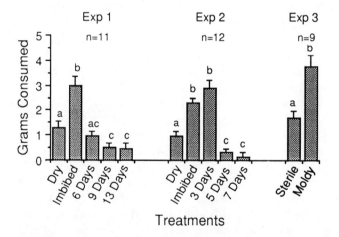

Figure 4.1   Histrograms indicating the mean (and standard error, vertical bars) amount of seeds eaten by banner-tailed kangaroo rats in three preference experiments. The abcissa indicates days of incubation and seeds that are dry, inbibed, sterile, or moldy. Imbibed and sterile seeds are moist but nonmoldy. Level of moldiness increases with days of incubation. The ordinate is the mean amount (g) consumed. Sample sizes are given at the top of the graph. Those bars within an experiment that share a lowercase letter are statistically indistinguishable at the .05 probability level. Redrawn from Reichman and Rebar 1985.

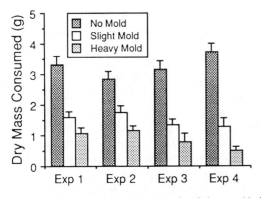

Figure 4.2   Histogram indicating the mean (±SE) dry mass (g) of seeds consumed by banner-tailed kangaroo rats in four feeding experiments. Each experiment tested the kangaroo rats' preference for seeds at three levels of moldiness: no mold, slight mold, and heavy mold. Exp 1, all seed treatments were equally moist. Exp 2, all seed treatments were equally moist and rodents were food deprived for 24 hours before the experiment. Exp 3, all seed treatments were dried after incubation. Exp 4, all seed treatments were dried after incubation, and rodents were food deprived for 24 hours. All bars within each experiment are significantly different at the $P = .05$ level. Redrawn from Frank 1988a.

capitalize on the beneficial metabolites of fungal growth while avoiding the harmful mycotoxins resulting from uncontrolled growth of some fungi. An alternative hypothesis to explain these results is that heteromyid rodents that inhabit arid regions consume slightly moldy seeds because these seeds have imbibed water from the atmosphere of their humid burrows. In other words, moldy seeds may be eaten in spite of, rather than because of, the fungi, the real goal being to capitalize on a source of preformed water. Frank (1988b) found that Merriam's kangaroo rats (*Dipodomys merriami*) preferred to eat the most moist seeds available and were able to detect very small differences in the moisture content of seeds. Further, Frank (1988a) determined that banner-tailed kangaroo rats strongly preferred eating nonmoldy seeds over both slightly moldy and heavily moldy seeds when the water content of all treatments was equal (fig. 4.2). One major advantage that desert rodents and harvester ants may gain by storing seeds in their burrows may be that the hygroscopic seeds serve as a means of "harvesting" water from the humid burrow (Morton and MacMillen 1982).

Do these rodents actually manage their seed stores to control the level of moldiness of seeds? Reichman, Fattaey, and Fattaey (1986) attempted to answer this question by providing banner-tailed kangaroo rats with a situation in which they could select one of three relative humidities (54%, 68%, and 95%) in which to store seeds. They found that about half (eight of seventeen trials) of the kangaroo rats moved sterile millet seeds into the highest humidities available, and in eleven of twelve trials, kangaroo rats moved moldy millet seeds into sites with intermediate humidity (nine trials) or low humidity (two trials). Although Reichman, Fattaey, and Fattaey (1986) did not demonstrate that these relative humidity regimes inhibited or enhanced fungal growth, the chambers to which the kangaroo rats moved seeds were consistent with the hypothesis that they managed seeds to promote the hygroscopic uptake of water while controlling the level of moldiness.

Heteromyid rodents have at least a ten million year history of seed storage (Voorhies 1974). Throughout this period, they have had to contend with the ubiquitous fungi that inhabit the soil (Reichman, Fattaey, and Fattaey 1986; Frisvad, Filtenborg, and Wicklow 1987) and that flourish in the rodents' cool, moist burrows (e.g., Kay and Whitford 1978). Although it has yet to be established that rodents derive any benefit from this interaction, they have adapted to the omnipresent fungal spores by managing the extent of fungal growth. This is not a simple problem for rodents. The significant short-term benefits of allowing seeds to imbibe water at the risk of limited fungal growth must be weighed against the long-term benefit of maintaining seed stores free of fungi. However, rodents may be able to maximize both short-term and long-term benefits by storing most seeds in chambers where the relative humidity is low enough to inhibit fungal growth while regularly moving small quantities of seeds to humid chambers more favorable to the uptake of water. Although there are no data that directly test this hypothesis, the occurrence of seeds stored by banner-tailed kangaroo rats in the field at two distinct depths and moisture levels (Reichman, Fattaey, and Fattaey 1986) is consistent with it.

Animal Storage

Some animals store other animals but avoid spoilage (as well as excessive drying and escape) by keeping prey alive and immobilized. For example, short-tailed shrews (*Blarina brevicauda*) produce a toxin in their submaxillary glands that, when transmitted in saliva, causes mice, earthworms, and insects to become comatose (Tomasi 1978; Martin 1981). These comatose prey may live for as long as 5 days (Martin 1981). Comatose prey neither escape nor spoil and retain nutritive value. Short-tailed shrew venom can be viewed as an adaptation to increase the "shelf life" of stored prey.

The European mole, *Talpa europaea,* stores earthworms during fall and early winter (Skoczen 1961). Before storage, anterior segments of most worms are eaten or mutilated. Worms are then pushed into hollows in runways and nest chamber walls and covered with dirt. Injuries inflicted on the worms, along with low winter temperatures, prevent escape.

Several food-hoarding birds also immobilize prey. Ligon (1968) found a live sphinx moth with its legs and one wing removed and two live crickets with bitten thoraces stored in nest cavities by elf owls (*Micrathene whitneyi*). Burrowing owls (*Athene cunicularia*) store long-horned beetles alive but incapacitated around their nest burrows (Rich and Trentlage 1983). These fresh, immobilized prey are available to adults and nestlings during the day when adults do not forage away from the nest. Shrikes sometimes impale live insects (Durango 1951; Beven and England 1969), and northwestern crows (*Corvus caurinus*) immobilize crabs before storing by removing their legs (James and Verbeek 1983). Crested tits

(*Parus cristatus*) mutilate the heads and anterior segments of insect larvae before storage (Haftorn 1954), and many of these larvae live for several days. Larvae often are stored in contact with lichens (77.7% of
larvae cached), which are purported to have antiseptic qualities and thus
may retard spoilage (Haftorn 1954).

Some araneid spiders immobilize prey by wrapping them in silk
(Robinson, Mirick, and Turner 1969). Live, quiescent prey are left hanging at the capture site (*Argiope*) or are transported to the hub of the web
and reattached by silk (*Nephila*).

Immobilization of invertebrate prey is widespread in mass-provision-
ing wasps (e.g., Rathmayer 1962; Evans 1966a; Piek and Simon Thomas
1969; Evans and Eberhard 1970; Alm and Kurczewski 1984). Wasps
sting prey at the capture site and then transport the comatose prey to the
nest. Effects of the paralysis are only temporary in some species of
wasps (Evans, Lin, and Yoshimoto 1953; Kurczewski and Spofford 1987),
but by the time prey regain mobility, they are trapped in a cell from which
they cannot escape (fig. 4.3). The qualitative effects of paralysis resulting
from the sting of mass provisioners often differ from those of progressive
provisioners. For example, most wasps of the genus *Bembix* are progressive provisioners, but *Bembix hinei* mass provisions cells with between
seven and thirteen flies. Flies paralyzed by *Bembix hinei* stay alive and
fresh far longer than those stung by other *Bembix* (Evans 1957b).

The ant *Cerapachys turneri* (Formicidae) raids colonies of other ant
species (e.g., *Pheidole* spp.), stings adults and larvae to immobilize
them, and carries them back to its nest (Hölldobler 1982). Stored larvae
remain in a state of "metabolic stasis" for up to 2 months and do not
pupate. Paralysis of prey also occurs in other species of ponerine ants
(Wheeler 1933; Maschwitz, Hahn, and Schönegge 1979).

The neotropical stingless bee *Trigona hypogea* is unusual in that it
stores meat from animal carcasses as a protein source for larvae. Young
worker bees predigest the meat, converting it to a brown viscous liquid
with a pH of 3.0–4.0 (Gilliam et al. 1985). During this process, the pre-
digested material is inoculated with several species of *Bacillus* bacteria
that apparently aid in converting the protein to a form metabolizable by
larvae and preserving the provisions through production of antibiotics.
This transformation permits *T. hypogea* to store a rich source of protein
for long periods in the humid tropics, where meat normally decomposes
rapidly.

Carnivorous birds and mammals may store prey that they capture
when they are not hungry or that are too large to be consumed all at once
(Miller 1931; Collins 1976; Macdonald 1976; Lyons and Mosher 1982;
Solheim 1984). Tigers (*Panthera tigris*), for example, hide carcasses of
large prey in brushy areas and cover them with leaves, grass, and litter
(Schaller 1967), which may keep them cool and thus retard microbial activity. Brown bears (*Ursus arctos*) also cover carcasses, which may slow
decomposition and maintain nutritional value (Elgmork 1982), as well as
make carcasses less accessible to flies and other carrion-feeding invertebrates (Mysterud 1975). In about half of the caches studied by Elgmork

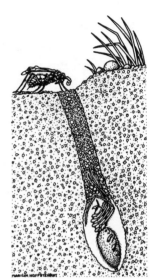

Figure 4.3
A paralyzed spider entombed in a
small chamber by the spider wasp
Anoplius apiculatus. Note the wasp
egg on the spider's abdomen. Original drawing by Marilyn Hoff
Stewart.

(1982), *Sphagnum* moss, which contains a phenolic compound that has bactericidal and fungicidal properties, may have slowed microbial growth. Mysterud (1973) has suggested that the moss and litter that bears rake over carcasses also may serve as visual and olfactory "filters" that decrease the probability that carcasses will be detected by competitors. For tigers, bears, and most other carnivores, storage of large carcasses is only possible for a short period before microbes render them unusable, usually less than 1 week (e.g., Balgooyen 1976; Janzen 1977; Phelan 1977; Walter 1979; Rijnsdorp et al. 1981; but see Maccarone and Montevecchi 1981), forcing most carnivores to use caches quickly.

During winter at high latitudes, prey stored by carnivores and raptors is likely to freeze rather than spoil, and thus prey can be stored for much longer periods (e.g., Solheim 1984). Once frozen, however, prey is not easily ingested. Bondrup-Nielsen (1977) found that captive boreal owls (*Aegolius funereus*) and saw-whet owls (*A. acadicus*), when given frozen mice, assume a posture on top of the mouse similar to that used during incubation. As portions of the prey thawed, they were immediately eaten. Thawing of prey has also been reported for great horned owls (*Bubo virginianus*) and least weasels (*Mustela nivalis*) (Criddle 1947); George and Sulski 1984).

Nectar and Honey Storage

Honey bees (*Apis mellifera*) prevent fermentation of stored nectar by altering its composition (White 1966, 1975), changing nectar into honey by converting sucrose to dextrose and levulose with the enzyme invertase. Water content of honey also is reduced during this curing process. With sucrose inverted, honey attains higher sugar concentrations, and the greater osmolality of concentrated sugars has an antibacterial effect. In addition, honey bees secrete glucose oxidase into nectar, which converts glucose to hydrogen peroxide and gluconic acid (White 1966). Peroxide has a strong antibacterial effect in uncured honey. White (1966) suggested that peroxide (also known as "inhibine") present during ripening of honey, along with high acidity due to gluconic acid, may preserve nectar until high sugar concentration begins to exert its antibacterial effects. In humid tropical forests, it is difficult for bees to reduce water content of honey to a level where osmolality exerts an antibacterial effect. Some neotropical stingless bees mix plant resins with wax used to build honey pots, and Roubik (1983) has suggested that these resins may prevent fermentation of the honey. Bacteria (primarily *Bacillus*) and yeasts occur in pollen stored by honey bees (Gilliam 1979). The role of these organisms is unknown. Bacteria and fungi also inhabit the nectar and pollen provisions of solitary bees. These organisms may cause the provisions to ferment, occasionally destroying some cells (Batra 1970, 1980). The cellophanelike cell lining of some soil-nesting bees may reduce infection of soil-dwelling microorganisms (Hefetz, Fales, and Batra 1979).

A bizarre form of food storage is found in honey ants. Certain workers called repletes store fluids in their extremely distended abdomens (see fig. 10.11). For example, repletes of the honey ant *Myrmecocystus mexicanus* of southwestern United States act as receptacles for extra-floral nectar and honeydew brought to the nest by foraging ants. These turgid repletes hang from the ceiling of special domed chambers and re-gurgitate honey to workers as it is needed. Burgett and Young (1974) found that some repletes stored lipids, apparently derived from insect prey. Honey ant repletes not only serve as convenient storage vessels in this arid and semiarid climate, but immunological systems of repletes probably act to protect stored liquids from microbes.

4.2   LOSS TO CACHE ROBBERS AND PARASITES

The host of animals that exploit the foods stored by hoarders fall into three groups. First, some individuals remove hoarded food from a cache site before the hoarder has an opportunity to retrieve it. These cache robbers are an important and widespread source of cache loss (e.g., Grinnell 1936; Murie 1936; Costello 1947; Long 1950; Ristich 1953; Free 1955; Ribbands 1955; Goodwin 1956; Sharp 1959; Gibb 1960; Hutchins 1966; Beven and England 1969; Robinson, Mirick, and Turner 1969; Rue 1969; Martin 1971; Clark and Comanor 1973; Haftorn 1974; Källander 1978; Parker 1980; Kendall 1983; Wille 1983; Olsson 1985; Alexander 1986; Clarkson et al. 1986; Hallwachs 1986; Koenig and Mumme 1987). Cache robbers may be members of other species (e.g., Pinkowski 1977; Smith and Balda 1979; Seeley, Seeley, and Akratanakul 1982), conspecifics (e.g., Butler and Free 1952; Elliott 1978; Peck and Forsyth 1982; James and Verbeek 1983; Baker et al. 1988), or even the mate of the hoarding animal (e.g., Goodwin 1956; Applegate 1977; Powlesland 1980). Some food hoarders seem to strongly prefer consuming food from a conspe-cific's larder over foraging for unstored food or consuming their own food stores (e.g., Wrazen and Wrazen 1982). Second, some animals consume stored food without removing it from the cache site. These include a vari-ety of cache symbionts such as earwigs, beetles, and microarthropods that infest and gradually degrade foods in the larders of mammals and some birds (e.g., B. Morris 1962, 1963; Elliott 1978; Seastedt, Reich-man, and Todd 1986). Third, some species parasitize nests provisioned with food, ovipositing on the provisions in the cell and using the food sup-ply solely for their own offspring. These nest parasites are primarily ar-thropods that preempt the provisions of wasps, bees, and dung beetles (e.g., Torchio and Youssef 1968; Eickwort and Eickwort 1972; Tengö and Bergström 1977; Gwynne and Dodson 1983; Alm and Kurczewski 1984; Vinson, Frankie, and Coville 1987).

The removal rate for artifically cached food items demonstrates the impact that cache robbers can have on food hoarders. Over 80% of clams cached by James and Verbeek (1983) to simulate caches by northwestern crows were gone after 7 days, and 84.6% of 500 horse chestnuts (*Aescu-lus hippocastanum*) cached by Thompson and Thompson (1980) were re-moved by gray squirrels over a 5-month period (fig. 4.4). Nutcrackerlike caches experienced predation rates of between 57% and 95% (Tomback

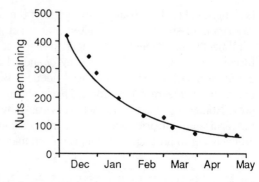

Figure 4.4
Rate of disappearance of 500 horse chestnuts experimentally buried under a thin layer of soil (≤ 5 mm) in late November. Most nuts were found by gray squirrels. Redrawn from Thompson and Thompson 1980.

1980; Hutchins and Lanner 1982). These observations demonstrate that many caches are highly vulnerable to robbery by other animals.

Food-hoarding animals exhibit a variety of protective behaviors that reduce the probability of cache loss. I have grouped these into four categories: (1) cache site selection, (2) cache preparation, (3) optimal cache spacing, and (4) aggressive cache defense.

Cache Site Selection

Some food-hoarding animals avoid loss to cache robbers by storing food in sites inaccessible to most or all potential robbers. This result is achieved by selecting a site where robbers do not have physical access to stored food or transporting food to sites where potential cache robbers are absent or uncommon.

Red-headed woodpeckers (*Melanerpes erythrocephalus*) tightly wedge many food items in crevices of tree branches and trunks and seal some stores in cavities (Kilham 1958a; MacRoberts 1975). These woodpeckers hammer moist splinters of partially rotten wood into the entrance until the splinters completely block access to the cache. As the splinters dry, they harden and become difficult to remove. Red-bellied woodpeckers (*Melanerpes carolinus*) achieve the same result as red-headed woodpeckers but in quite a different way. Red-bellied woodpeckers have long, protrusible tongues that are adept at manipulating objects at a distance (Kilham 1963). These birds store individual food items deep in crevices where competitors, including red-headed woodpeckers, cannot reach. In marked contrast to red-headed woodpeckers, red-bellied woodpeckers were never aggressive over stored food items, and Kilham attributed this difference to red-bellied woodpecker stores being less accessible to robbers. Solheim (1984) observed that pygmy owls (*Glaucidium gnoma*) cached prey in nest boxes with smaller entrances and roosted in boxes with larger holes. Caches in boxes with smaller entrances were safer from potential large-bodied cache robbers.

Solitary wasps and bees block the entrances to nests with soil, pebbles, plant debris, or resins to discourage robbing of provisions. The mud-dauber wasp *Trypoxylon xanthandrum* (Sphecidae), for example, constructs a large clay plug in the entrance to its nest, which deters cache robbing ants (Coville and Griswold 1983). An analog to this behavior is found in the Mongolian pika (*Ochotona pallasi*), which uses small pebbles to "wall in" entrances to underground chambers to prevent plundering of stored hay (Okunev and Zonov 1980).

Several species of food-hoarding birds transport seeds and nuts from areas where they were produced and where cache predators are numerous to other habitats where cache predators are few. Blue jays store many acorns in meadows and early successional habitats (Darley-Hill and Johnson 1981; Harrison and Werner 1984; Johnson and Adkisson 1985). By so doing, jays protect caches from tree squirrel (*Sciurus* spp.) predation, which can be very high in forests. Stapanian and Smith (1986), for example, found that fox squirrels (*Sciurus niger*) removed between 33.3% and 83.3% of 96 black walnuts buried to simulate caches in forests, but no nuts were removed from ninety-six caches more than 9 m from the forest edge.

Nutcrackers transport many pine seeds 5–15 km (and occasionally up to 22 km) and cache a portion of them in sites where rodent densities are likely to be low (Vander Wall and Balda 1981; Mattes 1982). Rodents often are numerous in the seed source area, and many actively search for pine seeds to make their own winter stores. The wind-swept ridges and cliffs used by nutcrackers, however, are harsh, open environments where rodents are few and cache loss is potentially much less. Although the pattern of caching by nutcrackers has usually been described as promoting accessibility of caches to birds while decreasing accessibility to rodents, nutcrackers also may benefit by caching some seeds in sites that will soon become inaccessible to themselves. In northwestern Wyoming, Clark's nutcrackers make thousands of caches on north-facing slopes and along ridges near treeline, sites that will be under cornices by early winter (Vander Wall and Hutchins 1983, unpubl. data), where snowpack may exceed 5 m. Rodents, with the possible exception of *Thomomys* spp., do not appear to be active under deep snowpack, so those seeds cached under deep snow are essentially locked away from nutcrackers and most cache predators until spring. Beginning in May and continuing through August, nutcrackers recover seeds at the edge of the receding snowpack within hours or days of snowmelt (Vander Wall and Hutchins 1983).

Kruuk (1972) observed spotted hyenas (*Crocuta crocuta*) carrying large chunks of bone or meat to margins of lakes and caching them in water about 30–50 cm deep. The water was muddy, and carcasses were not visible above the surface. By caching in water, spotted hyenas avoid loss of prey to a host of scavengers that find food by sight and smell.

Many scatter hoarders select cache sites that are readily accessible to potential cache robbers but gain a measure of protection by making cache sites inconspicuous and widely spaced (see below). Some scatter hoarders modify their selection of cache sites in the presence of competitors. Bank voles (*Clethrionomys glareolus*) allowed to cache food pellets in terraria, for example, redistributed them to "safer" sites after a conspecific was introduced into the terraria (Hansson 1986). Willow tits (*Parus montanus*) reduced overlap in cache site selection with Siberian tits (*Parus cinctus*) in areas of sympatry by reducing cache height (fig. 4.5) and altering the frequency of tree and substrate use (Alatalo and Carlson 1987). Stevens (cited in Sherry 1987) allowed captive marsh tits (*Parus palustris*) to store food in several substrates and then systematically removed stored items from one of these substrates. During future storing bouts, the tits avoided caching in substrates that earlier had been pilfered. This functional response to cache loss would seem to require that tits remember specific cache sites and keep some accounting of their foraging success within each substrate type when searching for stored food.

Food-hoarding animals often retrieve items they have stored and recache them elsewhere. Such behavior has been widely reported, but its extent and adaptive significance is poorly understood. DeGange et al. (1989) found that nearly all acorns that Florida scrub jays (*Aphelocoma coerulescens coerulescens*) retrieved from caches in the fall, when other

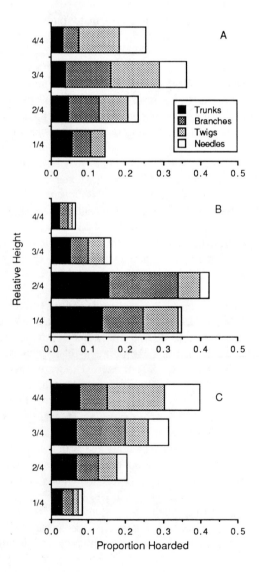

Figure 4.5
The distribution of cache sites (relative height and substrate) in spruce trees for willow tits at Norrbotten (A), willow tits at Lappland (B), and Siberian tits at Lappland (C), Sweden. The willow tits altered their selection of caching sites when sympatric with Siberian tits, a change that reduced cache site overlap with the dominant Siberian tit. After Alatalo and Carlson 1987.

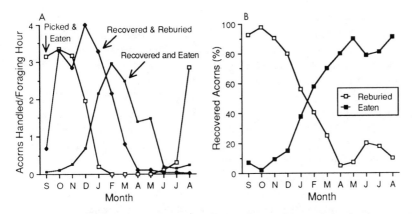

Figure 4.6   A, Monthly rates of acorn consumption (fresh and recovered from caches) and acorn re-burial by individual Florida scrub jays at the Archibold Biological Station in 1974–75. Points represent average number of acorns handled per hour of observed foraging behavior, pooling four focal individuals each month. B, Monthly changes in the percentages of all acorns recovered from caches by Florida scrub jays that were subsequently reburied or eaten. Redrawn from DeGange et al. 1989.

foods were plentiful, were restored but that the percentage recached gradually decreased during the winter and spring (fig. 4.6). Few other quantitative data seem to exist on recaching. At least five non-mutually exclusive explanations for the behavior have been offered. First, animals may retrieve and rehide food items if potential cache robbers are nearby. Collins (1976) observed this type of behavior in a captive great horned owl (*Bubo virginianus*), and I have observed captive Clark's nutcrackers retrieve and rebury pine seeds after a conspecific had unsuccessfully tried to find a recently made cache. Second, the retrieval, manipulation, and restorage of food may prevent or reduce microbial degradation of stored food by allowing items to dry more thoroughly. Bock (1970) suggested that Lewis' woodpeckers (*Melanerpes lewis*) may retrieve and restore almond fragments to reduce loss to fungi, and the movement of seeds by giant kangaroo rats from scattered surface caches to underground larders after the seeds had dried and become less susceptible to molds (Shaw 1934) was discussed earlier in this chapter. Third, as food items dry, they often shrink. Restorage may be a means of ensuring a tight fit of the food item in the cache site to avoid loss to cache robbers or to reduce the chance of dislodgement by physical forces. This appears to be why acorn woodpeckers move shrunken acorns to smaller holes (MacRoberts and MacRoberts 1976; Koenig and Mumme 1987). Fourth, when food is temporarily abundant, some animals rapidly cache it in convenient but relatively vulnerable sites and later, after the food is depleted, move cached items to more secure sites. Nutcrackers, for example, often rapidly cache pine seeds under source trees. Later in the fall, they dig up some of these seeds and transport them out of the forests to alpine areas (e.g., Kuznetsov 1959; Hutchins and Lanner 1982), where loss to rodent cache robbers is presumably less. This two-stage caching may permit nutcrackers to increase their share of a temporarily abundant resource.

And fifth, animals may locate and restore food from scattered caches as a means of monitoring the quality and quantity of food that they have stored. This feedback may be important for assessing losses to cache robbers and microbes. DeGange et al. (1989) suggested that Florida scrub jays may retrieve and restore acorns for this reason.

Cache Preparation

Two behaviors are important in reducing conspicuousness of cached food. First, because many cache robbers can find cached food only if they see caches being prepared, animals often are secretive when hoarding food to avoid attracting attention (e.g., Ewer 1965; Mueller 1974; Källander 1978; Powlesland 1980; Stone and Baker 1989). Goodwin (1956) found that prospective cache robbers watched caching jays attentively from a distance and then approached and searched the vicinity of the cache site carefully. If given a chance, jays would retrieve and recache food items that cache robbers were about to discover. Kilham (1963), Pinkowski (1977), Tomback (1978), and Brockmann and Barnard (1979) have made similar observations. Mud-daubers (Martin 1971) and digger wasps (Ristich 1953; Brockmann 1980) also may lose potential cache items to birds that observe wasps in the act of provisioning their young.

Källander (1978) noted that rooks (*Corvus frugilegus*) storing walnuts were not secretive if other nearby rooks also were actively storing walnuts. When a rook without a walnut in its bill landed near a rook burying a nut, the caching rook immediately picked up the nut and held it in its bill until the newcomer left. Birds burying food may not be a threat to other such birds, probably because an animal making a cache is "preoccupied" with its own caching activity. Eurasian jays and Clark's nutcrackers cannot find nuts stored by others if they do not watch the cache being made (Bossema 1979; Vander Wall 1982). Captive gray jays (*Perisoreus canadensis*) will cache food in the presence of another gray jay and Clark's nutcrackers, but not in the presence of Steller's jays (*Cyanocitta stelleri*) (Burnell and Tomback 1985). Both Steller's jays and blue jays (*Cyanocitta cristata*) are known to prey heavily on gray jay caches in the wild (Rutter 1969; Burnell and Tomback 1985). Northwestern crows will eat, rather than store, cachable items if they are being watched by another crow (James and Verbeek 1983); however, captive black-capped chickadees that were able to observe a conspecific store food did not have any greater success in finding that food than they did when foraging for food that they had not observed being hidden (Baker et al. 1988).

The second component of cache preparation that reduces conspicuousness of stored food is cache concealment. Cache concealment is particularly widespread among scatter hoarders, which hide single food items in a variety of situations: under lichens (Haftorn 1954), in crevices of bark and dead wood (Kilham 1963; McNair 1983), in soil (Goodwin 1956; Vander Wall and Balda 1977), in foliage (Haftorn 1956a; Dow 1965), and in moss banks and grass clumps (Collopy 1977; Cowie, Krebs, and Sherry 1981; James and Verbeek 1983). Many species that hide food cover storage sites with litter (e.g., Haftorn 1954; Kawamichi 1980;

Robinson and Brodie 1982; James and Verbeek 1983). Some solitary
wasps camouflage entrances to provisioned nests with leaves and twigs
(Evans 1966a). In contrast, South Island robins (*Petroica australis*) do not
cover scatter-hoarded earthworms (Powlesland 1980), and MacGregor's
bowerbirds do not conceal stored fruit (Pruett-Jones and Pruett-Jones
1985). The reason these species do not cover their food stores is not
clear.

Cache concealment reduces visual and, in some cases, olfactory signals, decreasing the success of naive foragers. James and Verbeek (1983)
found that clams "cached" on the ground disappeared significantly faster
than those cached under a layer of moss or grass in the manner of northwestern crows (fig. 4.7A). Bossema (1979) compared rates of disappearance of surface-cached acorns to jaylike caches, i.e., buried .5 cm deep
and covered with soil and litter. Exposed acorns disappeared 8.5 times
faster than buried acorns the first day: over half the surface "caches" disappeared on the first day, but half of the concealed caches were still present on the seventeenth day (fig. 4.7B). Scavengers found 60% of the
artificial red fox caches that were left exposed but only 13% of those that
were well camouflaged (Henry 1986). Vander Wall (1982) found that the
ability of nutcrackers to find caches made by other nutcrackers was
greater within 3 days of caching than after that time. Experiments demonstrated that small objects on the ground surface, including items placed
on caches by nutcrackers, were not essential elements of the nutcracker

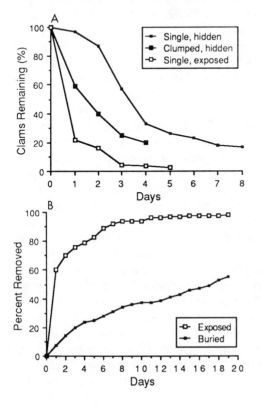

Figure 4.7
Removal rate of experimentally
cached food items. A, Percent of
clams remaining as a function of
time in three different types of experimental caches made on a slope
used as a caching area by northwestern crows. The single, hidden
clam caches are similar to those
made by northwestern crows. B,
Rate of disappearance of exposed
and buried acorns in a field in The
Netherlands. The buried acorns are
similar to caches made by Eurasian
jays. A, redrawn from James and
Verbeek 1983; B, redrawn from
Bossema 1979.

spatial memory of cache sites. Vander Wall interpreted these results to mean that objects placed on cache sites by Clark's nutcrackers obscured cache sites temporarily until weathering completely obliterated all visual sign of caches.

American kestrels often store rodents with the dorsal surface upward. This orientation of carcasses led Balgooyen (1976) and Collopy (1977) to suggest that the kestrel was taking advantage of its prey's natural protective coloration to help make stored prey more inconspicuous. Bildstein (1982) tested this hypothesis by conducting an experiment in which body orientation of two unicolor varieties of laboratory mice (white and black) cached by American kestrels were compared to orientation of countershaded white-footed mice (*Peromyscus leucopus*). Neither mouse color nor countershading significantly influenced the tendency of American kestrels to store mice dorsal side up. Bildstein (1982) concluded that the observed pattern in prey orientation was a result of species-characteristic, prey-handling behavior and not of an attempt on the part of kestrels to camouflage stored prey. Prey-handling behavior includes grasping prey by the dorsum while breaking its neck with the beak. Prey carried to a cache site is thus more likely to be cached dorsum upward. It is still possible, however, that the American kestrel's prey-handling behavior has evolved, in part, from selection based on conspicuousness of cached food.

Nest parasites can oviposit on a large proportion of the provisions of wasps, bees, and beetles. Parasitic anthophorid bees (*Mesoplia* sp.) laid eggs in 59% of the cells ($N = 22$) of their anthophorid bee hosts (*Centris flavofasciata*) (Vinson, Frankie, and Coville 1987), the bee *Sphecodes kathleenae* infested over 76% of the cells ($N = 21$) of the social sweat bee (*Dialictus umbripennis*) in some nests (Eickwort and Eickwort 1972), and satellite flies (Sarcophagidae; Miltogramminae) laid their eggs in 31% of the cells ($N = 29$) of the digger wasp (*Palmodes laeviventris*) (Gwynne and Dodson 1983). Beetles and flies can parasitize the provisions of dung beetles and burying beetles as they prepare the nest (Hammond 1976; Peck and Forsyth 1982; Wilson 1985). Satellite flies, named for following close behind prey-laden wasps, exhibit a range of behaviors: *Senotainia* larviposits on prey as the host wasp pauses at the nest entrance, *Metopia* locates open nests and deposits a larva on prey in the nest chamber, and *Phrosinella* digs into closed nests to larviposit on prey. Host wasps and bees defend against nest parasites in a variety of ways. Wasp flight behavior, for example, reduces the probability of contact between the wasp and parasites (McCorquodale 1986). At the nest, wasps carry excavated sand away from the entrance to make it less conspicuous, plug the entrance with soil to keep out intruders, or dig accessory (dead end) burrows near the entrance to confuse nest parasites (Evans 1966a; Gwynne and Dodson 1983). The empty but sealed cells that some mud-nesting wasps construct in their nests between fully provisioned cells have been interpreted as functioning to reduce the success of parasites (Tepedino et al. 1979). The sphecid wasp *Plenoculus davisi* places a few prey in a cell just outside fully provisioned cells, a behavior that may have evolved to prevent nest parasites from finding and ovipositing on the provisions

(Kurczewski 1968). Other wasps reduce nest parasitism by nesting in colonies where the density of provisioned nests satiates nest parasites, resulting in lower levels of nest parasitism (Wcislo 1984).

Optimal Cache Spacing

As pointed out in chapter 1, food-hoarding animals exhibit two general patterns of cache spacing or dispersion: scatter hoarding and larder hoarding. Each of these cache dispersion patterns is associated with several means of protecting hoarded food. Scatter hoarders, in addition to broad spacing of caches, rely on secretive hoarding behavior and inconspicuous caches to avoid predation; aggressive defense plays a secondary role or may be absent. On the other hand, larder hoarders must devote considerable time and energy to aggressive defense because the larder is attractive to potential cache robbers. Because of the amount of food at one place, it is usually difficult to make food inconspicuous to either visually or olfactorily oriented predators, although some burrowing mammals (e.g., some shrews, moles, rodents) and arthropods (e.g., some ants, wasps, dung beetles) effectively hide larders underground.

A number of environmental factors have constrained the evolution of these cache dispersion patterns. First, the period of activity of important cache robbers may strongly influence how a food hoarder can distribute stored food. Most food-hoarding birds, for example, are diurnal, whereas many cache-robbing mammals are nocturnal. Because of this difference in period of activity, most birds cannot effectively use aggressive defense to protect caches from cache-robbing mammals. This may be the primary reason why larder hoarding, which usually requires aggressive defense, is uncommon in birds (e.g., Pitelka, Tomich, and Treichel 1955; Mac-Roberts 1970; Bock 1970; Reese 1972; Collins 1976; Walter 1979). Second, aggressiveness of food-hoarding animals compared to important cache robbers also may influence use of larders. Species such as chickadees and nuthatches cannot defend a larder from formidable cache robbers such as chipmunks and squirrels and consequently must rely on widely spaced and inconspicuous caches to protect cached food. Third, climatic conditions during the period of food scarcity may influence the accessibility of scattered food. Deep winter snow may make scattered caches energetically expensive to recover, and if caches were scattered, it may force the hoarder to leave a burrow during inclement weather. Storing all food at a central point near the nest decreases energetic costs of obtaining food and reduces time spent outside the security of the nest. Species such as pikas, red squirrels, and chipmunks (C. C. Smith 1968; Elliott 1978; Conner 1983) may store food in larders primarily for this reason. And fourth, the way food is "packaged" may determine the feasibility of storing it in a larder or widely spaced caches. For example, C. C. Smith (1968) suggested that differences in hoarding behavior of red squirrels (larder hoarders) and tree squirrels (scatter hoarders) are largely due to the characteristics of the food items they store. Red squirrels keep conifer seeds inaccessible to many competitors by storing closed cones in cool, moist middens. The woody cones make the seeds an inefficient food source for large herbivores and nocturnally foraging mice. The acorns and nuts stored by tree squirrels, on the other hand, have relatively small amounts of woody tissue protecting the energy-rich contents. Because many species forage for acorns and nuts, it would be difficult for squirrels to defend them if they were stored in a larder. Burying individual nuts in widely spaced sites protects them from birds, which have a poorly developed sense of smell, and their inconspicuousness provides them a measure of protection from conspecifics and nocturnal rodents. Because nuts and acorns are difficult to defend, scatter hoarding is the most effective way of protecting them from cache robbers (C. C. Smith 1968).

The body of theory and experiments that addresses optimal cache dispersion for scatter hoarders can be traced to theory on spacing in camouflaged prey (Tinbergen, Impekoven, and Franck 1967; Croze 1970). This theory states that well-camouflaged prey species "live at interindividual distances which greatly exceed the distance from which predators usually detect them directly" (Tinbergen, Impekoven, and Franck 1967: 307). Wide spacing is favored by selection pressure exerted by predators that use a combination of "area-restricted search" and "search image" to find additional prey in the same vicinity following discovery of an initial prey. In this situation, prey in closely spaced populations suffer higher mean rates of predation than individuals in well-spaced populations. This theoretical framework applies equally well to cached food with one difference: there is a cost to the hoarder in spacing the food.

Stapanian and Smith (1978) were the first to develop a model that predicts the optimal dispersion of hidden food based on costs and benefits of spacing caches. They tested their model on fox squirrels burying walnuts, but the model applies equally well to any animal that caches food taken from a single, concentrated source. In the model, the benefit of caching food is equal to the hoarding animal's ability to retrieve it. Because successful cache robbers are less likely to find additional stored prey through a change in search behavior when caches are widely spaced, potential benefit to the hoarder increases with increasing intercache distance until an intercache distance is reached that completely discourages further search by predators. Cost of carrying a nut from a central location to a cache site is proportional to the distance nuts are carried. Changes in benefits and costs of scatter hoarding prey at different densities are shown in fig. 4.8. The mean intercache distance that results in maximum difference between cost and benefit is the optimal cache spacing. Recently an alternative model has been presented by Clarkson et al. (1986). In this model, the optimal dispersion of caches is a compromise between minimizing travel time and reducing the intensity of density-dependent cache loss. The model of Clarkson et al. is more general than the model of Stapanian and Smith, and in fact, Stapanian and Smith's model appears to be a special case of the Clarkson et al. model, where the period of storage is long and density-dependent cache loss is very high.

These models lead to two sets of predictions of how a scatter hoarder should distribute caches around a concentrated food source. The first set

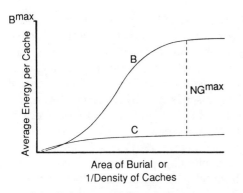

Figure 4.8   A model of the variation in cost (C) and benefit (B) of scatter hoarding nuts at different densities around a central source. The benefit is proportional to the probability of survival of a nut with B^{max} equaling the energy content of the nut. The cost is proportional to the mean distance a nut is carried. The survival of nuts is assumed to be related to predation by a population of predators with a normal distribution of the density at which individuals switch from finding all to none of the nuts. The optimum density is determined by the maximum net gain (NG^{max}), which is the density at which there is a maximum difference between cost and benefit. Redrawn from Stapanian and Smith 1978

deals with the survivability of caches at different densities relative to the optimal density. Hoarders that cache food too densely are likely to lose a high percentage of their caches, whereas hoarders that cache too sparsely waste time traveling. The second set of predictions concerns how a hoarder should establish the predicted array of caches, that is, the distance, direction, and sequence of caching events. The predictions of the Stapanian and Smith (1978) and Clarkson et al. (1986) models of caching behavior are summarized in table 4.1.

Optimal density models are difficult to test directly. The density of caches produced by an individual during a harvest period can be very difficult to determine without careful, time-consuming observations of marked individuals. To complicate the problem further, the "background" density of caches made by other hoarders often is unknown. If these problems can be overcome, yet another difficulty arises: it is very hard to determine whether the hoarder or some other animal retrieves a cached item, making it very difficult to determine the benefits of caching and how benefits vary with cache density. However, some indirect support for these models has come from field experiments that compare cache mortality at varying cache densities. Stapanian and Smith (1978) buried sets of sixty-four black walnuts in grids with internut distances of 2.4, 4.6, and 9.2 m. Wild fox squirrels were the primary nut predators. Walnut survivorship over 31 days increased with increasing internut distance, ranging from 4.7% for grids with 2.4 m spacing to 87.5% for grids with 9.2 m spacing. In addition, walnut densities after 31 days were similar on all grids, averaging .87 nuts per 100 m^2. This density is equivalent to an internut distance of 10.7 m, a value similar to the mean distance Stapanian and Smith observed in nuts scatter hoarded by wild fox squirrels (i.e., 9.9 m). In a later study, Stapanian and Smith (1984) demonstrated that the optimal density of buried nuts was inversely proportional to the food value of the nuts. Sherry, Avery, and Stevens (1982) provide further support for the model. They determined that spacing of sunflower seeds stored to simulate marsh tit (*Parus palustris*) caches significantly affected their survival. Grids of twenty-five seeds (five by five arrays) were laid out at intercache distances of 1.8, 3.6, 7.2, 10.8, and 14.4 m. Previous studies had shown that marsh tits store seeds at a mean intercache distance of 7.2 m (Cowie, Krebs, and Sherry 1981). Seeds spaced 1.8 m and 3.6 m apart were found more often by marsh tits than those stored 7.2 or more meters apart. These results suggest that at intercache distances used by marsh tits (7.2 m), survivorship of caches is near a maximum and any wider spacing would have no significant benefit. Waite (1988) found that simulated gray jay caches survived in inverse proportion to cache density (fig. 4.9). The pattern of cache predation suggested that caches at low density survived better, not because they were less likely to escape initial detection but because at low cache density, predators are less likely to persist in area-restricted search (Waite 1988).

Stapanian and Smith (1986) and Clarkson et al. (1986) also found that cache survival varied inversely with density of caches. But not all studies have found a relationship between cache density and cache removal rates. Bossema (1979), Kraus (1983), Jensen (1985), and Henry (1986) found no relationship between cache density and the rate of cache disappearance. In Kraus and Jensen's studies, virtually all of the food items were removed on all density treatments. Henry (1986) found no statistical difference in predation rates at 0 m (larder), 3 m, 15 m, and 75 m spacing for pieces of meat hidden to simulate red fox caches. This result is especially surprising, for if predation rates on scattered caches are the same as for larders, why should animals waste time and energy dispersing food items? The answer to this question, as Tinbergen (1965) and Ewer (1968) have suggested and Henry's data support, appears to be that by scattering food items in inconspicuous caches, scatter hoarders can re-

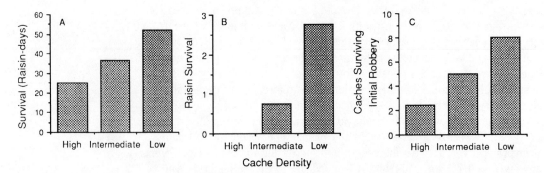

Figure 4.9 Mean survival of ten raisins artificially cached in white spruce foliage at three densities: high (about 2.12 m^{-3}), intermediate (about 0.24 m^{-3}), and low (about 0.06 m^{-3}). Cache sites were similar to those used by gray jays. A, Summation, over 10 days, of the number of raisins surviving at the end of each day. B, Number of raisins surviving at the end of the 10th day. C, Number of raisins surviving at the end of the first day on which at least one cache was removed. Redrawn from Waite 1988.

duce the variance in cache loss and thereby reduce the probability of wholesale loss of all of their food reserves. In other words, how losses are distributed is important. In the terminology of optimal foraging theory, these scatter hoarders are "risk-aversive," spacing their food caches to reduce the future variance in foraging success (Stephens and Krebs 1986). Loss of a portion of scattered food reserves may have relatively minor consequences compared to the complete loss of a larder, which, although less frequent, may be very detrimental to the hoarder.

How does the optimal dispersion of caches come about? Scatter hoarders generally begin caching a short distance from a food source, avoiding the area immediately surrounding the source, perhaps because of the concentration of other foragers there. There is only limited support for the prediction of Stapanian and Smith (1978; table 4.1, prediction 1) that subsequent caches should be placed at increasing distances from the source. Northwestern crows, hooded crows (*Corvus corone*), and most black-billed magpies (*Pica pica*) that have been observed follow the predicted pattern (Sonerud and Fjeld 1985; James and Verbeek 1985; Clarkson et al. 1986), but other species do not. Marsh tits first store seeds far from the source, placing later caches closer (Cowie, Krebs, and Sherry 1981; Sherry, Avery, and Stevens 1982), and some magpies exhibit a variable pattern (Clarkson et al. 1986). Fox squirrels observed by Stapanian and Smith (1978) did not carry successive nuts farther from the central food source as their model predicted. The density of caches has often been found to be higher near the food source and gradually decreases with distance from the source (Stapanian and Smith 1978; Cowie, Krebs, and Sherry 1981; Sherry, Avery, and Stevens 1982; James and Verbeek 1985; Clarkson et al. 1986), consistent with the prediction of Clarkson et al. (1986; table 4.1, prediction 2).

Fox squirrels and red squirrels carry items of greater food value greater distances and cache them at lower densities (Stapanian and Smith 1984; Hurly and Robertson 1987), in agreement with prediction 3 (table 4.1). This pattern is consistent with the findings of Stapanian and Smith (1984) that the food value of nuts influences the threshold density below which naive cache predators have no effect.

Caches are rarely uniformly distributed 360° around the source. Individual fox squirrels, gray squirrels, red squirrels, marsh tits, and black-billed magpies transport food items to only a portion of the area surrounding the food source resulting in cache dispersion that is strongly directional (Stapanian and Smith 1978; Cowie, Krebs, and Sherry 1981; Kraus 1983; Clarkson et al. 1986; Hurly and Robertson 1987). This pattern is consistent with predictions (table 4.1, prediction 4) if the "area suitable for caching" is affected by presence of other food hoarders in addition to characteristics of the substrate and habitat. In the case of fox squirrels, individuals carry walnuts toward their den trees (Stapanian and Smith 1978). Such a strategy may result in a suboptimal expenditure of travel time in the short term (compared to caching in a 360° arc around the source), but by placing food closer to the core area of the home range, the hoarder may be able to increase the number of food items it

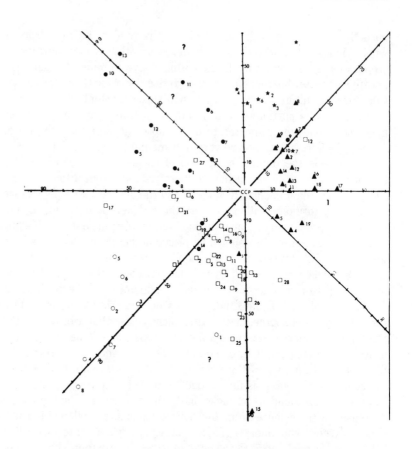

Figure 4.10　Burial sites of walnuts supplied at a common central point (CCP) for six fox squirrels in Sunset Cemetery, Manhattan, Kansas, in the fall of 1975. The axes are marked at 5-m intervals. Symbols indicate different squirrels, and numbers indicate the sequence of caching for each squirrel. From Stapanian and Smith 1978.

recovers over the long term. Further, in fox squirrels, the composite distribution of the local population approaches a more uniform distribution (fig. 4.10), suggesting that individuals avoid using overlapping caching areas. The scatter hoarding of individual red squirrels around a concentrated food source is neither uniform nor random with respect to direction (Hurly and Robertson 1987), and this result is the same when the composite cache dispersion pattern of all red squirrels caching around the food source is considered.

Predictions 5 and 6 (table 4.1) appear not to have been rigorously tested.

Despite the generally positive results of cache-spacing experiments, it is not clear how the exhibited cache-spacing pattern benefits caching animals. Stapanian and Smith (1978: 885) assumed that "memory of the animal caching the nut gives it an advantage over naive competitors in retrieving the nut," but experiments have yet to demonstrate that a nut- or seed-caching mammal can find its caches more easily than other individuals can. Experiments with marsh tits (Sherry 1982; Sherry et al. 1981, 1982; Shettleworth and Krebs 1982) and nutcrackers (Balda 1980b; Vander Wall 1982; Kamil and Balda 1985) demonstrate that, at least in some birds, the individual that scatter hoards prey has a greater probability of recovering its caches than do competitors. Another way a hoarder can

Table 4.1 Predictions of two models of how scatter hoarders should set about establishing the optimal dispersion of caches around a concentrated food source

Stapanian and Smith 1978	Clarkson et al. 1986
Premise:	
Optimal cache spacing is a compromise between the costs of transporting food and the benefits of caching food. The benefit of caching food, which is equal to the ability to retrieve it, will be proportional to the area in which it is cached up to that area which completely discourages further search by naive competitors.	Optimal cache spacing is a compromise between the costs of transporting food and the benefits of caching food. The benefits of caching depend on the strength of density-dependent cache loss.
Predictions:	
1. Caching should begin near the source, with subsequent caching occurring at increasing distances from the source.	Caching should begin near the source, with subsequent caching occurring near and at increasing distances from the source.
2. The density of caches for a food of a particular value should be uniform throughout the area used for caching (i.e., nearest-neighbor distances should be constant).	The density of caches for a food of a particular value should be highest near the source and decrease with increasing distance from the source (i.e., nearest-neighbor distances should increase as distance from the source increases).
3. The optimum density for storing items of low food value should be greater than that for storing items of greater food value (i.e., mean nearest neighbor distances should vary proportionately with the value of the food being cached).	Mean nearest neighbor distances should vary proportionately with the value of the food being cached.
4. Caching should occur in a 360° arc around the source unless some of the area is unsuitable for caching. As the arc of area suitable for caching around a food source decreases, the average distance food is carried from the source increases.	Caching should occur in a 360° arc around the source unless some of the area is unsuitable for caching. As the arc of area suitable for caching around a food source decreases, the average distance food is carried from the source increases.
5. An increase in the amount of food at the source increases the average distance food is transported.	No parallel prediction.
6. No parallel prediction.	If food is stored for short periods, caches should be hidden closer to the source.

increase the probability that it will be the one to retrieve its caches is to place caches near the center of its home range, where it has nearly exclusive access to the food. Both Stapanian and Smith (1978) and Clarkson et al. (1986) found that hoarders carried food in the direction of their territory or den tree; although tree squirrels are not territorial, this behavior may have placed the nuts more directly under their control.

Aggressive Cache Defense

Intensity of aggressive defense depends on attractiveness of hoarded food to other foragers. Attractiveness is influenced by the degree to which cached items are clumped and on food characteristics such as caloric value and nutritional qualities. Whether food is scattered or concentrated in a larder, this attractiveness affects the number of foragers attempting to acquire stored food and the intensity of robbing efforts.

Aggressive defense of hoarded food has been described in a variety of animals. Many long-term food hoarders, such as some nuthatches and woodpeckers, maintain winter territories and defend their food stores. Red-headed woodpeckers, for example, are highly aggressive toward birds and mammals that approach their food stores (Kilham 1958c). Larder-hoarding rodents such as giant kangaroo rats, red squirrels, and eastern chipmunks (*Tamias striatus*) chase conspecifics and other potential cache robbers away from their burrow larders (Shaw 1934; C. C. Smith 1968; Elliott 1978). Many large carnivores guard cached prey (Schaller 1967; Macdonald 1976; Elgmork 1982), and captive squirrel monkeys (*Saimiri sciureus*) defend stored food from conspecifics (Marriott and Salzen 1979). Group defense of larders is common among some social Hymenoptera and also occurs in some vertebrates. For example, groups of acorn woodpeckers (*Melanerpes formicivorus*) defend their granary trees by supplanting likely cache robbers that land near the food stores (MacRoberts 1970). Species such as tits and some tree squirrels (*Sciurus* spp.), on the other hand, make little or no attempt to defend their scattered, inconspicuous food stores.

The males of some species of wasps guard the nest entrance in the female's absence to prevent robbers and kleptoparasites from entering and stealing or ovipositing on stored provisions (Cross, Stith, and Bauman 1975; Brockmann 1988). In many of these species (e.g., *Pison* and *Trypoxylon*), the male remains inside the nest and attempts to copulate with the female when she visits (Evans and Eberhard 1970; Brockmann 1988). These guards may be relatively ineffective at excluding conspecific females that attempt to remove paralyzed prey because, under these circumstances, the males often are more concerned with copulating than guarding (e.g., Brockmann 1988).

Seeley, Seeley, and Akratanakul (1982) described a rich repertoire of protective behaviors in southeast Asian honey bees. *Apis florea* and *A. dorsata* build single combs in exposed sites but protect combs with three to six layers of worker bees. *A. florea* is a small bee with relatively painless sting. If large *Protaetia* beetles or other insects seeking honey or pollen land on the protective curtain, they are mobbed by bees, rolled down the side of the comb, and eventually fall to the ground. If the comb is discovered by a vertebrate predator, workers attempt to drive it away by stinging. *A. dorsata* is a large bee with a painful sting. If an intruder should approach, bees in the outer layer of the protective curtain extend their abdomens outward in synchronized waves that produce a "shimmering" effect, which discourages the intruder from landing. Vertebrates that try to reach the comb are faced with massive stinging attacks. A third species, *A. cerana,* gains considerable protection by placing the nest inside cavities of trees or caves. If an intruder succeeds in entering a nest, *A. cerana* workers can mount a moderately powerful stinging defense. All three species of Asian honey bees use alarm pheromones to focus defensive attacks (Seeley, Seeley, and Akratanakul 1982).

Honey bees (*Apis mellifera*) defend hives from wasps and other bees that attempt to steal honey (Butler and Free 1952; Free 1955; Ribbands 1955; Seeley 1985; Breed et al. 1989). Robber bees are identified by their behavior and are intercepted at the entrance by guard bees. These robbers are mauled and may be driven away or stung to death. Robbing tends to occur more frequently when little nectar is available in the field. During "honey flows," guards are seldom found at the hive entrance and robbers can enter the hive with impunity. Once a robber has been discovered within the hive, however, the hive quickly becomes alerted and guard bees station themselves at the entrance to intercept other intruders.

4.3 ESCAPE OF HOARDED PREY

Under favorable conditions, prey stored alive can escape from the cache site. This is a regular occurrence with seeds, which germinate and translocate nutrients and materials to seedlings, but also occurs with various underground plant parts (e.g., bulbs) that sprout, producing vegetative shoots. Also, some invertebrates that have been temporarily immobilized may leave the cache site before they are retrieved by the hoarder. Another phenomenon, the dislodgement of food items from a cache site by physical forces (e.g., wind, rain, etc.), will also be considered here.

Germination of Seeds and Sprouting of Bulbs

Several taxa of seed-hoarding animals avoid seed germination by storing seeds where germination is unlikely. Various woodpeckers, tits, and nuthatches store seeds above ground in foliage, bark, moss, and lichens (e.g., Haftorn 1954, 1956c; Kilham 1958a, 1963; Bock 1970; Mac-Roberts and MacRoberts 1976; Cowie, Krebs, and Sherry 1981; Moreno, Lundberg, and Carlson 1981; Sherry, Avery, and Stevens 1982; McNair 1983; but see Wong 1989), and numerous species of rodents and ants hoard seeds in nests deep underground (e.g., Shaw 1934; Quink, Abbott, and Mellen 1970; Elliott 1978; Shaffer 1980; Carroll and Risch 1984). For many of these species, protection from seed germination is a secondary benefit derived from selecting a cache site safe from predation or accessible during winter. Other animals appear to reduce the number of seeds germinating from larders by actively managing the seeds. Reichman, Wicklow, and Rebar (1985) found that many of the seeds in the burrow of banner-tailed kangaroo rats (*Dipodomys spectabilis*) germinated after the rodent was experimentally removed, suggesting that the rodent somehow prevented the seeds from germinating. Exactly how the rodents accomplish this is not clear. For animals that scatter hoard seeds in soil, germination of seeds may be a significant source of cache loss (see fig. 7.9). Some species that depend on these stored food reserves have traits that prevent seeds from germinating or that recoup some of the energy lost from those that do germinate.

One way that hoarders minimize loss of germinating seeds is to find seedlings breaking through the soil and eat what remains of the seed. Such behavior has been found in both corvids and rodents. Thirty-two of 163 whitebark pine (*Pinus albicaulis*) seeds caches located by Clark's nutcracker during summer contained one or more sprouting seeds (Vander Wall and Hutchins 1983). Nutcrackers seemed to search specifically for new seedlings as they broke through the soil surface by slowly and deliberately inspecting the ground, a foraging behavior strikingly different from that used to recover seeds from remembered cache sites (Vander Wall 1982). Nutcrackers located seedlings just as stems or seed hulls broke through the soil, detached seeds from seedlings, shelled them, and ingested the contents before they excavated the area to find ungerminated seeds. Jays (*Garrulus glandarius*) tear many of the acorns they eat during June and July from oak seedlings (Bossema 1968, 1979). Jays, which find acorns by tugging on the green stems of recently established oaks, seem to use young oaks as cues that edible acorns are buried close by. Oak seedlings are seldom damaged by this treatment. White-footed mice (*Peromyscus leucopus*) dig up and destroy caches containing young seedlings while searching for ungerminated seeds (Abbott and Quink 1970). Rodents also excavate caches of freshly emerging bitterbrush (*Purshia tridentata*) seedlings (West 1968).

Animals that use emerging seedlings to find caches may experience a certain loss in energetic value and possibly nutritional quality of the seeds. On the other hand, the grazing of seedlings may provide some beneficial nutrients not available in the seeds (Beatley 1969). The effects of this devaluation or enhancement of cached food on the hoarder have not been investigated. But, by exploiting recently germinated seeds, food-hoarding animals can salvage at least a portion of the stored food. Further, animals with weak olfactory abilities that cannot easily find concealed food can, once seeds germinate, exploit caches made by others.

The practice of excising embryos and shoots appears to be widespread in rodents. Elliott (1978) noted that virtually every beechnut (*Fagus grandifolia*) he found in eastern chipmunk burrows in spring was sprouting and that chipmunks had nipped off the sprouting radicle, an action that slowed seed degradation and may have provided some nutritional benefits to the rodent. Squirrels excise the embryo from some white oak (subgenus *Lepidobalanus*) acorns before they store them (Wood 1938; Fox 1982). The Cape mole-rat (*Georychus capensis*) of South Africa bites off buds of bulbs and tubers to keep them from sprouting (Shortridge 1934; Nowak and Paradiso 1983), as do mole-rats (*Spalax leucodon*) in Israel (Nevo 1961; Galil 1967). Davies and Jarvis (1986) failed to find any sign of disbudding of sprouts by mole-rats (*Bathyergus suillus* and *Cryptomys hottentotus*) in South Africa and suggested that removal of buds may be a seasonal phenomenon. The common vole in Lorraine (Europe) is known to disbud bulbs of *Ranunculus bulbosus* before storing them (Bourliere 1954).

Germination of stored seeds and nuts is of ecological importance because it forms the basis of some mutualistic—and in some cases co-evolved—seed dispersal systems (see chapter 7).

Animal Escape

Reports of animals escaping from caches are few, and no detailed studies have addressed this subject. Crested tits store sawfly (*Lophyrus* sp.) cocoons unharmed, and Haftorn (1954) suspected that some sawflies might emerge from cocoons not recovered by tits. Haftorn (1956a) described the escape of a slightly damaged aphid and an apparently undamaged wasp from two caches made by coal tits (*Parus ater*), and Rich and Trentlage (1983) described long-horned beetles walking away from burrowing owl caches. Earthworms injured by moles (to immobilize the earthworms during storage) may regenerate new segments during winter, and as soil temperatures rise in spring, those that have not been consumed by moles will disperse (Skoczen 1961). Undamaged snails crawl away from the larders of short-tailed shrews when spring temperatures rise (Shull 1907).

Compared to escape of seeds through germination, escape of animals from caches is rare and of limited ecological interest. It is possible that some invertebrates may be dispersed through hoarding activities of their predators—for example aphids may be dispersed from tree to tree by coal tits (Haftorn 1956a)—but no evidence has been presented to demonstrate that such behavior enhances dispersal to any significant degree.

Prevention of Dislodgement

Dislodgement of hoarded food items from storage sites is a possible source of cache loss, especially for animals that deposit prey in foliage and branches of trees. Haftorn (1954, 1974) described several types of "fixtures" used by tits to keep prey firmly in place. Mechanical fixture refers to wedging seeds or insects in bark crevices or behind lichens. Webs from spiders and insect cocoons, as well as plant down, also may be used to secure seeds. "Blood" fixture results from coagulation of body fluids of mutilated insects and spiders. Saliva fixture is used for seeds and sawfly cocoons. Copious amounts of saliva are secreted over these prey before being stored; the dried saliva acts as an adhesive, holding prey in place. Gray jays also store food items permeated with saliva (Dow 1965). The sticky saliva, secreted by the jay's enlarged salivary glands (Bock 1961), is used to fasten items to foliage and twigs.

The food caches of beavers can be destroyed when ice breaks up and flooding occurs in spring (e.g., Swenson and Knapp 1980). By this time, however, other foods become available so that loss of the cache probably has little effect on the beaver colony.

4.4 ACCESSIBILITY OF HOARDED PREY

In section 4.2, I described how food-hoarding animals limit accessibility of stored prey to potential robbers by selecting sites that robbers could not reach or where robbers were not common. Here I examine another facet of site selection, that which maintains accessibility of hoarded food to the hoarder. It is important to ensure that food stores remain accessible throughout the period during which they are needed. Inaccessible stores are no more beneficial than those consumed by cache robbers or spoiled by microbes.

Beavers (*Castor canadensis*) store tree branches and logs during late summer and fall and feed upon young wood, bark, and cambium in winter. They embed stems in bottom sediments and construct floating caches. A floating cache is started by constructing a raft of a low-preference species such as alder (*Alnus*) and nonfood items such as peeled aspen (*Populus tremuloides*) logs (Slough 1978). Beaver place high-preference food species such as aspen and willow (*Salix*) stems beneath the raft. The raft gradually becomes waterlogged, causing the cache below it to sink (Slough 1978). When beaver ponds freeze in winter, rafts containing low-preference and nonfood items become locked in ice, but preferred food species are accessible, lodged below the raft.

Steller's jays store acorns and pine seeds in the ground in Arizona (Brown 1963a; Vander Wall and Balda 1981), but in northwestern Wyoming, where winter snow is typically much deeper, they store seeds in trees (Hutchins and Lanner 1982), where they remain accessible. Gray jays also store food primarily in foliage and twigs of conifer trees, a habit that may enable them to inhabit boreal regions in winter (Dow 1965).

Crested tits concentrate stores in midsections of storage trees (Haftorn 1954). Prey captured in the canopy is carried downward to be stored, whereas prey collected from the ground is carried up into trees several meters above ground. Furthermore, nearly all caches are placed in bark or lichens under branches. Haftorn reasoned that cache site selection by crested tits maintained accessibility of caches in winter. Both tree crowns, ground, and lower branches are covered by a thick blanket of snow in Haftorn's boreal forest study site in eastern Norway. Red squirrels may place cone middens under large firs (*Abies*), which protect the cache from heavy snowfall (Shaw 1936).

Selection of well-drained soils and construction of burrows that drain away from food-storage and nest chambers are probably important to all burrowing rodents in areas prone to seasonal flooding. For example, common mole-rats (*Cryptomys hottentotus*) in Zimbabwe construct nest and food-storage chambers in termite mounds (Genelly 1965; see fig. 8.27) and mole-rats (*Spalax leucodon*) build mounds in high ground to avoid flooding (Nevo 1961).

Clark's nutcrackers, in addition to caching seeds under deep snow, cache many seeds in areas of scant snow in high mountains. Open ground at high elevations occurs predictably each winter along wind-swept ridges, on cliff faces, and on sunny, south-facing slopes. Nutcrackers use all three situations for seed caching (Vander Wall and Balda 1981; Tomback 1978). Because of the limited extent of such sites, nutcrackers intermix many caches on communal caching areas, although caches are not shared. Pinyon jays which store seeds at much lower elevations in ponderosa pine forests and piñon-juniper woodlands, cache seeds in the ground on south sides of trees, the first areas to lose snow cover after a storm (Balda and Bateman 1971).

4.5 CONCLUSIONS

The act of hoarding food does not ensure that stored prey will be available when the hoarder returns to retrieve it. For many food-hoarding species, stored food is vulnerable to loss from a variety of sources. The eastern chipmunks studied by Elliott (1978) exemplify the problems that many food-hoarding animals encounter. Beechnuts stored in underground burrows by eastern chipmunks were attacked by tenebrionid beetles, infected by a blue mold, and pilfered by other chipmunks. Many of the beechnuts that survived these hazards germinated. The effects of these losses were reduced quality of stored food and decreased effective storage time.

To meet these varied threats, food-hoarding species have acquired repertoires of behavioral tactics for protecting cached items. Protection is achieved primarily through special handling of prey during caching, and in some species, aggressive defense of caches also may be important. Handling of prey to be cached has been modified in various ways, including all aspects of the caching process (e.g., preparation, transportation, placement, concealment). Red-headed woodpeckers, for example, scatter hoard food to make it hard to find, wedge food in crevices or seal it in chambers to make it less accessible, and defend cached food from potential cache robbers (Kilham 1958a).

The various techniques that food-hoarding species employ can be readily combined into a few groups that depend on the type of food stored and the type of loss that threatens it.

Animal prey often are kept alive but immobilized during storage. Immobilization is achieved by mutilating the head or locomotory organs of the prey or by injecting it with a toxin that causes it to become comatose. As long as prey are alive, they are unlikely to spoil. Dead animal prey may be cached in situations where bacterial growth is suppressed, such as cool microsites or substrates (e.g., *Sphagnum* moss) that have bactericidal properties. Large animal prey, however, cannot be effectively protected from decomposers. Large carnivores have adapted to the poor storability of large prey by consuming prey in a short period of time. Plant material and fungi are susceptible to attack by molds and are protected from spoilage by drying. Fermentation of nectar is prevented by changing its chemical structure and composition.

Food-hoarding animals protect prey from vertebrate and invertebrate cache robbers by secretive caching and concealment of caches. Some prey are placed in accessible sites or are widely spaced so that cache robbers are discouraged from searching for caches. Larder hoarders and some scatter hoarders defend cached prey.

Seeds and some animal prey are able to escape from caches. Some seed-hoarding animals prevent escape (germination) by selecting cache sites where germination is unlikely or by excising the embryo from seeds. If germination occurs, part of the energy in the cache can be salvaged by excavating the cache and ingesting what remains of seeds. Escape of animal prey is usually precluded by killing or immobilizing them.

Accessibility of hoarded food during winter is crucial to many birds and mammals. Even though these species may cache their winter food supply long before the onset of winter, access to food stores is ensured by selecting cache sites that are typically free of ice and snow.

A number of food-storing animals have been observed to store much more food than they are likely to consume during the period of time when stored food is used (e.g., M. C. Smith 1968; Vander Wall and Balda 1977). Various explanations have been proposed to explain why excess storage should occur. One is that the length and severity of the lean period may not be predictable, and the hoarder stores in excess to cover any eventuality. But perhaps equally important, excess stored prey may serve as insurance against loss of cached food through spoilage, robbery, escape, or caches becoming inaccessible. In this view, the inevitable loss of a portion of the stored food is "anticipated" and, in a sense, over-storage is another type of defense against cache loss.

Our knowledge of how food-hoarding animals protect stored food is still fragmentary. Few experimental studies (e.g., Butler and Free 1952; Kilham 1958a; Stapanian and Smith 1978, 1984; Seeley, Seeley, and Akratanakul 1982; Sherry, Avery, and Stevens 1982) specifically examine how food-hoarding animals protect stored food, and basic questions concerning cache-protecting behavior have not yet been adequately answered. Specifically, it would be desirable to know the following: (1) What portion of stored food is recovered by the hoarder and how does this vary among individuals in a population? These fundamental questions have not been answered satisfactorily for any food-hoarding animal. (2) What is the relative importance of spoilage, robbery, escape, and other means of loss for different food-hoarding species, and how do these sources of loss differ for each type of food stored? (3) To what extent do food-hoarding animals monitor food stores to gain feedback on the quality of stored food and the extent of cache losses? This is of particular interest in scatter hoarders, which may not live as close to their food stores as do many larder hoarders. (4) Does the way a food-hoarding animal protects cached prey from loss change in different habitats or geographic areas where the types of loss threatening caches are different? (5) How do closely related food-hoarding species differ in food-hoarding behavior, and do these differences allow any insights into the evolution of prey-handling behavior in food-hoarding animals?

5
Factors That Influence Hoarding Behavior

Considerable inter- and intraspecific variation exists in the degree to which animals store food. Further, the extent to which an individual hoards may change depending on the season, social context, type of food available, sex, age, and a variety of other independent variables. To understand food-hoarding behavior, it is necessary to know what factors influence hoarding and how each of these makes its influence felt. Observational studies provide some data concerning how these factors might operate, but much progress will be made by carefully planned experiments. It also is desirable to know how certain environmental variables might influence the outcome of experiments.

I have divided the factors known to influence hoarding behavior into two groups: effects resulting from internal states of the animal and those caused by the external environment. This division is somewhat arbitrary. Obviously, some internal states are profoundly influenced by external stimuli, and environmental variables can affect hoarding only by changing the internal state of an organism.

More than half of the experimental analyses of hoarding behavior that have been conducted have used domesticated strains of the Norway rat (i.e., laboratory rats) as experimental subjects and commercial, pelletized food as the items hoarded. This work began with the pioneering research of Hunt and Willoughby (1939) and Wolfe (1939). The subsequent emphasis on the laboratory rat (*Rattus norvegicus*) as the research animal of choice is, in some regard, unfortunate because these rodents appear to store little food in the wild (e.g., Calhoun 1963; Takahashi and Lore 1980; Yabe 1981). Consequently, they may not be an appropriate model species for studying hoarding behavior (Beach 1950; Nyby and Thiessen 1980; but see Wallace 1988). It cannot be emphasized too strongly that the physiological regulation of hoarding, outlined in the pages that follow, may differ significantly in other food-storing animals. Food hoarding has evolved many times in many distantly related taxa. It is highly unlikely that the same physiological mechanism serves species as dissimilar as laboratory rats, deer mice, squirrels, beavers, insectivores, carnivores, raptors, woodpeckers, corvids, and tits (to say nothing of bees, beetles, and spiders). For this and other reasons, future work should concentrate on the diversity of other food-hoarding species, with emphasis on verifying and extending the work that has been done on laboratory rats.

5.1 EFFECTS OF INTERNAL STATES

Genetics of Hoarding

The hoarding performance of animals in both the wild and in controlled experiments is typified by great phenotypic variability. Pika (*Ochotona princeps*) hay piles at one site in the Rocky Mountains ranged in size from 400–6,000 g (Millar and Zwickel 1972a), and red squirrel (*Tamiasciurus hudsonicus*) middens contained from 280–4,360 cones (Gurnell 1984). Individual laboratory rats and Syrian golden hamsters (*Mesocricetus auratus*) differ so greatly in their propensity to store food that subjects often have been categorized as hoarders and nonhoarders (e.g., Morgan, Stellar, and Johnson 1943; Morgan 1945; Marx 1950b; Koski 1963; Herberg, Pye, and Blundell 1972). Variability in hoarding performance is partially the result of environmental influences (e.g., territory quality, developmental history, social interactions), but to what extent do genetic factors influence hoarding behavior?

Knowledge of the influence of the genotype on hoarding behavior has been gained by comparing the hoarding behavior of inbred strains of laboratory mice, laboratory rats, and races of honey bees reared under similar environmental conditions (Lindzey and Manosevitz 1964; Manosevitz and Lindzey 1970; Rinderer et al. 1986). Some lineages differ markedly in their propensity to hoard food. Food-deprived black-hooded rats begin to hoard pellets sooner, hoard more, and continue to hoard longer after ad libitum feeding resumes than do Irish rats (Stamm 1954) (fig. 5.1). A high-hoarding strain of laboratory mouse (JK) stores about four times more than a low-hoarding strain (C57BL/1) (Manosevitz 1965), and fast- and slow-hoarding lines of honey bees can be produced through artificial selection (Rothenbuhler, Kulincevic, and Thompson 1979; Hellmich and Rothenbuhler 1986a). European honey bees store significantly more sucrose solution than Africanized honey bees (*Apis mellifera scutellata*) (Rinderer et al. 1982, 1986). Hoarding of sucrose solution in the laboratory is significantly correlated with honey production in the field (Kulincevic and Rothenbuhler 1973; Kulincevic, Thompson, and Rothenbuhler 1974; Milne 1985). Because environmental conditions were the same for

Figure 5.1
Mean daily hoarding scores for genetically different black-hooded and Irish strains of laboratory rats. Redrawn from Stamm 1954.

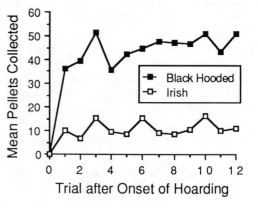

both strains or races, genotypic differences appear to be the cause of the differences in behavior.

The genetics of hoarding is virtually unknown in wild animals, and only an incomplete picture, derived from crossing inbred strains, is available in laboratory animals. Crossbreeding of the JK and C57BL/1 strains of laboratory mice produces F_1 and F_2 offspring that hoard at rates intermediate to those of the two parent strains (Manosevitz and Lindzey 1967). Offspring from backcrosses between the F_1 generation and the two parent strains also hoard at rates intermediate to those of their parents. Manosevitz and Lindzey (1967) concluded that these results are consistent with a polygenic model of hoarding; many genes contribute additively to the observed hoarding behavior. The fact that the F_1 offspring did not hoard like either parent strain indicated a lack of genetic dominance. Polygenic inheritance of hoarding behavior also has been proposed for honey bees (Rinderer and Sylvester 1978), but the genetic control of hoarding in rats may be different. Stamm (1956), on the basis of a partial genetic analysis, suggested that only a single gene is needed to explain the hoarding performance of inbred strains; however, he acknowledged that an alternative mechanism based on polygenic inheritance cannot be excluded.

Inbred strains are the result of repeated brother-sister matings for many (> 50) generations, resulting in a population of homozygous individuals that are nearly identical genetically. One result of such inbreeding is "inbreeding depression," a decline in general vigor often accompanied by reduced life span, lower reproductive potential, and a reduced ability to respond to changing conditions (Mayr 1970). Hybrids with a higher degree of heterozygosity often are more vigorous. In honey bees, the beneficial effects of heterosis extend to hoarding behavior. Workers from hybrid colonies store more honey than workers from inbred colonies in both field tests (Cale and Gowen 1956) and laboratory experiments (Bruckner 1980). However, inbreeding does not seem to depress hoarding behavior in inbred strains of laboratory mice in which hybrids hoarded at rates intermediate to the parent strains (Manosevitz 1965; Manosevitz and Lindzey 1967).

Manosevitz (1967) estimated that heritability, the ratio of additive genotypic variation to total phenotypic variation, of hoarding in mice ranged from 0.25–0.55. In honey bees, heritability ranges from 0.19 to as high as 0.92 (Collins et al. 1984; Milne 1985). The rest of the phenotypic variability is the result of environmental factors. The genetic component is large enough to be of considerable importance in understanding hoarding behavior.

The hoarding behavior observed in wild animals is the product of genetic endowment and environmental influences, both past and present. How do genes interact with the environment to produce the range of hoarding behavior exhibited in a population? Manosevitz and his colleagues (Manosevitz 1965, 1970; Manosevitz, Campenot, and Swencionis 1968) have examined this problem by exposing different strains of mice to specific environments and later comparing their hoarding behavior to that of control (standard cage environment) mice. "Fear" (induced by immersion in water for 10 seconds, 20 minutes before hoarding tests) more than doubled the number of pellets hoarded by males of JK mice (high-hoarding strain), whereas male C57BL/1 mice (low-hoarding strain) hoarded only half as many pellets as their controls (Manosevitz 1965). The hoarding by females of both strains was virtually unaffected. "Enriched" environments (large cages with ramps, platforms, columns, and tunnels) increase hoarding only in males of the high-hoarding strain and only in females of the low-hoarding strain (Manosevitz, Campenot, and Swencionis 1968). In addition, differences in the hoarding behavior of different strains may be associated with their responses to environmental variables. For example, AKR mice not only hoard more than C57BL/6 mice, but the two strains differ significantly in the ways they distribute hoarding activity during the light-dark cycle (Beigneux, Lassalle, and La Pape 1980). One interesting result of these studies is that a particular environmental effect influences hoarding differently in different strains. Researchers conducting experimental analyses of hoarding behavior should be aware of potential genetic effects and consider these effects in designing experiments and interpreting results.

Ontogeny of Hoarding Behavior

Hoarding is a complex and often stereotyped sequence of acts that, for the species thus far examined, seems to have its ontogenetic roots in food-handling behavior. Food storage has evolved many times; most frequently, it seems to arise from behavioral elements exhibited at feeding perches, from the transport of food to secure feeding stations, from delivery of food to a nest or other central place, or from protection of food from competitors (see section 3.2). In all of these hypothesized evolutionary precursors of hoarding, the way food is handled plays a critical role. The complexity of hoarding behavior and the ease with which hoarding can be elicited in experimental subjects makes it an excellent choice for studies in behavioral ontogeny. Future studies will no doubt illuminate the process of behavioral ontogeny and the evolution of hoarding. But despite this potential, detailed ontogenetic studies of hoarding seem to have been made on only three groups: hamsters, shrikes, and honey bees.

Food hoarding by Syrian golden hamsters follows a well-defined sequence of four behaviors, each occurring in a specific spatial context (Etienne, Emmanuelli, and Zinder 1982, 1983): (1) leaving the nest site to search for food, (2) filling the internal cheek pouches at the feeding site with food items picked up with the forepaws or directly by the mouth, (3) transporting food back to the nest site, and (4) emptying the contents of the pouches by opening the mouth and pressing the rear of the cheek with the forepaw. In adults, filling and emptying the pouch occur quickly, and movements to and from the food source are direct. However, the temporal and spatial contexts in which young hamsters perform these activities are unpredictable. Pups begin to manipulate food items in their mother's larder at about the age of 13 days. Pouch filling and emptying, first seen at ages 13–15 days, are uncoordinated and discontinuous.

Young may fill the pouch while chewing or press the cheek with the fore-paw without ejecting food from the pouch. Bouts of filling and emptying are short (one or a few items) and frequently interrupted by other behaviors (e.g., grooming). Furthermore, loading and emptying of the pouch occur at the same site without transport. By day 17–19, young hamsters begin to leave the nest, locate food, and transport it back to the nest, but efficient foraging excursions do not become frequent until about 30 days of age, when the hamster families begin to break up and young hamsters establish their own granaries (Etienne, Emmanuelli, and Zinder 1982). During this period, filling and emptying the pouches gradually becomes more coordinated and efficient, the behavior occurs only in certain spatial contexts, transport behavior becomes more direct, and the smooth temporal sequence of behaviors becomes established.

Researchers have been attracted to the study of behavioral development of shrikes (*Lanius* spp.) because of their unusual and complex behavior of impaling prey on sharp objects and wedging prey into crevices (Lorenz and von Saint Paul 1968; Wemmer 1969; Smith 1972). Impaling behavior evolved as a feeding adaptation. By impaling, shrikes immobilized prey so they could dismember and consume it. Subsequently, impaling became a means of storing excess prey. Loggerhead shrikes (*L. ludovicianus*) show incipient impaling movements at 20–25 days of age, when they grasp objects in the tip of their bill and repeatedly touch them to a branch or perch. Smith (1972) referred to this activity as *dabbing*. Within 2 days of initiation, this behavior gives way to *dragging*, as the shrike begins to pull objects along the perch toward itself (fig. 5.2). When dabbing and dragging first appear, the objects manipulated are often inedible, (e.g., bark, sticks, grass). Initially, dragging occurs along any perch or branch, but if thorns or forks are present, these movements gradually acquire directionality, and by 25–30 days of age shrikes begin to orient dragging toward thorns or forks. Successful wedging (securely immobilizing prey in a branch fork) typically first occurs on day 28–30 and the first successful impaling (pulling prey onto a sharp object) by day 33–35 (Smith 1972). Similar development times have been reported by Wemmer (1969). Straight, vertical objects stimulate young shrikes to im-

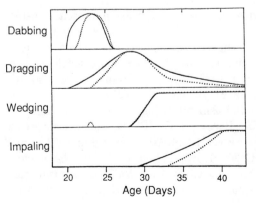

Figure 5.2
Ontogeny of impaling behavior in young loggerhead shrikes. The vertical axis represents an estimate of frequency of each behavior. Solid line, captive birds; dotted line, wild birds. After Smith 1972.

pale and wedge prey (Wemmer 1969), but initial impaling movements are inappropriate and inaccurate. For example, young loggerhead shrikes might jerk the head upward (instead of downward), missing an upward projecting nail by several centimeters. Within several days, however, the impaling movements become progressively more appropriate and accurate, with the point of impalement eventually being directed to the prey at the tip of the shrike's bill (Wemmer 1969).

The development of hoarding in shrikes and hamsters is similar in that early attempts at food handling are uncoordinated, but the path of development differs in some respects. In shrikes, the earliest components of hoarding (dabbing and dragging) are not components of the final adult behavior. These initial phases are replaced by new and different behaviors as the impaling act gradually attains its adult form. In hamsters, on the other hand, the earliest components (pouch filling and emptying) are part of the adult behavior. The earliest hoarding actions of young hamsters become more coordinated over time but are not lost. The major developmental change in hamsters concerns the spatial context and temporal sequencing of the behavior.

The development of hoarding in honey bees is markedly different from that of hamsters and shrikes. Among the three castes in a colony (drones, workers, and one queen), only workers store honey. In worker honey bees, adult behavior gradually changes in a predictable, deterministic way during the 6–10 weeks of life, a process called age polyethism (Free 1965; Seeley 1982). Seeley (1982) divided worker honey bees into four age subcastes based on predominant behavior—from 0–2 days of age, the newly emerged workers clean cells; from 2–11 days, they feed brood, cap the brood, and attend the queen; from 11–20 days, they receive nectar from foragers, store the nectar in cells, and pack pollen into cells; and from 20 days until death, they forage for nectar, pollen, water, resin, and other resources. In reality, these duties overlap somewhat, so that the changes in duties performed by a worker occur more gradually, but the most rapid changes in behavior occur at 2, 11, and 20 days. At each stage in the age polyethism schedule, the range of duties performed are spatially proximate tasks beginning in the center of the hive—the brood nest—and gradually moving to the periphery and eventually out of the hive.

Food-hoarding behavior in shrikes and hamsters reaches mature form within the first 2 months of life. The maturation of hoarding behavior in many other birds and mammals, although not well studied, seems to occur between the second and sixth months of life. For example, adult or near-adult levels of caching behavior occur by 75 days in fox squirrels (*Sciurus niger;* Cahalane 1942), 68 days in southern flying squirrels (*Glaucomys volans;* Hatt 1931), 2 months of age in bannertailed kangaroo rats (*Dipodomys spectabilis;* Bailey 1931), 101 days in African ground squirrels (*Xerus erythropus;* Ewer 1965), 20 weeks old in great horned owls (*Bubo virginianus;* Collins 1976), and not until 6 months of age in African giant rats (*Cricetomys gambianus;* Ewer 1967). Once a young animal has attained the typical adult hoarding pattern, change may still occur

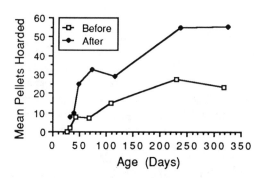

Figure 5.3
Increases in the hoarding rates
(mean food pellets per one-half
hour trial) with age of young laboratory rats before and after food
deprivation. Redrawn from Porter
et al. 1951.

in other aspects of the behavior. Laboratory rats, for example, gradually increase the number of pellets they hoard in one-half hour trials over the first 230 days of life (fig. 5.3). The extent to which such long-term developmental changes occur in other food hoarders has not been established.

Morgan (1947) and others maintained that hoarding in rats is largely an instinctive behavior. They based their argument on the fact that rats hoard spontaneously in species-typical form and that only food deprivation, not training, is necessary to elicit the behavior. But others (e.g., Marx 1950a, 1950b, 1951; Seitz 1954; Ross, Smith, and Woessner 1955; Guze 1958; Manosevitz, Campenot, and Swencionis 1968; Le Pape and Lassalle 1986) have argued that early experience contributes significantly to the form the behavior takes in adults. Laboratory rats from large litters eat less, weigh less, and hoard more than those from small litters when tested at 2 and 9 months of age (Seitz 1954; but see Lore and Moyer 1973). Infantile feeding frustration (food available for brief periods separated by long periods of deprivation) has been found to increase the propensity of some laboratory rats to hoard when later deprived of food as adults (Hunt 1941; Hunt et al. 1947; Albino and Long 1951), but the experimental results on this matter are inconsistent (McKelvey and Marx 1951; Marx 1952; Manosevitz 1970). Laboratory rats subjected to feeding frustration as adults also hoard significantly more when deprived (McCord 1941). Young rats that were undernourished when they were nursing from underfed mothers hoarded significantly fewer pellets when deprived as adults than the offspring of well-nourished mothers (Stephens 1982). Juvenile golden hamsters and laboratory rats with experience handling food pellets hoard significantly more as adults than individuals without experience (Holland 1954; Bevan and Grodsky 1958), but Smith and Ross (1953a) found that previous experience with pellets has no effect on the hoarding behavior of laboratory mice.

These two opposing points of view—instinct versus learning as the fundamental force behind the acquisition of hoarding behavior—can be reconciled by considering that the coordination of hoarding behavior, as with the coordination of many complex behavior patterns in vertebrates, is the result of interaction between developmental processes corresponding to the maturation of the nervous system and learning from individual experience. The relative contributions of these processes may differ among food-hoarding species. The earliest dabbing and dragging movements by shrikes appear to be unlearned, but early experience has important effects on shaping the hoarding behavior that eventually emerges in adults. Shrikes raised with certain types of impaling and wedging implements tended to use similar implements when later given a range of choices. For example, two subjects raised with vertical crotches and one with horizontal nails projecting 10° from the perch later used similar implements more than 99% of the time (Wemmer 1969). However, observational learning of young birds viewing adults or other young appear to play no role in the learning process (Smith 1972). Smith (1972) determined that the "critical period" for acquiring impaling experience occurred from 20–70 days old. Birds deprived of the option to impale and wedge for the first 90 days of life could not do so when first afforded an opportunity but learned to do so after 123–245 days of age (Smith 1972).

Etienne, Emmanuelli, and Zinder (1982) have argued that maturational processes more strongly influence the development of hoarding in golden hamsters than does experience. They determined that a pup's body mass was more strongly correlated with the onset of fully coordinated hoarding behavior than was its age. Etienne, Emmanuelli, and Zinder interpreted this to mean that development of hoarding behavior depends more on physical maturation than on the time the young hamster has had to perform the behavior. Further, they argue that "once mature hoarding behavior is clearly initiated, it quite suddenly appears in its final form." The sudden onset of the behavior, they believed, makes it unlikely that it occurred through learning processes. However, Etienne et al. (1982, 1983) did not conduct experiments to test for the possible contribution of early experience. Ewer (1965) concluded, although she did not provide experimental evidence, that the ability of African ground squirrels to fill the mouth with a large number of maize grains and carry them without dropping any was a learned skill but that the process of burying the seeds was not learned.

Hunger and Deprivation

The ingestion of food, a behavior closely associated with hoarding, is thought to be regulated at two levels (Herberg and Blundell 1967, 1970). Short-term regulation occurs as the result of a temporary satiating effect following feeding. Long-term regulation appears to be determined by the individual's nutritional status. The former responds to local conditions of hunger or satiety, whereas the latter responds to extended conditions of nutritional deficit or obesity. In both cases, the regulating response is the same: modification of the rate of food intake. But the responses differ markedly in the time scale over which they occur, their mechanism of physiological control, and impact on hoarding behavior.

There are at least three possible relationships between the short-term regulation of food intake and hoarding. First, when a hungry animal encounters food, feeding may compete with the motivation to hoard. In this case, the animal may eat, hoarding being temporarily postponed, causing the rate of hoarding to be less than that which occurs when the animal is sated. A second possibility is that hunger may actually stimulate or facilitate hoarding. Under this condition, feeding may or may not be postponed, but the overall rate of hoarding when hungry would be greater than that when satiated. Third, hunger may have no effect on the motivation to hoard; that is, eating and hoarding are independently motivated.

Considerable evidence supports the hypothesis that hoarding and eating are independently motivated. Morris (1962) found that the number of biscuits buried by green acouchis (*Myoprocta pratti*) under moderate levels of food deprivation had no relationship to the number eaten, a relative measure of hunger. African giant rats, hamsters, red foxes (*Vulpes vulpes*), Clark's nutcrackers (*Nucifraga columbiana*), marsh tits (*Parus palustris*), honey bees and many other species are known to hoard prodigious quantities when satiated (Ewer 1967; Free and Williams 1972; Macdonald 1976; Cowie, Krebs, and Sherry 1981; Vander Wall 1982; Etienne, Emmanuelli, and Zinder 1982; Seeley 1985). Under field conditions, animals have frequently been observed to hoard large quantities of food while eating little or none (e.g., George and Kimmel 1977; Baulu, Rossi, and Horrocks 1980). Wild American kestrels (*Falco sparverius*) may kill and cache up to twenty mice in a brief period, feeding only from the first kill (Nunn et al. 1976). The motivation to hoard persists undiminished long after satiation. Hoarding of large numbers of prey, despite the assertion of Nunn et al. (1976), is not the same as surplus killing (Kruuk 1972) because the prey are often used later by the predator (George and Kimmel 1977). But like surplus killing, hoarding illustrates that the motivations to search for, collect, transport, and hoard food items are disengaged from hunger and feeding. The independent motivation of feeding and hoarding is clearly advantageous because the payoff of eating is day-to-day survival, whereas the benefits of hoarding are often long delayed. Further, if hoarding was not independent of hunger, the accumulation of a food reserve during periods of surplus, when animals are likely to be satiated, would be unlikely to occur.

Despite the independent control of eating and hoarding, mild food deprivation does have some transient impacts on hoarding. Feeding may successfully compete with hoarding if the animal is hungry. Wemmer (1969) demonstrated this by providing shrikes with 0.6- and 1.3-g pieces of meat until they impaled a piece. Shrikes were most likely to impale meat after they had ingested four small pieces (2.4 g) or two large pieces (2.6 g) (fig. 5.4A). The consistency of shrike meal size suggests that they ingested sufficient food to fill their stomach before impaling. A close relationship exists between impaling probability and the quantity of meat recently ingested. Under the level of deprivation experienced by these shrikes, the tendency to eat exceeded the tendency to impale until the birds were satiated. Laboratory rats that have been deprived of food often feed more in the first half hour of a 1-hour hoarding trial and hoard proportionately more during the second half hour (e.g., Morgan, Stellar, and Johnson 1943). A brief period of feeding before hoarding trials increases the hoarding response rate by diminishing the motivation to eat (Nyby and Thiessen 1980). Hungry red-tailed chipmunks (*Tamias ruficaudus*) feed for 5–10 minutes following their first and sometimes

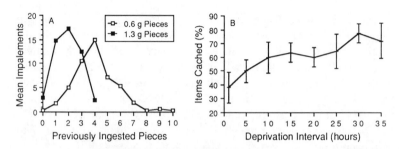

Figure 5.4 The effect of hunger on hoarding behavior. A, The impaling response of loggerhead shrikes as a function of previous ingestion of two different-sized pieces of meat. B, The frequency of caching by American kestrels in relation to deprivation interval. Vertical bars represent standard errors of the mean. A, redrawn from Wemmer 1969; B, redrawn from Mueller 1974.

second foraging trip after being deprived of food for 23 hours (Lockner 1972), again indicating a strong competition between the motivations to hoard and eat. But after feeding and when these chipmunks are satiated, they begin to store at rates significantly greater than subjects that had not been deprived. Thus, although the effect of hunger is at first inhibitory, the overall effect is that hunger stimulates hoarding. This effect, however, is short lived; the rate of hoarding gradually declines to satiation levels during the hoarding trial. Hunger also is known to facilitate hoarding in American kestrels (fig. 5.4B) and laboratory rats (Wolfe 1939; Mueller 1974; Borker and Gogate 1981). In American kestrels and red-tailed chipmunks, substantial hoarding occurred in nondeprived subjects, indicating that although hoarding may be stimulated by hunger, it is not dependent on hunger. Hunger may be sufficient, but it is not necessary for hoarding to occur.

Prolonged food deprivation (long-term regulation) that results in a loss of body mass stimulates hoarding in laboratory rats. Whereas hungry, mildly deprived rats hoard little or no food, severely deprived individuals are much more likely to hoard. This effect is distinct from that of short-term hunger (Morgan, Stellar, and Johnson 1943; Herberg and Blundell 1970). After food-deprived laboratory rats are allowed to refeed and reestablish normal body mass, hoarding intensity gradually declines over a period of several days to predeprivation levels. Hoarding in laboratory rats is so closely linked to deprivation that early researchers (e.g., Hunt and Willoughby 1939; Wolfe 1939; McCord 1941; Stellar and Morgan 1943; Smith and Ross 1950b) assumed that deprivation was necessary for hoarding. Deprivation is still considered by some to be necessary to elicit quantifiable levels of hoarding in laboratory rats (e.g., Herberg and Blundell 1970; Mendelson and Maul 1974; Fantino and Cabanac 1980), although others have concluded that deprivation is not necessary for hoarding (Bindra 1948a; Licklider and Licklider 1950; Porter, Webster, and Licklider 1951; McCain et al. 1964; Herberg, Pye, and Blundell 1972; Smith, Maybee, and Maybee 1979). Deprivation depresses hoarding in laboratory mice (Ross and Smith 1953; Smith and Ross 1953a,

1953b), an effect opposite that in laboratory rats. Hamsters and gerbils respond to deprivation much like rats, but hamsters and gerbils also store large quantities of food when satiated (Smith and Ross 1950a; Kumari and Khan 1979; Nyby and Thiessen 1980; Wong 1984). Many wild birds and mammals that have been studied store food when apparently satiated.

Body mass in animals often remains stable over long periods. One hypothesis for the control of body mass is that a regulatory response (i.e., food intake rate) is triggered when nutritional status is different from some set point. The set point is a hypothesized reference point that the hypothalamus (the center of body weight regulation in the brain) uses to maintain body mass at a certain level. This mechanism is what causes experimentally deprived subjects to compensate by "overeating" to bring body mass back to the predeprivation level. Rats, which have served as subjects for most of the experimental studies of postfast feeding compensation (hyperphagia following deprivation), store little food in the wild and thus their ability to compensate may be an important adaptation to survive periods of food scarcity. However, not all mammals fit this pattern. Hamsters are unusual among those mammals that have been tested in that they do not exhibit postfast feeding compensation, which would accelerate the return to normal body mass (Silverman and Zucker 1976; Wong 1984). Why hamsters fail to significantly increase food intake following periods of starvation is not yet clear, but it may be a consequence of their highly specialized food-hoarding habits. Hamsters store large quantities of food and feed from their larders frequently throughout the day and night. Unlike rats, whose nocturnal feeding habits result in a daily cycle of self-deprivation and compensation, hamsters have nearly constant access to a large, exclusive supply of food. When the cost of foraging is high (because of low food availability, predation risk, or other factors), size of the food store declines but body mass does not. In effect, hamsters behave so as to avoid ever having significant losses in body mass (Lea and Tarpy 1986). Perhaps their hoarding and frequent feeding have contributed to their inability to overeat to gain mass (Silverman and Zucker 1976).

The discovery that deprivation triggers hoarding in laboratory rats stimulated a series of illuminating investigations into the physiological control of hoarding behavior. Early work on the influence of deprivation suggested that the nutritional status of a subject controlled hoarding in some way. Depletion beyond a critical point elicited hoarding and, with reestablishment of nutritional balance, hoarding ceased. It was as if the depletion of some specific material in the blood or tissue during food deprivation elicited hoarding, and feeding to satiation replenished this material, eventually causing the hoarding response to diminish. Morgan, Stellar, and Johnson (1943: 293) formalized this concept in a hypothesis, later known as the "deficit hypothesis" (Bindra 1948b), which maintained that the onset of hoarding behavior was controlled by "the creation of some deficit in the body, which requires considerably more food-deprivation than does hunger for reaching a crucial level, and, similarly, takes a greater amount of food-intake to relieve than does hunger. It is possible that this deficit may be the result of the exhaustion of the supply of some specific material, such as sugar, or that it is a more general condition such as the lowering of body metabolism. . . . Although hoarding in these respects is independent of hunger, it is related to hunger in that food-deprivation produces both motivations and, when hoarding occurs in the presence of hunger, the two are in competition for the time of the animal."

The deficit hypothesis was attractive because a simple ebb and flow of some body chemical, such as sugar concentration in blood or tissues, might regulate hoarding. Deprivation causes a marked decrease in tissue sugar and moderate decrease in blood sugar compared to levels in a satiated individual. Stellar (1943) tested the deficit hypothesis by experimentally altering sugar metabolism by injecting glucose, epinephrin, or insulin into laboratory rats. If sugar concentration in blood, liver, or muscle tissues regulates the onset of hoarding, one would expect the glucose and epinephrine treatments to decrease hoarding and the insulin treatment to increase hoarding intensity. None of these treatments had the expected effect on hoarding. Following a similar line of reasoning, Bindra (1948a) fed rats diets deficient in various nutrients (e.g., protein, carbohydrate, water-soluble vitamins) but could find no evidence that depletion of any specific diet components controlled the onset of hoarding.

Undaunted, Stellar (1951) attacked the problem on another front. He reasoned that hoarding may be controlled not by some chemical substance but by the general metabolic rate of the rat. Stellar suggested that low metabolic rate might stimulate hoarding, whereas high metabolic rate might inhibit hoarding. He tested this hypothesis by lowering metabolic rates in two ways: by feeding rats thiouracil, a thyroid depressant, or by removing the thyroid gland. In addition, he raised metabolic rates of other rats with daily injections of thyroxine. None of these treatments significantly affected the onset of hoarding in the manner predicted, causing Stellar (1951: 297) to conclude that "general metabolic changes do not underlie the hoarding behavior of the laboratory rat."

Other developments make improbable the hypothesis that change in a single body component regulates the onset of hoarding. For example, rats deprived of water will hoard water in the form of water-soaked dental pledgets (Bindra 1947; Wallace 1983, 1984), but water deprivation does not result in food hoarding (Herberg and Stevens 1977). Rats deprived of vitamin B but otherwise given an ample and adequate diet hoarded food pellets high in vitamin B in preference to those lacking vitamin B (Gross and Cohn 1954); rats deprived of vitamin D likewise hoarded pellets high in vitamin D (Gross, Fisher, and Cohn 1955). These results demonstrate that the hoarding of rats can be specific to the items of which they are being deprived. Different physiological deficits are not necessarily equivalent. This being so, how could a change in a single body attribute, such as blood sugar or metabolic rate, provide the specificity needed to cause hoarding of certain items? Others found that rats stored food pellets based on characteristics other than their nutritive or caloric value, items irrelevant to any nutritional requirement. Licklider and

Licklider (1950) determined that rats store more foil-wrapped pellets than identical uncovered pellets. They concluded that "the factors that lead to hoarding and that determine what is hoarded are by no means entirely alimentary. The initiation of hoarding seems to be for the rat, as for the human being, a complex motivational problem to which sensory and perceptual factors, rather than blood chemistry, hold the key" (Licklider and Licklider 1950: 134). The direct nutritional impact of deprivation may be less important than how deprivation changes the rat's perception of food.

Recently the deficit hypothesis of Morgan, Stellar, and Johnson (1943) has been approached from a new perspective. Studies on the physiological control of hoarding have changed emphasis from attempting to find a biochemical trigger responsible for the onset of hoarding to understanding the role of hoarding in body weight regulation. Carefully controlled experiments have shown that elicitation of hoarding following food deprivation is not an all-or-none response, as claimed by Morgan, Stellar, and Johnson (1943) but is, in fact, a graded response proportional to change in body mass. Below a certain threshold level of body mass, the intensity of hoarding increases with decreasing body mass (fig. 5.5). This relationship is approximately linear and holds whether the subject is losing weight or gaining weight after having been deprived. In laboratory rats, the hoarding threshold (the x-intercept in fig. 5.5C) is the set point around which body mass is regulated (Fantino and Cabanac 1980; Herberg and Winn 1982; Fantino and Brinnel 1986; but see Borker and Gogate 1981). Thus, challenges to body mass—i.e., body mass below set point— trigger several responses in laboratory rats, including feeding and hoarding, an observation suggesting that hoarding in the rat is part of an adaptive syndrome that defends body mass. Food hoarding in Djungarian

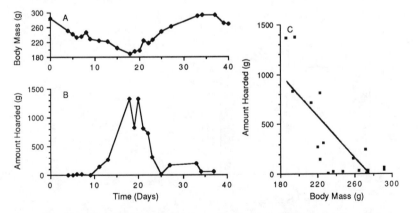

Figure 5.5 The relationship of hoarding behavior and body mass in a laboratory rat. Decrease in body mass following food deprivation (A) is associated with an increase in hoarding behavior (B). Food hoarding and body mass are significantly negatively correlated (C) below a threshold level in body mass that approximates the set point of body weight regulation. Redrawn with permission from *Physiol. Behav.* vol. 24, M. Fantino, and M. Cabanac, Body weight regulation with a proportional hoarding response in the rat, copyright 1980, Pergamon Press.

hamsters (*Phodopus sungorus*) also seems to be regulated in conjunction with body mass (Masuda and Oishi 1988).

Physiological control of body mass is served by the hypothalamus, which, among other things, regulates feeding behavior. Electrical stimulation of the lateral hypothalamic feeding area, in addition to stimulating eating, also elicits active hoarding in satiated laboratory rats (Herberg and Blundell 1967; Blundell and Herberg 1973). This effect on hoarding is similar to that resulting from prolonged food deprivation (fig. 5.6). Lesions of the ventromedial nucleus of the hypothalamus result in elevated ingestion (Herberg and Blundell 1970). The ventromedial nucleus is the seat of short-term satiety (i.e., the hunger response) and has not been shown to affect hoarding. These and other results led Herberg and Blundell (1970) to propose a simple model for the control of feeding and hoarding by the hypothalamus (fig. 5.7). This model states that hoarding in rats is regulated by long-term nutritional status monitored by the lateral hypothalamic feeding area. Nutritional deficiencies cause the lateral hypothalamus to stimulate feeding and hoarding as a defensive mechanism. Regulation of food intake is achieved by short-term inhibitory action of the ventromedial nucleus satiety area resulting from transient changes following ingestion of food. Hoarding is not subject to inhibition by the ventromedial nucleus. This simple model, if correct, explains the physiological basis for the independent effects of hunger and deprivation on hoarding.

What do deprivation-induced and naturally induced hoarding have in common? The model for the regulation of hoarding (fig. 5.7) in laboratory rats requires a body deficit to elicit hoarding, but hoarding in many wild rodents and birds occurs regularly in satiated animals. Does the model

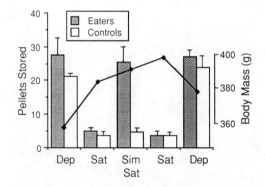

Figure 5.6   Hoarding scores and body masses obtained from eating and noneating (in response to electrical stimulation of the lateral hypothalamus) groups of laboratory rats under five conditions. Conditions: Dep, deprivation; Sat, satiation; Sim, electrical stimulation. Stimulation of satiated rats elicited intense hoarding, an effect similar to that of long-term food deprivation. The curve indicates changes in body mass in response to deprivation. Results shown under each condition represent the combined means of six consecutive measures from each subject. The small bars represent standard errors of the mean. Redrawn from Herberg and Blundell 1967.

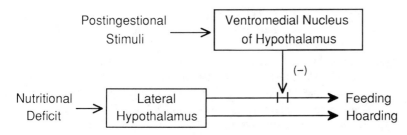

Figure 5.7
A model of the hypothalamic control of feeding and hoarding in the laboratory rat. See text for an explanation. Redrawn from Herberg and Blundell 1970.

proposed by Herberg and Blundell (1970) apply to these other species? Can this physiological perspective on hoarding be reconciled with the ecological view that animals hoard to avoid periods of food scarcity and that hoarding in wild animals is a proactive rather than a reactive response? Hoarding has evolved many times in taxonomically distant groups, so it is unrealistic to expect one physiological explanation to serve all species. But the two seemingly different hoarding responses in question are exhibited by members of the same rodent family. Nyby and Thiessen (1980) suggest that these differing responses can be reconciled at the physiological level if one considers seasonal changes in body weight regulation. Seasonal changes in body mass occur, and these are hypothesized to be the result of shifts in the set point caused by fluctuations in sex hormones elicited by environmental cues (e.g., photoperiod, temperature). At the end of the reproductive season, for example, sex hormones titers decline and as a result body mass set points are reset at higher levels. The hypothetical change in set point relative to body mass creates a nutritional deficit and the lateral hypothalamus responds to it by eliciting hyperphagia and hoarding, as in figure 5.7. Nutritional deficit still controls hoarding, as in the laboratory rat, but the deficit arises from an increase in set point rather than a decrease in nutritional status. Nyby and Thiessen (1980) point out that this hypothesis is supported by the fact that rodents often experience gonadal regression and hyperphagia at the same time that hoarding becomes prominent. Further, Herberg, Pye, and Blundell (1972) and Nyby et al. (1973) have demonstrated the inhibitory effect of sex hormones on food hoarding in rats and gerbils (see Sexual Differences in Hoarding). This hypothesis can be tested by determining whether it is consistent with patterns of weight regulation and hoarding behavior in other food hoarders.

Sexual Differences in Hoarding

Females of many rodent species hoard significantly more than do males. Female Syrian golden hamsters store twice as much as males (Smith and Ross 1950a; Wong and Jones 1985). Female laboratory rats usually store more than males (Wolfe 1939; Marx 1950a), and when deprived, females begin hoarding sooner than males (Morgan, Stellar, and Johnson 1943). Female Mongolian gerbils generally store more than males (Nyby et al. 1973; Thiessen and Yahr 1977), but Wong and Jones (1985) found no sexual differences in gerbil hoarding intensity. Male and female Eastern chipmunks (*Tamias striatus*) store similar quantities during summer (May to August), but at other seasons females hoard significantly more (Brenner and Lyle 1975). Female rice rats (*Oryzomys palustris*) and spiny pocket mice (*Heteromys desmarestianus* and *Liomys salvini*) also hoard more than males (Dewsbury 1970; Fleming and Brown 1975). There are exceptions to this pattern. For example, male southern grasshopper mice (*Onychomys torridus*) store considerably more than females (McCarty and Southwick 1975); male Indian gerbils (*Tatera indica*) hoard more food than females except when females have been provided with nest material (Kumari and Khan 1979; Sridhara and Srihari 1980); male Long Evans rats (a strain of laboratory rat) hoard more than do females (De Bruin 1988); and Namaqua gerbils (*Desmodillus auricularis*) exhibit no sexual differences in hoarding behavior (Christian, Enders, and Shump 1977). Despite these exceptions, there is a distinct trend for the females of many rodent species to hoard more than the males.

Sexual differences in the intensity of hoarding behavior have seldom been demonstrated in mammals outside the order Rodentia. The only example with which I am familiar is that female least shrews (*Cryptotis parva*) hoard significantly more than do males (Formanowicz, Bradley, and Brodie 1989).

Among arthropods, food hoarding is predominantly a female activity. Only female bees, wasps, and ants store food. Males usually take no active part in nest provisioning, although some male wasps may guard the nest and provisions while the female forages (e.g., Brockmann 1988). The males of some species of burying beetles and dung beetles assist the females in quickly burying food to avoid it being preempted by competitors (Halffter and Edmonds 1982; Wilson and Fudge 1984).

Few sexual differences in food-hoarding by birds have been documented. Only male MacGregor's bowerbirds store food (Pruett-Jones and Pruett-Jones 1985). Males of several species of shrikes, crows, and the New Zealand robin (*Petroica australis*) appear to store more food than females during the breeding season, when the females are occupied with incubation and brooding (Beven and England 1969; Applegate 1977; Powlesland 1980; James and Verbeek 1984; Kilham 1984b; McNair 1985). Most food-storing bird species are sexually monomorphic, so sexual differences in hoarding may have been overlooked.

Differences in hoarding intensity are not the only types of sexual differences exhibited. Differences in cache site selection, food type, and storing season, although much less frequently reported, also exist. For a single pair of European nuthatches (*Sitta europaea*) studied by Moreno et al. (1981), the male transported seeds significantly further than did the female, caching many seeds near the periphery of the territory. Otherwise, cache site selection was similar. Further study may demonstrate this to be an individual rather than a sexual difference. Male arctic ground squirrels (*Spermophilus parryii*) store mainly seeds in July and August, whereas females store primarily vegetation during June (McLean and Towns 1981).

Do these sexual differences in hoarding behavior have any adaptive significance? In rodents, higher levels of hoarding by females may be a consequence of their need for greater food reserves during the reproductive season, when energetic demands are great and time to forage is reduced. Lactation is energy demanding (Millar 1979), and the larder may facilitate successful reproduction (see Reproductive Benefits, chapter 2), although there appear to be few data on this problem. Permitting lactating hamsters to store food significantly increases their expression of maternal behavior (Miceli and Malsbury 1982). In situations where males store food in sites accessible to females during the reproductive season, the females often use this food to supplement their diets and that of their young (McCarty and Southwick 1975; Kilham 1984b). The different hoarding behaviors of male and female arctic ground squirrels are components of very different overwintering and reproductive strategies (McLean and Towns 1981). Male arctic ground squirrels enter hibernation later in the fall and emerge earlier in the spring than do females. Overwinter weight loss in males is less than in females, and McLean and Towns (1981) suggested that in their Yukon Territory population, males probably use stored food during periods of arousal from hibernation. The longer period of activity of males is thought to be advantageous because they can maintain their prehibernation fat reserves and enhance their access to females by being territorial in the fall and early spring. Females, on the other hand, avoid above-ground predation by entering hibernation earlier but lose more weight during the winter because of their longer period of hibernation and smaller food stores. As in arctic ground squirrels, sexual differences in hoarding of other rodent species may be well-integrated components of adaptive strategies that differ between sexes because of environmental contingencies, but for most species such strategies have not yet been demonstrated.

Hoarding behavior is controlled to some extent by the sex hormones. The general effect of sex hormones in laboratory rats, Syrian golden hamsters, and Mongolian gerbils is to suppress hoarding intensity. Nyby et al. (1973) demonstrated this suppressive effect by castrating male gerbils. Castrates—in which the secretion of gonadal testosterone was eliminated—hoarded significantly more than sham-operated controls (fig. 5.8). When castrates were injected subcutaneously with testosterone propionate in safflower oil, they hoarded significantly less than controls (castrates injected with only safflower oil) and at the same level as intact gerbils. Thus, testosterone regulates directly or indirectly (through its influence on other facets of behavior) the hoarding response of male gerbils.

The intensity of hoarding by female rats and hamsters is linked to the estrous cycle, which lasts 4–6 days. Hoarding is low on the day of estrus in golden hamsters (Estep, Lanier, and Dewsbury 1978; but see Bruce and Estep 1987) and laboratory rats (Herberg, Pye, and Blundell 1972;

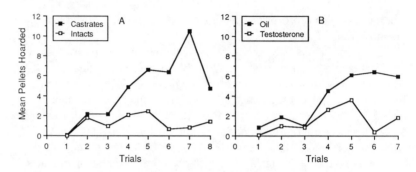

Figure 5.8   The effect of male sex hormones on hoarding behavior of Mongolian gerbils. A, Mean number of pellets hoarded in 30 minutes by castrated and intact male gerbils. B, Mean pellets hoarded in 30 minutes by castrated males following testosterone propionate treatment and oil controls. Redrawn from Nyby et al. 1973.

Fantino and Brinnel 1986) and significantly higher on all nonestrus days (fig. 5.9). Changes in body mass follow a similar pattern, fluctuating on average about 5 g between estrus and diestrus (Fantino and Brinnel 1986). Borker and Gogate (1984a, 1984b) and Borker, Dhume, and Gogate (1985), working with a different strain of rat and using different methods, found that hoarding was greatest on the day before estrus and declined steadily at estrus and the days following estrus. In any case, hoarding fluctuates in a predictable pattern during the estrous cycle. In hamsters and laboratory rats, estrogen and progesterone titers peak on the day of estrus (Estep, Lanier, and Dewsbury 1978; Fantino and Brinnel 1986). These data suggest that female sex hormones suppress food stor-

Figure 5.9
Influence of the ovarian cycle on hoarding behavior in female laboratory rats. 0, estrus day; 0 + 1, day after estrus (metestrus); 0 + 2 and 0 + 3, days of diestrus; and 0 − 1, day before estrus (proestrus). A, Change in amount of food hoarded. B, Change in body mass. All values of both parameters are significantly greater on nonestrus days than on estrus days. Redrawn with permission from *Physiol. Behav.*, vol. 35, M. Fantino, and H. Brinnel, Body weight set-point changes during the ovarian cycle: Experimental study of rats using hoarding behavior, copyright 1986, Pergamon Press.

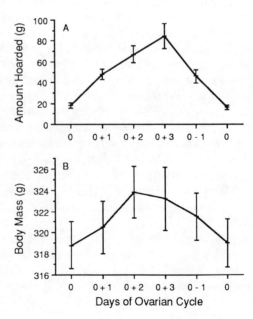

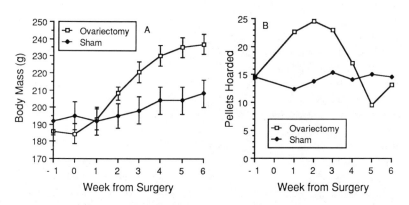

Figure 5.10   The effect of female sex hormones on hoarding behavior in laboratory rats. A, The effect of ovariectomy on body mass. B, Changes in hoarding behavior following ovariectomy. Redrawn with permission from *Physiol. Behav.*, vol. 29, J. G. Coling, and L. J. Herberg, Effects of ovarian and exogenous hormones on defended body weight, actual body weight, and the paradoxical hoarding of food by female rats, copyright 1982, Pergamon Press.

age behavior at a time when sexual receptivity and general activity are greatest. Building on this work, Coling and Herberg (1982) excised the ovaries of laboratory rats and compared their hoarding behavior to that of sham operants. This treatment eliminated the secretion of estrogen and progesterone in the treated rats. Ovariectomized rats gained mass and hoarded significantly more than the controls (fig. 5.10). Later, when body mass stabilized at a new, higher level, hoarding decreased to the preoperative level. To confirm the effect of the hormones, three groups of ovariectomized rats were injected with estradiol benzoate, progesterone, or testosterone propionate. Hoarding intensity and body mass of the estradiol-injected rats decreased significantly, but the progesterone- and testosterone-injected rats were unaffected. Thus, estrogen appears to regulate, directly or indirectly, the intensity of hoarding by female rats. A similar experiment on female eastern chipmunks, however, failed to demonstrate that estrogen has any significant influence on body mass or hoarding intensity (Bruce and Estep 1987).

An experiment by Fantino and Brinnel (1986) on the interacting effects of food deprivation and the ovarian cycle on hoarding intensity further supports the hypothesis that estrogen suppresses hoarding. The amount of food stored by female rats increased in proportion to the decrease in body mass below set point, as illustrated in figure 5.5C, but the amount of food stored at a given weight was significantly reduced during estrus compared to diestrus (fig. 5.11). Fantino and Brinnel (1986) interpreted this as evidence that body mass set point fluctuated with the ovarian cycle, the critical mass for the onset of food hoarding being 31.2 g lower at estrus than at diestrus.

Are sex hormones, through their suppressive effects on hoarding, partially responsible for females often storing more food than males? Increased hoarding of the ovariectomized rat (Coling and Herberg 1982) make it clear that the female's more intense hoarding is not caused solely

Figure 5.11
Regression of body mass against hoarding behavior computed at estrus and at diestrus for eight rats that continued to ovulate during food deprivation trials. Redrawn with permission from *Physiol. Behav.*, vol. 35, M. Fantino, and H. Brinnel, Body weight set-point changes during the ovarian cycle: Experimental study of rats using hoarding behavior, copyright 1986, Pergamon Press.

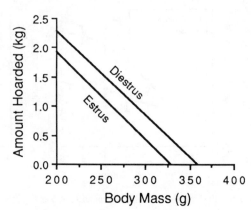

by ovarian hormones. Coling and Herberg (1982) hypothesized that the sexual difference may be in part the result of the different effects of estrogen and testosterone on weight regulation mediated through the lateral hypothalamus. Estrogen stimulates activity and inhibits feeding in female rats, thereby contributing to the lower body weights of female rats compared to males. In the absence of estrogen, the hyperphagia and weight gain that follow ovariectomy in rats may be caused by the establishment of a new, higher set point about which body weight is regulated. If estrogen does depress body mass set point in female rats, the hypothalamic model for the regulation of hoarding proposed by Herberg and Blundell (1970; fig. 5.7) would lead one to expect hoarding rates to be lower in females than males, not the reverse. Coling and Herberg (1982) suggested that the explanation for this seemingly paradoxical finding may be that estrogen also has a direct depressive effect of body mass independent of its effects on the set point. If this were the case, body mass would be maintained near or below the critical level at which hoarding would be elicited. Thus, despite their lower set points, undeprived females would be more likely to hoard than undeprived males. Even if this model is correct, Coling and Herberg (1982) stress that this cannot be the complete explanation for the sexual difference in hoarding because ovariectomized rats continued to hoard more than males after their body weights stabilized. Furthermore, gonadal hormones have been found to have little or no effect on body mass regulation in some species of wild rodents (Zucker and Boshes 1982; Bruce and Estep 1987). In these species, a somewhat different hypothesis to explain the tendency for females to hoard more than males would seem to be required.

5.2 INFLUENCE OF ENVIRONMENTAL FACTORS

Light Intensity

Hoarding intensity varies with time during the light-dark cycle. Two strains of laboratory mice conducted 83.3–99.5% of their hoarding activity during the night with peaks in activity occurring just before the light period begins (Beigneux, Lassalle, and Le Pape 1980; fig. 5.12). Most food hoarders confine all hoarding behavior to the light or the dark period. For these species, the lighting conditions may significantly influence the amount and placement of food stored. Stellar et al. (1952), for example, found that nondeprived albino rats in brightly illuminated alleys hoarded 15.3–39.4 pellets per one-half hour session in their darkened cages, whereas rats hoarded only 5.7–7.7 pellets per session when both the alley and home cage were dark. When alleys were dark or dimly lit, rats often loitered there, but under lighted conditions rats, which avoid bright light, minimized time in the illuminated alley and returned to the home cage immediately after getting each pellet. Minimizing time in the lighted area increases time available to carry pellets, and consequently hoarding increases. When Stellar et al. (1952) reversed the lighting conditions, making the home cage brightly lit and the alley dark, the rats sharply reduced hoarding in the home cage, hoarding pellets instead in the darkened alley. Stellar et al. (1952) and Charlton (1984) concluded that bright illumination was aversive and that it increased the "insecurity" and "anxiety" of foraging rats and hamsters. Wild honey bees, which place nests in darkened cavities, also store more when placed in the dark than when kept in the light (Free and Williams 1972).

Marx, Iwahara, and Brownstein (1957) found light intensity had no effect on the hoarding behavior of hooded rats. They interpreted their results to mean that hooded rats, which have normal eye pigmentation, are not as sensitive as albino rats to ambient light levels. However, hamsters and pine voles (*Pitymys pinetorum*), which have pigmented eyes, change hoarding intensity under different light regimes, as do albino rats (Waddell 1951; Geyer, Kornet, and Rogers 1984).

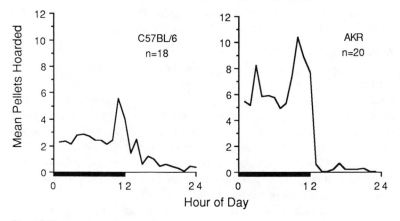

Figure 5.12
Mean number of food discs hoarded per hour over a 24-hour period for two strains of laboratory mice, C57BL/6 and AKR. There was a 12:12 LD cycle, the dark period is indicated by a black horizontal bar; 83.3% of hoarding was done at night by the C57BL/6 strain and 99.5% by the AKR strain. Redrawn with permission from *Physiol. Behav.*, vol. 24, F. Beignoux, et al., Hoarding behavior of AKR, C57BL/6 mice and their F1 in their home living space: An automatic recording technique, copyright 1980, Pergamon Press.

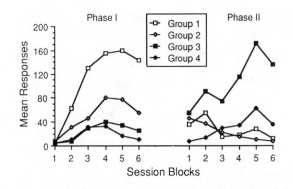

Figure 5.13   Mean number of lever press responses for food (most food was hoarded) in an operant conditioning chamber by four groups of Syrian golden hamsters under different conditions of chamber illumination and time of day. In phase I: Group 1, bright at night; group 2, bright during day; group 3, dark at night; and group 4, dark during day. During phase II, the conditions were reversed: group 1, dark at night; group 2, dark during day; group 3, bright at night; and group 4, bright during day. Redrawn from S. G. Charlton, 1984. Hoarding-induced lever pressing in golden hamsters (*Mesocricetus auratus*): Illumination, time of day, and shock as motivating operations. *J. Comp. Psychol.* 98 : 327–32. Copyright 1984 by the American Psychological Association. Adapted with permission.

The effect of illumination on food hoarding in hamsters varies with time of day (Waddell 1951; Charlton 1984). Hamsters exposed to bright alley illumination at night hoarded significantly more than those tested under bright conditions during the day (fig. 5.13). Furthermore, hoarding at night under darkened conditions (the situation pertaining in the wild) resulted in little food accumulation relative to brightly illuminated treatments. Presumably, illumination interacted with activity rhythms to produce these results. These studies suggest that illumination is not a neutral stimulus for the storing of food by certain animals.

Photoperiod

Many food-hoarding animals, particularly those living at mid to high latitudes, store food on a seasonal cycle. Peak hoarding activity usually occurs during fall, coinciding with peak availability of food items, such as nuts and seeds, and then declines to near zero in spring and summer (figs. 5.14 and 5.15; see also figs. 8.16 and 9.2). When food availability is experimentally maintained at a level far exceeding an individual's needs, the seasonal cycle of hoarding intensity persists (Muul 1970; Dufour 1978; Ludescher 1980). Seasonal changes in hoarding for such species thus are independent of food availability.

A short photoperiod has been found to stimulate hoarding in several species of rodents (Barry 1976; Masuda and Oishi 1988). White-footed mice (*Peromyscus leucopus*) maintained at 9 hours daylight at constant temperature (26°C) stored a mean of 22.4 g of soybeans per day, whereas conspecifics held at 16 hour day length stored only 5.5 g/day (Lynch et al. 1973). The number of hickory nuts stored by southern flying squirrels (*Glaucomys volans*) could be increased or decreased by shortening or lengthening (respectively) the photoperiod (Muul 1970). One possible

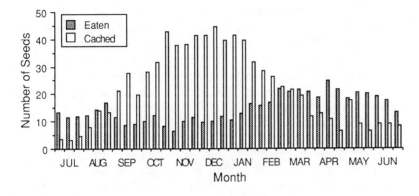

Figure 5.14
Seasonal pattern (5-year mean) of consuming and storing seeds by captive willow tits. Redrawn from Ludescher 1980.

explanation for these results is that nocturnal rodents store more food when exposed to a short photoperiod (long nights) simply because they have more time to forage. This explanation fails, however, because diurnal rodents, which have less foraging time as day length decreases, also increase hoarding intensity with decreasing photoperiod (e.g., Brenner and Lyle 1975). The similar reactions of diurnal and nocturnal animals suggest that shortening day length elevates the functional response to collect and hoard food.

The effect of photoperiod on hoarding behavior appears to be mediated by the sex hormones testosterone and estrogen. Blood titers of testosterone and estrogen increase as the gonads develop before breeding and then decline as the gonads regress following breeding. These changes are triggered by, among other things, increasing and decreasing day length. In addition to causing many of the physiological and behavioral changes exhibited by sexually active individuals through the reproductive

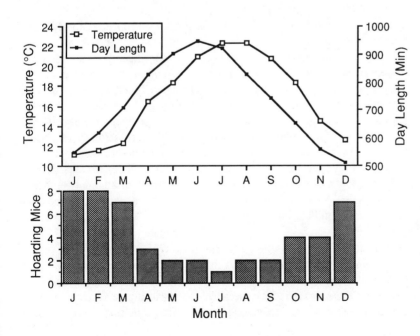

Figure 5.15
Seasonal change in hoarding (histogram), represented as number of wood mice in the sample that were hoarding, and its relationship to changes in day length and mean temperature. After Dufour 1978.

cycle, high levels of these hormones also are known to suppress hoarding activity in laboratory rats, Mongolian gerbils, and Syrian golden hamsters (Nyby et al. 1973; Estep, Lanier, and Dewsbury 1978; Coling and Herberg 1982; see Sexual Differences in Hoarding). These facts are consistent with the hypothesis that the onset of hoarding following the reproductive season is triggered by falling blood titers of sex hormones (Nyby and Thiessen 1980). It has not been established whether the hormonal regulation of hoarding behavior can explain seasonal hoarding in other mammal taxa and in birds.

Photoperiod is a reliable environmental cue that triggers behavioral and physiological responses—often well before the consequences of those responses can yield important advantages to the individual. This is especially evident in food hoarding, where the accumulation of food may begin several months before the food is actually needed. The importance of photoperiod can be expected to vary latitudinally with magnitude of the seasonal shift in day length and intensity of seasonal food scarcity. Photoperiod may be of little importance as a cue for seasonal hoarding in tropical regions, where photoperiod varies only slightly over the year. Further, the importance of photoperiod may be less in the southern portion of the temperate zone than in northern regions with more severe winters. Barry (1976) tested the hypothesis that the effectiveness of photoperiod as a stimulus for hoarding should increase with increasing latitude. He reported that northern subspecies of the white-footed mouse (*Peromyscus leucopus noveboracensis*) and deer mouse (*P. maniculatus bairdii*) from central Michigan stored more food pellets with short days (9:15 LD) and cold temperatures (7°C) than they did with long days (15:9 LD) and warm temperatures (27°C) (fig. 5.16). In the case of *P. m. bairdii,* the response was due largely to the effects of photoperiod, but for *P. l. noveboracensis,* temperature appeared to play a major role in causing the response. Conspecifics (*P. l. castaneus* and *P. m. blandus*) from more southerly locations (southern New Mexico) hoarded less food under the short-day, cold-temperature condition and did not differ significantly in hoarding intensity between the 2-day length treatments. *P. m. blandus* actually increased hoarding under warm temperatures, a result contrary to that expected if winterlike (short days and low temperatures) conditions stimulate hoarding. These results in general support the hypothesis that populations in more northern regions respond more strongly to photoperiod as an environmental indicator of future food shortage, but more studies are needed to confirm this pattern. Meanwhile, the hoarding of *P. l. noveboracensis* and *P. m. blandus* in response to ambient temperature suggests that other environmental variables also may trigger seasonal changes in hoarding intensity.

Temperature

Ambient temperature strongly influences the food-hoarding intensity of some animals. This is, of course, true for all food-storing arthropods, which for the most part lack the ability to internally regulate body temperature and so can be active only under warm conditions. Great golden

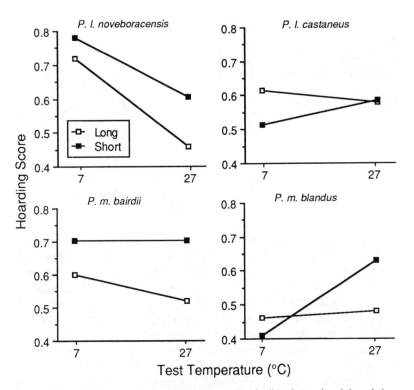

Figure 5.16   Mean hoarding scores (logarithm of the number of pellets plus one hoarded per day) under all possible combinations of test temperature (7°C and 27°C) and photoperiod (short day = 9:15 LD; long day = 15:9 LD) for two subspecies of white-footed mice (*Peromyscus leucopus*) and two subspecies of deer mice (*P. maniculatus*). *P. m. bairdii* and *P. l. noveboracensis* are northern subspecies from Michigan, and *P. m. blandus* and *P. l. castaneus* are southern subspecies from New Mexico. Redrawn from Barry 1976.

digger wasps (*Sphex ichneumoneus*), for example, deliver most prey to their nests at temperatures between 26°C and 35°C, and no prey are brought at temperatures below 20°C (Brockmann 1985). Honey bees kept at 35°C hoard more sucrose syrup in combs than bees kept at lower temperatures (Free and Williams 1972). For some rodents, on the other hand, hoarding intensity varies inversely with temperature. The hoarding responses of *Peromyscus leucopus noveboracensis,* a cricetid rodent of southern Michigan forests, *Liomys irroratus,* a heteromyid rodent of Mexican desert grassland, and *Apodemus sylvaticus,* a murid rodent of European deciduous forests, all increase at cooler temperatures (Barry 1976; Matson and Christian 1977; Dufour 1978). McCleary and Morgan (1946) found that the number of food pellets hoarded by laboratory rats was inversely proportional to temperature over a range of 9–34°C (fig. 5.17). The effect of deprivation (i.e., weight loss) on hoarding of laboratory rats is intensified at colder temperatures (fig. 5.18) (Fantino and Cabanac 1984), but this hoarding-temperature relationship is not consistent for all rodent taxa examined. *Peromyscus maniculatus bairdii* and flying squir-

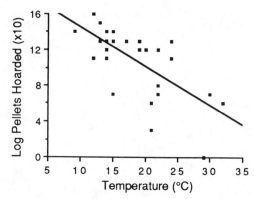

Figure 5.17
Mean hoarding scores of laboratory rats maintained at different ambient temperatures. The regression is significant (r = −0.71). Based on data from McCleary and Morgan 1946.

rels from southern Michigan do not change hoarding behavior in response to colder temperatures, and hoarding intensity of *Peromyscus maniculus blandus* of southern New Mexico, as already pointed out, increases in response to warm temperatures (Muul 1970; Barry 1976). Laboratory mice also appear to hoard more at higher temperatures (Ross and Smith 1953), but temperatures were not controlled in this experiment. The effect of temperature on avian food hoarding has not been explored.

There are a number of possible explanations for temperature-induced hoarding by rodents (e.g., Barry 1976), but the hypothesis most consistent with available information is that a change in temperature acts as an environmental cue to initiate or intensify hoarding in advance of some period of food shortage (McCleary and Morgan 1946; Barry 1976; Fantino and Cabanac 1984). In the temperate zone, approaching winter is heralded by low nighttime and daytime temperatures. Although temperature changes are less predictable and consequently less reliable zeitgebers than changes in photoperiod, low temperatures have significant metabolic consequences for foraging, nocturnal rodents; thus temperature may have a more immediate influence on behavior than does the gradual change in day length. The increased hoarding of *P. m. blandus* as temperature increases is atypical but may be explained if these mice hoard food for periods of inactivity during the hot summer months (Barry 1976).

The interaction of temperature and photoperiod as environmental cues is unclear. The studies of Muul (1970) and Barry (1976) suggest that a rodent taxon may be induced to hoard by low temperature or short photoperiod. More studies on a broader range of birds and mammals from various ecological contexts (e.g., alpine versus lowland, tropical versus temperate) are needed before the relative importance of temperature and photoperiod in influencing food-hoarding behavior can be established.

Familiarity with the Environment

In the wild, animals store food in their burrows or scatter items throughout an area that is intimately familiar. Familiarity with an area confers many advantages to animals; some of these advantages directly affect hoarding behavior. Knowledge of rich feeding areas arising from site familiarity may increase the amount of food an animal can gather. Fa-

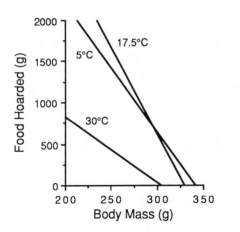

Figure 5.18
Amount of food hoarded during 3-hour trials as a function of body mass at three ambient temperatures. The slopes of the 5°C and 17.5°C relationships are not significantly different. Redrawn with permission from *Physiol. Behav.*, vol. 32, M. Fantino, and M. Cabanac, Effect of cold ambient temperature on the rat's food hoarding behavior, copyright 1984, Pergamon Press.

miliarity with refuges may allow hoarders, which often are vulnerable because they spend more time foraging, to avoid predators. Further, familiarity with an area and the site-related dominance that is often associated with residence may influence the ability of an individual to defend food stores. Consequently, familiarity may increase the amount of food stored and the proportion of items likely to be recovered.

During experiments, animals often are placed in unnatural and initially unfamiliar surroundings. What effect might these unfamiliar conditions have on hoarding behavior? In captive laboratory rats, lack of familiarity with any aspect of a restrictive environment will inhibit general activity, feeding behavior, and hoarding. When a new and unfamiliar cage is substituted for the old, familiar home cage, little or no food is carried into the new cage (Viek and Miller 1944). If the cage is familiar but the alley leading to the food source is unfamiliar, transportation of food pellets into the home cage occurs but is markedly less than in rats in familiar alleys. When the cage is unfamiliar but the alley familiar, rats do not store food pellets in the cage but carry the pellets down the familiar alley and leave them just outside the cage (Viek and Miller 1944). With repeated exposure to these unfamiliar surroundings, however, the rats gradually adapt to the new environment (i.e., the environment becomes familiar), and after several trials, hoarding rates equal those in familiar surroundings.

The cage characteristics that cause rats to respond to it as familiar have not been well studied. Miller and Viek (1944) compared hoarding of rats in environments that differed in visual (a new, different-appearing cage containing the rat's own scented shavings) and olfactory (an old familiar cage containing new shavings) cues and found that hoarding was greatest when the cage smelled familiar. However, rats initially hoarded less in both experimental conditions than in the control (familiar appearing and smelling) condition, indicating that visual cues also are important. It would be informative to observe how animals respond to visually familiar surroundings that harbored the odor of unfamiliar conspecifics.

Many species of mammals and birds readily hoard under experimental conditions, leaving the impression that familiarity with the experimental situation is quickly gained or unimportant. But the impact of unfamiliarity on animals is probably characterized by great inter- and intraspecific variation. In some studies, this variation is merely a statistical inconvenience, but in others—in which quantitative analyses of hoarding behavior of individuals or species is the goal—lack of control for familiarity of the surroundings may yield misleading results. For example, to what extent are studies that compare the hoarding response of two or more species (e.g., Lanier, Estep, and Dewsbury 1974; Barry 1976; Tannenbaum and Pivorum 1987) influenced by interspecific differences in hoarding response under novel conditions? Denenberg (1952), for example, observed that laboratory rats hoarded significantly more pellets when isolated than in the presence of another rat. He attributed the effect to social facilitation of eating under the group condition, a behavior that competes with hoarding. However, his experimental design, which had isolated rats hoard in their home cages and grouped rats hoard in neutral (unfamiliar) cages, may have confounded the results. It is impossible in reading Denenberg's experiment to establish whether reduced hoarding in groups was due to responses of rats to each other or to their strange surroundings.

Miller and Viek (1944) suggested that rats may hoard in the home cage because they regard the cage as a point of maximum security. This hypothesis, called the *security hypothesis* by Bindra (1948b), presumes that rats are relatively "insecure" and "anxious" when outside their home cage. Bindra (1948b) noted that the security offered by a particular situation depends on the nature of the organism and the characteristics of the situation. He termed the level of security felt by an animal in a particular situation "shyness" and went on to suggest that shyness may explain the variability in rats' hoarding responses frequently observed in experiments. Some rats do not hoard but eat in the food bin or alley, whereas other rats hoard large quantities of food in the home cage. These differences may be the results of differences in shyness. He proposed that shy individuals feel insecure outside the home cage and consequently hoard; less shy individuals feel more secure and do not accumulate food in the home cage.

Bindra (1948b) tested the security hypothesis by selecting rats that hoarded little or not at all (0–1.3 pellets per .5 hour) when deprived and in familiar (secure) surroundings. When tested in a less secure setting (open hoarding alley), these rats took longer to leave their cages (taken as a measure of their insecurity by Bindra) and hoarded large quantities of pellets (18.0–55.0 pellets per .5 hour), thus verifying his prediction. Subsequent experiments by Hess (1953), Smith and Powell (1955), and Manosevitz (1965) are consistent with Bindra's (1948b) hypothesis.

The security hypothesis has considerable appeal in explaining hoarding behavior of wild rodents because it is implicated in the evolution of food storage in burrow-inhabiting animals (see Security Hypothesis, chapter 3).

Food Type

The food items that animals store are not a random selection of those they encounter or those they eat. Certain foods are eaten immediately, others are stored, and yet others are ignored. How an animal treats a particular type of food depends on many characteristics, including size, composition, and perishability. For example, black-billed magpies (*Pica pica*) eat small pieces of bread and bury large pieces (Henty 1975), and laboratory rats (Long-Evans strain) eat small food pellets and hoard large pellets in their nests (Whishaw and Tomie 1989; fig. 5.19). Northern pocket gophers (*Thomomys talpoides*) consume underground plant parts that are high in protein as they encounter them and store low-protein items (Stuebe and Andersen 1985). Laboratory rats fed diets deficient in vitamin B or vitamin D preferentially store items high in these vitamins (Gross and Cohn 1954; Gross, Fisher, and Cohn 1955). Honey bees hoard nectars with high sugar contents (Sylvester 1978) or that differ in chemical composition (Rinderer and Baxter 1980b) and avoid sugar types (e.g., lactose, galactose) that are toxic (Sylvester 1979). An animal's preference for a food type may vary at a particular time, depending on the other types of food available. Eastern woodrats (*Neotoma floridana*), given a choice of acorns and commercial rodent chow, perferentially consumed and stored acorns (Reichman 1988). When the chow was paired with grapes, however, the woodrats stored mostly chow and ate the perishable grapes. Clearly, animals perceive the nutritional quality and perishability of food and make decisions on what to select and how to handle items based on these characteristics. Further evidence to support this notion comes from observations of how animals store different types of food. For example, red squirrels store conifer cones in moist litter, sites which keep the cone bracts from opening and the seeds from dispersing (Shaw 1936). When storing mushrooms, however, they place them in well-ventilated sites (e.g., branch forks in trees), where they quickly dry (Buller 1920). Alaska voles (*Microtus miurus*) store underground plant parts in subterranean chambers and cure hay on the ground surface or in low shrubs. In general, food items are stored under those

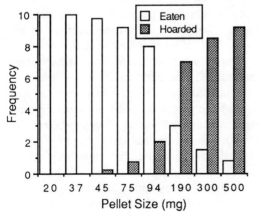

Figure 5.19
The effect of pellet size on the hoarding and feeding behavior of Long Evans laboratory rats. Pellet size is given in milligrams along the abscissa. After Whishaw and Tomie 1989. From *Psychobiology* 17:93–101. Reprinted by permission of Psychonomic Society, Inc.

conditions that minimize the chances that they will spoil. What we do not know is how animals allocate effort when storing an array of foods that differ in nutritional content or perishability.

A number of studies have examined the specific stimuli responsible for food selection or preferences, but the results are more confusing than revealing. We know, for example, that albino rats stored more food pellets wrapped in aluminum foil than otherwise identical but unwrapped pellets (Licklider and Licklider 1950). Rats also stored more saccharin-soaked (wet) chow than pelletized (dry) chow (Bindra 1948a). In paired preference tests, hamsters stored more bitter-tasting pellets soaked in citric acid than unmodified chow pellets, more sucrose pellets than citric acid–soaked chow pellets, and more glass beads than sucrose pellets (Scelfo and Hammer 1969; Hammer 1972)! Desert woodrats (*Neotoma lepida*) stored glass beads but not chow pellets (Clark and Gay 1976). Green acouchis (*Myoprocta acouchy*) stored more lightly pigmented (white, yellow) and smooth biscuits than darkly pigmented (brown, black) and rough biscuits of identical composition (Morris 1962). Pine voles, perhaps not surprisingly, stored more peanuts than similar-sized pieces of flavored, wooden dowels (Geyer, Kornet, and Rogers 1984), but they also stored pieces of wooden dowel soaked in corn oil, 25% sucrose solution, or 25% apple extract. In the case of laboratory rats, storage of inedible objects appears to be most strongly influenced by two characteristics: novelty and "partibility" (existing in carriable units) (Wallace 1978, 1979). Do these preferences have any bearing on how animals select edible foods for storage? And how are these nonfood items reinforcing? The answers are not yet known. Wallace (1983) states that the incentive properties of objects are crucial to the expression of hoarding behavior. But for many species, we have little knowledge of what food properties stimulate hoarding. One generalization that can be drawn from these studies is that rodents may manipulate and hoard items that stimulate visual, tactile, and gustatory senses regardless of the items' food value. However, the finding that blind or anosmic laboratory rats hoard as much or more than intact rats (Tigner and Wallace 1972) suggests that even this generalization may be invalid.

How might food type (i.e., its stimulus characteristics) influence the experimental analysis of hoarding behavior? Clearly, researchers should assume that any characteristic of a food could influence or bias hoarding behavior and therefore must consider all characteristics of a food when designing experiments on hoarding. Those using foods not normally part of a subject's "natural" diet should interpret their results with caution. Perhaps too much reliance has been placed on pelletized foods; an animal's natural hoarding behavior and food preferences need to be considered when formulating experiments. In comparative studies of hoarding behavior, Dewsbury (1970) and Lanier, Estep, and Dewsbury (1974) measured the hoarding of lab chow, corn, and sunflower seeds by eleven species of muroid rodents, including cotton rats (*Sigmodon hispidus*), white-footed mice, and Mongolian gerbils (*Meriones unguiculatus*). All three of these species failed to hoard significant quantities of any food, whereas hamsters and several species of voles did store. The authors tentatively concluded that "the ease of eliciting hoarding behavior in the laboratory appears roughly correlated with the tendency to hoard in the natural habitat." However, cotton rats construct hay piles of grasses and sedges (Nowak and Paradiso 1983), white-footed mice avidly store nuts and pine seeds (Hamilton 1943; Abbott and Quink 1970), and gerbils store large quantities of seeds, nuts and other foods (Naumov and Lobachev 1975). Further, Dewsbury's and Lanier, Estep, and Dewsbury's experiments were conducted under long-day (14 hour) photoperiod, which may have acted to inhibit hoarding in some species but not in others (see Photoperiod). The results of Dewsbury (1970) and Lanier, Estep, and Dewsbury (1974) indicate only that food pellets, corn, and sunflower seeds may be insufficient stimuli to elicit hoarding by these rodents under test conditions. To make valid interspecific comparisons of hoarding behavior, researchers are faced with the difficult problem of finding food items that are equally stimulating to all species being tested.

Social Facilitation and Inhibition

What effect might a conspecific have on an individual's food hoarding behavior? An animal present as another hoards food may (1) threaten the hoarder by retrieving the food after the hoarder has gone (see chapter 4, Loss to Cache Robbers); (2) compete for the limited unstored food available (see chapter 11, Food Hoarding and Competitive Ability); or (3) be a hoarding associate, an individual that participates in the preparation and consumption of a larder (see chapter 3, Communal Food Storage). In the first case, any animal, conspecific or otherwise, should be expected to inhibit hoarding behavior, whereas in the latter two cases another animal may stimulate hoarding.

A frequent consequence of hoarding in the presence of other animals is that the animal steals the food cache (e.g., Goodwin 1956; Källander 1978; Baulu et al. 1980; Burnell and Tomback 1985). Black-capped chickadees (*Parus atricapillus*) hoard less in the presence of several conspecifics than when alone (Stone and Baker 1989). Some laboratory rats hoard more when isolated than when placed in groups of two or four (Miller and Postman 1946; Denenberg 1952), but Cochrane (1968) found no significant difference in the hoarding rate of female laboratory rats in isolation or in group conditions. Among pairs of rats, the individual that hoarded the most was the dominant member (Ross, Smith, and Denenberg 1950; Weininger 1953). Presence of a member of the opposite sex may not affect the sexes similarly. Male pine voles (*Pitymys pinetorum*) hoard more when housed with females, but the female nearly stops hoarding when paired with a male (Geyer, Kornet, and Rogers 1984).

Rats do not cooperate in forming a common food cache, and there appears to be little social facilitation. When allowed to hoard in groups from a central food bin, each rat carries pellets back to its home cage (Miller and Postman 1946). There is a minimum of interaction; each rat hoards on its own. The presence of conspecifics has little effect on hoarding behavior with one exception. After each rat has accumulated a few pellets in its home cage, it then spends a portion of its time stealing pellets from the cages of other individuals. In an experiment by Miller and Postman (1946), 34% of all pellets hoarded came from cages of other individuals. Certain rats seemed to specialize in stealing from other rats, whereas others took pellets primarily from the central bin. Victims of pellet stealing showed no interest in the fact that their larders were being plundered, either by conspecifics or by experimenters (e. g., Miller 1945), a behavior in stark contrast to the aggressive protection of stored food by many species of wild rodents (see Loss to Cache Robbers and Parasites, chapter 4).

When animals compete for a limited resource, an individual may be stimulated to hoard food as a means of accumulating a disproportionate share of the food. Under these circumstances, food hoarding may become an important element in the competitive strategy of an individual or social group. This hypothesis has seldom been tested. Sanchez and Reichman (1987) found that white-footed mice stored significantly more when close to conspecifics they could smell and see than when totally isolated. Hansson (1986) found that bank voles (*Clethrionomys glareolus*) redistributed caches in the presence of conspecifics but did not report that they stored more food. Social groups of Mongolian gerbils competed aggressively for grain placed conspicuously near territorial borders, and more individuals became active in territorial defense when storable food was present. Partly as a consequence of their site-related dominance, territorial owners collected significantly more grain than did trespassers.

Communal hoarding is practiced by only a few species (e.g., honey bees, acorn woodpeckers, certain voles, Mongolian gerbils, beavers). Groups of thirty to fifty honey bees hoard significantly more per bee than groups of ten bees (Rinderer and Baxter 1978b). The cause of the difference was not established but may have been related to the frequency of interbee contact in this highly social species. Furthermore, uniform groups of worker bees hoard significantly more pollen in the presence of larvae (Hellmich and Rothenbuhler 1986b) and more honey if maintained with a queen (Free and Williams 1972). The queen is known to secrete substances that control certain aspects of colony behavior, and perhaps increased food storage is one indirect consequence of this control.

Stored Food as an Inhibitor of Hoarding
Regulation of the quantity of food animals store (larder size or the number of scattered caches) is a topic that seems to have drawn little attention. Some species collect and store several times more food than they can possibly consume during the course of a lean season (e.g., Clark's nutcracker, Vander Wall 1988; honey bees, Seeley 1985) whereas other species store only enough food to last a few days (see chapter 2). The former situation can be partially explained by the fact that a large proportion of the food stored may be consumed by cache robbers, degraded by decomposers, or simply lost. Although hoarding appears to be an expensive behavior with regard to allocation of time and increased risk to predation while collecting food, an animal faced with these environ-

mental uncertainties may evolve a strategy of hoarding food until the available supply has been exhausted—the "extra" food gathered serving, in effect, as an insurance policy against theft and loss and the vagaries of an unpredictable climate. In these species, the hoarding response may be elicited by the lowering of sexual hormone titers induced by photo-periodic or temperature cues or by some other means, but there may be no mechanism for inhibiting the behavior after a certain amount of food has been stored.

Although it is interesting to speculate about why some species seem to store more food than they can consume, it is also interesting to ask how some species regulate the size of their larders. One hypothesis is that a well-stocked larder curtails further hoarding behavior and, con-versely, an empty larder stimulates food collection. In other words, contact (visual, tactile, or olfactory) between the hoarder and its food re-serve results in feedback that acts to stimulate or inhibit food gathering. How this might happen is not clear. The inhibitory effect of a full larder has not been well studied, and in many species, the size of the food store seems to have no effect, either inhibitory or stimulatory, on further hoarding (e.g., Ewer 1967). Further, scatter hoarders, which typically hide food well so that its stimulus value is minimized, have not been shown to alter hoarding behavior in response to the quantity of food al-ready stored. But species that accumulate food in a conspicuous larder within a burrow or other site that they frequently visit can potentially be affected by the size of the stored food reserve. For example, the spiny pocket mouse (*Heteromys desmarestianus*) maintains a larder of about fourteen (range one to forty-two) palm (*Welfia georgii*) fruits in their bur-rows (Vandermeer 1979). If a captive spiny pocket mouse possesses a small hoard of palm seeds, it collects relatively few fruits when some are offered. If the larder is empty, the mouse collects nearly all *Welfia* fruits when fifteen or fewer are presented. When more than fifteen fruits are presented, the number stored does not increase proportionately (fig. 5.20). The reserve of seeds that spiny pocket mice maintain can sustain them for about 4–5 days. Similarly, captive red-tailed chipmunks (*Tamias*

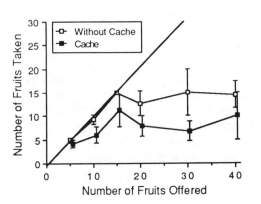

Figure 5.20
Removal rate of palm fruits by captive spiny pocket mice as a func-tion of number of fruits offered. The diagonal line represents hoard-ing of all fruits offered. Animals without an established larder hoarded more than those that al-ready possessed a larder, but after obtaining about fifteen fruits, they stopped hoarding. Vertical bars are 95% confidence intervals. Redrawn from Vandermeer 1979.

ruficaudus) stop hoarding if food pellets are allowed to accumulate in their cages (Lockner 1972).

An abundance of empty comb increases the rate of nectar gathering by foraging honey bees (*Apis mellifera*) during honey flows (Free and Williams 1972; Rinderer and Baxter 1978a, 1979, 1980a). Large amounts of filled and capped honey comb in the presence of little empty comb does not stimulate hoarding. The cause of the increased hoarding response when excess empty comb is available has not been established, but volatile chemicals, perhaps certain pheromones, that are incorporated into the comb by worker bees may be responsible (Rinderer 1981, 1982). The effect of this response may be to channel the activity of the workers toward comb construction and away from honey production to increase the long-term honey storage potential of the colony.

The benefits derived from hoarding should vary with size of the food store, but this important relationship has not been established for any food-hoarding animal. At what point does an incremental increase in larder size confer no further advantage on the hoarder? What are the costs of hoarding and how do they influence food-collecting behavior? What are the behavioral and physiological mechanisms that curtail hoarding? And what types of interactions between a species like the spiny pocket mouse and its environment cause it to accumulate enough food for a 4–5 day reserve instead of enough for a 10-day or 30-day reserve?

5.3 CONCLUSIONS

Although a wide variety of mammals, birds, and arthropods store food, most experimental investigations of the factors that regulate hoarding behavior have been conducted on a small subset of these species: domesticated strains of the Norway rat and house mouse, Syrian golden hamster, Mongolian gerbil, and the honey bee. Studies of the domesticated or laboratory rat accounts for about one-half of the experimental studies on hoarding behavior, but there is some reason to question whether the laboratory rat adequately represents food-hoarding rodents with regard to the factors that regulate hoarding behavior (Beach 1950; Lockard 1968; Nyby and Thiessen 1980), since Norway rats in the wild store almost no food (Calhoun 1963; Takahashi and Lore 1980; Yabe 1981). In the laboratory, hoarding is induced by depriving subjects of food, causing body mass to decrease well below ad libitum body mass. In contrast, many other hoarders store large quantities of food when satiated, with seasonal increases in food stores occurring when food availability is near its annual high. The "domestication" and intensive inbreeding of laboratory rats may have genetically disrupted species-typical behaviors such as hoarding (Lockard 1968). Despite these concerns for the laboratory rat in particular and the biased treatment of food-hoarding species in general, it seems desirable to draw some general conclusions concerning the regulation of hoarding behavior. How well these conclusions extend to other food-hoarding mammals, birds, and arthropods remains to be determined.

The food-hoarding behavior of an organism is the product of many factors that interact in complex ways. The genetic endowment of an organism, as well as its experience during formative stages of behavioral development, determines to a large degree its disposition to hoard as an adult. Much of the variation in the hoarding intensity of laboratory rats, laboratory mice, and honey bees has a genetic basis (Stamm 1954; Manosevitz and Lindzey 1967; Milne 1985), and ontogenetic changes are likely to augment this variability. It seems likely that the combination of genetic constitution and early experience would account for much of the intraspecific variation seen in other hoarders, but this has yet to be demonstrated. The physiological bases for these genetic and developmental effects is not well known. The hoarding behavior of some rodents appears to be a component of the physiological regulation of body mass controlled by the lateral hypothalamus. The nutritional state of the body is monitored by the hypothalamus and is compared to some set point. If a deficit (body mass below set point) occurs, the hypothalamus triggers a series of responses (including hyperphagia and hoarding) that result in defending body mass. How this hypothesis can be extended to explain hoarding in other species that store food in advance (from hours to months) of any possible food shortage to avert, in a sense, future deficits is not clear. Further, some species of animals hoard without noticeable hyperphagia (e.g., heteromyid rodents, all birds), and other species become hyperphagic without hoarding (e.g., certain ground squirrels). In fact, the occurrence of hyperphagia, fat deposition, and hoarding may vary among individuals of the same species (Wrazen and Wrazen 1982). Clearly, there are more pieces to the puzzle, and these pieces will only be found by conducting comparative studies on the regulation of hoarding and body mass in a wide array of species.

Superimposed on this "predisposition" to hoard are, at least in some mammals and birds, cyclic fluctuations in hoarding intensity. One identified cause for these cycles is fluctuating levels of sex hormones: testosterone and estrogen. The marked seasonal cycles that characterize the hoarding behavior of many species of mammals and birds in the temperate zone result from the suppressive effects of testosterone and estrogen on hoarding. The fall increase in hoarding occurs as the gonads are regressing, triggered by changes in photoperiod, temperature, and other environmental cues. In spring, gonadal development and the concomitant increase in sex hormone titers suppress hoarding. One might hypothesize that the reduced seasonality of hoarding in the tropics is due to the protracted reproductive season of many tropical organisms. The fact that tropical organisms are in breeding readiness for a longer period means a less marked fluctuation in sex hormone levels over the course of the year. Among some female rodents, hoarding intensity also cycles within the reproductive season in time with the estrous cycle. Hoarding decreases during behavioral estrus, when estrogen titers are high, and increases during diestrus after estrogen levels fall. This cycle, which apparently gains its rhythmicity from some internal biological clock, has a period of 4–6 days in laboratory rats and hamsters. The physiological effects of sex hormones on hoarding behavior in laboratory rats is thought to be through the reduction of body mass set point of the lateral hypothalamus. With the set point reduced, any physiological deficit is reduced or eliminated, and consequently the need to hoard is alleviated.

Within this framework of genetic and hormonal regulation of hoarding, hoarding intensity may be further modulated by stimuli impinging on the organism from the environment. Unfamiliar environments reduce hoarding, partially as a consequence of competing behaviors such as exploration. Bright illumination of the feeding areas or nest area, time during the light-dark cycle, and presence of conspecifics are known to influence an individual's disposition to store food. Characteristics of food including visual, tactile, and gustatory cues and traits like perishability and novelty influence an animal's choice of items to store, but for most species the factors that control food selection are not well known. In some species, the amount of food already stored may inhibit further collecting.

To what extent does the information gathered on the physiological and behavior control of hoarding apply to food hoarders in general? Is food hoarding in laboratory rats, kangaroo rats, beavers, jays, chickadees, and others regulated in essentially the same way or do these species have radically different regulatory mechanisms? The fact that laboratory rats respond so differently to food deprivation suggests that the latter case is a distinct possibility. More studies on other taxa (especially birds and nonrodent mammals) are needed to broaden our base of understanding and to either verify the conclusions derived from the study of laboratory rats or to determine what other mechanisms regulate hoarding.

6

Cache-Recovery Behavior

For scatter hoarding to be effective in a habitat rich in potential competitors for stored food, it is essential that caches be very inconspicuous. If visual, olfactory, or other sorts of cues are strongly associated with stored items, other animals might eventually learn where food is hidden, and excessive pilfering would ensue. But inconspicuousness of caches creates a problem for hoarders. How are hidden items to be relocated at some future time? This problem is compounded by the fact that many scatter hoarders store hundreds or thousands of items over large areas. Further, sites suitable for caching often are extremely abundant. The coal tit (*Parus ater*) that hides seeds among needles in a spruce forest would seem to have little chance of ever finding them if it relied on random search to locate them. The skepticism of some observers that animals could ever find food they have hidden is exemplified by Carl von Linné's statement that the nutcracker "as the farmer says takes the clouds as a mark" when storing nuts, and thus is not likely to find them again (cited in Swanberg 1951: 547).

How animals return to points in space also is of general interest to behaviorists. The means by which animals navigate by the sun, stars, magnetic fields, or other markers on regional or continental scales has received much attention. How animals return to specific points in space, which presumably depends on locating cues at or in the immediate vicinity of the point, has been studied relatively little. Food-hoarding animals are valuable resources in the endeavor to unravel this mystery because stored food motivates the cacher to return to specific points at some later time, a motivation that is sometimes difficult for experimenters to evoke in nonhoarding animals. Characteristics of a cache or its surroundings can be easily modified, permitting innovative researchers to elucidate the stimuli involved in relocating storage sites. Further, the large number of caches made by food hoarders can be used to the experimenter's advantage to replicate results. In a sense, food-hoarding animals can be "windows" into certain sensory and cognitive properties of animals.

The objective of this chapter is to review the literature on cache-recovery behavior, beginning with a survey of potential cache recovery mechanisms, hypotheses that have emerged as a result of numerous observational studies. Then the results of experimental studies on insects, birds, and mammals that support or refute these hypotheses will be presented. Much of the literature on cache recovery behavior has been reviewed by Shettleworth (1983, 1985), Sherry (1984b, 1985, 1987), and Balda et al. (1987).

6.1 HYPOTHESES OF CACHE RECOVERY

How might animals relocate hidden food? Vander Wall (1982) listed five hypotheses in a study of Clark's nutcracker (*Nucifraga columbiana*) cache recovery behavior, and these apply equally well to other animals that make inconspicuous caches. First, animals could find hidden food by using cues that emanate from the cache. Cues most likely to emanate from caches are chemicals (i.e., olfactory cues). Because the cues that emanate from a cache are most likely characteristic of the food and not the animal that stored it, one would predict that an animal that can use olfaction to recover stored food ought to be able to find food items that it did not hide. Even if animals scent-mark food items (e.g., Muul 1970) or storage sites (e.g., Smythe 1978) the scent should be detectable by conspecifics. One would further predict that if a food item is removed from a storage site (i.e., olfactory stimuli are removed), the hoarder would not return later to search at that site.

Second, animals might find hidden food using visible cues created during cache preparation. These cues include small disturbances of the substrate (e.g., soil, bark) created at cache sites, objects placed on cache sites by some hoarders, and, in some cases the partially exposed food item itself. Some of the disturbances will likely weather away within a short time, limiting their utility as cues. Note that this hypothesis does not require an animal to remember the spatial position of a cue or beacon but simply recognize one whenever encountered. If an animal used this type of information in cache recovery, one would predict that since memory of cues is not involved, individuals should be able to locate each other's caches, and that if the cues were disturbed, the animal could no longer find the stored food.

Third, animals might find hidden food by simply probing or digging at random. In the strictest sense, random search or trial-and-error search would be unaided by any general recollection of the area where food was stored or the types of cache sites used. Such a recovery technique would likely yield extremely low success rates. A modified form of this hypothesis (hypothesis 4) asserts that animals have a general knowledge of the areas and sites they used for storing food and that random search at these sites, which represent a very narrow subset of the possible sites, will produce a reasonable expectation of finding stored food. The means by which animals select suitable cache sites also may guide them in identifying likely storage sites when attempting to relocate food. This hypothesis has been proposed for several species of food-hoarding birds (e.g., Korelov 1948; Haftorn 1954; Goodwin 1956; Kishchinskii 1968; Rutter 1969; Källander 1978). For example, Haftorn (1956b) believed that willow tits (*Parus montanus*) had a "memory impression" of localities in a spruce forest where most storing had been conducted and that storage points at these localities were "highly standardized." These two factors guided willow tits in their search for stored food, which Haftorn believed to be a communally used resource. This hypothesis has two important predictions: (1) animals should store food in some very restricted subset of possible sites and (2) the success rate of random search at such sites can be easily predicted by knowing the proportion of suitable sites that are filled. Further, to the extent that conspecifics use similar sites, individuals should be able to find food stored by others.

A fifth frequently suggested mechanism of cache recovery is that animals remember the precise location of each cache site using visual cues. This conclusion has usually been based on the observation that animals go directly to cache sites "as if they know" exactly where they are or that animals have success rates too high to be explained by any other means (e.g., Swanberg 1951; Goodwin 1956; Tordoff 1955; Collopy 1977; Elliott 1978; Emmons 1980; Waite 1985). Mueller (1974), for example, observed captive American kestrels (*Falco sparverius*) attempt to retrieve stored prey by grasping at storage sites with their talons even though the prey had been previously removed. Murie (1936) and Scott (1943) inferred from red fox (*Vulpes vulpes*) tracks in snow and Henry (1986) concluded from carefully observing foxes that they remembered the exact location of hidden food. Animals often store prey near conspicuous objects, which might act as visual cues (e.g., Kruuk 1964). Two important predictions of the memory hypothesis, which contrast sharply with predictions for all other hypotheses, are (1) animals should find only the caches they prepare or observe other animals prepare and (2) if visual cues are remembered, systematic altering of these cues should result in animals searching in incorrect but predictable sites for food they have hidden.

Food-hoarding animals may use more than one of these hypothesized cache-retrieval mechanisms. In fact, redundant means of relocation are to be expected if other types of spatial behavior (e.g., navigation) are any guide. Cache-recovery behavior typical of a species has important implications for the evolution of food hoarding (Andersson and Krebs 1978) and may profoundly affect other aspects of its natural history. These concerns will be considered further at the end of this chapter.

6.2 RELOCATION OF PROVISIONED NESTS BY HYMENOPTERA

Among invertebrates, the problem of relocating stored food (provisioned nests) is most evident among solitary Hymenoptera. These wasps and bees leave their inconspicuous nests to forage, often at considerable distances, and then navigate back to the nest and find the entrance. Niko Tinbergen and his students, working with the bee wolf (*Philanthus triangulum*), wasps that inhabit heathlands in the Netherlands, were among the first to critically examine how wasps relocate their nests. Their series of experiments has become a classic study in animal orientation.

Bee wolves dig nests in the sandy soil of the heathland, leaving oval patches of sand about 10 cm in diameter next to entrances. Tinbergen (1935, 1972) noted that while constructing new nests wasps would occasionally emerge and hover directly over entrances. When they completed a nest, they blocked the entrance with sand and performed an elaborate departure flight during which they flew around the nest for up to 2 minutes before leaving to hunt for honey bees, the only prey they used to provision nests. Bee wolves would return several minutes to several hours later unerringly locate the entrance, dig out the sand, and enter the nest.

Tinbergen (1935, 1972) presumed, based on wasp behavior, that they relocated nests by visual cues around the entrance or olfactory cues emanating from the entrance. To test the visual orientation hypothesis, he arranged twenty pine cones, objects that littered the heathland and that may have been landmarks used by wasps to find nest sites, around a newly completed nest in a circle 30 cm in diameter. The wasp was allowed to enter and leave the nest several times and so had an opportunity to learn the arrangement of cones. After the wasp left on a foraging trip, Tinbergen constructed a sham nest about 30 cm to the side of the real nest and rearranged cones around the sham nest (fig. 6.1). When the wasp returned, Tinbergen recorded where it searched for the nest entrance. In all, seventeen wasps made 105 choices, all at sham nests surrounded with cones. Following these tests. Tinbergen replaced cones around real nests and allowed wasps to select again. The seventeen wasps selected the real nests eighty-six times of eighty-six choices. This experiment showed unequivocally that wasps were guided by the cone circles around nests.

Do sand patches at the entrance to nests act as beacons for returning wasps? Tinbergen and his followers investigated this possibility in several situations where excavated sand differed in contrast with the background. In an area where sand patches were identical in color to the surface sand, Tinbergen carefully blew away a sand patch and constructed a sham nest 20 cm to one side of the real nest. Wasps returning with prey ignored the sham nest and choose the actual nest in all of sixty-three choice situations; sand patches had no influence on wasps' ability to find nests. But similar tests in areas where pale sand patches contrasted sharply with the dark, brownish black substrate had different results. Five of six wasps (twenty-five of thirty choices) oriented consistently to sham nests. A further test in which neighboring nests also were shifted indicated that when sand patches contrast with the background, bee wolves may use those in the neighborhood of their nests as orientation cues. Tinbergen could find no evidence that olfactory cues facilitated nest relocation, but olfaction may be important in nest identification in other Hymenoptera (e.g., Steinmann 1976, cited in Wehner 1981; Anzenberger 1986).

In a long series of experiments that followed, Tinbergen and his colleagues compared the bee wolf's response to different types of objects arranged around nest entrances. Following training with two sets of objects forming a single or double ring around a nest, each set of objects was moved to sham nests on either side of the real nest and the wasps' choices observed. Tinbergen found that three-dimensional landmarks with detailed patterns were used more often than plain or flat objects. Objects that contrasted with the substrate were responded to more

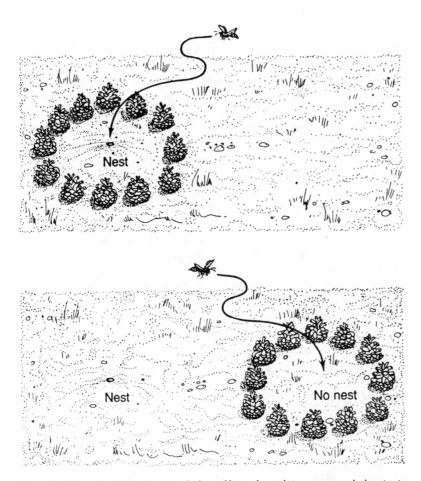

Figure 6.1   Bee wolves (*Philanthus triangulum*) are able to relocate their nest entrance by learning its position relative to landmarks in the vicinity. Tinbergen demonstrated this by placing a ring of pine cones around a wasp nest. After the wasp had learned these landmarks, Tinbergen shifted them 30 cm to one side, causing the wasp to search for its nest at the center of the ring. From Mitchell, Mutchmore, and Dolphin 1988. *Zoology,* figure 13.4. Copyright 1988 by Benjamin/Cumming Publishing Company.

strongly than those matching it. Close objects were more influential than distant objects, and large objects had a greater effect than small objects of the same height. Irregularly shaped objects evoked a stronger response that those with smooth surfaces, but height of objects was the most important character influencing their use as landmarks. Presumably, natural objects on the landscape surrounding nests that embody these traits are most likely to be used as landmarks by bee wolves. In another series of experiments using similar test procedures, van Beusekom (1948) demonstrated that choice of *Philanthus triangulum* is affected by configuration as well as characteristics of landmarks, with the former having greater influence. In fact, specific landmarks can be replaced with dissimilar objects in the exact configuration and a wasp can still relocate its nest. Only a few studies have investigated cue use by other solitary wasps and bees (e.g., Fenton 1923, cited in Wehner 1981; Tepedino, Loar, and Stanton 1979; Steinmann 1985; Anzenberger 1986), so it is premature to generalize about cue use in other nest-provisioning Hymenoptera.

Finding the precise location of an inconspicuous, closed nest entrance would appear to involve estimating distances and angles to landmarks. Cartwright and Collett (1982) suggested that honey bees remember objects around food sites in a form similar to a two-dimensional photograph. Bees may relocate points in space by constantly comparing images on the retina with the remembered image. Extending this idea, navigation through space may be guided by a succession of remembered visual panoramas. Studies of bee visual perception suggest that bees use retinal images to discriminate among different-sized objects, not physical size per se (Wehner 1981). Cartwright and Collett (1979) experimented with this concept, training honey bees to orient to a food source in a white room with a single black cylinder as a landmark. When the landmark was replaced by a smaller black cylinder of identical proportions, some bees searched closer to the landmark. When a larger landmark was used, some bees searched farther away from the landmark. Cartwright and Collett (1979) interpreted these results to mean that honey bees moved to a point where images on the retina and the snapshot were the same

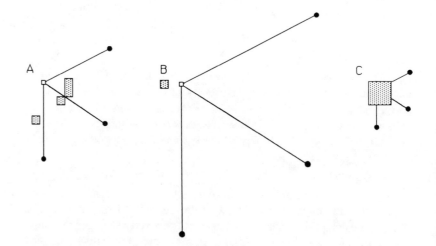

Figure 6.2   An experiment demonstrating that angles to landmarks are more important than distances to landmarks in the orientation of honey bees. A, Training configuration. A food source is placed at the intersection of the three lines (60° angles), 76 cm from three landmarks (cylinders 5 cm in diameter × 50 cm high). Shaded area indicates where bees spent most time searching when food was absent. When the distance to landmarks is doubled (B) or halved (C), bees spend most time searching (shaded area) at a position where they maintain a 60° angle between landmarks. Redrawn from B. A. Cartwright, and T. S. Collett 1982 and reprinted by permission from *Nature* 295:560–4. Copyright © 1982 Macmillan Magazines Ltd.

size. Not all bees responded to changes in landmark size the same way. The behavior of some bees suggested that they relied on motion parallax to determine how far they were from the single landmark.

In more complex situations, where several landmarks are involved, maintaining correct angles to cues seems to be more important than accurately estimating distance. Cartwright and Collett (1982) demonstrated this in honey bees by manipulating three black 5- × 50-cm cylinders near a food source. In training tests, the cylinders were arranged at 60° angles equidistant (76 cm) from a food source (fig. 6.2). In one treatment, distances from objects to a food source were doubled while keeping angles at 60°. In another experiment, distances were halved. In both experiments, foraging bees ignored distance and moved to a position where angles between landmarks were about 60° (the site where food was predicted to occur based on cue angle). Cartwright and Collett (1982) concluded that the most important aspect of landmark arrays to bees in their experiments was retinal position of landmarks as seen from the food source (i.e., angle between landmarks); apparent size of landmarks played a secondary role. Although these experiments dealt with relocating foraging sites, it seems likely that the same perceptual mechanism may be functioning when bees and wasps relocate stored food. Further experiments in which insects (hoarding and nonhoarding) apparently match remembered and retinal images are described by Wehner (1981).

Learning spatial arrangements of nest entrances and surrounding objects occurs during "locality studies": short, low, looping and zigzag flights in the vicinity of the nest entrance. Such orientation flights are characteristic of all bees and wasps that relocate their nest by landmarks (Wehner 1981). Van Iersel and van den Assem (1964), who studied orientation behavior of a digger wasp (*Bembix rostrata*), which orients to its inconspicuous nests in the same manner as bee wolves, demonstrated experimentally that site learning occurs primarily during locality studies when digger wasps leave their nests and less so during preentering flights when they return (fig. 6.3). There are brief periods during which

Figure 6.3   An experiment demonstrating that digger wasps *Bembix rostrata* learn landmarks at the time they leave the nest burrow. During preentering flights a black ring is positioned to the left of the nest entrance (A). While a wasp is in the nest, the ring is moved to the right side of the nest (B). After repeating this procedure several times, the ring is placed so that it encircles the nest entrance (C). When the wasps return, they usually land on the spot marked with a + (the "correct" site if landmarks are learned on preleaving flights), but sometimes on the nest (N). From van Iersel and van den Assem 1964.

bee wolves learn landmarks quickly: when nests are first being excavated and after long periods of inclement weather when wasps have not left their nests in several days (Tinbergen 1972). Following the latter situation, bee wolves could learn positions of new cues placed around their nests during one orientation flight that lasted as little as 6 seconds. At other times, two or more orientation flights were needed to learn positions of new objects placed around nests.

Objects placed near a nest while a wasp is away foraging cause the returning wasp to conduct reorientation flights. These flights, which occur when wasps next leave the nest, update landmark memories. If landmarks are disturbed when the wasp is in the nest, the wasp does not conduct a reorientation flight when it leaves, indicating that discrepancies in landmark position when entering a nest are what triggers reorientation flights (Wehner 1981). Reorientation time is proportional to degree of disturbance of landmarks around nests (van Iersel 1975). When very large objects placed near the nest block out many landmarks, digger wasps are able, after short reorientation flights, to find the nest site using only a subset of the original remembered landmarks (van Iersel and van den Assem 1964). Thus, digger wasps are able to attend to and to use the undisturbed cues.

Distant cues also may play a role in nest relocation (Wehner 1981). Van Iersel and van den Assem (1964) demonstrated this by placing visual barriers 120 cm long by 50 cm high 60 cm behind nests of *Bembix rostrata*. Returning wasps searched for more than a minute for entrances even though the vicinity (area within 60 cm) of nests was undisturbed

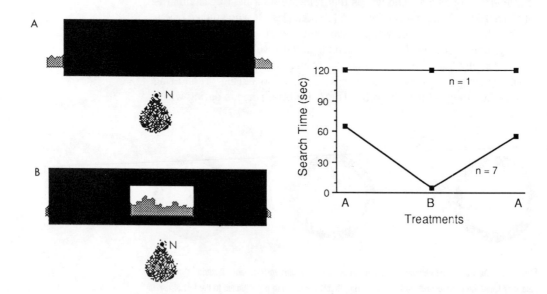

Figure 6.4   Experiment to show that *Bembix rostrata,* when flying in the immediate vicinity of its nest (N), uses points located in the far distance (e.g., the horizon) to aid orientation to the nest. Search time of seven of eight subjects was significantly greater than normal when the horizon directly behind the nest entrance was obscured (A) but returned to nearly normal when a window was made in the obstruction (B). After van Iersel and van den Assem 1964.

(fig. 6.4). When a window (30 × 50 cm) through which part of the horizon could be seen was made in the barrier (the new barrier was enlarged at the ends to have surface area equivalent to that of the original barrier), returning wasps usually found their nests immediately. Thus, it appears that although *Philanthus* may prefer vertical objects near nests as location markers, *Bembix* may also use the outline of the visible horizon from nest entrances as location cues. These results may be more consistent than they at first appear, since vertical objects near a nest are also likely to project on the horizon.

6.3 CACHE RECOVERY BY BIRDS

Among birds, nutcrackers, jays, tits, and chickadees have been focal points of inquiry on cache recovery behavior because they readily hoard in captivity and hide items in discrete sites that can be manipulated easily be experimenters. The two groups differ in that corvids hoard many thousands of seeds for up to nearly a year (e.g., Vander Wall and Hutchins 1983), whereas the tits whose recovery behavior has been most thoroughly studied store tens to hundreds of seeds from periods ranging from hours to days (e.g., Cowie, Krebs, and Sherry 1981). Despite these differences, a number of recent studies have demonstrated that both groups are able to relocate stored food using spatial memory of the cache site. But, before we consider the evidence for spatial memory, let us first assess the evidence for and against other potential cache recovery mechanisms.

Mechanisms of Cache Recovery

Olfactory cues seem to be unimportant in cache recovery of food-hoarding birds. Nutcrackers and jays seldom find food hidden by others, even though they may frequently pass close by storage sites (e.g., Bossema 1979; Vander Wall 1982; Bunch and Tomback 1986), and tits find seeds stored by an experimenter much more slowly than seeds they hide themselves (Cowie, Krebs, and Sherry 1981; Shettleworth and Krebs 1982). Yet these same birds will search intensely for seeds that have been removed in their absence and for which there is no apparent associated olfactory stimulus (Balda 1980b; Sherry et al. 1981; Shettleworth and Krebs 1982). Black-billed magpies (*Pica pica*) are able to find hidden food scented with cod liver oil (Buitron and Nuechterlein 1985), but their olfactory sense, which seems to be specific to aromatic oils and decomposing flesh, has not been shown to be effective in recovering items they cache themselves. Northwestern crows (*Corvus caurinus*) did not find decomposing, strong-smelling clams significantly more often than false clams (clam shells filled with wax or putty) that had no detectable odor to humans (James and Verbeek 1983). Gray jays (*Perisoreus canadensis*) found pungent-smelling caches of food (cheese and bread soaked with castor oil) at rates similar to that predicted from random search, which suggests no significant olfactory ability to detect these foods (Bunch and Tomback 1986). Birds generally have poorly developed senses of smell, and olfactory senses are especially poorly developed in omnivorous and granivorous species (Suthers 1978), trophic groups to which many scatter-hoarding birds belong.

Cache recovery rates (caches located/recovery attempts) in both field and experimental situations are usually high, ranging from about 50–99% (Bossema 1979; Balda 1980b; Shettleworth and Krebs 1982; Vander Wall 1982; James and Verbeek 1983, 1985; Bunch and Tomback 1986). These success rates are far too high to be explained by random search. Balda (1980b) predicted that if a Eurasian nutcracker (*Nucifraga caryocatactes*) in an experimental enclosure found caches using random search, it would need to make 680 probes to find twelve caches; the bird made only fifteen probes. Marsh tits (*Parus palustris*) relocated seeds they had stored in small receptacles in trees at rates intermediate to those expected from perfect knowledge of sites and random search (fig. 6.5). Further, as already mentioned, birds seldom find food hidden by others, a fact inconsistent with the random search hypothesis. This, of course, does not mean that birds cannot find stored items by random foraging (e.g., Waite 1985)—only that random search alone cannot explain the high success rates that have been observed.

If birds store food in specific kinds of sites, then searching randomly at similar sites should raise their expected success rate. Balda (1980b), for example, considered whether Eurasian nutcrackers used a "template"— a spatiotemporal pattern set relative to structural cues in the environment surrounding caches—when storing seeds. If the template was followed closely, birds could recover hidden food by simply using the same template to identify likely storage sites. Sherry, Krebs, and Cowie (1981) expressed this same idea as a set of rules about which types of sites marsh tits use for storage. Considerable data demonstrate that food hoarders do prefer certain types of sites for storage (e.g., Haftorn 1954, 1956a, 1956b; Bossema 1979; Balda 1980b; Cowie, Krebs, and Sherry

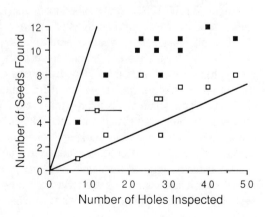

Figure 6.5
Number of stored seeds encountered versus number of holes inspected (excluding revisits) by a captive marsh tit. Filled symbols represent the bird's performance. Open symbols represent results of a simulation search. The lower diagonal line represents expected results of random search within the bird's "acceptable subset" of holes. The upper diagonal line represents errorless performance. The horizontal line gives the range of the bird's data for the first five seeds encountered. Redrawn from Shettleworth and Krebs 1982.

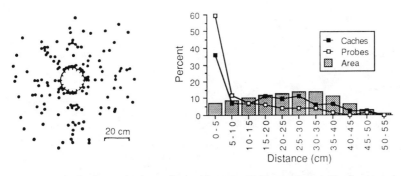

Figure 6.6   The distribution of caches made in soil by captive Clark's nutcrackers. A, The direction and distance from an individual's caches to the nearest large object. The radial distribution is not significantly different from random. B, The distribution of probes (made by birds that had not stored or observed other birds store seeds), caches, and ground surface area as a function of distance to the nearest large object. Nutcrackers stored seeds and searched for seeds significantly closer to large objects than one would expect by chance. After Vander Wall 1982.

1981; Moreno, Lundberg, and Carlson 1981; Vander Wall 1982; Bunch and Tomback 1986) and that individuals may have different preferences or change preferences rapidly over time (Moreno, Lundberg, and Carlson 1981; Sherry, Krebs, and Cowie 1981). Experimental data show, however, that potential advantages gained by storing in these select sites and then searching randomly at similar sites is slight and inadequate to explain the high success rates mentioned in the preceding paragraph. For example, Vander Wall (1982) found that predicted success rates of Clark's nutcrackers searching within 5 cm of an object, sites they strongly preferred for seed storage (fig. 6.6), was 2.7%, up from 0.4% for a bird foraging completely at random. Success rates of two nutcrackers that had stored seeds were 52.3% and 68.0%. Two nutcrackers that had not stored seeds in the enclosure searched most frequently near objects and thereby achieved success rates (1.6%) slightly above that predicted by totally random search. Cowie, Krebs, and Sherry (1981) used the radioisotope [99]technetium to locate sunflower seeds cached by marsh tits and then hid two sunflower seeds in identical-appearing sites 10 and 100 cm away from each marsh tit cache. Caches made by marsh tits survived a mean of 7.7 hours, whereas control caches survived means of 13.5 hours (10 cm) and 20.4 hours (100 cm) (fig. 6.7). The difference in survivorship rate between marsh tits' caches and the 100-cm controls was significant. Slower removal of the control caches hidden in identical-appearing situations suggests that marsh tits were not using simple rules or templates to recover caches. Using a rule constructed to take advantage of the preferences demonstrated by individual marsh tits, Shettleworth and Krebs (1982) found that simulated success rates were lower than the birds' performance.

There is no evidence that birds "mark" cache sites to assist future relocation. A Eurasian nutcracker searched at correct sites after Balda (1980b) raked and smoothed the soil surface over the caches, destroying

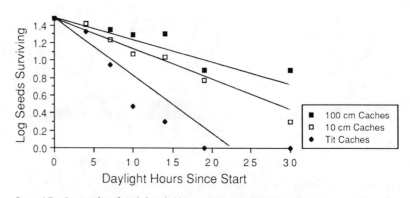

Figure 6.7 Survivorship of seeds hoarded by marsh tits and artificially cached seeds placed 10 cm and 100 cm away from the marsh tit cache. The fitted lines are constrained to meet on the ordinate at a point representing the original number of seeds stored. Redrawn from Cowie et al. 1981.

any potential cues. Vander Wall (1982) rearranged all soil features on one-half of an experimental enclosure and found that recovery success of Clark's nutcrackers on the treatment and control sides of the enclosure were not significantly different. However, success rates of two nutcrackers that had not stored food dropped from 8.0–8.8% on the control area to 0.7–2.9% where the cues had been erased, suggesting that birds that search for caches of other birds may use soil disturbances as cues. Success of these birds in finding recent (less than 3 days old) caches in the control area was significantly greater than for old (more than 3 days old) caches, indicating that weathering processes may quickly erase cues, limiting their use to within a few days of cache preparation.

Under certain circumstances, seedlings emerging from the ground during spring and early summer may serve as visual cues. In this case, animals may learn to associate seedlings with the possible presence of buried seeds. Several species of rodents, jays, and nutcrackers have been shown to use such cues (Bossema 1968; West 1968; Abbott and Quink 1970; Vander Wall and Hutchins 1983).

Spatial Memory

As stated earlier, memory of cache sites by birds has been supported by several recent experiments. Bossema and Pot (1974) and Bossema (1979) displaced visual cues near jay caches with the result that jays could not find cached food but dug correctly relative to the visual cues. I elaborated on this approach to test the impact of shifting large objects (potential visual cues) on the search behavior of Clark's nutcrackers (Vander Wall 1982). In separate sessions, two nutcrackers were allowed to cache piñon pine nuts on an oval arena while I carefully mapped cache locations using a Cartesian coordinate system with origin on a rock near the left end of the arena (fig. 6.8). Following caching, the birds were removed, and one end of the arena was extended 20 cm. All objects at the extended end of the arena were shifted 20 cm to the right so that they were in the same position relative to the margin of the arena. When nutcrackers searched for seeds 3 days later, I measured the x and y components of any errors in searching, i.e., x and y components of distance from center of cache to initial probe. Nutcrackers accurately located many of their caches at the left end of the arena, where potential visual cues were stationary, but they missed most caches at the shifted end of the arena by about 20 cm (mean error = 20.5 ± 3.2 cm). Error along the Y axis (ΔY) was only 1.5 ± 2.7 cm, not significantly different from the ΔY on the control half of the arena (0.9 ± 1.6 cm). This experiment demonstrated that nutcrackers remembered the position of the stored seeds relative to the large objects and margin of the arena, which served as spatial reference points.

Further important insights into nutcracker recovery behavior have been gained by constraining caching into a small, investigator-defined subset of potential sites. Kamil and Balda (1985) trained nutcrackers to store and search for seeds in a small room containing 180 sand-filled cups set in a 15×12 array in the floor. During caching sessions, cups were either available (open and filled with sand) or unavailable (capped with a wooden plug). Thus, experimenters could determine which sites were available for caching and which were available during searching, thereby gaining manipulative control over certain facets of caching behavior. For example, Kamil and Balda (1985) eliminated the effects of site preferences and systematic path selection on ability to recover seeds. They found that with these confounding effects eliminated, nutcracker cache recovery rates still greatly exceeded chance levels. Errors that did occur were usually revisits to previously emptied sites or probes in holes adjacent to cache sites. Success rates gradually decreased within and between sessions, a pattern also observed in marsh tits (Shettleworth and Krebs 1982). Taken together, these results suggest that nutcrackers relocate stored food using spatial memory.

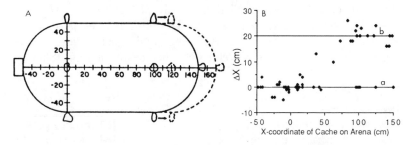

Figure 6.8   The effect of moving spatial cues on the searching behavior of Clark's nutcrackers. A, A 1 $\times$ 2 m oval caching arena where Clark's nutcrackers were allowed to cache seeds. Solid lines indicate arena perimeter and positions of objects during seed caching. During the cache recovery phase, the right array of four objects and the right margin of the arena were shifted 20 cm to the right (broken lines). The x- and y-coordinates of each cache and probe were determined in centimeters with reference to the imaginary axes. B, Distance along the x-axis from a cache to an initial probe directed at that cache (ΔX) as a function of the x-coordinate of the cache on the arena. Line a indicates cache recovery; line b indicates the expected ΔX if nutcrackers used the shifted objects as visual cues. The stationary array of objects occupies the area from -50 cm to 0 cm, the shifted array from 100 cm to 150 cm. Redrawn from Vander Wall 1982.

Tomback (1980) estimated cache-recovery success rates of Clark's nutcrackers by checking for presence of broken seed coats around excavations that appeared to be made by nutcrackers (birds were not observed excavating the seeds). Seventy-two percent of the excavations had shells nearby in spring, and 30% had shells nearby in summer, much higher than success rates likely if animals searched randomly. Tomback (1980) went on to measure nearest-neighbor distances among these holes and found that the distribution of successful-unsuccessful nearest neighbor distances ranged from 2–242 cm. She predicted before the study that such pronounced variation would suggest that nutcrackers rely on trial and error search to find caches. However, based largely on the presumed high success rates, Tomback eventually concluded that nutcrackers must remember individual cache sites. Without knowledge of the initial distribution of all caches in the study plots and lacking convincing evidence that nutcrackers were in fact the animals excavating all of the caches, the distribution of holes in the ground tells one nothing about the sensory and cognitive abilities of the animals that made the holes. Furthermore, high success rates per se do not prove that memory was involved in the recovery process.

A large number of experimental studies indicate that marsh tits and black-capped chickadees remember precise locations of storage sites (Cowie, Krebs, and Sherry 1981; Sherry, Krebs, and Cowie 1981; Sherry 1984a; Shettleworth and Krebs 1982, 1986; Baker et al. 1988). Seed retrieval by tits and chickadees is characterized by relatively few errors during recovery (fig. 6.9). Experimental birds often found the first three or four seeds for which they searched at the beginning of a session without making any errors. Success rates usually decreased to about two or three errors per seed found as more seeds were recovered. This increase in errors is consistent with the memory hypothesis if one assumes that birds do not remember all sites equally well and tits recover food from the best-remembered sites first. Tits preferred certain types of storage sites, but high recovery rates were not due just to these preferences or to following fixed search paths when searching (e.g., Shettle-

Figure 6.9
Mean number of visits by captive black-capped chickadees to sites where they had hoarded food and sites they had not previously used for storage in an arena. During search trials (when the stored food items had been removed by the experimenter), the subjects visited storage sites significantly more frequently than sites not used for storage. Redrawn from Sherry 1984a.

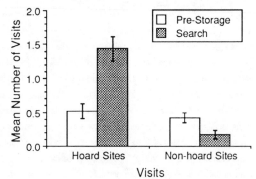

worth and Krebs 1986). Shettleworth and Krebs (1982) put preferences for specific sites in opposition to memory by allowing marsh tits to store more seeds just hours after storing a first batch. While making the second set of caches, marsh tits avoided filled sites, which suggests they knew where the first caches were. During recovery, tits relocated seeds from first and second batches equally well, suggesting that any possible site preference used when storing the first batch of seeds yielded no advantage when searching. In further experiments, marsh tits and black-capped chickadees avoided visiting sites they had already emptied, and black-capped chickadees did not return to sites they had found pilfered (Sherry 1984a). Black-capped chickadees sometimes recovered preferred items (sunflower seeds) before retrieving less preferred items (safflower seeds), suggesting that they remembered not only the location but the contents of cache sites (Sherry 1984a). In a field study, Stevens and Krebs (1986) used tiny magnetic switches (Hall-plate detectors) placed near cache sites to demonstrate that individual marsh tits (which carried magnetic leg bands that tripped the switch when the tit approached to recover a seed) recovered about 30% of the seeds that they had stored. None of these observations by itself demonstrates that tits and chickadees remember storage sites, but the quantitative accuracy, as well as qualitative properties, of cache recovery behavior strongly suggest that tits and chickadees rely on spatial memory to find food they have hidden.

More conclusive evidence comes from a study by Sherry, Krebs, and Cowie (1981), in which they used an innovative approach to demonstrate the importance of vision in cache recovery of marsh tits. Their experiment used the phenomenon of interocular transfer; that is, the ability or, in many cases, inability of the nervous system to transfer visual information from one brain hemisphere to the other. If there is little transfer of visual information gathered during caching and if visual information is essential to cache recovery, a bird that stored food with one eye covered would be able to find that cache only if the same eye was used when searching for caches. Differences in recovery performance using the "experienced" versus "naive" eye would indicate whether cache sites must be viewed (and remembered) for efficient recovery. Sherry, Krebs, and Cowie (1981) used eyelash adhesive to attach an opaque, plastic cap over one eye of marsh tits. Each tit was allowed to cache sunflower seeds in three metal trays filled with moss in a small enclosure. Observers noted exact cache sites and the number of visits and proportion of time spent in each quadrant of the trays. Following caching, the marsh tits were removed from the enclosure, and seeds were located and removed from the trays so that there could be no olfactory or direct visual cues. The plastic cap was removed and either glued back on the same or on the contralateral eye. When searching for seeds 3 hours later, birds with the same eye covered as during caching spent significantly more time searching in quadrants where they had stored seeds than did marsh tits that had the cap switched to the other eye (fig. 6.10). This failure of interocular trans-

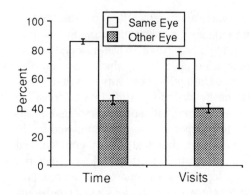

Figure 6.10
The percent of time spent and visits made by marsh tits (during the first 2 minutes of the recovery phase) in quadrants of a rectangular arena where they had stored seeds. Subjects had one eye covered with an opaque cup during caching and recovery trials. *Same eye* refers to trials where the same eye was unobscured during caching and recovery. *Other eye* refers to trials where the eye that was unobscured during caching was covered during the recovery trial. Vertical bars are standard errors of the mean. Redrawn from Sherry et al. 1981.

fer demonstrates that visual information is collected during caching and that this information must be remembered for relocation of stored food to occur.

How similar are memory retention intervals demonstrated in experiments to those likely to occur in the wild? Cache duration in experimental studies of tits and chickadees has ranged from 2–48 hours (Sherry, Krebs, and Cowie 1981; Sherry 1984a; Shettleworth and Krebs 1982, 1986). Although some tits store seeds for much longer periods (e.g., Haftorn 1956c; Alatalo and Carlson 1987), black-capped chickadees and marsh tits usually recover seeds within 2 days. Thus the duration of caches in experiments is representative of those stored in the wild. On the other hand, nutcrackers and Eurasian jays store seeds for many months (e.g., Bossema 1979; Vander Wall and Hutchins 1983), and for practical reasons, tests of memory retention have been much shorter than the birds' apparent capabilities. Tests of jay memory were conducted over only 1 day (Bossema 1979), whereas nutcrackers have successfully retrieved seeds after periods ranging from 3–31 days (Balda 1980b; Vander Wall 1982; Balda and Turek 1984; Kamil and Balda 1985).

Spatial information appears to be processed by the hippocampal complex of the forebrain. Perhaps the first experimental evidence implicating the hippocampus in spatial memory of cache sites was provided by Krushinskaya (1966, 1970). Krushinskaya's goal was to elucidate the functional role of the hippocampal cortex, a region of the brain thought to be involved with memory of complex food-searching reactions in birds. He chose as subjects Eurasian nutcrackers because of their purported highly developed memory of cache sites. Krushinskaya (1970) removed the hippocampus of six nutcrackers and, as controls, removed the hyperstriatum from three subjects, the neostriatum from three subjects, and left four other subjects intact. Following recuperation, each bird was allowed to store pine seeds in a 3.0- × 5.5-m aviary. Krushinskaya mapped locations of each cache (fig. 6.11). Several hours after hoarding sessions, the recovery success of birds was determined. All three control groups recovered seeds effectively with success rates ranging from 78–91% (number of caches found/number of caches made). Subjects with the hippocampus removed, however, had success rates of only 13%. These birds went to the general areas where they had stored seeds and dug repeatedly but with little success. Areas where digging occurred were significantly correlated with where seeds were cached, suggesting that a general memory of cache locations was retained. Recently, Sherry and Vaccarino (1989) have performed similar experiments on black-capped chickadees with nearly identical results. Chickadees with a portion of the hippocampus surgically aspirated found seeds that they had stored at about chance levels, significantly lower than controls (hyperstriatum accessorium aspirated and unoperated subjects). However, experimental birds stored seeds, ate seeds, and searched for seeds at levels similar to controls. Further, removal of the hippocampus did not influence the chickadees' ability to perform remembered nonspatial tasks. Thus, loss of the hippocampus renders nutcrackers and chickadees unable to perform precise spatial tasks.

To summarize thus far, experimental data on several species of avian food hoarders suggest that spatial memory of visual stimuli is the primary means of cache recovery in these birds, although other methods may also be used. Non-seed-hoarding Clark's nutcrackers, for example, which lack the option of finding seeds using spatial memory, nevertheless located some cached seeds. Hypotheses consistent with the behavior of these nutcrackers are that they used visual cues generated when the cache was made (i.e., soil disturbances) and searched randomly at "likely" sites (Vander Wall 1982). These techniques, however, yielded low success rates. Soil disturbance can be used effectively only during a short period after caching, and the quantity of food located by birds using these meth-

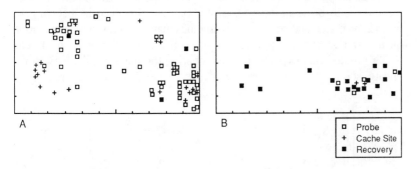

Figure 6.11
Experiment demonstrating the importance of the hippocampus in cache recovery in the Eurasian nutcracker. A, Cache recovery success for a subject that has had its hippocampal cortex surgically removed. Probes are assumed to be unsuccessful cache recovery attempts. B, Success rate of control (sham operated) nutcrackers. Redrawn from Krushinskaya 1966.

ods alone is probably insufficient to sustain a nutcracker during winter at subalpine elevations. Marsh tits also are able to find stored seeds without recourse to memory (e.g., Cowie, Krebs, and Sherry 1981). Pravosudov (1986) has suggested that the experimental demonstration of spatial memory of relatively few sites for short periods by tits does not adequately explain how they can relocate thousands of caches over many weeks or months, and further he suggests that individualistic caching and foraging patterns may be very important in cache retrieval. Jays and nutcrackers may locate many seeds in spring by seeking out seedlings as they emerge through the soil (Bossema 1968; Vander Wall and Hutchins 1983), so although spatial memory may be the primary means of retrieving hoarded seeds, other means of locating seeds are available.

Spatial Orientation

Compared to the Hymenoptera described earlier in this chapter, birds conduct subtle "locality studies," and learning of sites must occur very quickly. Tits and chickadees cock their heads quickly from side to side after hiding a food item as if to "fix" a visual image of the site with each eye (Sherry, Krebs, and Cowie 1981; Shettleworth and Krebs 1986). Nutcrackers appear simply to glance at a few nearby objects during and following caching. Although no experiments have addressed the function of these head movements, it seems probable that birds gather some important spatial information—perhaps a visual image—at this time. If so, one would predict that body orientation of a bird recovering a cache would be similar to that adopted when making the cache, so that the same visual field could be observed each time. The body axis of jays when recovering caches (fig. 6.12) is consistent with this hypothesis (Bossema and Pot 1974). This pattern holds even though jays may approach a site from different directions during caching and recovery (Bossema 1979). The interocular transfer experiment of Sherry, Krebs, and Cowie (1981) also demonstrates that some sort of visual image plays a role in returning to the cache site.

The preparation of a cache may impart a special status to the site that permits it to be remembered. Motor patterns involved in inserting an item in a crevice and covering it with moss may help fix visual information in a way that simply viewing the site cannot. Shettleworth and Krebs (1986) tested this hypothesis by comparing the marsh tits' ability to relocate seeds they had stored with seeds they had encountered when

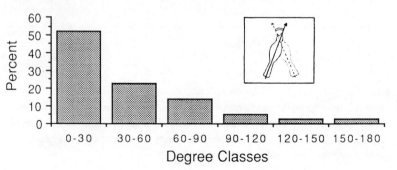

Figure 6.12
Deviation in the body axis of Eurasian jays between cache preparation and cache recovery. Redrawn with permission from Bossema and Pot 1974.

searching for storage sites. Encountered seeds were placed in storage sites by the experimenters and covered with a Plexiglass window so that tits could see but not remove the food item. Shettleworth and Krebs (1986) termed such encounters *window-shopping*. Both stored and encountered seeds were recovered equally well, suggesting that motor patterns used while making a cache (other than the head movements mentioned previously) do not play a role in fixing a cache site and that viewing the site is all that is necessary. Individuals that have the opportunity to observe caching from a distance, however, do not necessarily increase their subsequent searching success for those caches (Baker et al. 1988).

Landmarks or spatial cues used by corvids to find cache sites include large, conspicuous objects. Vander Wall (1982) found rocks, logs, tree trunks, and shrubs to be suitable, but any permanent, conspicuous feature of the environment will probably do. Very small items (i.e., pebbles, twigs), also may be used, but removal of these items does not significantly impair cache relocation. Given a choice, Eurasian jays will search preferentially (more digging and better accuracy) at vertical objects compared to identical horizontally oriented objects (fig. 6.13). Some evidence also suggests that objects nearest a cache site are preferred or relied on more heavily as landmarks than more distant objects (see fig. 6.8), but despite the importance of close objects, the wide environment around a cache site seems to be very important in the bird's ability to orient properly to a storage site (Bossema and Pot 1974). The qualities of landmarks used by tits and chickadees have not been investigated.

How do birds use spatial cues to orient to cache sites? Limited data suggest that jays and nutcrackers relocate cache sites by a process analogous to triangulation. Vander Wall (1982) found that in a few cases, when shifted and stationary objects produced conflicting information on cache location, nutcrackers dug at sites approximately halfway between the sites predicted by the two sets of markers (fig. 6.8). This result is consistent with a cache recovery mechanism in which nutcrackers triangulated the position of the cache using one or more objects from each set. Bossema (1979) approached this question by training jays to dig up artificial caches buried midway between, and 30 cm in front of, two vertical markers spaced 60 cm apart (fig. 6.14A). The markers and buried food were moved around in the experimental enclosure but were always oriented in the same direction, so that the jays would learn not to rely on other potential spatial cues available. After training, markers were stationed in the absence of food in one of two situations, either 30 or 90 cm apart. If jays relocate remembered sites by recalling distances and angles from landmarks, the exact point where the jays dig should indicate something about how the triangulation system works. Initial probe sites of the jays under these conditions were measured and compared to the control (60 cm) condition. In both test situations, the y-coordinate of the initial probe was not significantly different than that during control tests (fig. 6.14). The x-coordinate of the mean initial probe under the 30-cm condition was 16.7 ± 6.3 cm, very close to the midpoint of the two markers

hiding

recovery

Figure 6.13
Design of an experiment testing the relative importance of horizontal sticks (represented by paired straight lines) and vertical sticks (paired dots) of the same size on a caching arena. During the hiding phase, both types of objects were present. During the recovery phase, only one type of visual cue was present. The jays had significantly greater cache recovery success when the vertical sticks were present. Redrawn from Bossema 1979.

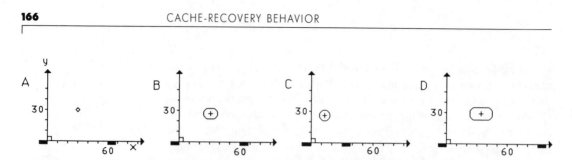

Figure 6.14   Jays were trained to search for a buried piece of food 30 cm to one side and midway between two stakes spaced 60 cm apart (at the diamond in graph A). After training, the location of probing for food was measured in the absence of hidden food (B). Mean position of the center of digging along the x and y axes is denoted by a cross, and the standard deviation about the mean is approximately delineated by the octagon. In condition (C), the stakes (dark rectangles) were repositioned 30 cm apart, and the jay shifted its area of search to the area midway between the stakes and nearly 30 cm in the y direction. In condition (D) the stakes were repositioned 90 cm apart. The jay's search behavior did not differ significantly from the training condition (B) but was shifted slightly to the right. Based on data from Bossema 1979.

(15 cm) and significantly less ($P < 0.01$) than the x-coordinate of the control condition (33.0 ± 11.1 cm). Under the 90-cm condition, the x-coordinate of the initial probe was 37.8 ± 16.9 cm, not significantly different from the control situation (26.6 ± 10.9 cm), but also the standard deviation about the mean of the initial probe included the predicted search point (45 cm) if jays triangulated using the two vertical markers. These results suggest that the jays used at least one fixed relationship in the spatial array, i.e., the perpendicular distance between the hidden food and the line connecting vertical markers. The x-coordinate of the search point (and consequently the angle between the vertical markers as seen from the search point) varied, and at least in the 30-cm trial, was midway between the vertical markers. These limited data suggest that absolute and relative distances are more important than angles.

A major difficulty of the memory hypothesis is explaining the continued high accuracy of birds when many of the spatial reference cues around a cache have been obscured, as happens following heavy snowfalls. Nutcrackers, jays, and crows have been reported digging through enough snow to completely obliterate all surface features, maintaining success rates of greater than 80% (Swanberg 1951; Crocq 1977; Bossema 1979; Mattes 1982; Waite 1986). Although definitive experiments have yet to be conducted, it seems highly likely that a few landmarks must project above the snow surface and that they are sufficient for the birds to accurately relocate a cache. The experiment of Bossema (1979) that indicated a preference for vertical objects when the complete constellation of cues was disrupted is consistent with this explanation. However, the ability of corvids to find cache sites hidden under snow indicates that relocation of storage sites is far more complex than the simple matching of remembered visual images with retinal images, as has been proposed for some Hymenoptera. When digging through deep snow, these birds are able to pinpoint a storage site based on only a few disjunct elements of a remembered visual field and then dig to that point, often at an angle, when unable to see the visual cues above the snow! This suggests that nutcrackers have the ability to recollect cognitive maps of storage sites based on only a few visible landmarks and that they can be very precise in judging angles and distances.

These preliminary findings emphasize how little we know about how birds use the spatial cues they remember at each cache site. I suspect that many important contributions to understanding spatial memory in vertebrates will result from experiments on food-hoarding species.

Sequential Patterns of Recovery

The sequence in which birds recover caches may indicate something about how information on storage sites is organized in memory. There are many possible sequential patterns, but two seem particularly likely. First, birds may recover food in the same order in which they store items, i.e., a primacy effect. This could be because the first few caches made in a sequence are remembered better or that chains of performance may be used to assist cache recovery. Second, the most recent cache prepared may be the first one retrieved, i.e., a recency effect. For a given type of food stored in a particular type of site, loss of stored food is a function of time. The longer an item is stored, the greater the chance that it will be gone when the hoarder returns to retrieve it. Furthermore, memory decay (forgetting) may differentially influence recovery of older caches. Thus, recovery in the reverse order of caching should maximize the number of cached items retrieved (Shettleworth and Krebs 1982; Sherry 1984a).

The sequence of recovery has been examined in numerous studies and has seldom been found to be different from random (Balda 1980b; Sherry 1982, 1984a; Sherry, Avery, and Stevens 1982; Balda and Turek 1984; Baker et al. 1988; Balda and Kamil 1989; and see Reichman, Wicklow, and Rebar 1985 for a rodent example). Bossema (1979) found that even with two sets of caches, one up to 1 hour old and the other up to 14 days old, jays do not seem to discriminate among them. Cowie, Krebs, and Sherry (1981) found that marsh tits tended to recover their earliest caches first, but since sequence of recovery was not observed precisely, this result needs to be confirmed. The strongest evidence for a primacy effect has been found in northwestern crows (James and Verbeek 1985). The crow's pattern of recovery may be due to the fact that clams, the food type stored most frequently, decompose quickly over relatively short storage periods, making it profitable to search for the oldest caches first. On the other hand, Shettleworth and Krebs (1982) detected a slight but significant recency effect in the retrieval behavior of two marsh tits. Given this relatively weak support and the lack of serial order effects in numerous studies, it is necessary to conclude that if sequential patterns are important in cache recovery, few experiments have been designed to demonstrate them.

Other patterns are possible. For example, birds may recover stored items in a way that minimizes travel time between storage sites. This is especially likely for species that store numerous items over a wide area and when storage periods are very long—both situations reducing the advantage of recovering caches in either serial order. Under these circumstances, the energetically least expensive pattern of recovery would be to harvest caches near each other, ignoring the sequence of caching. A captive Eurasian nutcracker observed by Balda (1980b) retrieved seeds in a way that minimized travel time between sites, but Clark's nutcrackers studied by Balda and Turek (1984) and northwestern crows studied by James and Verbeek (1985) did not. Nutcrackers store food in complex spatial patterns. Clark's nutcrackers, for example, may store each pouch-load of seeds in a dozen or so sites scattered over large cache areas used jointly with other nutcrackers. After several thousand caches have been prepared, the temporospatial pattern of caches may be extremely difficult for a bird to recall and, if it could do so, would be a very inefficient foraging pattern to have to replicate. A more efficient use of foraging time would be to retrieve caches that are near each other. That such a pattern is not more evident in laboratory studies may be due to scale. Arrays of caches are so compressed in experiments (compared to the wild) that spatial patterns of recovery are unlikely to be energetically profitable.

Observing patterns of recovery may yield important information concerning cache recovery behavior, but further studies are needed in situations designed to enhance the likelihood of these patterns being expressed before conclusions can be drawn about sequential patterns of recovery. Important elements of such designs are temporal and spatial patterns that approximate those used by the birds in the field; however, it is already evident that sequential patterns of search are not required to relocate caches.

Revisits

A bird that retrieves stored items using memory also should have a way of distinguishing visited sites from sites still likely to harbor food. This could be accomplished by dropping sites from memory (i.e., actively forgetting) or by remembering new information about visited sites, so that a bird would not waste time revisiting them in the future. A bird that could not handle this information storage problem would eventually suffer low effectiveness in recovering food it had hidden because the number of empty remembered sites would greatly outnumber filled remembered sites.

Some data suggest that tits and nutcrackers differ in their tendency to revisit empty cache sites. Marsh tits appear to avoid sites they have emptied several hours earlier, visiting these sites at levels at or below those predicted by chance encounter (Sherry, Avery, and Stevens 1982; Shettleworth and Krebs 1982). The Eurasian nutcracker studied by Balda (1980b) occasionally visited emptied cache sites, and the four Clark's nutcrackers used by Kamil and Balda (1985) revisited cache sites much more often than one would expect by chance. The percentage of unsuccessful probes resulting from revisits by the four Clark's nutcrackers are not given by Kamil and Balda (1985), but these can be calculated from their data (their tables 1 and 5): Adolph, 58%; Johann, 22%; Marcel, 17%, and

Uli, 31%. Overall, 25% of their errors resulted from searching at sites where they had already removed food. Based on these high proportions of revisits, Kamil and Balda (1985: 103) concluded that "nutcrackers do not appear to distinguish between cache sites that have been emptied and cache sites that have not yet been visited." However, in a study I conducted (Vander Wall 1982) in a more natural setting, two Clark's nutcrackers rarely revisited empty cache sites.

What is the possibility that the high revisit frequencies observed in some nutcracker studies are due to artifacts of the experimental procedure? Balda (1980b) removed seeds from the Eurasian nutcracker caches, and this effect alone could have contributed to revisiting. The nutcracker could have been searching again for the seeds it did not find during the first visit to the probe site. Kamil and Balda (1985) constrained nutcracker caching to a "pigeonhole" style matrix of sand-filled cups in which they altered the number of seeds in caches and smoothed the soil over cache sites following each recovery session. Balda, Kamil, and Grim (1986) tested the effects of these manipulations in two experiments. In one experiment, they reconstructed the dig marks in each cup from which nutcrackers previously removed seeds, but as before, they reduced cache size to two seeds. In another experiment, they left cache size unaltered but dug up and later replaced seeds, thus altering their exact placement in the cup. They also filled in excavations. Although two of their four experimental subjects revisited less frequently when dig marks were left at cache sites, they concluded that neither of these manipulations affected the tendency of nutcrackers to revisit emptied cache sites. It is still uncertain, however, whether the pigeonhold caching apparatus or other alterations of the cache sites might have influenced the frequency of revisits. They concluded that either nutcrackers are not able to remember which cache sites have been emptied or that they can distinguish empty cache sites but continue to revisit them for some unknown reason, perhaps related to their experimental procedures. The different tendencies of tits and nutcrackers to revisit caches in laboratory experiments is one of the major characteristics that distinguish their recovery behaviors, but this difference has not been shown to exist in captive subjects hoarding in more natural situations.

Species Differences in Memory Ability

To what extent do the memory capabilities of scatter-hoarding birds differ? It may be that food-hoarding animals have the same degree of spatial memory possessed by any forager but that this memory is simply more evident in hoarders that store extensively. A possible reason for this presumed lack of interspecific differences is that all foragers need precise spatial memory (e.g., Gill and Wolf 1977; Kamil 1978; Shettleworth et al. 1989). Some (e.g., Shettleworth 1983, 1985; Kamil and Balda 1985; Shettleworth et al. 1990), however, have hypothesized that the habits of animals that store and recover large quantities of food may have led to the evolution of adaptive specializations in memory ability.

One rationale for this hypothesis is that food-hoarding animals differ in their degree of behavioral and morphological specialization to store food. Specialized hoarders store more food, recover the food over longer periods, and depend more on their stored food reserves. One example of such specialization in hoarding is the gradation in behavior and morphology of conifer seed–caching corvids (Vander Wall and Balda 1981) discussed in chapter 3. Another is the difference among tits that never or rarely hoard (e.g., great tits and blue tits); tits that store food for short periods (e.g., marsh tits and black-capped chickadees); and tits that apparently store for long periods (e.g., crested tits, willow tits, and coal tits) (Haftorn 1954, 1956c; Sherry 1984a, 1989, Sherry, Krebs, and Cowie 1981; Shettleworth and Krebs 1986; Alatalo and Carlson 1987; Nakamura and Wako 1988). Recent studies suggest that specializations in memory ability are part of the suite of traits that produce some highly specialized food hoarders. Krebs et al. (1989) have shown that the hippocampal complex of food-storing passerine birds is larger than those of similar-sized nonhoarding passerines. This result holds even when storing and nonstoring members of the same bird family are compared. They interpreted the large hippocampi of food-hoarding species as adaptive specializations associated with increased memory capacity. Balda and Kamil (1989) found that Clark's nutcrackers and pinyon jays, highly specialized seed-caching corvids, were significantly more accurate in finding their caches than were scrub jays, a relatively unspecialized seed-caching corvid. They caution, however, that these species differences are not necessarily due to differences in memory ability.

More studies are needed to compare memory abilities of food hoarders that differ in the amount of food they store, as are studies that compare memories of closely related hoarders and nonhoarders. The challenge in the latter case is to devise tests that make meaningful comparisons between species. The window-shopping technique of Shettleworth and Krebs (1986) may be one means of comparing hoarding and nonhoarding tits. It also is important to attempt to compare memory abilities of distantly related taxa. Shettleworth (1985) has summarized some of the results on memory abilities of white rats and pigeons and compared these to food-hoarding species. Balda and Kamil (1988) have compared memory of nutcrackers with that of pigeons by testing nutcrackers in an analog of the radial-arm maze. In these experiments, nutcracker memory far exceeded that of pigeons, but much more information is needed before any general conclusion can be drawn.

6.4   RETRIEVAL OF STORED FOOD BY MAMMALS

The olfactory sense of mammals is well developed compared to that of birds. It is not surprising, therefore, that olfaction has frequently proven to be important in finding stored food in rodents and canids. Most experiments that examine mammalian cache recovery behavior have been designed to test the importance of olfaction, but there is some evidence that spatial memory also plays a role in cache recovery in some species.

Importance of Olfaction

Evidence for the use of olfactory cues comes mostly from experiments in which mammals (usually rodents) have been allowed to search for food buried by an experimenter. Because these caches are artificial, spatial memory of cache sites cannot assist recovery. Captive deer mice (*Peromyscus maniculatus*) can find nearly all conifer seeds and many cereal grains buried 5 cm deep in peat by experimenters (Howard and Cole 1967; Howard, Marsh, and Cole 1968). Seven of eight desert rodents tested in a small, indoor arena by Johnson and Jorgensen (1981) were able to find seeds buried in dry sand (see table 7.1). In field tests, nuts, seeds, and other food items in artificial caches are readily found by squirrels (*Sciurus*), kangaroo rats (*Dipodomys*), pocket mice (*Perognathus*), agoutis (*Dasyprocta punctata*), and red foxes (*Vulpes vulpes*) (Cahalane 1942; Reynolds 1958; Richards 1958; Tinbergen 1965; Lockard and Lockard 1971; Murie 1977; Reichman and Oberstein 1977; Thompson and Thompson 1980). Tests of optimal dispersion of scattered caches (Stapanian and Smith 1978, 1984; Kraus 1983; Jensen 1985; see chapter 5), although designed to examine other aspects of hoarding behavior, also demonstrate that rodents can readily find seeds and nuts they did not hide.

Behavior of mammals searching for hidden food usually suggests that olfaction is important in locating caches (e.g., Nichols 1927; Moore 1957; Kawamichi 1980). Rodents sniff the soil, move slowly until over a cache, and then dig directly to it. Tinbergen (1965) followed fox trails and concluded that they detected buried eggs as they passed close to a site on the downwind side. Number of caches found as a proportion of cache recovery attempts (digs) was not reported in many of the previously cited studies. Nevertheless, the precision with which mammals locate hidden food (e.g., Richards 1958; Tinbergen 1965; Murie 1977) clearly indicates they are not digging at random.

Visual cues have often been reported as unimportant in cache recovery by rodents. Howard et al. (1968) found that deer mice located just as many buried seeds when tested in complete darkness as they did in dim light, when landmarks were visible. These results seem consistent with the fact that deer mice and many other scatter-hoarding rodents are nocturnal and so might seldom see clearly enough to rely on visual cues. But for diurnal rodents, visual cues may be more important. Agoutis, for example, dug significantly more frequently at artificial caches containing corn than at dummy caches where soil was disturbed but no food buried (Murie 1977), but the fact that they dug at dummy caches at all suggests that disturbed soil was a visual cue used by the agoutis to identify sites where buried food might be found.

Several experiments show that detectability of buried seeds varies with strength of the olfactory signal emanating from caches. High moisture content of soil facilitates diffusion of chemical signals; thus moisture is an important (but relatively unstudied) variable in how well rodents are able to find hidden food. Rodents that quickly find buried nuts in moist substrates are much less accurate when searching for nuts or seeds buried in very dry substrates (Dice 1927; Cahalane 1942; Johnson and Jorgensen

1981). The relatively poor success of desert rodents in finding buried
seeds in Johnson and Jorgensen's (1981) experiment may have been due
to their use of dry sand. The decreased ability of rodents to find buried
seeds with increasing depth of the cache (e.g., Cahalane 1942; Reichman
and Oberstein 1977; Schmidt 1979; Johnson and Jorgensen 1981) is a predicted result for an olfactorily-oriented predator.

Modifying the odor of buried seeds and nuts also influences the likelihood that they will be found. For example, grain treated with safflower
or lecithin mineral oil was more frequently found by deer mice than untreated seeds (Howard, Marsh, and Cole 1968). On the other hand,
cleaning nuts to reduce the odor emanating from them reduces rodents'
retrieval success. Nielsen (1973) removed the husk and washed and
scrubbed the hull of 100 black walnuts and buried them 5 cm deep in rows
15 cm apart. Nearby he buried 50 walnuts with husks. Within 1 week, gray
squirrels had found all 50 walnuts with husks but only about 10% of the
cleaned walnuts. Nielsen's (1973) results are not consistent with all other
results on dehusking of walnuts (Phares, Funk, and Nixon 1974) but
nevertheless demonstrate that recovery success of squirrels is related to
strength of olfactory signals.

Although there is as yet no firm evidence, it seems plausible that
mammals could use glandular secretions or urine to scent-mark food or
storage sites to facilitate later relocation of caches (Steiner 1975; Kivett,
Murie, and Steiner 1976; Smythe 1978). Korytin and Solomin (1969,
cited in Harrington 1981) hypothesized that urine marking of cache sites
by canids may serve as location cues that assist in cache recovery. However, this hypothesis was not supported by Henry (1977) and Harrington
(1981), who found that urine marking of cache sites by wolves and red
foxes usually occurred after food was retrieved rather than after it was
cached. Smythe (1978) suggested that agoutis may scent-mark cache
sites by sitting with the perineum in contact with soil while they bury
food, but this has not been tested. Avenoso (1968) and Muul (1968,
1970) suggested that southern flying squirrels marked stored nuts with
an oily secretion, perhaps from modified sebaceous or sudoriferous glands
on the lips (Quay 1965). Muul (1970) came to this conclusion after ob-

Figure 6.15
Preferences of flying squirrels for
hickory nuts that had not been
previously handled by a squirrel
and those that had been previously
stored. The strong preference shown
for "new" nuts can be diminished
by washing previously stored nuts
in soapy water or carbon tetrachloride. The sample size is given
above each bar. Redrawn from
Muul 1968.

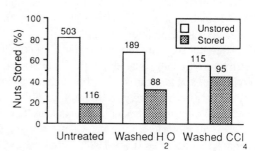

serving that flying squirrels strongly preferred "fresh" hickory nuts to those previously handled and stored by squirrels. He found that if nuts previously handled by squirrels were washed in carbon tetrachloride, a lipid solvent, the difference in acceptability of fresh and previously stored nuts disappeared (fig. 6.15). The effect of the hypothesized secretions on the behavior of captive flying squirrels was to inhibit the storer or other squirrels from recovering already stored nuts. The presence of this secretion and its role in flying squirrel food storage behavior needs to be established.

Spatial Memory

The experiments described previously demonstrate conclusively that olfaction is an important means of finding stored food by various rodents and by the red fox. This result is not surprising in light of the demonstrated reliance of these mammals on olfaction in foraging, social behavior, and other contexts (e.g., Eisenberg and Kleiman 1972; Reichman 1981). It is perhaps simply the consistency of these cache-recovery results with known olfactory capabilities that has caused few researchers to probe further for additional cache retrieval mechanisms. The fact that rodents can and do use olfaction effectively to find hidden food does not preclude them from using spatial memory or other means to find food they have stored. Clark's nutcrackers, for example, used very different modes of search when "expecting" to find their own versus other birds' stored seeds (Vander Wall 1982). McQuade, Williams, and Eichenbaum (1986) have demonstrated that gray squirrels weigh visual and positional cues more heavily than extrinsic olfactory cues when attempting to locate hidden food for which they have been trained to search. Although the study of McQuade, Williams, and Eichenbaum (1986) does not prove that spatial memory is used to recover cached food, it does demonstrate that gray squirrels can remember the position of hidden food relative to visual cues. To test whether mammals also can use spatial memory to find food they have stored, it is necessary to observe individuals attempting to recover food items that they have stored. Few experimenters, to my knowledge, have attempted this.

Macdonald (1976) has tested the hypothesis that red foxes retrieve food they have hidden by remembering the precise location of caches. He began by walking a semitame vixen on a leash around the perimeter of an agricultural field. She found mice previously dropped along the trail by Macdonald and either ate or cached them. One or 2 days later Macdonald walked the food-deprived fox on the same route, allowing her to search for buried food. The fox walked directly to forty-eight of fifty mice it had stored. Retrieval was characterized by a lack of exploratory behavior, suggesting that the vixen knew the exact location of each buried mouse. In a subsequent experiment, Macdonald (1976) allowed the fox to hide mice and then, after removing her, hid another mouse in a similar site within 3 m of the fox's storage site. On recovery trials, the fox walked directly to her own stored prey, and found most of them. She found only two of about twenty artificially stored mice, however. The fox's behavior when finding these two mice was markedly different than when relocating her own caches. In both cases, she nearly stepped on the cache, paused, sniffed and then proceeded to seek out the hidden mouse. It was apparent from her behavior that she used olfaction to find the two mice.

In a third experiment, Macdonald (1976) moved mice stored by the vixen to a similar site 1 m away. She returned subsequently to her empty cache sites and and dug extensively. Eventually she began sniffing around the cache site but found the recached mouse at only 25% of the sites.

In a fourth and final experiment, Macdonald (1976) tested whether foxes could find each others' hidden food. When tested singly, each of three foxes could accurately relocate its own caches, but when Macdonald allowed the three foxes to search for each others' stored mice, they walked past hidden mice, apparently unaware of their presence. In one representative trial, one fox stored nine mice in eight caches along the trail. The next day two other foxes walking along the trail failed to find any of the cached food. Over the next couple of days, the original cacher found and ate all nine mice. These experiments suggest that, although foxes have keen senses of smell and are able to find inconspicuous food items using olfaction (Tinbergen 1965), the fox that makes a cache has a decided advantage over other foxes in retrieving its stored food. Results of experiments 2 and 3 are consistent with the assertion that red foxes remember the location of caches precisely.

Gray squirrels and Merriam's kangaroo rats (*Dipodomys merriami*) also seem to rely on spatial memory to retrieve stored nuts in some situations. Jacobs (1989) allowed gray squirrels to store hazelnuts in a large outdoor arena. After each squirrel had finished, Jacobs made an equal number of control caches. Two to 4 days later squirrels found significantly more of their own nuts than those stored by the experimenter. The differential recovery of nuts stored by the squirrel suggests that more than olfactory cues were used to find nuts. This result is consistent with the spatial memory hypothesis, but further tests are needed to confirm the use of spatial memory in cache recovery. Stapanian and Smith (1978) reported that red squirrels (*Tamiasciurus hudsonicus*) unerringly returned to pine trees in which they had earlier stored fungi. Fungi could not be seen by red squirrels approaching on the ground, and fungi were high enough that olfactory cues could not have been important in tree selection. Stapanian and Smith (1978) suggested that the squirrels likely remembered the trees in which fungi had been stored.

Several mammal species thus seem to remember exactly where they have stored food and rely on this memory to find storage sites. Further experiments are needed to determine how widespread the use of spatial memory is in mammals.

6.5   CONCLUSIONS

Nest-provisioning Hymenoptera and scatter-hoarding birds use spatial memory of visual cues to precisely locate storage sites. Visual cues include proximal cues (features near sites), which facilitate precise relocation of storage points (e.g., Tinbergen 1972; Vander Wall 1982), and distal cues (e.g., surrounding landscape), by which animals orient to the proximal cues (Chmurzynski 1964; van Iersel and van den Assem 1964; Wehner 1981). Some birds appear to remember qualitative characteristics (e.g., seed type) of stored food (Sherry 1984a) in addition to precise location. Despite the general similarity of spatial memory in certain birds, wasps, and bees, there also are many important differences. Wasps and bees perform elaborate locality studies over sites to be remembered, a behavior reduced in birds to subtle head turning or no apparent orientation movements at all. Number of sites simultaneously remembered by birds ranges from tens to thousands, but wasps and bees need remember only one or a few sites. Birds return to cache sites days or months after having prepared them, whereas the time between visits to nest sites by wasps and bees ranges from minutes to hours. Thus the memory task confronting food-hoarding birds appears much more complex than that performed by Hymenoptera. Finally, although wasps and bees respond similarly to displaced visual cues, it is not known whether they perceive storage sites as birds do.

Some mammals use spatial memory to relocate cache sites, but these and many other species have been found to rely heavily on olfaction to find hidden food. Consequently their food recovery behavior is fundamentally different than that of birds and wasps. Mammals can rely on stimuli intrinsic to stored food, whereas wasps and birds use features of the environment extrinsic to stored food, features that have attained stimulus value due to their spatial association with storage sites. Future studies of mammals may demonstrate that memory plays a larger role in cache recovery than is currently known. Birds and mammals also behave opportunistically, exploiting other possibilities to locate hidden food. These include random search at preferred storage points and direct visual cues (e.g., partially buried nuts, seedlings from stored nuts breaking through the soil). Although these techniques may at times provide foragers with rich rewards, they are ineffective means of sustaining hoarders over prolonged periods of food scarcity.

That some vertebrates have redundant cache recovery mechanisms is not surprising. In variable environments, natural selection should favor behavioral plasticity so that activities important to the persistence of an animal can be carried out if one means of completing the activity should temporarily fail. An avian migrant, for example, may be able to navigate by sun compass, magnetic compass, or perhaps general memory of migratory routes to ensure that an overcast sky or some other unforeseen event will not prevent the necessary migration. Among food-hoarding animals, red foxes, tits, and nutcrackers seem to rely primarily on spatial memory to relocate stored food (Macdonald 1976; Sherry, Krebs, and Cowie 1981; Shettleworth and Krebs 1982; Vander Wall 1982; Kamil and Balda 1985), but foxes also can effectively find hidden food using olfaction (Tinbergen 1965; Henry 1986). In addition, tits recover many food items by searching likely sites (e.g., Shettleworth and Krebs 1986), and nutcrackers will search at preferred storage sites to retrieve some hidden seeds (Vander Wall 1982). Gray squirrel may use memory to relocate food (Jacobs 1989) but also find many nuts using only olfaction (e.g., Thompson and Thompson 1980).

The behavioral mechanism or mechanisms by which an animal locates hidden food has important implications for the evolution of food-hoarding behavior (Andersson and Krebs 1978). Recovery of stored food using spatial memory is a "private" system; only the hoarder knows the extrinsic stimuli that can be used to retrieve the food. Olfaction, random rearch, and direct visual modes of cache recovery, on the other hand, can be thought of as "public" systems because all foraging animals with appropriate sensory modalities can recognize the stimuli intrinsic to the stored food and thereby locate the food. If limited to these means of recovery, hoarders would have little or no advantage in recovering food they have stored. In such situations, cheating—the exploitation of stored food by nonhoarding conspecifics—may become a profitable behavioral option. Some of the conditions under which cheating may be expected to occur have been discussed by Andersson and Krebs (1978). For food hoarding to evolve, the cacher must have a significant advantage over unrelated animals in cache recovery (Smith and Reichman 1984). Spatial memory of cache sites is one way that this advantage can be gained, but other means of ensuring that hoarders recover most of their stored food are possible. For example, olfaction-mediated cache recovery coupled with aggressive defense of storage areas could provide a hoarder with a nearly exclusive stored food supply. Such a system might possibly be used by scatter-hoarding kangaroo rats. Thus, spatial memory of cache sites or behavior that protects the storage area may have had to evolve concurrently with the evolution of scatter hoarding.

Although there is strong agreement concerning the importance of spatial memory in cache recovery behavior, the experiments thus far conducted are not without shortcomings and limitations. Enclosures and arenas in which studies have been conducted only remotely resemble field situations in which birds and mammals store food. Storage sites in the field often are continuous and virtually limitless, whereas in experimental settings hoarders may be constrained to use a few discrete sites. This often causes storage sites to be spatially compressed instead of broadly spaced, as is usually the case in the wild. Hoarders are allowed to cache and recover during "sessions" and are held in separate "home cages" between sessions, a practice that greatly oversimplifies the temporal complexity of caching. For experiments that employ discrete storage sites, reuse of storage sites is frequent, again a practice not seen in the wild, probably because repeatedly used sites would eventually attract cache robbers. Furthermore, cache contents often are removed or altered by experimenters. The consequences of these restrictions are not well understood, but several effects seem likely. Motivation to store and recover food may be absent or reduced because of lack of reinforcement (e.g., Marx 1951, 1957; Marx and Brownstein 1957; Balda 1980b). Not finding food where it was hidden may increase exploration in the enclosure, which may have several effects, including reduced recovery success rates and increased incidence of revisiting empty cache sites. Spatial and temporal constraints on caching may obscure serial position effects. Many of these constraints on experimental design cannot be remedied for practical reasons, but it is important to realize how such studies are constrained and to better understand how those constraints might influence experimental results.

7

Food-Hoarding Animals as Dispersers of Plants

A number of food-hoarding animals have influenced the structure and functioning of plant communities as a direct consequence of their storing of seeds, nuts, or vegetative structures in soil. Those propagules that animals bury in favorable seedbeds, at appropriate depths, in hospitable microsites, and that are overlooked by the hoarder and escape detection by robbers and microbes can germinate or sprout under appropriate conditions. The portion of propagules buried in such sites and that do not succumb to predation is, in many cases, relatively small, yet this means of dispersal is effective to the degree that some plant species have evolved propagules and supporting structure that attract food-hoarding animals and appear to depend on them as primary vectors of seed dispersal.

The role of food-hoarding animals in plant dispersal has until recent years been largely overlooked. Several recent reviews have touched on the influence of food hoarders in seed dispersal (Howe and Smallwood 1982; Smith and Reichman 1984), but except for Price and Jenkin's (1986) review of seed dispersal by rodents, no thorough treatment of the literature has thus far been undertaken. In this chapter, I examine how the food-storing activities of birds, mammals, and insects influence the dispersal and establishment of plants. I first describe the types of food hoarding that contribute to plant establishment and the animals responsible. Second, I examine the benefits and costs of seed dispersal by food-hoarding animals. Third, I survey the types of plants that benefit from this form of dispersal and the adaptations of their seeds, nuts, cones, and other supporting structures that promote harvesting and storage by food hoarders. And fourth, I describe how food hoarders have affected the physiognomy of the plants they disperse and how they have influenced plant succession and other types of changes in plant communities.

7.1 FOOD HOARDERS THAT DISPERSE PLANT PROPAGULES

The ways in which food hoarders handle the items they store influence their effectiveness as agents of plant dispersal. Animals that bury plant propagules in shallow depressions in the soil place those propagules in environments potentially favorable to establishment and so may serve as agents of dispersal and as predators. In contrast, animals that accumulate food in larders seldom place plant propagules in situations that encourage establishment and consequently are predators whether or not they eventually recover the stored food. Larders are not only located in protected sites (often deep underground) where there is little chance for the elongating epicotyl or shoot to reach the soil surface, but the large number of propagules simultaneously emerging from a larder results in intense competition. Further, although seeds, bulbs, and other propagules stored in these sites often sprout (e.g., Hawbecker 1940; Griffin 1971; Elliott 1978), they are usually prevented from establishing because watchful food hoarders selectively graze the sprouting propagules (e.g., Galil 1967). Although there are a few well-documented instances of plants establishing from rodent larders—for example, oaks from rodent burrows (Olmsted 1937), *Oxalis* from mole-rat storerooms (Nevo 1961), and fir from red squirrel middens (Hutchins and Lanner 1982)—most plants that establish from food caches appear to do so from propagules scatter hoarded under a thin layer of soil or litter.

Among mammals, dispersal of propagules is confined to several rodent families. Tree squirrels (Sciuridae) and agoutis (Dasyproctidae) scatter hoard nuts and seeds in soil. Chipmunks and ground squirrels (Sciuridae), kangaroo rats (Heteromyidae), and mice (Cricetidae, Muridae) are primarily larder hoarders but also store seeds in shallow surface caches. Pocket gophers (Geomyidae) and mole-rats (Spalacidae and Bathyergidae) are fossorial larder hoarders that are seldom responsible for dispersing seeds, but vegetative structures (root cuttings, bulbs, and corms) have been known to establish from larders near the soil surface (Cook 1939; Nevo 1961; Galil 1967).

Among birds, several species of nutcrackers and jays account for most of the observed seed dispersal (e.g., Vander Wall and Balda 1977; Tomback 1978, 1982; Bossema 1979; Mattes 1982; Johnson and Adkisson 1985; Nilsson 1985). These birds scatter hoard conifer seeds, softshelled nuts, and acorns in shallow caches in soil. A couple of nuthatch species (Richards 1958; Bromley, Kostenko, and Okhotina 1974; Nilsson 1985) and perhaps a few species of chickadees or tits (Gibb 1954; Owen and Owen 1956; Cowie, Krebs, and Sherry 1981; Nilsson 1985) also disperse seeds, but this has not been well documented. Other seed- and fruit-storing birds hide items above ground (MacRoberts and MacRoberts 1976; Moreno, Lundberg, and Carlson 1981; Pruett-Jones and Pruett-Jones 1985), and although germination can occur (e.g., MacRoberts and MacRoberts 1985; Wong 1989), there is little chance for establishment.

The only insects known to store seeds are harvester ants (e.g., Carroll and Risch 1984; MacKay and MacKay 1984) and the larvae of several species of carabid beetle (Kirk 1972; Alcock 1976). These insects depsoit seeds in deep chambers and monitor them closely. There is little chance for seedlings to establish from ant nests because seeds that germinate often are carried to refuse heaps, and seedlings that do emerge from nest mounds often are quickly cut down and removed by some species of harvester ants. Only seeds that ants remove from nests and discard in refuse piles regularly establish plants (e.g., Wheeler 1910; Cole 1934; MacKay 1981). This form of seed dispersal is very interesting, but it is peripheral to food storage. Another type of seed dispersal effected by ants, the gathering of elaiosome-bearing seeds that the ants eventually dump in refuse piles or discard in underground chambers where they may eventually germinate (Bond and Slingsby 1984; Beattie 1985), is not food storage (although it affects plant establishment in a similar fashion) and will not be considered further. Dung beetles are potential seed dispersers, but this form of seed dispersal is only just beginning to be explored (e.g., Wicklow, Kumar, and Lloyd 1984). Dung beetles bury the feces of medium and large mammals as provisions for their larvae (Halffter and Edmonds 1982). Viable seeds are common constituents of feces (e.g., Wicklow and Zak 1983; Janzen 1984), and the potential for dung beetles to facilitate seedling establishment seems great. This interaction is in need of study.

The efficacy of dispersal performed by these food-hoarding animals varies markedly. A number of food hoarders harvest large quantities of seeds whenever they are available and place a portion of these in sites that favor plant establishment. Squirrel (*Sciurus* spp.) dispersal of walnuts (*Juglans*) and hickory (*Carya*) nuts, pinyon jay (*Gymnorhinus cyanocephalus*) dispersal of piñon pine (*Pinus edulis*), and nutcracker (*Nucifraga* spp.) dispersal of the stone pines of Asia, Europe, and western North America are some familiar examples. As detailed below, the seeds of some of these plants appear to be highly adapted for dispersal by these animals. Other species store relatively few plant propagules and place many of these in sites that do not favor establishment.

7.2 BENEFITS OF DISPERSAL BY SCATTER HOARDERS

The benefits of seed dispersal by food-hoarding animals lie in the large number of nuts and seeds cached and the quality of the sites selected. Three stages in the harvesting and storing process have important consequences for plants: (1) selection of seeds, (2) transport of seeds away from source crowns, and (3) burial of one or more seeds in a substrate. As already outlined in chapter 4, food-hoarding animals use a variety of protective measures that reduce the probability that cached food will be lost. Many of these behaviors benefit plants because they minimize the depredations of other foragers after seeds are buried.

Number of Seeds and Nuts Stored

Studies that have quantified the number of seeds and nuts stored by scatter hoarders vary greatly in the importance they ascribe to these animals as dispersal vectors, with estimates ranging from virtually no seeds stored (e.g., Sork, Stacey, and Averett 1983) to millions of seeds stored (e.g., Vander Wall and Balda 1977). Variability in the harvesting activity of food hoarders results from variation in the attractiveness of seeds, the number of seeds produced, the number and types of individuals storing seeds, the number and types of seed-predators competing for seeds, and a variety of site-specific variables.

Jays are noted for their ability to store large quantities of acorns and nuts. A group of thirty-five Eurasian jays (*Garrulus glandarius*), for example, stored nearly 200,000 acorns during the fall of 1951 in the Hainault Forest, Essex, England (Chettleburgh 1952), and sixty-five Eurasian jays studied by Schuster (1950) in Vogelsberg, West Germany, buried an estimated 300,000 acorns during a 1-month harvest period. A population of blue jays (*Cyanocitta cristata*) stored 133,000 pin oak (*Quercus palustris*) acorns during the fall in Blacksburg, Virginia, which represented 54% of the mast crop (Darley-Hill and Johnson 1981). And near Saukville, Wisconsin, about seventy-five blue jays transported approximately 100,000 American beech (*Fagus grandifolia*) nuts away from a woodlot during September (Johnson and Adkisson 1985). Each pinyon jay can store between 18,000 and 21,500 piñon pine (*P. edulis*) nuts following a good nut crop (Ligon 1978; Balda 1980a). A flock of 200 pinyon jays (an average colony size for this species) could then store 3.6–4.3 million seeds. Based on a mean cache size of 1.4 seeds (Vander Wall and Balda 1981), from 2.6–3.0 million caches are likely made in a colony's traditional home range of several hundred hectares.

Individual Clark's nutcrackers (*Nucifraga columbiana*) have been estimated to store 32,000 whitebark pine (*Pinus albicaulis*) seeds in the Sierra Nevada, California (Tomback 1982), and 98,000 whitebark pine seeds in Squaw Basin, Wyoming (Hutchins and Lanner 1982). In the latter study, nutcrackers were estimated to have stored or consumed 36% of the seed crop. Lanner and Vander Wall (1980) estimated that a population of Clark's nutcrackers in the Raft River Mountains of northwestern Utah harvested 70–98% of the limber pine (*Pinus flexilis*) seed crop and stored 30,000 seeds/ha on a windswept ridge that served as a shared cache area. Following heavy seed crops, Clark's nutcrackers may store the equivalent of 1.8–3.3 times their likely energetic needs for the coming winter and breeding season (Vander Wall and Balda 1977; Tomback 1982; Vander Wall 1988). In the Ural Mountains of the Soviet Union, Eurasian nutcrackers (*N. caryocatactes*) made 3,334 caches/ha of Siberian stone pine (*P. sibirica*) seeds in Siberian stone pine forests and 1,665 caches/ha in mountain tundra (Bibikov 1948). Eurasian nutcrackers in the Swiss Alps each stored about 47,000 Swiss stone pine (*Pinus cembra*) seeds following a "scanty" nut crop in 1975 and 109,000 seeds following the "medium" nut crop of 1974 (Mattes 1982).

Quantitative estimates of the number of seeds and nuts scatter hoarded by rodents are surprisingly few. Individual squirrels bury hundreds and perhaps thousands of nuts and acorns each fall (Cahalane 1942; Stapanian and Smith 1978; Thompson and Thompson 1980). A single giant kangaroo rat (*Dipodomys ingens*) made 875 surface caches around its burrow entrance (Shaw 1934). Heteromyid rodents made 1.4 and 5.8 seed caches/m^2 (13,700 and 57,600 caches/ha) in two studies conducted near Reno, Nevada (La Tourrette, Young, and Evans 1971; McAdoo et al. 1983). Chipmunks (*Tamias* spp.) and mice make dozens of seed caches within their home ranges (Abbott and Quink 1970; Elliott 1978; Kawamichi 1980). When viewed at the population or community level, the number of seeds and nuts stored by rodents must often exceed thousands per hectare.

Seed and Nut Selection

Food-hoarding animals are often highly selective when collecting nuts and seeds (e.g., Dennis 1930; Mailliard 1931; Ligon and Martin 1974; Vander Wall and Balda 1977; Bossema 1979; Darley-Hill and Johnson 1981; McAdoo et al. 1983; Sork 1983a). Many of the nuts and seeds borne in the crowns of plants or that fall to the ground are inedible. Some appear sound externally but may be empty, occupied by insect larvae, or infested with microbes. Discrimination of seed quality is complicated by the fact that seeds and nuts are stored whole, the contents usually being assessed without breaking the hull.

Cues used by corvids to discriminate between edible and inedible seeds include such factors as color, density, and sound when manipulated in the bill (Ligon and Martin 1974; Vander Wall and Balda 1977; Bossema 1979; Johnson, Marzluff, and Balda 1987). Olfactory cues appear to be most important in nut discrimination by rodents, although visual and tactile cues also may be important (Dennis 1930; Mailliard 1931; Lloyd 1968). In combination, these cues can be used by animals to sort seeds and nuts with a high degree of accuracy, even when edible and inedible nuts may appear identical to a human observer. Johnson and Adkisson (1985), for example, found that 100% of beechnuts carried by blue jays were edible, whereas only 11% of the nuts they sampled from source crowns were edible. One should not jump to the conclusion, however, that animals always achieve such high levels of accurate discrimination. Stiles and Dobi (1987), for example, found that 11% of the 136 horse chestnuts (*Aesculus hippocastanum*) buried by gray squirrels (*Sciurus carolinensis*) in New Jersey were completely rotten, and 57% were partly rotten.

To the extent that viability is correlated with edibility of seeds (e.g., Johnson and Adkisson 1985), this selection process benefits plants. Quickly removing viable propagules from source crowns, where seed mortality by seed predators usually is very high (Janzen 1971), increases seed survivorship. But discriminate foraging has another, and perhaps equally important, influence on plants. Preferences of food-hoarding animals for viable seeds of varying characteristics may have important consequences for evolution of seed and nut morphology and fruiting phenology. If dispersal agents are in limited supply and if they discriminate among seeds with different characteristics, then harvesters could act as strong selective agents on seed size, seed composition, hull thickness, timing of nut production, and other traits. Although several studies have described suites of coadapted (and perhaps coevolved) characters between dispersal agents and propagules (Vander Wall and Balda 1977; Bossema 1979; Lanner 1982a), few studies (Bossema 1979; Hallwachs 1986; Johnson, Marzluff, and Balda 1987) have adequately documented that these harvesters discriminate among nuts or trees so as to act as selective agents. Such studies are badly needed.

Transport

Stage two of the harvest is movement of seeds and nuts away from source crowns to scattered locations. The distances food-hoarding taxa transport items vary greatly. Rodents carry seeds relatively short distances. Squirrels usually carry nuts mean distances of 10–30 m to storage sites (Heaney and Thorington 1978; Emmons 1980; Kato 1985; Stiles and Dobi 1987), but they carry some seeds over 100 m (e.g., Stapanian and Smith 1978) and Miyaki (1987) reported that Eurasian red squirrels (*Sciurus vulgaris*) may transport Korean stone pine (*Pinus koraiensis*) seeds as far as 600 m. Sork (1984) used metal-tagged acorns to determine the distances that rodents carry red oak (*Q. rubra*) acorns. Distances from the source at which metal pieces were found (points where they stored or ate acorns) fit a negative exponential curve with a maximum recovery distance of 19 m. Mice, chipmunks, kangaroo rats, and agoutis seldom carry seeds more than 100 m (Hormay 1943; R. Hardy 1949; Reynolds 1958; Abbott and Quink 1970; Elliott 1978; Imaizumi 1979; Kawamichi 1980; Hallwachs 1986; Miyaki and Kikuzawa 1988). Yellow-necked mice (*Apodemus flavicollis*), for example, bury most beechnuts within 10 m of the source (fig. 7.1). Birds, on the other hand, carry seeds much further. Jays, rooks, and nutcrackers are known for their long transport flights. Eurasian jays and blue jays routinely carry acorns and other nuts 4 or 5 km to storage sites (Schuster 1950; Johnson and Adkisson 1985), although distances are often much shorter (e.g., Chettleburgh 1952; Bossema 1979). Nutcrackers carry pine seeds up to 22 km (Vander Wall and Balda 1977), but distances of 5–15 km seem to be much more usual for both species of nutcracker (Turcek and Kelso 1968; Vander Wall and Balda 1977; Tomback 1978; Mattes 1982; Vander Wall 1988). Pinyon jays typically transport seeds 1–5 km (Balda and Bateman 1971), but a distance of 10 km has been reported (Vander Wall and Balda 1981). Some corvids, such as scrub jays (*Aphelocoma*

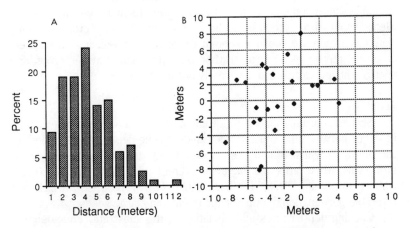

Figure 7.1   Dispersal distances (A) and dispersion (B) of radioactively labeled beechnuts carried from a single source (0,0) by wood mice and possibly bank voles in Denmark. Redrawn from Jensen 1985.

coerulescens) and black-billed magpies (*Pica pica*), usually transport seeds less than 500 m (Vander Wall and Balda 1981; Clarkson et al. 1986), and tits carry seeds less than 100 m (Haftorn 1956c; Sherry, Avery, and Stevens 1982). As impressive as some of these distances are, food hoarders cannot approach the dispersal distances attained by some migratory birds and mammals that transport seeds in their guts or in their feathers and fur for hundreds of kilometers.

Animals carry seeds from source crowns for a number of reasons (Becker, Leighton, and Payne 1985). One of the driving forces responsible for the evolution of scatter hoarding appears to be that the loss of buried seeds and nuts to foragers is directly proportional to the density of cached seeds (see section 4.2). Further, loss of buried seeds may decrease inversely with the distance from the cache to the source area where seed predators forage more intensely. The combined movement and scattering of seeds away from source crowns benefits the hoarder as well as the seeds. Propagules that remain under the source crown experience extremely high mortality rates resulting from seed predators and intense competition with the parent plant (Watt 1923; Janzen 1971; Howe and Smallwood 1982; Hutchins and Lanner 1982; Nilsson 1985; Hallwachs 1986). Nuts cached away from the parent plant will, on average, experience less competition than those germinating under the source crown. Since many food-hoarding animals store nuts and seeds in open sites (Pivnik 1960; Harrison and Werner 1984; Johnson and Adkisson 1985; Stapanian and Smith 1986), a portion of the nuts carried away may be cached in tree-fall gaps or on the edge of clearings, sites favorable for establishment and growth.

The distance that food hoarders transport seeds and nuts may be to some extent under the influence of the plant. Stapanian and Smith (1978) and Stapanian (1986) have suggested that if scatter hoarders reduce cache loss by maintaining uniform nearest-neighbor distances between buried food items, then very large crops of seeds at intervals of several years (mast seeding), as occurs in many conifers and nut-bearing trees, should result in some propagules being carried further from the source. Mean dispersal distance should be proportional to seed crop size, and plants can increase mean dispersal distance by producing large seed crops at infrequent intervals rather than producing smaller crops annually. This prediction has apparently not been tested.

Burial

Scatter hoarders deposit seeds and nuts in shallow subsurface caches in soil or between the soil and litter layer. Cache depth is constrained by the depth to which rodents can economically dig and by the length of birds' bills. Most birds and mammals cache food 0.5–10 cm deep (e.g., Abbott and Quink 1970; Tomback 1978; Hallwachs 1986), which are favorable germination depths for many nuts and large seeds.

Caching depth varies with qualities of the substrate. Fox squirrels (*Sciurus niger*) and gray squirrels in cemeteries, parks, and other urban settings, where several important studies have been conducted, bury nuts

Table 7.1 Discovery rates (percent) of eight rodent species for clumps of 100 Indian ricegrass seeds placed at three depths in dry sand

Species	Depth of Buried Seeds		
	Surface	0.6 cm	1.3 cm
Reithrodontomys megalotis	100.0	0.0	0.0
Peromyscus maniculatus	100.0	0.0	2.5
Dipodomys ordii	100.0	5.0	5.0
Perognathus longimembris	100.0	12.5	10.0
Dipodomys microps	100.0	37.5	10.0
Perognathus parvus	100.0	42.5	17.5
Microdipodops megacephalus	100.0	50.0	37.5
Perognathus formosus	100.0	57.5	27.5

Source: After Johnson and Jorgensen 1981.

in very shallow caches, often 0–2 cm deep (Cahalane 1942; Thompson and Thompson 1980; Stiles and Dobi 1987). Many nuts may be partially exposed. In these sites, the soil is often highly compacted, and grass forms a dense mat. The shallowness of caches no doubt contributes to the very high mortality rates reported for cached nuts in some of these studies (e.g., Cahalane 1942). In forests with undisturbed understories and more friable soils, squirrels often bury nuts deeper, 3–8 cm below the surface (e.g., Madson 1964; Sviridenko 1971).

Burial of seeds by food hoarders has at least three beneficial effects on seeds. First, burial decreases the probability of seeds being located and eaten by seed predators. Predation rates of large seeds and nuts on the soil surface often are near 100% but decrease as depth of seeds in soil or litter increases. This general pattern has been demonstrated in a number of experimental studies (Watt 1923; Cahalane 1942; Reynolds 1958; Lockard and Lockard 1971; Perkins, Bienek, and Klikoff 1976; Floyd 1982; Hutchins and Lanner 1982; Evans et al. 1983; Brown and Batzli 1985; Kikuzawa 1988; Borchert et al. 1989). Johnson and Jorgensen (1981), for example, placed caches of 100 Indian ricegrass (*Oryzopsis hymenoides*) seeds at three depths (surface, 0.6, and 1.3 cm) in dry sand and allowed eight species of rodents to search for seeds in a 1.2- × 2.4-m experimental arena. Species differed in their ability to detect buried seeds, but all species exhibited a significant decrease in cache recovery ability with increasing depth (table 7.1). In an experiment that considered both cache size and depth on Merriam's kangaroo rat's (*Dipodomys merriami*) ability to locate seeds, Reichman and Oberstein (1977) found a significant negative relationship between the ratio of depth to size of a seed packet (mixed bird seed) and the probability that kangaroo rats dug for seeds (fig. 7.2).

Figure 7.2
The proportion of Merriam kangaroo rats that dug for seeds (i.e., probability of digging) as a function of the ratio of burial depth (cm) to size (g) of seed packets (commercial bird seed). The least square regression is significant ($P < 0.01$). Redrawn from Reichman and Oberstein 1977.

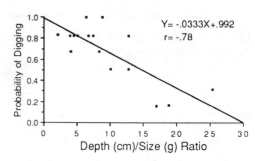

$Y = -.0333X + .992$
$r = -.78$

Probability of Digging

Depth (cm)/Size (g) Ratio

The relationship between seed depth and predation rate exists for several reasons. First, seeds on the soil surface are available to both visually and olfactorily oriented predators, but buried seeds are available only to olfactorily oriented predators (except if the animal that made the cache is still guided by visual cues when relocating a cache). Second, the strength of an olfactory signal emanating from a seed cache decreases proportional to the inverse of the square of the distance from the source (Reichman 1981). As seeds are buried more deeply, the odor at the soil surface becomes weaker and the area within which seeds can be detected becomes smaller (fig. 7.3). At a certain depth (specific to the seed predator, type of seeds, and physical characteristics of the substrate) the seeds can no longer be detected. For visually oriented corvids that remember cache sites (Bossema 1979; Vander Wall 1982; Kamil and Balda 1985), depth of seeds has not been shown to affect the storer's ability to relocate a cache. Third, even if buried seeds are detected, seed predators may not attempt to retrieve them if the quantity of seeds is too small or the position too deep to repay the effort. This explanation was proposed by Lockard and Lockard (1971) for desert kangaroo rats (*Dipodomys deserti*) that dug small pits directly over seeds but then did not completely excavate caches buried between 4 and 20 cm below the surface.

The second benefit of burial is that the cool, moist environment experienced by a buried nut or seed may help maintain viability. Acorns are particularly vulnerable to damage from unsuitable conditions such as excessive drying, intense sunlight, or low temperatures (Watt 1919; Jones 1959; Griffin 1971; Borchert et al. 1989). Chestnuts lose viability within weeks of maturing unless stored in a cool, moist medium such as soil (Schopmeyer 1974). The more moderate underground environment may influence viability of other nuts and seeds similarly.

Certain handling techniques of the hoarder may break seed dormancy. Merriam's kangaroo rats gnaw the seed coats of mesquite (*Prosopis juliflora*) seeds to test their quality. Mesquite seeds similarly treated in the laboratory had 90% germination compared to 6% for control seeds (Reynolds and Glendening 1949). Removal of the lemma and palea of Indian ricegrass seeds by heteromyid rodents significantly increases seed germinability (McAdoo et al. 1983). These effects, however, are not nec-

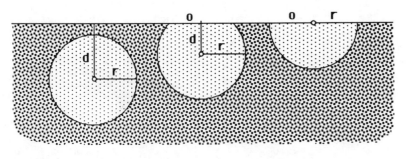

Figure 7.3
The detectable odor emanating from a seed or clump of seeds is represented (schematically) as a sphere of radius r. Seed clumps of equal size (equal r) buried at different depths (d) will vary in detectability in proportion to the chord (o) of the interception of the odor sphere by the ground surface. For seeds on the ground surface, o is equal to $2r$. When d exceeds r, the seeds cannot be detected. Redrawn from Reichman 1981.

essarily advantageous to the plant, for they may cause the seed to germinate at an inappropriate time.

Third, buried seeds frequently have a higher probability of establishing than unburied seeds. The depth of burial is critical. A large body of literature supports the notion that there is an optimal range of depths in soil or litter at which germination of seeds and subsequent emergence of seedlings is maximized (e.g., Watt 1923; Korstian 1927; Barrett 1931; Wood 1938; Heydecker 1956; Kinsinger 1962; Saito 1983b; Lal, Nautiyal, and Sharma 1984; Fenner 1985). Germination of seeds on the ground surface, especially large seeds such as nuts and acorns likely to be cached by animals, is infrequent, but germination increases markedly once propagules are buried (fig. 7.4). Major exceptions to this pattern are white oaks (subgenus *Lepidobalanus*) and live oaks, which produce nondormant acorns that germinate within a few weeks of maturing, often on the ground beneath source trees (e.g., Griffin 1971). When germination on the soil surface occurs, the elongating radicle has difficulty penetrating the soil and quickly dries out, resulting in death of the seed. Seeds beneath a protective layer of soil or litter experience a much more moderate environment where temperature and moisture fluctuate within narrower limits than on the soil surface (Korstian 1927). With all surfaces of the seed contacting the substrate, moisture content of the seed before germination is higher and more uniform, and germination is more successful. Seeds buried deeply, on the other hand, may establish poorly because they lack sufficient stored reserves to permit the epicotyl to elongate to the ground surface, oxygen may be limiting at greater depths, or because they cannot push their way through a thick layer of soil. These germinants become etiolated, and most eventually die (fig. 7.4A). Thus, in general stored seeds and nuts have the greatest chance of producing seedlings if buried at some intermediate depth.

Seed size, number of seeds in a cache, and characteristics of the soil influence the depths from which seedlings can successfully emerge.

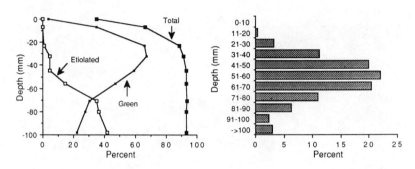

Figure 7.4

Probability of establishment as a function of seed depth. A, Changes in percent total germination, percent healthy green seedlings, and percent etiolated seedlings as burial depth of chestnut oak acorns in loose litter increases. Etiolated plus green seedlings sums to less than total germinated seedlings due to early seedling mortality from insects and other agencies. B, Depths from which Indian ricegrass seedlings establish under natural conditions in sandy soil. A, redrawn from Barrett 1931; B, based on data from Kinsinger 1962.

Large seeds contain more stored food reserves and can usually success-fully establish from greater depths than can small seeds. Clumps of seeds germinating simultaneously are better able to push through the soil and thus can establish from greater depths than single seeds. Saito (1983b), for example, found that single seeds of Japanese stone pine (*Pinus pumila*) could successfully emerge only when buried less than 4.5 cm deep, whereas groups of twenty seeds could emerge from as deep as 10 cm. Similar results have been found for antelope bitterbrush seeds (Basile and Holmgren 1957).

7.3   COSTS OF DISPERSAL BY SCATTER HOARDERS

Fruits whose seeds are dispersed through the hoarding activities of ani-mals differ markedly from fruits whose seeds are dispersed by frugivores that ingest the seeds. Fruits eaten by frugivores typically contain one or more seeds surrounded by a layer of fleshy tissue that serves as a nutri-tional or energetic reward for the animal that consumes them (Howe and Smallwood 1982). Dispersal occurs when seeds are eventually eliminated in feces or are regurgitated away from the parent plant. Although the distinction is not always clear, the reward and the seeds are separate components of the propagule. For fruits adapted for dispersal by food hoarders, the nutritious seeds or nuts often are produced within an in-edible woody, papery, or leathery husk or cone. Exceptions to this pat-tern occur in the tropics, where some seeds buried by food hoarders are surrounded by an edible flesh (primary reward) that the hoarder con-sumes before storing the seed (secondary reward) (Smythe 1970, 1978; Hallwachs 1986). In those fruits for which the supportive structure has no nutritive value to the food hoarder, the seed or nut itself is the attrac-tant. Animals eat some seeds at the food source and carry others off to store. A portion of the seed crop serves as a "reward" to foragers for dispersing the rest of the crop. Seeds of this type need to be both attractive energy sources—large and easily digestible—and readily accessible. As a result of these and other traits both hoarders (seed predators and potential dispersal agents) and nonhoarders (seed predators) are attracted to seed sources in large numbers.

From the plant's perspective, the cost of seed dispersal by seed-caching animals is the proportion of seeds and seedlings that die as a result of the process. I divide these costs into three classes: (1) predispersal seed mortality, (2) postdispersal seed mortality, and (3) postgermination seedling mortality.

Predispersal Seed Mortality

Predispersal seed mortality can be caused by animals that disperse some seeds to suitable germination sites and by a host of nondisperser organisms. The great variety of animals attracted to these seeds and nuts is due in part to masting—the production of abundant seed crops at irregular intervals that act to satiate vertebrate and invertebrate seed predators (e.g., Silvertown 1980)—which has prevented the most efficient seed predators and dispersers from competitively excluding others (Smith and Balda 1979). A seed crop may be attacked by several types of insects as it develops, often resulting in 50% or more of the crop being destroyed (e.g., Korstian 1927; Roy 1957; Ovington and Murray 1964; Gysel 1971; Nielsen 1977; Kessler 1979; Nixon, McClain, and Hansen 1980; Darley-Hill and Johnson 1981; Sork 1983b). As the seeds mature, other insects and vertebrates, such as weevils, deer, peccaries, turkeys, pigeons, and woodpeckers (Shaw 1968b; McDonald 1969; MacRoberts and MacRoberts 1976; Nilsson 1985; Hallwachs 1986), which act solely as seed predators, begin feeding on the seeds. Although such depredation is well known, its extent has been estimated for only a few of the plants dispersed by food hoarders. Hutchins and Lanner (1982), for example, estimated that 64% of the whitebark pine seeds handled by vertebrates on their Squaw Basin study site in 1980 were consumed or stored by animals that acted solely or primarily as seed predators.

Further seed mortality may be caused by dispersal agents that consume a portion of the seed crop. Blue jays consumed 20% of an acorn crop and stored 54%, the remaining 26% being consumed by insects (Darley-Hill and Johnson 1981). Thirty-nine percent of the limber pine seeds handled by nutcrackers at one study site in 1978 and 21% of the piñon pine (*P. monophylla*) seeds handled in 1980 were immediately eaten (fig. 7.5). Nutcrackers placed the rest of the seeds in their sublingual pouches and transported them to cache areas (Vander Wall 1988).

Postdispersal Seed Mortality

During the interval from caching to germination, which may last 6 or more months, seeds must endure threats from microbes that can penetrate seed coats and from vertebrates that search for the caches. Cool, moist soil is an ideal medium for fungi and other microbes, and some seeds fall prey to them. I have, for example, excavated clumps of recently germinated seedlings of singleleaf piñon pine (June) and whitebark

Figure 7.5
Seasonal changes in number of limber pine (A) and singleleaf piñon pine (B) seeds eaten at source trees and placed in sublingual pouches by Clark's nutcrackers (to be stored later) during the fall seed harvests of 1978 and 1980 in the Raft River Mountains, Utah. Redrawn from Vander Wall 1988.

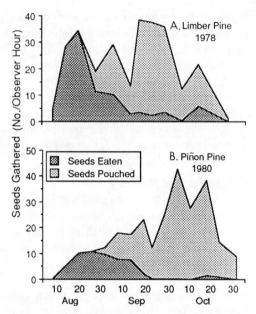

pine (July) and found that 7% and 27%, respectively, of the seeds initially cached were attacked by fungi or were discolored. Clumps of antelope bitterbrush (*Purshia tridentata*) seedlings excavated by Evans et al. (1983) contained a mean of 12.5 individuals plus 8 ungerminated seeds, many of which apparently were nonviable before they were cached. Fungal infection of acorns experimentally buried under natural conditions was less than 0.5% (Korstian 1927).

Vertebrate cache robbers usually are rodents that find buried seeds and nuts using their highly developed olfactory sense. Rodents may destroy as much as 80% of Eurasian nutcracker caches of Japanese stone pine seeds (Pivnik 1960). Experiments in which researchers have simulated animal caches by burying seeds and nuts at depths and sites similar to those used by food-storing animals demonstrate that disappearance rates can be very high, often in the range of 50–100% (Cahalane 1942; Richards 1958; Stapanian and Smith 1978, 1984; Thompson and Thompson 1980; Tomback 1980; Kraus 1983). Several points must be kept in mind, however, when interpreting these results. First, in some of these studies, the period during which seeds and nuts were exposed to foragers was much shorter than their residence time in the soil under natural conditions. In this regard, these studies probably underestimate actual removal of seeds and nuts. Second, these are artificial caches, and the experimenters may not have mimicked techniques that food-hoarding animals use to reduce cache loss to seed predators. For example, small wooden stakes sometimes are used by experimenters to relocate caches (e.g., Cahalane 1942), and these may provide additional cues to cache location for cache robbers. Third, many of the animals that find stored seeds are food hoarders themselves. It is likely that these animals rebury some of the nuts and seeds (e.g., Cahalane 1942; Kamil and Balda 1985;

DeGange et al. 1989). Disappearance cannot be equated to predation. More of the seeds and nuts may survive to germinate than is indicated by experiments using artificial caches.

Some seeds are "killed" by rodents before they store them. Squirrels and other rodents notch the apex of white oak acorns (subgenus *Lepidobalanus*), often excising the embryo, which prevents the seed from germinating (Barnett 1977; Fox 1982). Wood (1938) determined that the germination rate of chestnut oak (*Q. montana*) acorns notched by gray squirrels was 18%, whereas 94% of undamaged acorns from the same tree germinated. Mole-rats disbud some of the bulbs they accumulate in larders to prevent them from sprouting (Shortridge 1934; Nevo 1961).

The most effective predators of cached seeds often are the individuals that store the seeds. Food hoarders are often highly dependent on stored seeds as a winter food supply and, not surprisingly, have acquired very effective means of recovering the food. Reliable estimates of the proportion of seed stored in soil that an animal recovers itself are exceedingly difficult to obtain, but there are some estimates of seed recovery by members of the population. White-footed mice (*Peromyscus leucopus*) and perhaps other rodents recovered all the eastern white pine seeds from 85 of 129 caches (Abbott and Quink 1970). Of 251 nuts stored by western fox squirrels during fall, 249 were retrieved before germination could occur in spring (Cahalane 1942). Large Japanese field mice (*Apodemus speciosus*) and other rodents retrieved before snowfall 47% of the acorns they had hoarded during the fall, and they removed an additional 52% before germination could occur in spring (Miyaki and Kikuzawa 1988). Pre- and postdispersal seed mortality often eliminates most of the seed crop. Comparisons of seedling yield immediately following germination in the spring with numbers of nuts or seeds produced the previous fall give an estimation of overall seed mortality. For sessile oak (*Q. petraea*), seedling yield has been estimated at 0.6–1.5% (Shaw 1968a) and 0.1–0.5% (Ovington and Murray 1964). These values are probably underestimates, however, because these workers did not account for acorns possibly carried away from source crowns by jays. Reynolds (1958) found that 1.8% of the velvet mesquite seeds he placed in feeders emerged from Merriam kangaroo rat caches. Only 0.02% of Indian ricegrass seed produced during a favorable moisture year emerge as seedlings from rodent caches (McAdoo et al. 1983). Twelve of 485 radioactively labeled acorns (2.5%) taken from feeding stations by small rodents were found to have germinated (Jensen and Nielsen 1986), but this should be considered a minimum estimate of survivorship because the radioactive label may have reduced viability (e.g., May and Posey 1958).

Postgermination Seedling Mortality

Seedling mortality can be attributed to a host of factors, but I consider here only those sources of mortality that result from the fact that seeds had been stored. Postgermination mortality that can be attributed to hoarding animals includes (1) seedling death caused by seeds being stored in sites inappropriate for establishment, (2) seedlings damaged by animals digging up the germinants, and (3) attrition within a clump of seedlings caused by competition for water or other resources. Hoarders store many seeds in sites inappropriate for germination or subsequent seedling survival. Clark's nutcrackers carry many piñon pine seeds to higher elevations and store them on cliff faces, slopes, and wind-blown ridges, where only a small proportion ever establish and where few of the trees that reach maturity produce viable seeds (Vander Wall and Balda 1977; Lanner and Vander Wall 1980; Lanner, Hutchins, and Lanner 1984). Tropical squirrels store many nuts at the base of trees (Heaney and Thorington 1978), which must in most cases be unsuitable sites for establishment.

At the time of seedling emergence and for a brief period thereafter, seed hoarders and other animals feed on the seedlings and buried seeds. Food-hoarding rodents that discover germinating nuts may eat the radicle and rebury the nut (Walton 1903; Cahalane 1942). Rodents uproot clumps of seedlings in search of ungerminated seeds (Hormay 1943; West 1968; Abbott and Quink 1970; McAdoo et al. 1983). Seventeen percent of germinating beechnuts in Aberdeenshire, England, were eaten by mice as cotyledons emerged through the soil (Watt 1923). During June and July, nutcrackers search for germinating seeds breaking through the soil surface. They detach seeds and eat what remains of the contents before excavating the cache site in search of other seeds (Bibikov 1948; Vander Wall and Hutchins 1983). The effect on the population of germinating seedlings may be small, however, because the critical period during which germinating seeds are eaten by nutcrackers appears to last only a few days.

Many of the acorns eaten by Eurasian jays in June and July are taken from English oak (*Q. robur*) seedlings (Bossema 1968, 1979). The cotyledons within these germinated acorns are of some nutritional value to the seed predator long after seedlings emerge. Jays find the acorns by searching out young seedlings with greenish stems. They tug on the stem to raise the buried acorn slightly and then dig up the acorn and tear it from the stem. Bossema (1968) suggested that this action may have little detrimental effect on the seedlings but later stated that such treatment often harmed young seedlings (Bossema 1979). Korstian (1927) has shown experimentally that removal of white oak (*Q. alba*) cotyledons within the first month after germination significantly reduces first-year seedling weights of both roots and shoots. This effect decreases with the age at which cotyledons are removed (table 7.2). The cotyledons of sessile oak acorns are thought to serve as nutrient reserves, covering losses from defoliation and annual shedding of leaves during the first year or two of seedling life (Ovington and MacRae 1960; Shaw 1974). Removal of sessile oak acorns from seedlings may have a great impact on seedling growth, and this impact may be increased on nutrient-deficient soils.

Further seedling mortality may result from competition among seedlings germinating from a cache. Number of stems per clump has been found to be negatively related to clump age for eastern white pine, ante-

Table 7.2 Influence of removal of cotyledons on development of white oak (*Quercus alba*) seedlings

Stage at Which Cotyledons Were Removed	Days after Pumule Emerged	Mean Size of Seedlings on 5 June, about 47 Days after Germination		
		Height (cm)	Shoot Weight (g)	Root Weight (g)
Plumule emerged from acorn	0	*	*	*
Plumule 12.5 mm tall	5	4.25	0.17	0.43
Leaves unfold	12	6.00	0.34	0.75
8 days after leaves unfold	20	9.50	0.70	1.24
18 days after leaves unfold	30	9.25	1.03	1.91
Control, cotyledons not removed	47	10.25	1.02	1.76

Source: After Korstian 1927.
*Immediate death.

lope bitterbrush, and limber pine (Abbott and Quink 1970; Sherman and Chilcote 1972; Woodmansee 1977). Careful excavation of seedling clumps of singleleaf piñon pine and whitebark pine often reveals seed and seedling mortality within the first growing season (fig. 7.6). Species that occupy arid sites may experience especially severe competition for moisture. Mature piñon pines, for example, are rarely found in multiple-stemmed clumps like whitebark pine and limber pine, which occupy relatively more mesic sites. Clumps of singleleaf piñon pine seedlings usually are reduced to a single seedling or are eliminated entirely within a few growing seasons.

7.4 TYPES OF PLANTS ESTABLISHED BY FOOD-HOARDING ANIMALS

Plants dispersed by food hoarders can be divided into four ecological groups based on the size and structure of their propagules: (1) conifers, (2) nut-bearing, broad-leaved trees, (3) shrubs and grasses, and (4) bulbs, tubers, and roots of geophytes. The fourth category is distinct in that reproduction is by vegetative rather sexual means. Although there is in some cases broad overlap, each group of propagules is stored by different types of food hoarders and buried in different ways.

Conifers

Establishment of conifer seedlings from rodent caches has received only superficial study. At two sites in Oregon 15% and 85% of ponderosa pine seedlings resulted from rodent caches (Munger 1917; West 1968). Abbott and Quink (1970) demonstrated experimentally that caches made by white-footed mice establish eastern white pine seedlings, but the mice

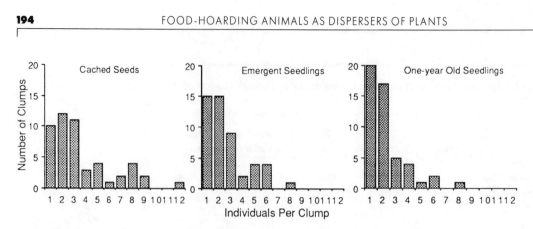

Figure 7.6

Size of fifty whitebark pine seedling clumps excavated 19 June 1981, about 1 year after germination, at Dunraven Pass, Wyoming, an area used as a caching ground by Clark's nutcrackers. Original cache sizes and original number of emergent seedlings were determined by adding number of ungerminated seeds and dead seedlings, respectively, to number of healthy 1-year-old seedlings (Vander Wall and Hutchins, unpublished data).

also caused much seed and seedling mortality. Seedlings from caches occurred in clusters of twenty-five to thirty individuals. Establishment of Jeffrey pine (*P. jeffreyi*) and sugar pine (*P. lambertiana*) from rodent caches is sometimes extensive (Hormay 1943; Tevis 1953; pers. observ.). Eurasian red squirrels (*S. vulgaris*) scatter hoard Korean stone pine (*P. koraiensis*) seeds and are thought to be an important disperser of this tree in Hokkaido, Japan (Miyaki 1987). I have observed a cluster of eighty Engelmann spruce (*Picea engelmanii*) seedlings (fifty-two individuals survived to age 4 years) in western Wyoming (Vander Wall and Hutchins, unpubl. data), and clumps of twenty to thirty Utah juniper (*Juniperus osteosperma*) seedlings in northeastern Utah. Little seems to be known about the importance of these caches in producing mature trees. Rodents dig up and destroy many of the clumps, and attrition within the few clumps monitored by Abbott and Quink (1970) was high. Mature, multistemmed ponderosa pines are not uncommon in some areas (West 1968; pers. obs.), but it is uncertain whether they originated from rodent caches. The prevailing opinion of forest managers has been and continues to be that rodents' role as dispersers of conifer seeds is greatly outweighed by their impact as seed predators (Krauch 1945; Smith and Aldous 1947; Shaw 1954; Abbott 1962; Radvanyi 1966; Lindsey 1977; Sullivan 1978). However, seeds dispersed and buried by rodents may actually have a much greater chance of establishing because the rodents inoculate seeds with ectomycorrhizal fungal spores (e.g., Piroznski and Malloch 1988), and this interaction may have very important implications for the dynamics of some forests.

The establishment of pines (especially those with wingless seeds) through the activities of jays and nutcrackers is well known (Formozov 1933; Reimers 1953; Vander Wall and Balda 1977; Ligon 1978; Tomback 1978; Mattes 1982; Hutchins and Lanner 1982; Saito 1983b). Most wingless-seeded pines belong to the subgenus *Strobus* (Critchfield and Little 1966), the soft pines. Twenty-one members of this group of about thirty-three species have relatively large seeds with little or no wing (table 7.3). Corvids, including Clark's nutcrackers, Eurasian nutcrackers, pinyon jays, Steller's jays (*Cyanocitta stelleri*), and scrub jays, are known to disperse nine of these species. All of these nutcrackers and jays bury single seeds or small groups of seeds (two to fifteen but as many as forty-eight [Egorov 1961]) in soil at depths of 1–5 cm. The composite geographic range of these five corvids overlaps broadly the composite range of the wingless-seeded soft pines (Lanner 1982b).

All other pines, about sixty-two species, are called hard pines (subgenus *Pinus*). Only a few members of this group have large wingless seeds. One of these is Italian stone pine (*P. pinea*), whose seeds are collected and stored by azure-winged magpies (*Cyanopica cyana*) (Turcek and Kelso 1968).

The wingless-seeded pines have a number of traits that make them attractive to seed-eating animals (Vander Wall and Balda 1977; Tomback 1978; Lanner 1982a, 1982b; Mattes 1982; see Coevolution of Plant Propagules and Food Hoarders, chapter 3), traits not shared with wing-seeded pines. The large size (fig. 7.7) and high caloric content of the seeds make them attractive food sources. Wingless pine seeds also ac-

Table 7.3 Winged (plain) and wingless (**bold**) seeded pine species of the subgenus Strobus arranged by subsection

Subsection				
Cembrae	Balfourianae	Strobi	Cembroides	Gerardianae
P. koraiensis	P. balfouriana	P. strobus	**P. cembroides**	*P. gerardiana*
P. cembra	P. aristata	P. monticola	**P. edulis**	**P. bungeana**
P. pumila	P. longaeva	P. lambertiana	**P. monophylla**	
P. sibirica		**P. flexilis**	**P. johannis**	
P. albicaulis		**P. strobiformis**	**P. juarezensis**	
		P. ayacahuite[a]	**P. culminicola**	
		P. peuce	**P. maximartinezii**	
		P. armandii	**P. pinceana**	
		P. griffithii	**P. nelsonii**	
		P. dalatensis		
		P. parviflora		
		P. morrisonicola		
		P. fenzeliana		
		P. wangii		

Source: After Lanner 1982b.

[a] One variety of *P. ayacahuite* has wingless seeds.

Figure 7.7   Size (seeds per pound) of winged and wingless seeds of some soft pines. 1, Korean stone pine, *Pinus koraiensis;* 2, Parry piñon pine, *P. quadrifolia* var. *juarezensis;* 3, Mexican piñon pine, *P. cembroides;* 4, chilgoza pine, *P. gerardiana;* 5, singleleaf piñon pine, *P. monophylla;* 6, Armand pine, *P. armandii;* 7, Siberian stone pine, *P. sibirica;* 8, Colorado piñon pine, *P. edulis;* 9, Swiss stone pine, *P. cembra;* 10, sugar pine, *P. lambertiana;* 11, whitebark pine, *P. albicaulis;* 12, southwestern white pine, *P. strobiformis;* 13, Japanese white pine, *P. parviflora;* 14, limber pine, *P. flexilis;* 15, blue pine, *P. griffithii;* 16, Japanese stone pine, *P. pumila;* 17, Balkan pine, *P. peuce;* 18, bristlecone pine, *P. aristata;* 19, eastern white pine, *P. strobus;* 20, western white pine, *P. monticola.* Redrawn from Lanner 1982b; data from Schopmeyer 1974.

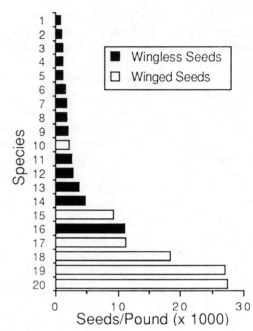

count for a relatively high proportion of the energy invested by the plant in reproductive tissue. Whitebark pine seeds, for example, comprise 17–46% of total cone energy, whereas ponderosa and lodgepole pine (*P. contorta*) seeds make up only 8% and 1.6%, respectively, of the energy in cones (Smith 1970; Lanner 1982a). Further, animals can extract the seeds of bird-dispersed pines much more quickly than they can extract seeds from the cones of pines with wind-dispersed seeds. This is because most wingless-seeded pines retain seeds in cones where they can be easily found and extracted. Whitebark pine and other stone pines have cones conspicuously arranged on upswept branches. Cones open only slightly or not at all when mature because they lack tracheids on the adaxial surface of the bracts (Tomback 1978; Lanner 1982a). The absence of tracheids also makes the cone bracts weak and easy for a foraging nutcracker to sever. Bracts break just distal to the midpoint of the seeds, leaving each seed exposed but wedged between the base of its bract and the core of the cone (Lanner 1982a) (fig. 7.8). Piñon pine achieves a similar result in a very different way. The cones open when mature, but seeds are not shed from cones (see fig. 3.14). The seeds are set in deep pockets of the cone bracts and are embraced by broad flanges of the interior surface of the bracts (Vander Wall and Balda 1977). The relatively short cone axis, wide bract angles, and cone orientation on branches (upward and outward) result in a highly visible seed display. Piñon seed coat color—most edible seeds are rich chocolate brown and most aborted seeds are pale yellow brown—is used by Clark's nutcrackers and pinyon jays as a cue to seed quality (Ligon and Martin 1974; Vander Wall and Balda 1977).

Limber pine and southwestern white pine (*P. strobiformis*), members of the white pine subsection, lack any obvious structures that retain seeds in cones, but nutcrackers foraging on limber pine extract most of the seeds before they fall from the cones (Lanner and Vander Wall 1980; Benkman, Balda, and Smith 1984). Wingless pine seeds often have relatively thin hulls that nutcrackers and jays can crack easily. Further, cones of wingless-seeded pines lack massive woody bracts and spines, which deter seed predators from foraging on cones. Structural modifications of *P. armandii, P. bungeana, P. gerardiana,* and *P. parviflora* cones that accommodate seed dispersal by birds have not been formally described, but the rudimentary wing adheres to the bract surface in all four species, retarding seed fall from cones (R. M. Lanner, pers. comm.).

Large seeds also have some obvious benefits for the pines. The energy reserves contained in large seeds serve an important role in the establishment of these pines, most of which occur on arid sites where seedlings often experience heavy mortality.

As with many wing-seeded pines, wingless-seeded pines usually produce large seed crops at intervals of 2–5 years, with relatively few seeds produced in intervening years (Formozov 1933; Schopmeyer 1974). The great variation in seed crop size over time makes the seed source difficult for seed predators to exploit (Smith 1968; Smith and Balda 1979), but the highly synchronized seed crops, by producing many more seeds than can be immediately consumed, also ensure that a great excess of seeds will be stored and thereby increase the proportion of seeds that eventually establish (Abbott and Quink 1970; Ligon 1978).

Conifer seed–caching corvids also have acquired a number of mor-

Figure 7.8
Whitebark pine cones. A Clark's nutcracker has broken off some of the bracts and removed some of the seeds from the cone on the left. The base of the cone bracts (still visible) hold the seeds in place, allowing nutcrackers to harvest them more efficiently. Photograph by author.

phological and behavioral adaptations to harvest seeds (see Evolution of
Specialized Food Hoarding, chapter 3). These include a large, pointed bill
for prying into cones, food-transporting structures for carrying seeds,
and high degree of reproductive dependence on stored food (Ligon 1978;
Vander Wall and Balda 1981). Probable coevolutionary scenarios of these
corvids and pines are offered in chapter 3.

Nut-bearing, Broad-leaved Trees

Many of the large nuts commonly harvested and stored by animals
are familiar to us because they form the basis of the commercial nut industry. These include walnuts (*Juglans*), hickory nuts and pecans (*Carya*),
hazelnuts (*Corylus*), beechnuts (*Fagus*), and chestnuts (*Castanea*). Other
less economically important nuts include chinquapin (*Castanopsis*), acorns
(*Quercus* and *Lithocarpus*), and horse chestnuts (*Aesculus*). The distribution, characteristics, and hoarders of these nuts are summarized in
table 7.4. A large variety of other nuts, such as pistachios (*Pistacia*),
brazil nuts (*Bertholletia*), paradise nuts (*Lecythis*), almonds (*Prunus*),
cashews (*Anacardium*), and macadamia nuts (*Macadamia*), may be dispersed through the activities of hoarding animals, but as yet there seems
to be no evidence that this is so. These genera represent eight families of
plants, suggesting that nuts have evolved independently many times.
In tropical areas, the nuts of palms (*Scheelea* and *Astrocaryum*) and various trees (e.g., *Spondias, Oleiocarpon, Dipteryx, Gustavia, Persea* and
Hymenaea) are scatter hoarded by rodents (Smythe 1970, 1978; Janzen
1971; Smith 1975; Heaney and Thorington 1978; Smythe, Glanz, and
Leigh 1982; Hallwachs 1986).

It is well established that animals, principally tree squirrels, flying
squirrels, agoutis, chipmunks, mice, jays, nutcrackers, rooks, crows,
and tits, bury nuts (table 7.4), and it is widely accepted that these animals
are the primary dispersers of many nut-bearing trees and shrubs. However, exactly what happens to the nuts after they have been buried is
poorly understood. For most nut-bearing trees, we have little information
on what proportion of the nut crop is stored and even less information on
what proportion germinates from the stores. For example, I can find no
documentation that hickory nuts establish from animal caches. Postgermination mortality caused by the actions of food hoarders (e.g., English oak mortality as a consequence of jay foraging; Bossema 1968,
1979) has seldom been examined. Parent-offspring distances are virtually
unknown, a factor that may have important consequences for genetics of
nut-bearing forest trees (Hallwachs 1986). A few studies have linked animal nut storage with colonization of nonforest habitats such as old fields
(Olmsted 1937; Grey and Naughton 1971; Harrison and Werner 1984;
Stapanian and Smith 1986), but even these studies have just presumed
that animals were the dispersal vectors. Part of the difficulty stems from
the fact that nuts, unlike smaller seeds, are usually stored singly. The
multistemmed clumps of conifers, shrubs, grasses, and geophytes make
the actions of hoarding animals easier to recognize.

Table 7.4 Distribution, characteristics, and dispersers of temperate zone nut-bearing trees and shrubs

Genus	Number of Species	Distribution*	Nut Characteristics*	Important Scatter Hoarders
Juglans	15–20	North America, north-western South America, West Indies, southern Europe, southern and eastern Asia	Fruit spherical to pyriform; nut 20–40 mm in diameter, rugose to smooth, enclosed in a semi-fleshy, four-valved, dehiscent or indehiscent husk; hull thick and very hard to thin; nuts fall September to November; cotyledons sweet, oily; large crops produced at 1- to 3-year intervals	*Sciurus carolinensis* (Brown and Yeager 1945; Smith and Follmer 1972); *Sciurus niger* (Brown and Yeager 1945; Smith and Follmer 1972; Stapanian and Smith 1978); *Sciurus vulgaris* (Sviridenko 1971); *Sciurus lis* (Kato 1985); *Corvus frugilegus* (Källander 1978; Purchas 1980)
Carya	18	Eastern and southern North America, Mexico, south-eastern Asia; previously also in Europe, north Africa, Asia	Fruit spherical to lenticular; nut 20–45 mm in diameter; enclosed in a fiborous, dehiscent to partially dehiscent, four-valved husk; hull smooth or ribbed and moderately thick; nuts fall September to December; cotyledons sweet or bitter; large nut crops produced at 1- to 5-year intervals	*Sciurus niger* (Cahalane 1942; Brown and Yeager 1945; *Sciurus carolinensis* (Brown and Yeager 1945; Thompson and Thompson 1980); *Glaucomys volans* (Muul 1970); *Tamiasciurus hudsonicus* (Layne 1954); *Corvus brachyrhychos* (Conner and Williamson 1984)
Corylus	15	Temperate North America, Europe, Asia	Nuts ovoid, 15–20 mm in diameter, enclosed or partially enclosed in pubescent, papery involucral bracts; hull thick and hard; falling August to October; cotyledons sweet; large nut crops produced at 2- to 3-year intervals	*Sciurus vulgaris* (Tonkin 1983; Wästljung 1989); *Tamiasciurus douglasii* (Mailliard 1931); *Corvus frugilegus* (Källander 1978; *Nucifraga caryocatactes* (Swanberg 1951, 1956; Volker and Rudat 1978; Wästljung 1989)
Fagus	10	Temperate eastern Asia, Europe, eastern North America	Fruit a spiny, dry, four-valved dehiscent burr enclosing two to three cordate nutlets 12–18 mm in diameter; hull thin and leathery; nuts falling September to November; cotyledons sweet; large nut crops produced at 2- to 20-year intervals	*Cyanocitta cristata* (Johnson and Adkisson 1985); *Tamias striatus* (Elliott 1978); *Apodemus flavicollis* (Jensen 1985); *Apodemus sylvaticus* (Ashby 1967); *Parus ater* (Richards 1958)
Castanea	11	Southwestern and eastern Asia, southern Europe, northern Africa, eastern North America	Fruit a spiny, dry, spherical burr that encloses one to three nuts; husk dehiscent, four-valved; nuts oval, flattened, 20–35 mm long; hull thin, leathery; nuts fall August to October; cotyledons sweet	?

Table 7.4 (*continued*)

Genus	Number of Species	Distribution	Nut Characteristics	Important Scatter Hoarders
Castanopsis	110	Southern and eastern Asia, western North America	Fruit a dry, spiny, spherical burr enclosing one to three nuts; husk dehiscent, four-valved; nuts pyriform, 10–15 mm long; hull hard and woody; nuts fall September to October of second year; cotyledons sweet	?
Lithocarpus	275	Southeastern Asia, Indonesia, western North America	Fruit an acorn, borne singly or in clusters of two or three; scaly cup encloses basal end of acorn; acorn oblong, 25–50 mm long; hull woody and hard; cotyledons bitter; acorns fall September to October of second year; large nut crops produced in alternate years, but some acorns produced almost every year	?
Quercus	350	North America to northern South America, Europe, northern Africa, Asia, Indonesia	Fruit an acorn, borne singly or in clusters of two to five; scaly cup encloses basal end of acorn; nut oblong to nearly spherical, 6–37 mm long; hull thin and relatively soft; cotyledons sweet to bitter; maturing in one or two seasons, falling August to December; large acorn crops produced at intervals of 1–10 years	*Garrulus glandarius* (Schuster 1950; Turcek 1951; Chettleburgh 1952; Bossema 1968, 1979); *Cyanocitta cristata* (Darley-Hill and Johnson 1981); *Aphelocoma coerulescens* (Grinnell 1936; McDonald 1969; DeGange et al. 1989); *Sciurus* spp. (McDonald 1969); *Tamiasciurus hudsonicus* (Hatt 1929); *Apodemus* spp. (Jensen and Nielsen 1986; Miyaki and Kikuzawa 1988)
Aesculus	25	North America, Central America, southeastern Europe, eastern and southeastern Asia	Fruit a leathery, spherical, three-celled capsule either smooth or with weak spines; husk dehiscent, three-valved; one to 3 nuts per capsule; nuts oval to flattened, 20–50 mm in diameter; hull chocolate brown, thin, relatively soft; cotyledons bitter; nuts fall September to November; large crops produced at intervals of 1–2 years	*Sciurus carolinensis* (Thompson and Thompson 1980)

*Compiled from Schopmeyer 1974 and Harlow et al. 1979.

Nevertheless, several lines of evidence suggest a high degree of interdependence between nut-bearing trees and nut-burying animals. The geographic distribution of nut-bearing trees and dispersers often coincide (e.g., Turcek 1951). Nuts share a set of traits that increase their likelihood of being harvested by food-burying animals. Nuts are large and usually palatable. Nuts dispersed by animals range in size from beechnuts (7-mm diameter) to horse chestnuts (50 mm). Nuts usually are relatively high in lipid and protein (Botkin and Shires 1948; Woodroof 1979), traits that result in highly nutritious propagules that attract potential dispersal agents from a wide area. Of course, the large energy reserves of nuts also are beneficial to young seedlings competing with other vegetation, often in heavy shade (Korstian 1927; McComb 1934; Jarvis 1963; Smythe 1978). Some nuts, such as most acorns, horse chestnuts, and bitternut hickory (*Carya cordiformis*), contain tannins and other compounds that give the seeds astringent flavors to humans. These chemicals may impart added protection to the nuts from insects during development.

Many types of nuts are enclosed in a fibrous, semiwoody husk or pericarp. The husk in many cases is derived from the floral involucre and, in some cases, is homologous with the fleshy pericarp of fruits (see section 3.4). The pericarp serves various functions. The nuts of beeches, chestnuts, and chinquapins are produced in a burr armed with sharp spines that may deter some seed predators. The unripe husk of cashews contains an oil that causes painful blistering if contacted (Janick et al. 1974). The three- or four-valved husks of horse chestnuts, beeches, chinquapins, chestnuts, and some walnuts and hickories dehisce at maturity, releasing the nut. The fruits of some tropical plants have an edible pericarp, which the hoarder eats immediately; this pericarp surrounds a tough seed coat that sometimes contains toxins that discourage immediate consumption (Smythe 1978). Agoutis and other animals store these nuts and recover them months later, perhaps after the toxins have leached from the seeds or deteriorated.

Nut drop is a widespread trait of animal-dispersed nuts. Nuts fall shortly after the hull dehisces, as in horse chestnuts, or may be retained for a short period, as in the cupule of beech. Other nuts fall still enclosed in husks or within the fleshy pericarp. The adaptive significance of nut drop is not always clear. Nut drop may increase the foraging efficiency of food-hoarding animals, which before the nuts fall harvest nuts directly from tree canopies. After nuts fall, however, they become vulnerable to deer, peccaries, turkeys, and other animals that act exclusively as nut predators. The fact that nuts drop suggests that they are adapted for dispersal by ground foragers (i.e., terrestrial rodents).

The woody hull or endocarp of nuts act as a filter, determining which animals gain access to the nutritious embryo and cotyledons within. Hulls range in quality from the thick and very hard shell of black walnut (*Juglans nigra*), which is nearly impenetrable to all animals except squirrels (*Sciurus* spp.), to the thin and relatively soft covering of acorns and beechnuts (table 7.4). Consequently, acorns and beechnuts are eaten and dispersed by a larger array of animals than are black walnuts. The durability of nut hulls apparently is a compromise between excluding seed predators while maintaining the appeal of the nut to dispersers (Smith and Follmer 1972).

Mast fruiting, the production of large crops of nuts at intervals of 2 or more years (Waller 1979; Silvertown 1980; Sork 1983b; Smith and Scarlett 1987), is characteristic of nut-bearing trees (table 7.4). Nut crops of each species usually are synchronized over wide geographic areas. In general, mast seeding results in satiation of seed predators during mast years; during nonmast years, these animals must switch to other foods, emigrate, or starve. This results in a resource that is difficult for predators (especially those with short generation times, such as insects) to track through time, with the result that a high proportion of seeds escape predispersal predation during mast years (Smythe 1970; Smith 1975; Nilsson 1985; Wästljung 1989). An additional benefit may accrue if nut hoarders store more nuts than they can consume during mast years, leaving a greater number of nuts to germinate (e.g., Ligon 1978; Jensen 1985).

Nondormancy of some white oak and live oak acorns is yet another adaptation of nuts for being buried by animals (Barnett 1977; Fox 1982). Once buried, the greatest threat to an acorn is being found. Certain white and live oak acorns avoid this problem by germinating in the fall within weeks of maturity. Many acorns germinate on the ground surface; those that have been buried by animals germinate under more favorable conditions. Nutrients and stored energy are quickly translocated to the swollen taproot (Lewis 1911). The future above-ground portion of the seedling also exits the acorn but does not elongate, remaining only a few millimeters long and protected below ground (see fig. 3.17). If the acorn is discovered after these changes have occurred, the slender connection between the acorn and the seedling breaks, leaving the seedling relatively unharmed, although with reserves depleted.

Shrubs and Grasses

A variety of shrubs, grasses, and forbs are known to establish from caches; however, only three species (antelope bitterbrush, Indian ricegrass, and velvet mesquite) have been studied. The dozen or so other species that have been found germinating from rodent caches include a variety of shrubs (snowbush, *Ceanothus velutinus;* squawcarpet, *Ceanothus prostratus;* green rabbitbrush, *Chrysothamnus viscidiflorus;* desert bitterbrush, *Purshia glandulosa;* snowberry, *Symphoricarpos albus;* manzanita, *Arctostaphylos patula;* wax currant, *Ribes cereum*), several species of native and introduced grasses (Idaho fescue, *Festuca idahoensis;* squirreltail, *Sitanion hystrix;* needlegrass, *Stipa occidentalis;* cheatgrass, *Bromus tectorum*), and various species of cultivated grains (wheat, barley, rye, and oats) (Scheffer 1938; Hawbecker 1940; Hormay 1943; R. Hardy 1949; Reynolds and Glendening 1949; Tevis 1953; Reynolds 1958; West 1968; La Tourrette, Young, and Evans, 1971; Everett and Kulla 1977). Some of these plants can also establish in clumps by having several seeds dropped in one place in the feces of birds or mammals (e.g., Howe 1989), so studies are needed to determine how often these

Figure 7.9
Clumps of plants arising from scattered caches: Upper left, antelope bitterbrush seedlings emerging from a chipmunk cache (partially excavated to show the individual stems); upper right, singleleaf piñon pine seedlings established from a Clark's nutcracker cache; lower left, Jeffery pine seedlings escaping from a cache; lower right, mature whitebark pine trees often grow in clumps arising from animal caches. Photographs by author.

plants establish from rodent food caches. The rodents responsible for these caches have seldom been identified, but kangaroo rats, pocket mice (*Perognathus* spp.), deer mice, golden-mantled squirrels (*Spermophilus lateralis*), and chipmunks are the species most often suspected.

Dispersal of forb seeds has seldom been observed to occur through the hoarding activities of food-storing birds. Thistle, corn, and sunflower have been known to establish from scrub jay and blue jay caches (Stearns 1882; Laskey 1943; Michener and Michener 1945). Haftorn (1956c) suggested that marsh tits (*Parus palustris*) and willow tits (*P. montanus*) dispersed the seeds of *Galeopsis* by dropping seeds after transport flights, but such dispersal was peripheral to seed storage.

Keen interest has been shown in bitterbrush because of its importance as big game forage and the importance of rodent caching in its establishment. At sites in California and Oregon, frequency of clumped stems indicated that between 50% and 58% of all bitterbrush shrubs derived from rodent caches (Hormay 1943; West 1968). The actual proportion of shrubs coming from rodent caches may be even higher because death of all seedlings except one in a clump would result in the plant being scored as an individual rather than a cache. At germination, clumps contained an average of about twelve plants (West 1968; Sherman and Chilcote 1972) and ranged in size from 2 to 139 stems (Hormay 1943) (fig. 7.9). Following good seed crops, from 1,300–3,600 clumps of germinants have been found per acre (3,250–9,000 clumps/ha) (Sanderson 1962; Evans et al. 1983). Many of these occur in interspaces between plants, where establishment is most likely to be successful. Clumps are long lived (greater than 30 years), but because of their shrubby growth form, close inspection is required to distinguish clumps from plants growing singly.

Hormay (1943) concluded it is doubtful that bitterbrush could perpetuate itself without rodents; however, even with rodents, bitterbrush is not establishing well on some sites. Competition for moisture with exotic grasses (Sanderson 1962) and accumulation of litter because of fire suppression (Sherman and Chilcote 1972) strongly influence bitterbrush establishment. These observations demonstrate that simply caching seeds in a suitable seedbed is insufficient to ensure establishment.

Rodents may have a major influence on Indian ricegrass establishment. Germination of untreated seed is very low (at least initially) but increases markedly if rodents remove the lemma and palea (McAdoo et al. 1983). Seedling establishment has been attributed to several species of heteromyid rodents, which bury seeds in large clusters averaging 250 seeds at mean depths of 5.5 cm. Most successful germination of Indian ricegrass is from seeds 3–8 cm deep (Kinsinger 1962) (fig. 7.4B). McAdoo et al. (1983) estimated that 8% of the seed crop was stored by rodents, or about 3.4 million seeds per hectare. However, only a small proportion (about 0.02%) of the seeds establish. Indian ricegrass is a bunch grass, making it impossible to distinguish clumps from solitary plants without excavating the bunches. One of the bunches dug up by McAdoo et al. (1983) was composed of twenty-one individuals.

Velvet mesquite has invaded southwestern rangelands, partially through the actions of Merriam's kangaroo rats (Reynolds and Glendening 1949; Paulsen 1950; Reynolds 1954, 1958). The kangaroo rats scatter hoard small clusters of mesquite seeds within 50 m of parent plants. The role of the kangaroo rat in large-scale range extension, however, seems minor. Reynolds (1954) calculated the maximum dispersal rates due to hoarding by kangaroo rats to be 1 mile in 500 years, a fraction of the rate actually observed on the Santa Rita Experimental Range in southern Arizona. Cattle, which disperse seeds much greater distances in their feces, are thought to be much more effective long-range dispersers. Merriam's kangaroo rats may be important in increasing mesquite density on a local scale.

It seems likely that other species of shrubs and grasses rely on rodent seed caching as an important, if not primary, means of dispersal. Problems in identifying these plants arise from the fact that their growth

Table 7.5 Dispersal of bulbs, tubers, and root cuttings by fossorial mammals

Rodent Species	Plant Species	Cache Size	Depth (cm)	Reference
Geomyidae				
Thomomys sp.	*Cirsium* sp.	Up to 200 root cuttings	20	Cook 1939
Spalacidae				
Spalax leucodon	*Oxalis cernua*	Up to 250 bulbs	15–50	Galil 1967
	Arisarum sp.	Bulbs	Shallow	
	Bellevalia sp.	Bulbs	Shallow	
Bathyergidae				
Bathyergus suillus	*Othonna* sp.	11 tubers	34	Davies and Jarvis 1986
Georychus capensis	*Micranthus junceus*	Scattered corms	?	Lovegrove and Jarvis 1986
Cryptomys hottentotus	*Micranthus junceus*	Scattered corms	?	Lovegrove and Jarvis 1986

forms tend to obscure their clumped dispersion once they have passed the seedling stage. Future research should focus on identifying those species that grow in clumps and determining the impacts of rodents in dispersing them.

Bulbs, Tubers, and Roots of Geophytes

Bulbs, tubers, and root cuttings stored by fossorial rodents sometimes sprout and establish plants. Only a few instances of this sort of vegetative reproduction have been reported, but it seems probable that this form of dispersal is more common than we realize.

Six families of rodents are known to store underground plant parts: pocket gophers, Geomyidae; mole-rats, Spalacidae, Bathyergidae, and Rhizomyidae; octodonts, Octodontidae; and voles and muskrats, Arvicolidae. Establishment of plants from caches has been documented in the first three families (table 7.5). Rodents of all three families are highly specialized burrowers. These rodents collect underground plant parts in foraging galleries and store them in small chambers at the end of short lateral tunnels—either in the foraging gallery or near the nest (Smith 1948; Galil 1967; Lovegrove and Jarvis 1986). Storage chambers in the foraging galleries often are sealed off from the main burrow with a plug of earth. Roots and bulbs in these caches are most likely to sprout and establish plants, probably because the chambers are usually near the ground surface (the foraging galleries are dug in the root zone of plants), and they are monitored less frequently than food stored near the nest site.

Rodents that store bulbs, corms, and other vegetative structures decrease the likelihood that these organs will establish plants in two ways. First, they frequently excise the buds of bulbs before they store them. Mole-rats (*Spalax leucodon*), for example, disbud bulbs of *Narcissus, Muscari,* and *Bellevalia* (Nevo 1961). Cape mole-rats (*Georychus capensis*) bite off the buds from tubers and bulbs to prevent them from sprouting (Shortridge 1934). Second, if bulbs, tubers, or root cuttings sprout in the cache, the rodents graze on the shoots. Mole-rats may clip the shoots of *Oxalis cernua* repeatedly (Galil 1967).

For some plant species that have established from rodent caches, such as Canadian thistle (*Cirsium* sp.) sprouting from pocket gopher caches (Cook 1939), this mode of propagation may be a relatively unimportant mode of dispersal. For plants such as *Oxalis cernua*, however, collection and storage of underground vegetative organs by rodents may be the principal mode of dispersal. *O. cernua* is widely scattered by the mole-rat *S. leucodon* in Israel. Clumps of *O. cernua* are conspicuous and frequently observed plants on the coastal plain (Galil 1967). The plant does not set seed, so seed dispersal is not a means of propagation, yet clumps of *Oxalis* may be 10–20 m apart. Galil (1967) established that bulbs of *O. cernua* are a favored food of *S. leucodon* and are stored in large numbers. Many of the clumps have been excavated and found to consist of numerous individuals sprouting from bulbs in neglected mole-rat caches. These bulbs are able to establish from a depth of more than 40 cm.

Establishment of plants from stored underground plant parts seems a fertile area for future research and of considerable importance, since some of the plants are noxious weeds. Monitoring of chambers containing underground plant parts following experimental removal of the resident rodent would be informative.

7.5 PATTERNS OF PLANT ESTABLISHMENT

Dozens of reports describe the dispersal and establishment of plants by seed caching animals. Few studies, however, have done more than speculate on the possible impacts that seed caching has had on plant populations and plant communities. Such impacts include the effects of clumped dispersion of seedlings arising from a cache on members of the clump, the role of seed-caching animals in plant succession, and the influence of dispersers on changes in plant distributional ranges.

Clumped Seedling Dispersion

Clumping of seedlings is a common pattern observed in plants dispersed by food-hoarding animals (Galil 1967; West 1968; Woodmansee 1977; Sherman and Chilcote 1972; Lanner 1982b, 1988; see also Howe 1989) (table 7.6, figs. 7.9 and 7.10). It has been long realized that seedling clusters result from animal caches (e.g., Munger 1917; Olmsted 1937; Hormay 1943; Reynolds and Glendening 1949; Kondratov 1953), but only recently has the ecological significance of the clumped dispersion patterns been examined. Clump size is obviously related to number of seeds stored per cache. For most rodents, each cache consists of one load of seeds carried in the mouth, jaws, or cheek pouches. Thus, maximum cache size is usually constrained by maximum load size. Corvids, on the other hand, typically divide a load among numerous caches, each containing several seeds. Birds and mammals carry large nuts singly, and consequently nut-bearing trees establishing from caches are usually single-stemmed.

Table 7.6 Frequency of stem clumping in various plants dispersed by scatter-hoarding birds and mammals

Plant Species[a]	Sample Size	Percentage of Sample in Multi-stem Clumps	Stems per Clump		Reference and Site
			Mean	Range	
Indian ricegrass (seedlings)			1.4	1–17	McAdoo et al. 1983; Nevada
Indian ricegrass (mature)			1.2	1–21	McAdoo et al. 1983; Nevada
Antelope bitterbrush		58		2–139	Hormay 1943; California
Antelope bitterbrush	122	52	11.6[b]	2–50	West 1968; Oregon
Antelope bitterbrush	494	66	5.2	1–36	Vander Wall unpubl.; Nevada
Velvet mesquite	913	27		4–13	Reynolds and Glendening 1949; Arizona
Ponderosa pine		85		2–50+	Munger 1917; Oregon
Jeffrey pine (seedlings)	566	27	1.9	1–54	Vander Wall unpubl.; Nevada
Limber pine (seedlings)	470	56	3.1	1–15	Woodmansee 1977; Colorado
Limber pine (seedlings)	94[c]	11[c]	1.2[c]	1–6	Lanner & Vander Wall 1980; Utah
Whitebark pine	1270	47		1–8	Lanner 1982b; Wyoming
Japanese stone pine	30	77	4.8	1–23	Saito 1983b; Japan
Swiss stone pine	345	31		1–7+	Mattes 1982; Switzerland
Swiss stone pine	214	39		1–7+	Mattes 1982; Switzerland
Swiss stone pine	54	67	3.1	1–11	Crocq 1978; France
Bristlecone pine	511	75	1.8	1–6	Lanner 1988; Nevada
Bristlecone pine	352	71	1.8	1–9	Lanner 1988; Utah
Bristlecone pine	122	69	2.7	1–11	Lanner 1988; California

[a] Stems of all ages and sizes included in sample unless otherwise stated.
[b] Does not include 112 single-stemmed plants; adjusted mean = 6.5 stems per clump.
[c] Calculated from data in Lanner and Vander Wall 1980.

Clumping of seedlings also can result in other ways. For example, wind-blown seeds can accumulate in a hoof print, frugivorous animals may eliminate numerous seeds in their feces, and runoff can cause seeds to collect in depressions, where they germinate in clumps. Browse damage to woody plants may result in lateral branches forming new vertical stems that superficially resemble a clump of plants. Although the effect of these agents is usually distinguishable from that of clumps arising from cached seeds, caution must be exercised when interpreting the causes for clumped dispersion patterns.

Some species are relatively tolerant of crowding, and clumps thus

Figure 7.10
The distribution of clump sizes and individuals within clumps of 209 singleleaf piñon pine seedlings in June 1986 in piñon-juniper woodland, shortly after emergence. Most, but probably not all, seedlings established from scrub jay and Clark's nutcracker seed caches (Vander Wall, Mesch, and Crusifulli, unpublished data).

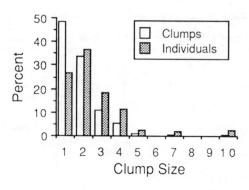

persist through maturity. Antelope bitterbrush, limber pine, and white-bark pine are notable examples (Sherman and Chilcote 1972; Wood-mansee 1977; Hutchins and Lanner 1982). It is often possible to find stem clumps each representing different age classes in a population. The oldest clumps are undoubtedly those of bristlecone pine (*Pinus longaeva*) reported by Lanner, Hutchins, and Lanner (1984) and Lanner (1988). Electrophoretic studies have shown whitebark pine stems in clumps to be genetically distinct (Linhart and Tomback 1985; Furnier et al. 1987).

The clumped dispersion of plants established from caches has several important impacts on plant biology. Clumps of seeds may enjoy certain advantages in increased early survivorship and growth (Munger 1917; Ferguson 1962; Ferguson and Basile 1967; Bullock 1981; Saito 1983b). This effect is presumably the result of mutual shading. In addition, plants that are not self-compatible will benefit by having a neighbor with compatible pollen close at hand. This may be especially important when plants colonize new sites and other sources of pollen are distant. Clumped plants may act as small breeding neighborhoods and may exhibit greater adaptability in other ecological relations (Bullock 1981). For many species, however, the clumping of seedlings may have negative consequences (Howe 1989). Individuals in clumps are more likely to contract communicable diseases or parasites, and caching establishes a situation in which competition among seedlings may be severe. This competition results in some or all of the seedlings in a cluster dying (McAdoo et al. 1983). Those stems that persist often differ in height and diameter, one or a few individuals being significantly larger than the rest (Olmsted 1937; Lanner 1982b). At maturity, the number of reproductive individuals in a clump often is reduced to less than half the original germinants (fig. 7.11). Seedling competition may have played an important role in the evolution of seed characteristics. The evolution of large seeds in wingless-seeded pines, for example, may be due in part to large seeds faring better in intracache competition.

Plant Succession

Nut-bearing trees and other plants dispersed by food hoarders often dominate late successional and climax plant communities. Many of these plants depend almost completely on scatter hoarders for maintaining the

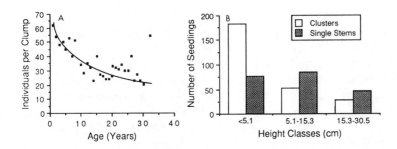

Figure 7.11
Attrition within clumps of plants
establishing from animal caches. A,
Reduction in number of bitterbrush
stems per clump with increasing
clump age. B, Number of limber
pine seedlings growing in clusters
or singly in three height classes.
A, Redrawn from Sherman and
Chilcote 1972; B, redrawn from
Woodmansee 1977.

climax communities and moving their propagules into early successional habitats. Stand maintenance results primarily from seeds stored between the canopies of mature plants and in small disturbances, such as tree-fall gaps, where seedlings receive adequate sunlight and less competition for moisture and nutrients (e.g., Mellanby 1968). It is significant, in this regard, that many food-hoarding animals transport food items beyond the canopies of the source plants (Reynolds and Glendening 1949; La Tourrette, Young, and Evans 1971; Sherman and Chilcote 1972; Bossema 1979; Stapanian and Smith 1984).

One propensity of seed-hoarding animals that diversifies their impacts on plant communities is their tendency to transport food well away from concentrated food sources and often into nonforested habitats. Nut and seed hoarders benefit by caching in early successional habitats because cache loss to cache robbers is likely to be low in these sites (Ligon 1978; Hutchins and Lanner 1982; Vander Wall and Smith 1987; Stapanian and Smith 1986; Kraus and Smith 1987). The potential for rodents and birds to move propagules to other habitats differs markedly. Plant establishment from rodent seed caches will generally be within 100 m of the parent plant and thus influences recruitment into and maintenance of the local population. Movement of nuts into disturbed areas and old fields extends the influence of rodents beyond the forest but for only relatively short distances. Dispersal by corvids, which are able to cross such barriers as streams, large patches of unsuitable habitat, and mountain ridges, not only influences plant establishment on the local scene but is likely to colonize disjunct patches of suitable habitat. Bibikov (1948), for example, found Siberian stone pine seed cached by Eurasian nutcrackers in the Ural Mountains in Siberian stone pine forests (3,334 caches/ha), alpine tundra (1,665 caches/ha), fluvial mixed forest (833 caches /ha), and fir-birch open forest (417 caches/ha). Vander Wall and Balda (1977) found that piñon pine had the broadest elevational range of any conifer in the San Francisco Mountains of northern Arizona. Further, since pollen dispersal is often effective over only short distances (Levin and Kerster 1974), long-range dispersal of seeds by corvids may be an important source of gene flow between scattered populations.

Movement of propagules into other plant communities by scatter hoarders can occur very early in the successional process. Establishment of black oak (*Q. velutina*) and white oak seedlings in a Michigan old field, presumably the result of blue jay hoarding, began 15 years after abandon-

ment (Harrison and Werner 1984). At year 30, 401 seedlings and saplings occupied the 0.64-ha study plot. Most young oak seedlings occurred
in patches of mineral soil with sparse lichens and mosses, cache sites preferred by blue jays. These sites were rapidly colonized by grasses and
forbs, which acted to increase oak survivorship. In Denmark, acorn-
storing by wood mice (*Apodemus sylvaticus*), yellow-necked mice, and
perhaps bank voles (*Clethrionomys glareolus*) has caused the invasion of
oaks (*Q. robur* and *Q. petraea*) into heathlands (Jensen and Nielsen 1986).
The rate of invasion is very slow, about 300 m in the past 100 years, and
the hoarding behavior of the wood mice adequately accounts for the
colonization of the oaks. Stapanian and Smith (1986) surveyed native
prairie along 1.45 km of hardwood forest edge and found 385 seedlings
and saplings of black walnut, bur oak (*Q. macrocarpa*), chinquapin oak
(*Q. muehlenbergii*), and bitternut hickory. These seedlings and saplings
occurred up to 151 m (species means = 3.8–16.8 m) from the forest
edge. Stapanian and Smith (1986) attributed the colonists to nut-caching
squirrels. Squirrels disperse acorns onto the sand plains of Connecticut
up to 60 m from mature trees (Olmsted 1937). Incidently, the "squirrels" responsible for the oak dispersal in Olmsted's study were probably
not tree squirrels (*Sciurus*) because the acorns germinated from relatively deep underground galleries, sites that tree squirrels have rarely
been reported using (see Tree Squirrels, chapter 8). As mentioned earlier, Merriam's kangaroo rats contributed to the rapid encroachment of
velvet mesquite into southern Arizona rangeland (Reynolds and Glendening 1949).

Food-hoarding animals do not seem to avoid disturbed areas and perhaps are even attracted to them. Nutcrackers frequently store pine seeds
in sparsely vegetated sites and thus initiate restocking of burned sites
within a few years of a fire (Reimers 1953, 1958; Lanner and Vander Wall
1980; Tomback 1986). Pinyon jays store millions of piñon pine seeds in
large tracts of land where humans have eradicated this tree to "improve"
grazing land (Ligon 1978). These areas are excellent sites for germination and growth of piñon pine. Antelope bitterbrush seedlings germinating from rodent caches established better in a rototilled area than in
undisturbed sites, apparently because reduced competition for water
with grasses and forbs (Sanderson 1962).

The benefit to hoarders of transporting nuts and seeds out of the forest (e.g., Stapanian and Smith 1984, 1986) is countered to some extent
by higher probabilities of predation on the hoarders in habitats with little
cover. In some cases, this requirement for cover has slowed colonization
of nut-bearing trees into old fields until other woody plants establish
(Bard 1952). For example, Grey and Naughton (1971) observed that black
walnut often invaded pastures after osage-orange (*Maclura pomifera*) had
established. Osage-orange is dispersed in the feces of cattle and fruit eaters. Once it is established in the pasture, fox squirrels venture out to the
osage-orange thickets to feed on fruits and to store walnuts under its protective canopy. Walnut saplings grow well in shade and overtop the osage-
orange in several years. Invasion of oaks into pine and spruce forests

through the actions of acorn-storing Eurasian jays has been documented by Turcek (1951) and Nilsson (1985), as has the invasion of piñon pine into oak forests by Floyd (1982). Once mature, isolated nut-bearing trees serve as seed sources for further colonization of successional habitats (Olmsted 1937) and further add to the complexity of plant communities.

Grinnell (1936) was one of the first to implicate nut-storing animals as agents that act to overcome gravity, resulting in movement of nuts uphill. This effect is especially noticeable at timberline, where upward dispersal of conifer seeds primarily by nutcrackers is responsible for the persistent encroachment of these trees into alpine zones (Reimers 1953; Oswald 1956; Mattes 1982; Lanner, Hutchins, and Lanner 1984; Daly and Shankman 1985; Lanner 1988). Low viability of high alpine conifer seeds (Tranquillini 1979) suggests that these trees may not be able to maintain timberline forests and colonize alpine tundra without nutcrackers continually bringing in seeds from lower elevations. The interaction of nutcracker cache site selection and the rigors of the physical environment at high elevations often results in predictable patterns of tree establishment along ridges, rocky cliffs, and south-facing slopes (Oswald 1956; Clausen 1965; Mattes 1982; Tranquillini 1979). Swiss stone pines established at alpine sites in the Swiss Alps through the actions of Eurasian nutcrackers serve as natural avalanche barriers, and Mattes (1982) argued that destruction of some of these high-elevation forests during the Middle Ages is still being paid for today through the establishment of artificial avalanche walls.

Seed-caching animals also play an important role in plant establishment in habitats where seedlings are subject to moisture stress (Woodmansee 1977). This effect is probably widespread but seems especially evident in plants not adapted for dispersal by seed hoarders. Pines with winged seeds, for example, appear to establish from animal caches more frequently on stressful sites (Munger 1917; Lanner, Hutchins, and Lanner 1984; Lanner 1988). The reason appears to be that pine seeds buried several centimeters deep are more likely to experience favorable moisture regimes during establishment than surface broadcast seeds that on average germinate nearer the soil surface.

Establishment of antelope bitterbrush is adversely affected by thick mats of pine litter in ponderosa pine forests. Fire control has increased litter depth on some sites to the extent that few sites are now suitable for bitterbrush establishment. As a result, current establishment on some sites is occurring only in the interspaces between pines where litter is sparse (Sherman and Chilcote 1972).

Long-range dispersal of large nuts and seeds allows plants to colonize widely separated patches of suitable habitat. The importance of this ability has become increasingly apparent since humans have altered land use patterns and fragmented natural landscapes. The reduction of large tracts of deciduous forest in the eastern United States to small woodlots isolated by tracts of agricultural crops and pastures virtually eliminates any interpatch nut dispersal by rodents. Darley-Hill and Johnson (1981) and Johnson and Adkisson (1985, 1986), however, have demonstrated

that blue jays effectively transport acorns and beechnuts across fragmented landscapes and thus play a major role in maintaining populations
of members of the Fagaceae. Such dispersal of heavy nuts to scattered
sites also may have been important in the past (e.g., Webb 1987).

Impacts of Seed Hoarders on Plant Distribution during the Holocene

During the Quaternary, plant communities in the north-temperate
zone have seldom maintained a constant species composition for more
than a few thousand years. With each glacial advance, climatic change has
forced plant species southward into refugia, and with each glacial retreat,
they have recolonized northern areas. During the most recent glacial epi-
sode, some species have undergone displacement as great as 2,000 km
(Davis 1981). This phenomenon has been especially well documented
in eastern North America and western Europe using pollen profiles
from lake and pond sediments and plant macrofossils (van der Hammen,
Wijmstra, and Zagwijn 1971; Davis 1981; Delcourt and Delcourt 1984).
Reconstruction of the Quaternary history of coniferous forests in western North America is based largely on macrofossils from woodrat (*Neotoma*) dens (Van Devender and Spaulding 1979; Wells 1983).

Nut-bearing hardwoods and wingless-seeded conifers participated in
these migrations. Rates of northward migration of some of these tree
species after the most recent glacial retreat, 16,000–10,000 years ago,
have been estimated at between 100 and 350 m/yr (Davis 1981) (table
7.7), which far exceed that which could be accomplished by physical factors such as gravity and wind. To what extent did animals ancestral to
those that disperse these plants today facilitate these shifts in distributional range?

Table 7.7 Northward dispersal rates of certain nut-bearing hardwood trees during the
Holocene following glacial retreat

Nut-bearing Plants		Primary Dispersers		Estimated Dispersal Rates	
Taxon	Minimum Generation Time (yr)	Taxon	Maximum Dispersal Distance (m)	Maximum Predicted[a] (m/yr)	Observed[b] (m/yr)
Oak	20–35	Blue jay Squirrels	4,000 200	115–200	350
Hickory	25–40	Squirrels	200	5–8	200–250
Beech	40	Blue jays	4,000	100	200
Chestnut	8	Squirrels	200	25	100

[a]Maximum dispersal distance divided by minimum generation time.
[b]Based on pollen records (Davis 1981).

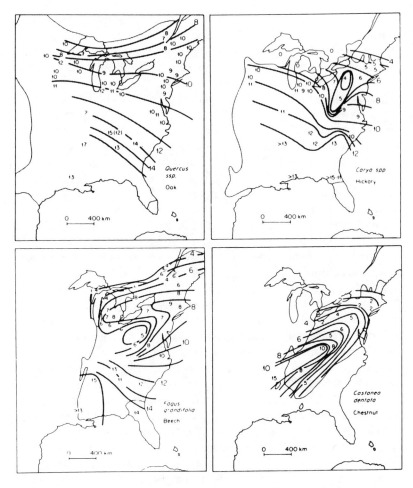

Figure 7.12
Migration maps for oak, hickory, beech, and chestnut during the Holocene. Numbers refer to radiocarbon ages (in thousands of years) of the first appearance of each group at the site after 15,000 years ago. The first appearance was determined from increased pollen abundance or presence of macrofossils. Isopleths connect points of similar age and represent the leading edge of northward migrating populations. The modern ranges of each plant group are outlined. From Davis 1981.

The nut-bearing trees in eastern North America that have well-documented migrational histories are the oaks, beech, hickories, and American chestnut (*Castanea dentata*) (Davis 1981; Bennett 1985; Webb 1987; Woods and Davis 1989). During the Wisconsin glacial maximum, before 16,000 years ago, oaks were widespread in what is now the southeastern United States, whereas beech, hickories, and American chestnut occupied the lower Mississippi River Valley and adjacent regions. Starting about 15,000–12,000 years ago, these trees began dispersing northward in response to the amelioration of climate that initiated the present interglacial, reaching the northern edges of their current geographic ranges between 10,000 and 2,000 years ago. Each group appears to have migrated at a different rate and to have taken different routes to reach their present range (fig. 7.12). Oaks moved the fastest, completing the 1,200–1,600 km range extension in less than 5,000 years, a rate of about 350 m/year (Davis 1981). American chestnut was the slowest, moving only about 100 m/year. Migration rates of beech and hickory were intermediate (table 7.7; Bennett 1985). Interestingly, the migration rates of some animal-dispersed trees (i.e., oaks, beech, and hickories) appeared to have been much faster than those of wind-dispersed trees (i.e., elms, firs, hemlock, spruces, and pines) of eastern North America that also participated in the northward migration.

To estimate the potential maximum dispersal rates of nut-storing animals, we need to know maximum dispersal distances of the animals and minimum generation times (age at first fruiting) of the trees. The former were given earlier in this chapter, and the latter are presented in table 7.7. Based on these data, blue jays could have moved oak at a rate of 115–200 m/yr and beech at a rate of 100 m/yr. These rates are about half those predicted from pollen data. American chestnut and hickories, the nuts of which are stored primarily by rodents, were apparently transported much more slowly. Assuming a maximum northward movement of 200 m per tree generation, rodents could have moved hickory northward 5–8 m/yr and American chestnut 25 m/yr. These rates are about 4% and 20%, respectively, of mean rates of spreading Davis (1981) estimated from pollen data. Johnson and Webb (1989), however, suggested that American chestnut also may have been dispersed by blue jays and that the jays easily could have dispersed the trees the necessary 1.2 km per generation to account for the estimated observed rate of spreading.

There are three general reasons why migration rates of nut-bearing trees predicted from pollen profiles and nut dispersal rates do not agree more closely. First, dispersal rates by nut-hoarding animals may be underestimated. Small errors in minimum generation time or maximum dispersal distance could cause these predicted rates to differ significantly from actual maximum dispersal rates. Second, it is possible that the dispersal rates predicted from pollen analyses (table 7.7) are overestimates of the actual rates of nut-tree dispersal. This could occur if, for example, isolated individuals or small stands of these trees existed on favorable sites north of their presumed range during glacial maxima. Such isolated individuals may not have left any detectable pollen record (Bennett 1985). Occurrence of oak and hickory macrofossils at Meadowcroft Rockshelter in western Pennsylvania about 14,500 years before the present (Adovasio et al. 1983), far north of their presumed range at that time, supports this argument. Isolated individuals and small stands of these trees could have acted as centers of dispersal when climate ameliorated, significantly shortening the necessary annual migration rate. Further studies of pollen profiles and plant macrofossil discoveries that provide greater resolution to Wisconsin glacial plant distribution may help resolve this question. Third, other nut dispersal vectors may have assisted the northward movement of nut trees. Chestnuts, beech nuts, and hickory nuts may have been transported many kilometers in the crops of turkeys (*Meleagris gallopavo*) or passenger pigeons (*Ectopistes migratorius*) before these birds were killed by predators and the contents of their crops scattered (Webb 1986). Wood pigeons (*Columba palumbus*) and mallards (*Anas platyrhynchos*) have been suggested as likely dispersers of beech seeds in northern Europe (Hemberg 1918, cited in Nilsson 1985). Such long-range dispersal may have been a rare event, but it would have greatly accelerated the northward movement of nut-bearing trees. Native Americans, who are known to have inhabited eastern North America at the time of glacial retreat (Adovasio et al. 1983), also may have accelerated the northward migration by transporting, trading, and even planting the nuts. Nut trees established by these agents may have acted as centers of dispersal by nut-burying rodents and jays.

Corvids have influenced the migrations of wingless-seeded conifers in Europe and western North America. Eurasian nutcrackers are thought to have been the vectors of Swiss stone pine in Europe as it repeatedly moved up and down mountain slopes during the Pleistocene (Mattes 1982). In the western United States 12,000–8,000 years before the present, singleleaf piñon pine occupied much of what is now the Mojave Desert of southern Nevada and southeastern California (Van Devender and Spaulding 1979; Wells 1983). Today piñon pine is one of the dominant conifers on the lower slopes of mountains throughout much of the Great Basin, as much as 640 km north of its former range. In the Mojave Desert, piñon pine still occupies isolated mountaintops. Based on the conservative assumption of a 100-year generation time and a 13-km maximum dispersal distance, Lanner (1983) estimated that nutcrackers could disperse singleleaf piñon pine from the Sheep Range of southern Nevada to the northern edge of the Humboldt Range of northeastern Nevada, 459 km to the north, in just 2,354 years. This estimate agrees with that made by Wells (1983). Thus, seed dispersal by nutcrackers can easily account for Holocene movements of singleleaf piñon pine. Nutcrackers also may have played a role in transporting ponderosa pine, limber pine, bristlecone pine, and Douglas-fir between mountain ranges in the Great Basin, but Wells (1983) was incorrect to state that nutcrackers and pinyon jays also might have dispersed the seeds of white fir (*Abies concolor*) and berries of Utah juniper. Clark's nutcracker has never been reported to eat or store the seeds of white firs or spruces; juniper berries are dispersed by many birds and mammals that consume the fruits, but nutcrackers and pinyon jays rarely eat the berries.

7.6 CONCLUSIONS

Food hoarders that effectively disperse plant propagules have had far-reaching effects on plant and animal communities. The most obvious ecological effect is that hoarders of seeds, nuts, and underground plant parts often establish new individuals beyond the crown of the parent plant. In situations where dispersal and establishment are sufficiently frequent, the activity of food hoarders can account for regeneration of plants within a stand and the colonization of new favorable sites. Over long periods, seed and nut hoarders assist plants in responding to changing environments by acting as the vehicle for directional shifts in distributional range. These ecological effects are worthy of study in themselves, but the impacts of some food hoarders extend well beyond plant dispersal; food hoarders can have pervasive influences on the evolutionary biology of the plants they disperse. Some hoarders have acted as strong selective agents on nut and seed characteristics that enhance their attractiveness to foragers. These include nut size, nutritional content, conspicuousness and other traits. The relatively large energy content of hoarder-dispersed nuts and seeds also increases the probability that seedlings will establish. The benefits of large, energy-packed nuts, both to the seedling and to the hoarder, sets up a strong positive feedback that promotes further increases in nut size and energy content, as well as greater dependence of hoarders on stored nuts (see fig. 3.18). The evolutionary response of seed and nut hoarders has taken the form of morphological and behavioral specialization for harvesting and transporting seeds and reproductive dependence on the stored food (e.g., Vander Wall and Balda 1981; Ligon 1978; Vander Wall and Hutchins 1983).

Dispersal of propagules by food hoarders has been a very successful adaptive syndrome, at least in north-temperate regions. Many plant communities are dominated by nut-bearing trees or wingless-seeded pines. To cite only a few examples, piñon-juniper woodlands and oak woodlands occupy millions of hectares in southwestern North America; oak-hickory forests are dominant plant communities in parts of the southeastern United States; beech forests are widespread in eastern North America, Europe, and Great Britain; and Siberian stone pine is the dominant member of the taiga of northern Asia. Despite the importance of seed hoarding as a means of plant dispersal, our knowledge of seed dispersal by hoarders is far from adequate, especially in the tropics and south-temperate regions, where the dispersal of nut-bearing genera like *Nothofagus* in New Zealand, *Araucaria* and *Macadamia* in Australia, *Lithocarpus* in southeast Asia, and *Bertholletia* in South America, to name only a few, has yet to be thoroughly studied. In the north-temperate zone, where interest in food-hoarding animals has increased greatly in the past 20 years, many interesting questions still need to be addressed. What are the fates of hoarder-dispersed propagules during various stages of production, dispersal, and establishment? What are the shapes of hoarder-generated seed shadows and how does hoarder cache-site selection (e.g., depth, substrate, distance from parent plant, presence of shade) affect the probability of seedling establishment? How do nut characteristics affect seedling growth and establishment? To what degree are hoarders and the plants they disperse dependent on each other? Are there as yet unidentified grasses, shrubs, trees, and geophytes that are dispersed through the hoarding activities of animals? We have learned in recent years that food-hoarding animals can be effective dispersers of plants, but we know very little about how this mutualistic interaction influences the population biologies of these plants.

8

Food-Hoarding Mammals

Food-hoarding has been reported in six orders and thirty families of mammals (table 8.1). Most food-hoarding mammals are members of the orders Rodentia and Carnivora. The array of food-storing behavior in the class is very diverse, encompassing food items as varied as seeds, nuts, conifer cones, foliage, twigs, fruit, bulbs, lichens, mushrooms, earthworms, vertebrates, and carrion. Mammals store these foods in sites ranging from deep below ground to the crowns of tall trees. Many mammals have well-developed hoarding behavior that is essential to their survival during periods of food scarcity. In the pages that follow, I emphasize how and where food is stored, the types and amounts of food hoarded, seasonal patterns in food storing, and how stored food is eventually used. I follow the taxonomy of Honacki, Kinman, and Koeppl (1982). A list of the food-hoarding mammals mentioned can be found in appendix I.

8.1 MARSUPIALS

Pygmy Possums (Burramyidae)

The mountain pygmy possum (*Burramys parvus*) is the only marsupial known to store food (fig. 8.1). This small, uncommon possum, for many years known only from fossil bones and teeth, was rediscovered in the mountains of southeastern Australia in 1966. Much of what is known about its behavior is based on observations of captive individuals. In captivity, mountain pygmy possums accumulate sunflower seeds and peanuts in their nest boxes (Calaby, Dimpel, and McTaggart Cown 1971; Kerle 1984). Invertebrates comprise much of the possum's diet, but they are not stored even if they are the only food items available (Dimpel and Calaby 1972). In the wild, mountain pygmy possums inhabit boulder-strewn, snow gum (*Eucalyptus niphophila*) forests. The possums nest under large boulders (Dixon 1971), which makes examination of nests and food stores very difficult.

The mountain pygmy possum is the only mammal restricted to the alpine and subalpine zones in Australia (Calaby, Dimpel, and McTaggart Cown 1971). In regions where the possum has been trapped, snow covers the ground for about 3 months each winter. In captivity, one mountain pygmy possum entered hibernation when ambient temperature fell below 13°C but awoke occasionally and fed (Dimple and Calaby 1972). Stored

Table 8.1 Mammalian taxa known to store food with a summary of their food-storing behavior

Order Family	Dispersion[a]	Food Type[b]	Substrate/ Location[c]	Storage Duration[d]
Marsupials				
Pygmy possums (Burramyidae)	L	S	N	L
Insectivores				
Shrews (Soricidae)	L	I,SM,A,Fi,S	N,B	L
Moles (Talpidae)	L	I	B,N	L
Primates				
Squirrel monkey (Cebidae)	S?	Mi	?	S
Green monkey (Cercopithecidae)	L?	Fr	C	S
Chimpanzee (Pongidae)	S?	Me	?	S?
Carnivores				
Foxes, wolves, et al. (Canidae)	S,L	SM,MM,Bi,E,Re	S,Sn,C	S,L
Bears (Ursidae)	S,L	LM	S,L	S
Weasles, mink, et al. (Mustelidae)	L	SM	B,N,S,C	S,L
Hyenas (Hyaenidae)	S	Ca	W	S
Tigers, bobcats, et al. (Felidae)	S	MM,LM	S,L,Sn,T	S
Rodents				
Mountain beaver (Aplodontidae)	L	V	B	S,L
Squirrels and chipmunks (Sciuridae)				
Chipmunks	L,S	S,Nu,Bu	B,N,S	L
Red squirrels	L	Co,Fu	L,F,T	L
Tree squirrels	S	Nu,Fr,Fu	S,T	L
Ground squirrels	L,S	S,Nu,V	B,S	L
Flying squirrels	S,L	Nu	S,L,T	L
Pocket gophers (Geomyidae)	L	Bu,Ro	B,Sn	L
Kangaroo rats, et al. (Heteromyidae)	L,S	S	B,S	L
Beavers (Castoridae)	L	WV	W	L
Mice, hamsters, et al. (Cricetidae)				
New World mice	L,S	S,Nu,I	B,S	L
Woodrats	L	V,S,Nu,Fr	N	L
Hamsters	L	S	B	L
Gerbils	L	S,Nu,Ro,V	B	L
Mole-rats (Spalacidae)	L	Bu,Ro,V	B	L
Mole-rats (Rhizomyidae)	L	Bu,Ro,V	N	?
Voles and muskrats (Arvicolidae)	L	V,Bu,Ro,WV	B,G,Sn,F	L
Old World rats and mice (Muridae)	L,S	S,Nu	B,S	L
Dormice (Gliridae)	L	Nu,Fr,Mi	C	L
Jerboas (Dipodidae)	L?	?	B	L?
Old World porcupines (Hystricidae)	S	?	S	?
Agoutis and acouchis (Dasyproctidae)	S	Nu,Fr	S	L
Octodonts (Octodontidae)	L	Bu	B	L
Tuco-tucos (Ctenomyidae)	L	Bu	B	L?
African mole-rats (Bathyergidae)	L	Ro,Bu	B	L
Lagomorphs				
Pikas (Ochotonidae)	L	V	G,B	L

Note: See text for references and exceptions to the general patterns within taxa.

[a]Dispersion patterns: L, larder; S, scattered.

Figure 8.1
A mountain pygmy possum in Kosciusko National Park, Australia. The shrub is *Podocarpus lawrenceii,* the seeds of which are a major food of these pygmy possums. Photograph courtesy Linda Broome.

nuts and seeds may act as a food reserve during the period of winter dormancy.

8.2 INSECTIVORES

Shrews (Soricidae)

Shrews comprise a large group of small insectivores that often feed as carnivores. Honacki, Kinman, and Koeppl (1982) list twenty-one genera and 288 species of shrews, but only a few species are known to hoard food. The paucity of known food-hoarding species in the group may be because shrews are inconspicuous animals that, with few exceptions, have not been well studied. The best-known food hoarder in the family is the short-tailed shrew (*Blarina brevicauda*).

Short-tailed shrews, like most shrews, are semifossorial predators that burrow through plant litter and loose soil in search of invertebrate and small vertebrate prey. Shrews viciously attack virutally any animal they encounter. Venom secreted from their submaxillary salivary gland immobilizes but does not kill prey (DeMeules 1954; Tomasi 1978; Martin 1981). Shrews also immobilize some snails by cracking their shells (Shull 1907). When an abundance of prey is presented to a captive shrew, it eats the first few items it captures and then caches subsequent prey in its nest (fig. 8.2). Within the burrow, shrews store food in a larder in the nest chamber, in a specially excavated food-storage chamber, or along one of several runways connecting the shallow nest chamber to the surface

[b]Food types: A, amphibians; Bi, birds; Bu, bulbs; Ca, carrion; Co, cones; E, eggs; Fi, fishes; Fr, fruit; Fu, fungi; I, invertebrates; LM, large mammals; Me, meat; Mi, miscellaneous; MM, medium mammals; Nu, nuts; Re, reptiles; Ro, roots; S, seeds; SM, small mammals; V, green vegetation; WV, woody vegetation.
[c]Substrates and locations: B, burrow chambers; C, cavity or chamber (not in burrow); F, foliage; G, ground surface; L, litter; N, nest; Sn, snow; S, soil; T, tree trunk and branches; W, water.
[d]Storage duration: S, short-term (generally <10 days); L, long-term (generally >10 days).

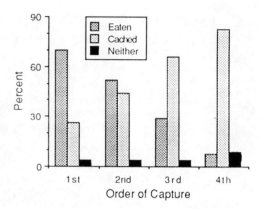

Figure 8.2
The influence of order of capture on the probability that short-tailed shrews will eat, cache, or neither eat nor capture prey. Redrawn from Robinson and Brodie 1982.

(Shull 1907; Ingram 1942; Martin 1984; Merritt 1986). Short-tailed shrews will use the nests of rodents for prey storage (Platt 1976). Captive shrews cache prey in the most protected sites available. Short-tailed shrews and water shrews (*Sorex palustris*) piled prey in their nest chambers (small jars) (Sorenson 1962; Robinson and Brodie 1982). After the nests were filled, they stored food on the ground surface under a layer of soil or plant litter near the nest.

Animal prey cached by short-tailed shrews includes insects, snails, slugs, and mice (Shull 1907; Ingram 1942, 1944). Captive short-tailed shrews also cached mealworms, dog food, raisins, frogs, salamanders, sunflower seeds, oatmeal, and beechnuts (Hamilton 1930; Robinson and Brodie 1982; Martin 1984). Despite the importance of seeds in the winter diets of shrews (Criddle 1973), seeds have not been found in caches in the wild. Captive water shrews stored suet, aquatic insects, and several species of fishes (Sorenson 1962). Water shrews cached fishes, which they caught in active underwater pursuit or by pouncing on them in shallow water, in hollow logs or in plant litter. Water shrews given the opportunity to kill and hoard mice did not do so. Captive least shrews (*Cryptotis parva*) stored mealworms and crickets (Formanowicz, Bradley, and Brodie 1989).

Maintenance of caches is characterized by frequent checking and repositioning of prey items (Robinson and Brodie 1982). Short-tailed shrews carry live snails to their burrow entrances in winter and pile them in the snow on the surface (Shull 1907; Hamilton 1930). Shull monitored several piles of snails during late winter and early spring and determined that the number of snails left at burrow openings each day fluctuated inversely with temperature; on cold days snails appear at the burrow openings and on relatively warm winter days shrews take snails below ground. Shull (1907) suggested that the shrews kept snails in the coldest places available. Excavation of the burrows in spring revealed numerous empty shells embedded in the walls of the burrows and filling small chambers off the main burrows. Some chambers contained both live snails and empty shells. Ingram (1942) found that short-tailed shrews carry the shells of snails on which they have fed to discard chambers.

Captive water shrews seldom defend territories or food caches

(Sorenson 1962). Water shrews use each other's nest sites and eat indiscriminately from each other's food stores. Such disregard for the ownership of larders has not been reported in short-tailed shrews. Two short-tailed shrews housed in the same terrarium did not store food, whereas isolated individuals under similar conditions did hoard (Martin 1984). After one of the pair died, the second individual began to hoard food. Short-tailed shrews often defecate and may also urinate on prey when they deposit the item in the larder and again when they visit the larder (Robinson and Brodie 1982), but it is not clear whether this behavior is "territorial" marking. The effects of the presence of conspecifics on food hoarding by shrews require further study.

The shrew's hoarding of prey appears to be almost insatiable. Robinson and Brodie (1982) observed a captive short-tailed shrew kill eighty-two green frogs (*Rana clamitans*) and store eighty of them in its nest in an hour. A captive water shrew that was given over 100 aquatic insects quickly ate several and cached the rest in a hollow log (Sorenson 1962). But Buckner (1964) found that at very high prey abundances the number of items hoarded by shrews began to level off. Buckner presented four species of shrew (masked shrew, *Sorex cinereus;* arctic shrew, *Sorex arcticus;* pygmy shrew, *Microsorex hoyi;* and short-tailed shrew) with different quantities of sawfly (*Pristiphora erichsonii*) cocoons ranging from 200–8,000 cocoons per day and counted the number of cocoons hoarded during 24 hours at each prey density. The four shrew species responded differently to the change in prey density (table 8.2), but each species increased the number of cocoons stored with increasing cocoon availability. Masked and arctic shrews, the species for which the most data are available, stored about 170 and 220 cocoons/day, respectively, at availabilities of 2,000 or more.

Summer and winter hoarding by captive short-tailed shrews differ both qualitatively and quantitatively (Martin 1984). Summer stores consist of a few items scattered about the nest chamber, which were usually retrieved and eaten within a few days. Winter stores consist of a much

Table 8.2 Relation between larch sawfly cocoon abundance and mean number of cocoons hoarded during a 24-hour period by four species of shrews

Species	N	Cocoons Required per Day	Number of Cocoons Presented								
			200	400	600	800	1,000	1,200	1,500	2,000	4,000
Sorex cinereus	10	87	10	85	146	151	156	164	165	173	166
Sorex arcticus	5	99	0	75	134	172	196	209	216	219	217
Microsorex hoyi	2	95	6	62	—	—	121	—	—	180	—
Blarina brevicauda	5	138	0	0	64	190	214	263	—	—	—

Source: After Buckner 1964.

larger quantity of food. Captive short-tailed shrews under natural conditions begin accumulating winter food in October and increase hoarding intensity until each shrew has accumulated a large larder by late November. These large stores are maintained until February, when hoarding begins to wane and most large larders are gone by the end of April. Martin's data do not make it clear whether temperature, photoperiod, or some other variable caused the seasonal pattern in hoarding activity, but captive shrews maintained under summer conditions did not show this seasonal pattern in hoarding behavior.

Short-tailed shrews select certain types of food for storage, and this selectivity gradually weakens as winter approaches. In preparing the winter caches, they first stored sunflower seeds, included dog food in later October, insects by mid-November, and mice by late November. They removed food from the larder in reverse order: first mice, then insects, dog food, and finally seeds. Martin (1984) suggested two possible reasons for this sequence of prey caching and recovery: (1) shrews eat more desirable prey (mice and insects) and store less undesirable prey (seeds and dog food), and (2) shrews store first those items that have greater storability. In winter, short-tailed shrews forage and nest in protected microsites to conserve thermoregulatory energy and the food stored in the larder (Merritt 1986).

Moles (Talpidae)

Moles are fossorial insectivores indigenous to those portions of Europe, Asia, and North America that have moist, friable soils. Food storage has been reported only in the Old World genus *Talpa,* and nearly all of the published records pertain to the European mole (*Talpa europaea*). Burrow systems, which are usually inhabited by a single mole, consist of a chamber containing a nest of dry grass or leaves and galleries 5–30 cm below ground. Moles construct some nests or "fortresses" under unusually large molehills that may measure 100 cm across and 30 or more centimeters in height. Food storage, which occurs from late fall to early spring, always occurs in the fortress and adjoining galleries (Skoczen 1961).

European moles store large worms and insect larvae in the walls of galleries near the winter nest and in chambers (fig. 8.3). The moles prevent earthworms from leaving caches by eating or mutilating the two to seven anteriormost segments of the head region (Degerbøl 1927; MacDougall 1942; Skoczen 1961, 1970; Raw 1966; Funmilayo 1979). Some large worms appear undamaged, but close examination of many of these reveals minute lacerations to the anterior segments where the mole has cleanly bitten the head (Skoczen 1961). These mutilations render worms immobile, apparently because loss of or damage to anterior segments interferes with burrowing. Under the cool temperatures that obtain in soil during winter, injured worms remain fresh and otherwise healthy for months, until the moles retrieve and consume them. Worms, however, slowly regenerate damaged parts and may eventually regain full health (Evans 1948). The degree of regeneration can be used to determine the approximate time of storage (Skoczen 1970). If worms are not consumed by moles, they may disperse as the soil warms in spring (Skoczen 1961).

Storage of worms in gallery walls can be so extensive that virtually all of the galleries within a meter or more of the winter nest and the walls of the winter nest chamber itself are underlain with stored worms (Skoczen 1961). The greatest concentrations of worms often occur in the bottom walls of the galleries. When confronted with excess worms, captive moles excitedly inspected the worms and then set about digging small cavities in soil (Degerbøl 1927; Skoczen 1961). They then dragged up to ten worms, one at a time, to the prepared site, and there pushed the worms into the hole with their forepaws and snout. Finally, they pushed soil over the caches with the forepaws and chest and firmly packed the soil with their snout and forebody.

Moles also accumulate worms and larvae in a specially excavated chamber located near the winter nest (fig. 8.3). Such a chamber examined by Skoczen (1961) was 30 cm in diameter, 50 cm below the ground surface, and contained 146 earthworms. Another larder contained 160 worms and two moth larvae, the worms weighing 380 g (Skoczen 1961). One of the largest caches found by Dahl (cited in MacDougall 1942) contained 1,280 earthworms, which weighed 2.1 kg, and eighteen cockchafer grubs in a chamber and embedded in the wall of a 1.5-m segment of tunnel near the winter nest.

Worms most often found in stores of European moles are the large *Lumbricus terrestris* and *Octolasium cyaneum* and the medium-sized *Allolobophora caligniosa*. Other worms encountered in mole stores are listed in table 8.3. Insect larvae found in caches include those of cockchafers, noctuid moths, and carabid beetles. Siberian moles (*Talpa altaica*) also store worms extensively (Yudin 1972). European moles seem to include in their stores all the earthworm species that occur in the vicinity of their foraging galleries, but they do not store them in the proportions encountered (table 8.3). Rather, they cache more large and medium-sized worms (Evans 1948; Skoczen 1961; Funmilayo 1979). Small worms encountered by moles apparently are more often eaten or possibly ignored. Skoczen suggested that small worms are tastier and therefore moles eat them immediately, whereas large worms are less desirable and therefore moles defer consuming them. An alternative explanation is that large worms are no less tasty but are large loads and consequently worth the energetic cost of transport.

Unlike many other food-hoarding animals, moles do not collect food in late summer or fall to consume during winter. Moles remain active throughout the winter, foraging in galleries and feeding from or adding to food stores. Stored earthworms have been found in mole burrows from October through April, with the largest stores found in late winter (MacDougall 1942; Evans 1948; Skoczen 1961; Funmilayo 1979). Especially large accumulations of worms have been noted immediately following severe frosts (Dahl 1891; Skoczen 1961). Skoczen suggested that storing intensifies at such times because low soil temperatures cause worms to become inactive and hence easy prey for moles. As a consequence, moles are thought to have an abundance of prey during cold weather.

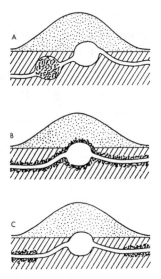

Figure 8.3
Three types of food storage distributions in the European mole. A, Worms accumulated in a chamber near the nest. B, Worms imbedded in the walls of the nest chamber and galleries leading from the nest. C, Worms imbedded in the walls of galleries at a distance from the nest. From Skoczen 1961.

Table 8.3 Number and condition of worms stored in five fortresses of the European mole (*Talpa europaea*) and the proportional occurrence of earthworm species in mole stores and in soil

Worm Species	Number			Condition		Proportion	
	Adults	Immatures	Total	Injured (%)	Dead (%)	Stores	Soil
Allolobophora caligniosa	209	61	270	81.5	0.4	83.1	45.7
Allolobophora chlorotica	1	—	1	—	—	0.3	9.5
Dendrobaena rubida	1	—	1	—	—	0.3	1.7
Eisenia rosea	2	—	2	—	—	0.6	4.2
Lumbricus castaneus	1	—	1	—	100.0	0.3	0.6
Lumbricus rubellus	5	2	7	85.7	—	2.2	32.8
Lumbricus terrestris	12	10	22	90.9	—	6.8	1.7
Octolasium cyaneum	17	4	21	95.2	—	6.1	1.7
All species	248	77	325	82.2	0.6	100.0	100.0

Source: After Funmilayo 1979.

Some workers have been skeptical of this interpretation (e. g., Raw 1966). Yudin (1972) examined the influence of earthworm storing on the overwintering of Siberian moles.

8.3 PRIMATES

Monkeys and Chimpanzees (Cebidae, Cercopithecidae, Pongidae)

Most nonhuman primates do not store food. The few species that have been observed storing food are not closely related and represent both Old World and New World forms. Captive squirrel monkeys (Cebidae: *Saimiri sciureus*) store food and "toys" (Marriott and Salzen 1979). Squirrel monkeys, which were housed three to a cage, appeared to hide food to prevent other individuals from stealing it. Preferred storage sites were those that could not be viewed from within the cage, such as opaque plastic cups outside the cage and the top of the cage. Items they hid most frequently were those that were likely to elicit stealing attempts by other monkeys. Monkeys aggressively defended food caches, and eventually they retrieved and ate the food—actions that distinguished hoarding from play behavior.

Baulu, Rossi, and Horrocks (1980) observed a subadult Barbados green monkey (Cercopithecidae: *Cercopithecus aethiops*) store two golden apples, one at a time, in a deep hole among rocks at the base of an apple tree. This occurred when the crop of golden apples, a favorite food of the monkeys, was nearly depleted. The monkey retrieved the first apple minutes later only to have it taken by a dominant group member. The hoarding monkey deferred retrieving the second apple until after the other monkeys had left the area.

Jolly (1972: 87) implies that chimpanzees (Pongidae: *Pan troglodytes*) store meat but does not provide any details.

The lack of hoarding in primates is surprising given the propensity of aboriginal and modern man (Hominidae: *Homo sapiens*) to store a diverse array of food types. Our proclivity to store food would suggest that primate ancestors may have stored food to varying degrees and that this habit would have been preserved in many extant species, however, this does not seem to have been the case. Only a few observations of food storing have been made despite thousands of hours of careful behavioral observations in the field by numerous primate biologists.

8.4 CARNIVORES

Carnivores exhibit several strategies for handling excess prey (Elgmork 1982). First, individuals of some species (e.g., wolves [*Canis lupus*] and African wild dogs [*Lycaon pictus*]) gorge to ensure themselves the largest portion of a kill. Second, some large predators (e.g., mountain lions [*Felis concolor*], tigers [*Panthera tigris*], and bears [*Ursus* spp.]) guard prey carcasses from scavengers. Third, all or part of the prey may be hidden to be eaten later. Finally, a kill simply may be abandoned, with no attempt to conserve it for future use, as occurs with cheetahs (*Acinonyx jubatus*) (Eltringham 1979; G. Frame pers. comm.).

There have been few detailed studies of food caching in the order Carnivora. Except for several fine studies on the red fox (*Vulpes vulpes*) (Kruuk 1964; Tinbergen 1965, 1972; Macdonald 1976; Henry 1986), most accounts of food caching by carnivores are anecdotal, making it necessary to piece together bits of information from various scattered sources. This paucity of information is no doubt the result of the extreme difficulty and vast individual effort required to observe predatory behavior in carnivores. Some of these difficulties have been overcome by observing free-roaming, semitame individuals (Krott 1960; Macdonald 1976; Henry 1977), by conducting extensive tracking studies (Murie 1936; Kruuk 1964; Tinbergen 1965; Nellis and Keith 1968), and by studying captive individuals (e.g., Harrington 1981, 1982).

Members of five families of terrestrial carnivores are known to cache prey (table 8.4): Canidae (foxes, wolves), Ursidae (bears), Mustelidae (weasels, mink, wolverines), Hyaenidae (hyenas), and Felidae (tigers, bobcats, leopards). Food caching is unknown in aquatic carnivores (seals, sea lions, and walruses) and four families of terrestrial carnivores: Procyonidae (raccoons, ringtails, coatis), Viverridae (civets, genets), Herpestidae (mongooses, meerkats), and Protelidae (aardwolves). Many members of these noncaching terrestrial carnivores feed on insects, fruits, and small vertebrates in warm or tropical regions. There may be no advantage for these species to cache small, perishable food items (Macdonald 1976).

Within the five families that cache food, hiding of prey follows one of several distinctive patterns (Elgmork 1982). Canids typically bury prey in shallow surface pits, whereas felids and bears never dig a hole in which to

Table 8.4 Some carnivores known to store food

Family Common Name	Scientific Name	References
Canidae		
Arctic fox	*Alopex lagopus*	Braestrup 1941; Pedersen 1966
Red fox	*Vulpes vulpes*	Kruuk 1964; Tinbergen 1965, 1972; Macdonald 1976; Henry 1986
Fennec fox	*Vulpes zerda*	Gauthier-Pitters 1962
Crab-eating fox	*Dusicyon thous*	Macdonald 1976
African wild dog	*Lycaon pictus*	Malcolm 1980
Coyote	*Canis latrans*	Sooter 1946; Harrington 1982
Wolf	*Canis lupus*	Murie 1944; Mech 1970; Magoun 1979; Harrington 1981
Golden jackal	*Canis aureus*	Wyman 1967; H. and J. van Lawick-Goodall 1970
Black-backed jackal	*Canis mesomelas*	Wyman 1967; Kruuk 1972
Ursidae		
Brown bear	*Ursus arctos*	Holzworth 1930; Mysterud 1973, 1975; Magoun 1979; Elgmork, 1982, 1983
Black bear	*Ursus americanus*	MacDonald 1965
Polar bear	*Ursus maritimus*	Elgmork 1982
Mustelidae		
Mink	*Mustela vison*	Bailey 1926; Yeager 1943; Erlinge 1969
Least weasel	*Mustela nivalis*	Criddle 1947
Long-tailed weasel	*Mustela frenata*	Hamilton 1933; Svendsen 1982
Short-tailed weasel	*Mustela erminea*	Johnsen 1969; Erlinge et al. 1974; Oksanen 1983
Polecat	*Mustela putorius*	Räber 1944
Striped polecat	*Ictonyx striatus*	Ewer 1973
African striped weasel	*Poecilogale albinucha*	Alexander and Ewer 1959; Ansell 1960
Wolverine	*Gulo gulo*	Ognev 1935; Krott 1960
River otter	*Lutra lutra*	Harper and Jenkins 1982
Badger	*Taxidea taxus*	Snead and Hendrickson 1942
European badger	*Meles meles*	Myllek 1986
Stone martin	*Martes foina*	Schmidt 1934
Pine martin	*Martes martes*	De Monte and Roeder 1987
Fisher	*Martes pennanti*	Powell 1978
Hyaenidae		
Spotted hyena	*Crocuta crocuta*	Kruuk 1972
Felidae		
Canadian lynx	*Lynx canadensis*	Nellis and Keith 1968
European lynx	*Lynx lynx*	Haglund 1966; Pulliainen and Hyypia 1975
Bobcat	*Lynx rufus*	McCord and Cardoza 1982
Mountain lion	*Felis concolor*	Wright 1934; Hornocker 1970
Tiger	*Panthera tigris*	Brander 1923; Schaller 1967
Leopard	*Panthera pardus*	Kurt and Jayasuriya 1968; Eltringham 1979; Stuart 1986

bury food but rake soil and plant litter over it. Bears differ from felids mainly in that they use much larger quantities of debris to cover prey. Mustelids cache prey in their dens and burrows or bury it in shallow surface pits like canids. Hyenas submerge prey in water.

Foxes, Wolves, and Their Relatives (Canidae)

Food caching by foxes, wolves, and other canids involves a stereotyped sequence of behaviors. An individual preparing a cache begins by digging a small hole with the forepaws just large enough to accommodate the prey item (fig. 8.4). The prey, which is carried in the jaws or regurgitated, is placed in the hole and nudged into the bottom of the hole with the nose. The material excavated from the hole is pushed over the prey with the bridge of the nose to cover the prey to a depth of several centimeters and compacted by softly pressing it with the tip of the nose. Canids rarely use their feet or mouth to place material over the cache site. Such behavior has been described for red foxes (Scott 1943; Fisher 1951; Kruuk 1964; Tinbergen 1965, 1972; Rue 1969; Macdonald 1976, 1978); fennec foxes (*Vulpes zerda*) (Gauthier-Pitters 1962); coyotes (*Canis latrans*) (Harrington 1982); golden and black-backed jackals (*Canis aureus* and *C. mesomelus*) (Wyman 1967); arctic foxes (*Alopex lagopus*) (Degerbøl and Freuchen 1935); African wild dogs (Malcolm 1980); and wolves (Murie 1944; Mech 1970; Magoun 1979). Canids make caches in loose, friable soil, sand, snow, and plant litter. Prey caught in the open are often carried to tall grass or dense vegetation to be cached (Fisher 1951; Macdonald 1976). Finished caches usually are well concealed and difficult for a human observer to find.

Figure 8.4
Red fox digging a small hole in which to bury food that it is carrying in its mouth. Reproduced with permission from J. D. Henry 1986: *Red fox: The catlike canine.* Washington, D.C.: Smithsonian Institution Press.

Occasionally canids store prey in other situations. Red foxes sometimes cache prey on the ground among dense vegetation (Tinbergen 1972). Maccarone and Montevecchi (1981) found a cache of seabirds inside an active red fox den and adjoining burrow. Elsewhere among carnivores, caching in dens and burrows has been noted only among weasels. Degerbøl and Freuchen (1935), Pedersen (1966), and others have found larders of the arctic fox in sheltered rock crevices. Wolves and red foxes sometimes cache prey under temporary beds in the snow (Mech 1967, 1970).

Canids store a large variety of prey, as might be expected from their opportunistic feeding habits. Red foxes, for example, cache rodents, cottontail rabbits (*Sylvilagus* spp.), snowshoe hares (*Lepus americanus*), weasels, raccoons (*Procyon lotor*), moles, gulls, terns, ducks, seabirds, snakes, frogs, insects, and bird eggs (Scott 1943; Sooter 1946; Fisher 1951; Kruuk 1964; Tinbergen 1965, 1972; Mech 1967; Maccarone and Montevecchi 1981). Arctic foxes cache birds, bird eggs, and rodents (Osgood, Preble, and Parker 1915; Tuck 1960; Pedersen 1966; Chesemore 1975). Wolves feed to satiation on large ungulate prey (e.g., moose, elk, deer) and then cache large chunks of meat and bones (Cowan 1947; Mech 1970). Jackals cache scraps of meat that they scavenge from animals killed by larger predators (Wyman 1967).

Seasonal changes in caching intensity have not been documented in canids (and most other carnivores). Caching of prey may occur at any time of year and seems to be more closely related to availability of prey and nutritional status of the predator than to season (Scott 1943; Macdonald 1976).

Canids, as a group, are scatter hoarders (Kruuk 1964; van Lawick-Goodall and van Lawick-Goodall 1970; Macdonald 1976; Tinbergen 1965; Magoun 1979). The typical pattern is especially well illustrated by the scattering of black-headed gull (*Larus ridibundus*) eggs by red foxes in sand dunes near Ravenglass, Scotland, as studied by Kruuk (1964) and Tinbergen (1965, 1972). Foxes carried eggs distances of 3–80 m in different directions from the nest so that the three resulting caches were widely separated (fig. 8.5). Some canids add prey to existing caches and even construct large larders. Fisher (1951) watched red foxes make several visits to cache sites. An examination of one of the caches revealed three voles (*Microtus* sp.), two house mice (*Mus musculus*), part of a garter snake (*Thamnophis* sp.), and six large grasshoppers. Maccarone and Montevecchi (1981) located several very large red fox caches in Newfoundland. One cache in an active den contained thirty Leach's storm petrels (*Oceanodroma leucorhoa*), three Atlantic puffins (*Fratercula arctica*), three black-legged kittiwakes (*Rissa tridactyla*), a common murre (*Uria aalge*), and a fox sparrow (*Passerella iliaca*). Pedersen (1966) found an arctic fox larder containing thirty-six little auks, two guillemots, four snow buntings, and numerous little auk eggs. Other arctic fox larders have been described by Gibson (1922), Degerbøl and Freuchen (1935), and Braestrup (1941).

Red foxes, wolves, and coyotes urinate to mark cache sites (Henry 1977; Harrington 1981, 1982). All three species direct urine at the cache

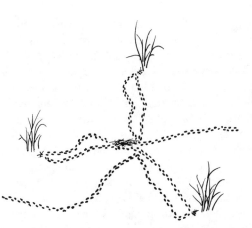

Figure 8.5
Tracks of a red fox burying three
eggs taken from a gull nest. The
fox scatters the eggs widely and
usually buries them near some vegetation. Redrawn from Kruuk 1964.

site, not the cached prey, and urine marking often occurs shortly after
food is excavated from a cache site. These canids rarely urine mark when
they prepare caches, and they seldom urine mark when they visit caches
but do not remove food (table 8.5). Juvenile red foxes do not urine mark
cache sites (Jeselnik and Brisbin 1980). Henry (1977) and Harrington
(1981, 1982) found that red foxes, wolves, and coyotes spend significantly less time investigating empty cache sites that have been urine
marked than those that have not. They interpreted these results to mean
that urine marking serves as a sign that food is no longer present, even
though food odors may still linger. However, in light of Macdonald's
(1976) findings that red foxes remember the exact locations of caches in
the wild (see Retrieval of Stored Food by Mammals, chapter 6) it is
unclear why such a "bookkeeping" system is necessary. Furthermore,
the restriction of urine marking of cache sites in coyotes and wolves to
dominant members of social units suggests that social signals also may be
communicated by marking caches.

Foxes and other canids usually recover cached meat within a day
(Osgood, Preble, and Parker 1915; Murie 1936; Scott 1943; Fisher

Table 8.5 Urine marking of cache sites by wolves, coyotes, and red foxes

Species	Natural Cache Sites Urine Marked When Created		Experimental Caches Sites Urine Marked When Found		Experimental Caches Sites Urine Marked after Recovery	
	%	N	%	N	%	N
Wolf	0.0	67	2.0	102	67.0	102
Coyote	0.0	244	13.2	53	58.9	73
Red fox	0.9	225	—	—	72.0	90

Source: Data from Henry 1977; Harrington 1981, 1982.

1951; Wyman 1967; Malcolm 1980). Short storage periods are no doubt a consequence of meat spoiling quickly. Nevertheless, there are several reports of long-term meat storage by foxes in cool climates. Beddard (1902), Gibson (1922), Pedersen (1966), and Chesemore (1975) reported that during winter, arctic foxes depended for a portion of their diet on prey they cached in cool sites during summer. Maccarone and Montevecchi (1981) implied that seabirds cached by red foxes in summer were used as a winter food supply. Kruuk (1964) and Tinbergen (1965, 1972) learned by periodically checking inconspicuously marked caches that red foxes recovered many gull eggs they had stored in spring 1–2 months later—after gulls had left the breeding area and other foods were scarce. At another site in Scotland, Frank (1979) found eggshells in red fox feces in winter and concluded that the eggs must have come from caches. If this conclusion is correct, the eggs would have been stored for approximately 6 months.

Captive canids often recover food cached by other individuals. Based on such observations, Jeselnik and Brisbin (1980) suggested that defense of an individual's caches is not an important trait of wild red foxes and that food is cached more for general use by the population than for the advantage it may confer on the caching individual. However, experiments by Macdonald (1976) on semitame red foxes in a natural setting demonstrated that a fox's caches significantly benefited only itself. Foxes sometimes defend cache sites and, if hidden food is in danger of being discovered, the fox that made the cache quickly digs it up and recaches it elsewhere.

The propensity to cache food varies considerably among canid species, and the species differences may result from other aspects of their behavior and environment (Macdonald 1976). For example, species that occupy a relatively small home range may be more likely to cache excess prey than wide-ranging species because sedentary species are likely to be near the cached prey when it is needed. Large or social predators are more capable of defending their kills and so may be more likely to defend than to hide excess prey. Thus, in general, species such as foxes, which may have difficulty defending prey and which occupy relatively small home ranges, store considerable quantities of food, whereas the large, social, far-ranging wolf seems to cache food sparingly. African wild dogs, which until recently were thought not to cache food at all (Kruuk 1972; Macdonald 1976; Andersson and Krebs 1978), cache sparingly during the breeding season, when the pack stays near a den, but they have not been observed to cache during their nomadic phase (Malcolm 1980). Clarification of the extent to which these and other ecological factors influence caching in canids awaits more rigorous quantification of caching intensity and the ecological context in which caching occurs.

Bears (Ursidae)

Bears conceal the remains of prey carcasses under large quantities of soil, vegetation, plant litter, and snow (fig. 8.6). When covering prey, bears stand on or beside the carcass facing away and rake material onto

Figure 8.6
Brown bear raking debris on top
of a moose carcass. Original drawing by Marilyn Hoff Stewart, based
on photographs and descriptions
in Elgmork 1983.

the carcass with backward movements of a forepaw (Magoun 1979; Elgmork 1982, 1983). Prey of brown bears (*Ursus arctos*) are often large (e.g., elk, moose, domestic sheep), and considerable quantities of material are used to cover the carcasses (Holzworth 1930; Semenov-Tian-Shanskii 1972; Mysterud 1973, 1975). Elgmork (1982) reported that scraped areas around carcasses in Norway averaged 43 m². The volume of debris heaped on four carcasses ranged from 0.5–1.5 m³, but Elgmork (1982) suggested that these figures may be underestimates. Polar bears (*Ursus maritimus*) cache seals on the sea ice under piles of ice and snow (Elgmork 1982).

Bears usually bury carcasses singly (Semenov-Tian-Shanskii 1972; Magoun 1979), but of sixteen caches of sheep studied by Elgmork (1982), three contained two sheep and one cache contained three sheep. The clumping of prey in these cases was probably done because sheep occur in flocks and are easy prey for brown bears. Brown bears will move large prey (e.g., moose) into dense trees before covering them (Mysterud 1973).

Magoun (1979) observed that brown bears in Alaska were most likely to cover carcasses when other scavengers were present and during rest periods. Covering of carcasses occurred most often in summer (July and August) in Norway (Elgmork 1982) and in fall in Wyoming (Craighead and Craighead 1972). Elgmork (1983) found brown bear beds and scats around caches (fig. 8.7), indicating that bears remain near cache sites, perhaps to guard them.

Weasels, Mink, Wolverines, and Their Relatives (Mustelidae)

Food caches of weasels and mink (*Mustela* spp.) consist of a collection of prey in a natural cavity, usually near a nest (Oksanen 1983). A mink (*Mustela vison*), for example, cached thirteen muskrats, two mal-

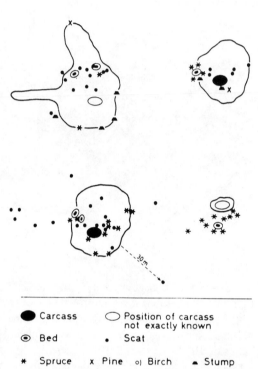

Figure 8.7
Four selected caching sites of brown bears in the Vassfaret area of Finland showing form and size of scraped areas (outlined), position of the carcasses, trees in the immediate surroundings as well as beds and scats in the vicinity. From Elgmork 1983.

● Carcass   ○ Position of carcass
                     not exactly known
◉ Bed   ● Scat

✳ Spruce   x Pine   o) Birch   ▲ Stump

lards (*Anas platyrhynchos*), and a coot (*Fulica americana*) in a hollow log that also contained its den (Yeager 1943). Yeager found other mink caches under rocks, logs, and overhanging banks. In riparian habitats in southern Sweden, minks stored crayfish, water beetles, and waterfowl (Erlinge 1969). Least weasels (*M. nivalis*) often appropriate rodent nests and store prey in the nests, nearby chambers, or adjoining tunnels (Criddle 1947). A long-tailed weasel (*M. frenata*) cached nine young golden-mantled ground squirrels in an unoccupied pocket gopher burrow 175 m from its den (Svendsen 1982). The weasel cached food at this site on three occasions over a 2-week period. Captive African striped weasels (*Poecilogale albinucha*) stockpiled mice and chunks of meat in their nest boxes (Alexander and Ewer 1959; Ansell 1960). Some weasels make very large larders. For example, Johnsen (1969, cited in Oksanen 1983) found 153 lemmings and a shrew in a cache made by short-tailed weasels (*M. erminea*).

Other mustelids usually bury single food items under a layer of soil or snow. When burying food, a captive male river otter (*Lutra lutra*) excavated a hole in soft sand, placed food items in the hole, and pushed sand over the hole with its nose (Harper and Jenkins 1982). The otter sometimes used its forepaws to flatten and compress the sand. It placed other food items in crevices, apparently exposed to view. Caching has not been described for wild otters. Badgers (*Taxidea taxus*) bury prey 15–25 cm deep in loose soil around the entrance to their burrows (Snead and Hendrickson 1942). Wolverines (*Gulo gulo*) in Sweden bury prey remains in deep holes in snow or soft ground or carry prey up into trees and place them in the angles formed by branches and trunk (Krott 1960). In northern Siberia, wolverines cache arctic foxes and ptarmigans in the snow (Ognev 1935). Unlike weasel caches, those of badger, otter, and wolverines often contain only one prey item.

Seasonal changes in caching intensity have not been well documented. Snead and Hendrickson (1942) found that badgers in Iowa stored food from mid-April to mid-May. Yeager (1943) stated that mink cache more commonly in winter.

Hyenas (Hyaenidae)

Hyenas studied by Kruuk (1972) in Ngorongoro Crater and the Serengeti submerged large chunks of meat or bones in 30–50 cm of still, muddy water. Hyenas usually returned to retrieve the cached food within a day, and in the hot climate of East Africa, storage is not likely to be effective much longer. Kruuk observed hyenas, which apparently remember the general location of cache sites using visual landmarks along the shore, recover carrion by repeatedly plunging their heads deep into water around cache sites. Some attempts to find submerged carrion were unsuccessful, and when the water holes dried up and lake margins receded, many bones were exposed, some probably remnants of hyena caches. Kruuk suggested that carrion cached underwater is safe from the hordes of terrestrial scavengers that search for meat by sight and smell.

Tigers, Bobcats, Leopards, and Their Relatives (Felidae)

Several species of predatory cats cover the remains of prey with soil, vegetation, snow, or other materials. Mountain lions (*Felis concolor*) cover the carcasses of elk (*Cervis canadensis*), mule deer (*Odocoileus hemionus*), and livestock with plant litter, snow, and fur of the prey (Wright 1934; Hibben 1937; Young and Goldman 1946; Robinette, Gashwiler, and Morris 1959; Hornocker 1970; Stuart 1986). Tigers (*Panthera tigris*) often cover carcasses with leaves and grass before they leave the kill (Brander 1923; Schaller 1967). Bobcats (*Lynx rufus*) and lynx (*Lynx canadensis*) cover prey with snow, leaves, and fur from the kill (fig. 8.8) (McCord and Cardoza 1982; Nellis and Keith 1968). Lions (*Panthera leo*) cover parts or all of a kill with soil and vegetation (Schaller 1972), but Schaller did not interpret this behavior as caching.

Covering behavior is similar in all species of felids known to cover prey. The predator faces away from the remains of a kill and rakes whatever material is available in the direction of the kill with backward movements of a forepaw (fig. 8.9). Covering behavior of young, captive-born mountain lions is identical to that of adults (Baudy 1976; Bogue and Ferrari 1976). Tigers sometimes bite grass off and cover the prey with it (Brander 1923; Schaller 1967). Felids have not been reported to dig holes in which to bury prey, as is typically done by canids, but they may use natural depressions as cache sites. Schaller (1967), for example, found

Figure 8.8
White-tailed deer covered with snow and fur by a bobcat. Photograph courtesy C. M. McCord, Division of Fisheries and Wildlife, Massachusetts.

the remains of a cow hidden in a narrow erosion gully overgrown with grasses. Felids often carry or drag prey heavier than themselves into shelter of dense vegetation before feeding on or covering them (e.g., Young and Goldman 1946).

Covering prey is generally thought to be an attempt by predators to conceal the carcass from scavengers. Covering also may function to keep prey cool and thereby retard spoilage.

Felids usually store prey items individually. Wright (1934) described an exception to this pattern. He found a mountain lion guarding two mule deer fawns that it had buried together in leaf litter. The fawns were of different ages, indicating that they were not siblings. Other evidence sug-

Figure 8.9
A tigress covering the remains of her kill by sweeping earth and grass over it with a forepaw. Photograph courtesy George Schaller.

Figure 8.10
Remains of an impala (*Aepyceros melampus*) that had been carried into a tree by a panther. Photograph courtesy James A. MacMahon.

gested that the lion made two separate kills and cached them together, perhaps to guard them more effectively.

Leopards (*Panthera pardus*) drag prey into the upper branches of trees for storage (Kurt and Jayasuriya 1968; Guggisberg 1975; but see Stuart 1986), a storing behavior quite different than that of other felids (fig. 8.10). Eltringham (1979) and Eisenberg and Lockhard (1972) suggested that this behavior results from the leopards' inability to protect kills from the hordes of scavengers that might quickly find carcasses stored on the ground.

Nellis and Keith (1968) found that caching of hares (*Lepus americanus*) by lynx was related to hunting success. In years when hares were easily caught, twelve caches were found along 194.8 km of lynx trail. Six caches contained fresh kills, and six others contained old kills relocated and eaten by the lynx. In a year when lynx hunting success was low, no hares were found cached along 225.4 km of lynx trail.

8.5 RODENTS

Mountain Beaver (Aplodontidae)

The mountain beaver (*Aplodontia rufa*) is the sole member of the Aplodontidae, the most primitive living family of rodents. It is restricted to humid forests of the Pacific Northwest from southwestern British Columbia to central California. Mountain beavers feed on forbs, ferns, and bark of woody plants and live in extensive burrows, often in moist soils along streams.

During nocturnal feeding excursions, mountain beavers clip vegetation and stack it in small piles on logs, on rocks, under logs, and on low branches of shrubs and trees around burrow openings (Taylor and Shaw 1927; Scheffer 1929; Voth 1968). Mountain beavers harvest and store a wide variety of plant species (Camp 1918; Hubbard 1922; Scheffer 1929; Voth 1968), but they often cache western sword fern (*Polystichum munitum*) and bracken fern (*Pteridium aquilinum*), plants that most herbivores will not eat. Some piles are composed of cuttings of a single plant species.

Mountain beavers pile vegetation throughout the year, but most harvesting activity occurs in summer, when green vegetation is readily available (Voth 1968). Composition of the caches changes with the seasons, apparently reflecting changing availability. For example, Scheffer (1929) observed that mountain beaver caches near Puget Sound contained few evergreen plants in summer, but evergreens predominated in caches during winter.

The fate of these piles of vegetation has been the subject of much speculation. Many early natural historians believed that mountain beavers dried vegetation in the "hay piles" and later stored the hay in special chambers as a winter food supply (Taylor and Shaw 1927; Morrison 1946). Others maintained that all or a portion of the dried vegetation was intended not as winter food but as bedding for the nest (Camp 1918; Scheffer 1929). It is now evident that mountain beavers move the piles of wilted vegetation into chambers near the nest when they still contain considerable moisture (Camp 1918; Cahalane 1947; Voth 1968; Martin 1971). They plug these chambers with hard balls of dirt and rocks. The only dried vegetation to be found in mountain beaver burrows is the nest lining, which is not derived from vegetation caches but from naturally dried vegetation collected from the forest floor (Voth 1968). When full, each chamber holds about 2.2–4.4 l of wilted vegetation, as well as roots of ferns and other plants (Camp 1918; Cahalane 1947). Mountain beavers feed extensively in these food-storage chambers during all seasons (Camp 1918; Voth 1968). Because of high moisture content of food stores and high relative humidity of burrows, stored vegetation molds quickly and has to be replaced within a week or two (Scheffer 1929; Cahalane 1947; Voth 1968). Decayed vegetation may be taken to refuse chambers or dumped outside a burrow entrance.

Chipmunks and Squirrels (Sciuridae)

Sciurid rodents are among the most familiar food hoarders. For ease of coverage, I divide this large and diverse family into five groups: (1) chipmunks, (2) red squirrels and Douglas squirrels, (3) tree squirrels, (4) ground squirrels, and (5) flying squirrels. Members of each of these groups store different types of food in distinctly different ways.

Chipmunks. Chipmunks (*Tamias*) are common inhabitants of forests and shrublands throughout the Holarctic region; twenty-three species are recognized in North America, and one species occurs in eastern Europe and northern Asia. Chipmunks forage on the ground and in low shrubs and trees for seeds, underground plant parts, fungi, and invertebrates. Food hoarding has been observed in many species, and all species of chipmunk probably store food extensively. Food is transported in spacious cheek pouches (see fig. 3.3).

Chipmunks store food in larders and scattered surface caches. Larder hoarding seems to be the predominant mode of food storage, but several recent studies suggest that scatter hoarding may be more common than previously suspected. For example, Kawamichi (1980) reported that 49% of 1,596 hoarding events of Siberian chipmunks (*T. sibiricus*) that he observed were at scattered sites.

Larders consist of nuts, seeds, bulbs, and corms. Chipmunks establish larders in one or more chambers near their nest of dry vegetation in the ground or in a hollow tree (fig. 8.11). Usually only one chipmunk inhabits each burrow. Storage chambers typically measure 10–35 cm in diameter and 10–15 cm in height, although the size and number of chambers varies among species and geographically within species (Walker 1923; Panuska and Wade 1956; Broadbooks 1958; Thomas 1974; Elliott 1978; Kawamichi 1980). A large cache of seeds also may be made in the lining of the nest (fig. 8.12) (Criddle 1943; Engels 1947; Broadbooks 1958; Elliott 1978). Small quantities of food are sometimes stored in auxiliary refuge burrows and in cavities in hollow trees (Thomas 1974; Kawamichi 1980).

Cache composition varies greatly among habitats, chipmunk species, and individuals, reflecting local preferences and availability. Eastern chipmunks (*T. striatus*) in the Adirondack Mountains, for example, cached primarily beech (*Fagus grandifolia*) nuts, maple (*Acer saccharum, A. pensylvanicum,* and *A. rubrum*) samaras, and trout lily (*Erythronium americanum*) bulbs (Elliott 1978). In contrast, eastern chipmunks in Louisiana stored large numbers of water oak acorns (*Q. nigra*) and bitternut hickory (*Carya cordiformis*) nuts (Thomas 1974). Similarly, Siberian chipmunks

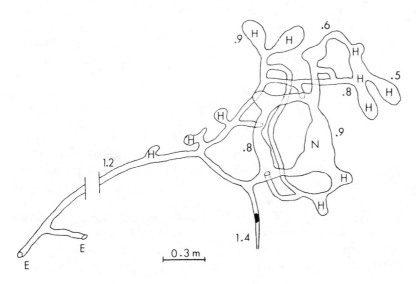

Figure 8.11
Floor plan (viewed from above) of an eastern chipmunk burrow showing food storage chambers (H), a nest (N), and entrances (E). Numbers indicate the depth of the burrow in feet. The dark spot is a rock blocking a deep passage. Redrawn from Elliott 1978.

Figure 8.12
A side view of a yellow pine chipmunk nest showing storage of seeds
in the nest lining. From Broadbooks 1958.

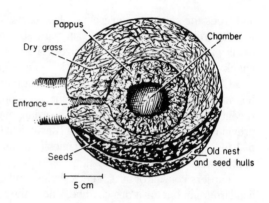

in western Sayan, USSR, stored Siberian stone pine seeds (Stilmark
1963), whereas in Hokkaido, Japan, they stored primarily acorns (*Q.
mongolica* and *Q. dentata*) (Kawamichi 1980). Broadbooks (1958) found
considerable variation in composition of yellow pine chipmunk (*T. amoenus*) larders.

The total food reserves in a single burrow may be substantial. The
largest cache in fifty eastern chipmunk burrows excavated by Thomas
(1974) in Louisiana contained 392 water oak acorns and two bitternut
hickory nuts and weighed a total of 1,073 g (wet weight). This was a late
winter cache (19 February); the same cache in the fall was probably
larger. Bachman and Audubon (1849, cited in Howell 1929) described an
eastern chipmunk burrow that contained a gill (118 ml) of wheat and
buckwheat in the nest chamber and a quart (1.1 l) of hazelnuts (*Corylus
rostrata*), a peck (8.8 l) of acorns, and 2 quarts (2.2 l) of buckwheat in
other chambers. Some other quantitative measures of chipmunk food
stores are given by Howell (1929), Allen (1938), Criddle (1943), Broadbooks (1958), and Elliott (1978) (see Section 2.3).

Scatter hoarding has been observed in eastern, least (*T. minimus*),
Siberian, red-tailed (*T. ruficaudus*), alpine (*T. alpinus*), yellow pine,
lodgepole (*T. speciosus*), and cliff (*T. dorsalis*) chipmunks (Swarth 1919;
Grinnell and Storer 1924; Criddle 1943; Hart 1971; Lockner 1972; Yahner
1975; States 1976; Elliott 1978; Kawamichi 1980; Kobayashi and Watanabe 1980; Shaffer 1980). A chipmunk prepares a surface cache by digging
a hole about 2–5 cm deep and 2–3 cm in diameter. The contents of the
cheek pouches are expelled into the hole and nudged with the nose firmly
into the bottom of the excavation (fig. 8.13). Next, the animal rakes soil
into the hole with a forepaw or pushes soil forward into the hole with both
forepaws and pats the surface of the cache with both forepaws simultaneously. Finally, the chipmunk camouflages the cache site by placing plant
litter over it. The finished cache is difficult for a human observer to find.
Scattered caches contain the same food types as do larders but sometimes also include perishable buds and insects (e.g., Kawamichi 1980).

Surface caches can be scattered throughout the home range (Hart
1971; Yahner 1975; Elliott 1978; Kawamichi 1980), but some eastern
chipmunks concentrate caches near the burrow entrance, an area they

aggressively defended (Shaffer 1980). Cliff chipmunks make surface caches between foraging sites and nest burrows (Hart 1971). Kawamichi (1980) concluded that the direction and distance between foraging sites and cache sites were random and that hoarding sites of Siberian chipmunks overlapped broadly with those of other individuals.

The longevity of surface caches made by eastern chipmunks appears to be relatively short. Chipmunks recovered twelve of sixteen caches monitored by Elliott (1978) within 2 days. Elliott (1978) and Shaffer (1980) observed eastern chipmunks retrieve numerous caches and carry the food into their nest burrow during summer. Kawamichi (1980) found that Siberian chipmunks retrieved a portion of their caches in fall and transported the contents to the nest burrow; they did not recover other caches until the following spring.

The intensity of chipmunk food hoarding changes over the seasons. Kawamichi (1980) reported a bimodal distribution of larder-hoarding activity in the Siberian chipmunk (fig. 8.14). A small peak in hoarding activity occurred in May comprised primarily of "prerest" hoarding (a single hoarding event as the chipmunk enters the nest for the evening). A second, larger peak in August through November occurred as chipmunks laid away a large food supply to sustain themselves during winter hibernation. The temporal distribution of scatter hoarding is similar to that of larder hoarding except that there is no July peak. Eastern chipmunks scatter hoard extensively in summer, but they store most food in the nest burrows during fall (Yahner 1975; Elliott 1978). This contrasts with the behavior of Siberian chipmunks (Kawamichi 1980). Cliff chipmunks scatter hoarded seeds from June through October in northern Utah (Hart 1971). Yellow pine chipmunks in eastern Washington stored food in the nest burrow during a brief period from late October to mid-November; no scatter hoarding was noted (Broadbooks 1958).

The burrow larders are critical for winter survival, as they are the

Figure 8.13
A yellow pine chipmunk preparing a cache of antelope bitterbrush seeds carried in its cheek pouches. Original drawing by Marilyn Hoff Stewart.

Figure 8.14

Seasonal change in hoarding activity of Siberian chipmunks in forests in Hokkaido, Japan. From 45–647 hours of observations contributed to each bimonthly sample period. Redrawn from Kawamichi 1980.

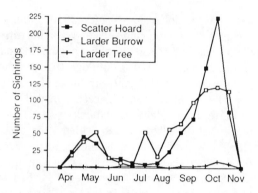

sole source of energy for chipmunks when they periodically arouse from hibernation during the winter (Panuska 1959; Cade 1963; Jameson 1964; Jaeger 1969; Brenner and Lyle 1975; States 1976; Wrazen and Wrazen 1982). Fall food storage also may have important consequence for the spring activities of chipmunks. Kawamichi (1980), for example, found that when the acorn crop failed, resulting in a small fall seed harvest, spring reproduction (measured as number of juveniles trapped) decreased (Kawamichi 1980). Also, stored food is used during late summer lulls in above-ground activity (Wrazen 1980) and during inclement weather.

The functional significance of scatter hoards is unclear. Hart (1971) suggested that cliff chipmunks scatter hoarded to increase foraging efficiency by reducing travel time during periods of profitable foraging. Scattered caches that persist through winter may be an important food source in spring (Kawamichi 1980). Yahner (1975) suggested that scatter hoarding in *Tamias* is a vestigial fixed-action pattern—a nonadaptive rudiment of ancestral scatter hoarding behavior. There is, however, no support for the assumption that the ancestral form of food storage in the group was scatter hoarding. Shaffer (1980) observed that scattered caches were quickly retrieved and carried into the nest burrow by their owners after pilfering of den storerooms by neighboring chipmunks. He concluded that scatter hoards served as an efficient means of replacing food stolen from the larder. This may be true but it seems unlikely that this is the primary functional significance of scatter hoarding. Other hypotheses, such as that scatter hoarding "cures" seeds, as may be the case for heteromyid rodents (e.g., Shaw 1934), or that it prevents infestation of seeds by insects, which may infect seeds stored in dens (Elliott 1978), or that it increases foraging efficiency (i.e., Hart 1971) need to be tested.

Red Squirrels and Douglas Squirrels. Red squirrels (*Tamiasciurus hudsonicus*), common inhabitants of coniferous forests throughout much of North America, and the closely related Douglas squirrel (*T. douglasii*) of the Pacific Northwest are among the most thoroughly studied of food-hoarding mammals. Largely because of their conspicuous and accessible middens in which they store large quantities of conifer cones, red and Douglas squirrels are favorite subjects for studies of territoriality, foraging behavior, and energetics (e.g., C. C. Smith 1968, 1981; M. C. Smith

1968; Finley 1969; Kemp and Keith 1970; Rusch and Reeder 1978; Vahle and Patton 1983; Gurnell 1984; Hurly and Robertson 1987). Further, foresters and nurserymen have taken great interest in these squirrels because their middens are sources of commercially valuable seeds (e.g., Korstian and Baker 1925; Baldwin 1942; Lavender and Engstrom 1956).

Middens are deposits of cone debris accumulated beneath traditional feeding perches where squirrels dismantle cones as they search for seeds. The squirrels frequently dig in these deposits, incorporating fresh cone bracts and cores with partly decomposed cone debris deposited many years earlier. This squirrel activity results in a thick, loose mat of litter, denuded of vegetative cover (fig. 8.15). Well-established middens typically are irregular-shaped areas 7–10 m in diameter and 40 cm or more thick (Hatt 1943; Finley 1969; Hutchins and Lanner 1982; Gurnell 1984; Patton and Vahle 1986). Middens usually are located beneath one or more large conifers that shade the midden and shelter the squirrel (e.g., Vahle and Patton 1983). One or more nests constructed of grass and foliage are usually located on sheltered tree branches or in hollow trees within 5–10 m of the midden (Yeager 1937; Hatt 1943; Patton and Vahle 1986). The midden and nest form the central point of the territory, which except for a brief period during the breeding season, is occupied by a single adult individual (C. C. Smith 1968).

The midden is of value to the squirrel because of its moisture-retaining properties (Shaw 1936; Finley 1969). The midden absorbs moisture during spring runoff and summer rains. During dry periods, the upper few centimeters of the midden may dry out, but deeper, where squirrels store most cones, the litter remains cool and moist. As long as

Figure 8.15
A midden of the red squirrel in a lodgepole pine forests in Colorado. Photograph courtesy Robert Finley, Jr.

cones remain moist, they will not open, and the seeds within will remain viable for years (Shaw 1936; Finley 1969). To increase moisture levels, red squirrels frequently establish middens in boggy areas or springs, where a constant supply of moisture is ensured.

Red squirrels begin collecting and storing cones in late July or early August and continue for 4–8 weeks coincident with ripening and depletion or opening of cones. Red squirrels usually collect cones by scurrying through the tops of trees from one branch tip to another, cutting individual cones or branchlets bearing clusters of cones and letting them fall to the ground (Shaw 1936; Shellhammer 1966; Smith 1981). Later, red squirrels descend to collect the cones and, after cutting off any branchlets, carry the cones one at a time to the midden. Douglas squirrels may gnaw off the distal portions of the scales of large cones, such as those of white fir (*Abies concolor*) and ponderosa pine (*P. ponderosa*), to make them light enough to be carried (Smith 1981). Clarke (1939) estimated that red squirrels can collect and store about 1,000 red pine (*P. resinosa*) cones in a day, but they clip and store the large cones of limber pine (*P. flexilis*) at rates of only about twenty-nine to thirty-two cones per day (Benkman, Balda, and Smith 1984).

Squirrels usually bury cones in groups of thirty or more in large pits up to 20 cm deep dug into the soft cone debris (Hatt 1943; Kendall 1980). Alternatively, red squirrels may simply pile cones on the surface of the midden or around the bases of trees (Dice 1921; Yeager 1937; Finley 1969; Kendall 1980; Hurly and Robertson 1987). Middens usually contain 2–4 bushels (70–141 l) of cones, but instances of 8–15 bushels (282–528 l) being taken by cone collectors have been reported (Cox 1911; Korstian and Baker 1925; Yeager 1937; Baldwin 1942). M. C. Smith (1968) estimated that red squirrels stored between 12,000 and 16,000 white spruce (*Picea glauca*) cones in a 6-week harvest period in interior Alaska. Gurnell (1984) found a mean of 2,187 lodgepole pine cones in nine middens (range = 280–4,360 cones per midden).

Red and Douglas squirrels cache numerous types of conifer cones, including pine, spruce, fir, Douglas-fir (*Pseudotsuga menziesii*), hemlock, and sequoia (*Sequoiadendron giganteum*). Finley (1969), C. C. Smith (1968), and others discuss the preferences of red squirrels for the cones of certain conifer species, but because most species of conifers do not produce large cone crops each year, types of cones cached in a given year depend largely on availability. Energy content of the seeds and toughness of the cones are important determinants of cone preference.

Red squirrels also may store cones in other situations. In Colorado, Finley (1969) found accumulations of cones in pools of streams 5–15 cm deep and attributed them to red squirrels. In these wet sites, the cones stay tightly closed; however, red squirrels have not been observed making these submerged caches, and behavioral studies are needed to determine whether squirrels are capable of retrieving cones from under water. Clarke (1939) observed red squirrels store red pine cones in branch forks and crotches of trees. Gurnell (1984) found 500 lodgepole pine cones in a hollow log. During the fall of 1973, I observed a red squirrel in the San Francisco Peaks, Arizona, that had filled a large, hollow tree with limber pine cones. Many of these cones had opened, and nuthatches visited the tree frequently to take seeds from the squirrel's larder.

Besides cones, red squirrels store various nuts, seeds, fruits, and some meat. Nuts stored include walnuts, hickory nuts (Layne 1954), chestnuts (Audubon and Bachman 1846, cited by Hatt 1929), beechnuts (Klugh 1927), and hazelnuts (Mailliard 1931). The squirrels hide nuts at scattered sites within their territories, including cavities in hollow trees or logs, crevices in bark, or buried in the soil (Audubon and Bachman 1849; Klugh 1927). Burton (1930) attributed to red squirrels accumulations of boxelder (*Acer negundo*) samaras found at the bases of trees, in hollow trunks, and in the crotches of tree branches. One of these caches contained more than a bushel (35 l) of seeds. Red squirrels occasionally store fruit, including cranberries (*Viburnum pauciflorum*), apples, hawthorn fruits, kinnikinnick (*Arctostaphylos* sp.) berries, wild cherries (*Prunus pensylvanica*), and currants (*Ribes cereum*) (Murie 1927; Layne 1954; Finley 1969). Red squirrels also may cache nestling birds in forks of branches or in bark crevices (Klugh 1927; Hatt 1929).

When scatter hoarding food items in the soil, a red squirrel digs a small hole with its forepaws, drops the nut into the depression, presses the nut into the hole with its nose, and then covers it by raking soil and litter into the hole with its forepaws (Klugh 1927). Squirrels cache nuts singly or in small clusters. Cache sites are well camouflaged and difficult for a human observer to find. Kendall (1980) located 201 whitebark pine seed caches on or near middens in Yellowstone National Park, Wyoming. These contained a mean of 17.3 seeds (range = 3–176 seeds). In pine plantations in Ontario, Canada, red squirrels scatter hoard cones (Hurly and Robertson 1987).

Bones and antlers are frequently encountered on middens, and it seems that these items are actively collected and hoarded by red squirrels (Hatt 1943). Gnawing bones may satisfy requirements for calcium or phosphorus not met by eating conifer seeds and fungi.

Mushrooms and other fungi are important constituents of red squirrel diets (McKeever 1964) and are extensively collected and stored (Buller 1920; Cram 1924; G. A. Hardy 1949). Squirrels place mushrooms on logs or carry them into trees and shrubs, where they lodge them in crevices, forks of branches, or in abandoned bird nests. Later, when mushrooms are dry, squirrels collect and recache some of them in hollow tree trunks. A cache found by G. A. Hardy (1949) in a hollow Douglas-fir stump in British Columbia, Canada, contained fifty-nine specimens belonging to fourteen species and occupied a volume of about 1 gallon (4.4 l). Most of the specimens were *Hymenogaster* and *Russula*. Buller (1920), Hatt (1929), and G. A. Hardy (1949) list the species of fungi that red squirrels are known to store (and presumably eat), as well as some that they apparently will not eat.

The stored food supply forms an important supplement to the squirrel's winter diet of shoots, cambium, and lichens. Without the cones stored in the midden, it is doubtful whether red squirrels could survive the winter in much of their range.

Tree Squirrels. Over 100 species of tree squirrels inhabit forests and woodlands throughout the world, except Australia and Madagascar. Many species of tree squirrels, like ground squirrels, forage on the ground, but they nest in trees and, if startled, seek shelter in trees. They are excellent climbers and often move through forests on arboreal travel lanes. The largest and best known genus of tree squirrels, *Sciurus,* is common and conspicuous in north-temperate forests. Food-hoarding behavior of most tropical tree squirrels, including numerous species of *Sciurus,* is poorly known.

The predominant mode of storage by tree squirrels is the scattering of single nuts in shallow subterranean caches (Dennis 1930; Cahalane 1942; Eibl-Eibesfeldt 1951; Moore 1957; MacClintock 1970; Stapanian and Smith 1978; Thompson and Thompson 1980; Moller 1982). Gray (*S. carolinensis*), fox (*S. niger*), and Eurasian red (*S. vulgaris*) squirrels make caches by first carrying an edible nut away from the source tree to a suitable burial site. With the nut still in its mouth, the squirrel quickly digs with its forepaws a hole slightly larger than the nut. The squirrel next lowers the nut into the hole and, with several rapid thrusts of the body, forces the nut tightly into the hole. After releasing the nut, the squirrel may strike it with several sharp blows with the upper incisors to seat it in the hole. It then pushes dirt into the hole with its forepaws. Nuts may be buried so that the top of the nut is level with the ground surface (Cahalane 1942; Thompson and Thompson 1980; Stiles and Dobi 1987) or as much as 8 cm deep (Sviridenko 1971). Finally, the squirrel scratches litter over the cache site.

Nuts with thick, woody seed coats resistant to decay are most often stored by tree squirrels. Gray and fox squirrels commonly cache black walnuts (*Juglans nigra*), various types of hickory nuts and pecans (*Carya* spp.), horse chestnuts (*Aesculus hippocastanum*), beechnuts (*Fagus* spp.), chestnuts (*Castanea dentata*), and acorns (*Quercus* spp.) (Allen 1943; Brown and Yeager 1945; Richards 1958; Stapanian and Smith 1978; Thompson and Thompson 1980). Tassel-eared squirrels (*S. aberti*) on the Kaibab Plateau, Arizona, which store relatively little food compared to gray and fox squirrels, bury piñon pine (*P. edulis*) seeds and whole conifer cones (Hoffmeister 1971; Hall 1981). Japanese squirrels (*S. lis*) also store whole cones (Kato 1985). Eurasian red squirrels store walnuts, hazelnuts (*Corylus avellana*), conifer cones, and acorns (Sviridenko 1971; Moller 1982; Tonkin 1983). African palm squirrels (*Epixerus ebii*) cache recently fallen *Panda oleosa* nuts with green husks but eat those nuts that are old with the husk rotted away (Emmons 1980). Squirrels can crack old nuts more easily than fresh nuts, so burial may facilitate opening the nuts. Red-tailed squirrels (*S. granatensis*) bury the woody nuts of *Scheelea* and *Astrocaryum* palms (Heaney and Thorington 1978; Glanz et al. 1982).

Squirrels cache nuts throughout their home ranges. Among temperate zone *Sciurus,* cache areas are not defended; two or more individuals may cache in the same area without conflict (Sharp 1959; C. C. Smith 1968; Stapanian and Smith 1978). Nuts are carried away from source trees, where they are usually abundant, to interspaces between nut-bearing trees. The more uniform spacing of nuts reduces nut loss to cache robbers (see section 4.2). Mean transport distances usually range from 10–30 m (Heaney and Thorington 1978; Emmons 1980; Kato 1985; Stiles and Dobi 1987), but when nuts are concentrated at one site the distances that squirrels transport nuts may be much greater (see fig. 4.10) (Stapanian and Smith 1978; Stiles and Dobi 1987). The even dispersion of caches suggests that squirrels avoid previous cache sites. Tree squirrels often cache nuts in the open, but red-tailed squirrels cache most often near objects such as the base of trees (Heaney and Thorington 1978).

Some tree squirrels scatter hoard nuts in trees, as well as in the ground. The African striped squirrel (*Funisciurus anerythrus*), for example, wedges *Raphia* palm fruits between leaflets of palm fronds, in branch forks, and among bracts of epiphytes (Emmons 1980). Fruits are usually cached in isolated trees away from arboreal travel lanes. Emmons found mean height of six cached fruits to be 7 m. Red-tailed squirrels on Barro Colorado Island place nuts and fruits, such as *Gustavia superba* and *Inga* sp., in branch forks and on top of lianas, 1–12 m high (Heaney and Thorington 1978). These arboreally cached nuts and fruits are not always well concealed. Japanese squirrels observed by Kato (1985) cached thirty-two of eighty-two items (mostly walnuts and pine cones) in trees. Prevost's squirrel (*Callosciurus prevosti*) in Malaysia wedges ripe fruit into cracks of branches (Becker, Leighton, and Payne 1985).

Occasionally, temperate zone *Sciurus* larder hoard numerous food items in tree cavities or scatter items in foliage. Raspopov and Isakov (1935, cited in Moller 1983) found two cavities containing 650 and 214 acorns stored by Eurasian red squirrels. Gray squirrels stored about 4.4 l of sweet gum (*Liquidambar styraciflua*) capsules in a hollow at the base of a tree (Brown and Yeager 1945). Gray squirrels in Wisconsin hid butternuts (*Juglans cinerea*) in the uppermost branches of eastern white (*P. strobus*) and red (*P. resinosa*) pine saplings, 2–4 m above the ground (Habeck 1960). The squirrels recovered many of these scattered nuts during early winter.

Eight of nine species of African rain forest squirrels studied by Emmons (1980) in Gabon were observed storing food. These were the African palm squirrel, sun squirrel (*Heliosciurus rufobrachium*), bush squirrel (*Paraxerus poensis*), oil-palm squirrel (*Protoxerus stangeri*), and four species of striped squirrel (*Funisciurus isabella, F. pyrrhopus, F. anerythrus,* and *F. lemniscatus*). The sun squirrel was observed storing only prized food items such as dead birds. Emmons (1980) did not observe Pygmy squirrels (*Myosciurus pumilio*) caching food.

The number of nuts stored by individual squirrels has not been well documented. There are numerous statements about squirrels storing most of the available nut crop (e.g., Allen 1943; Thompson and Thompson 1980). Acorn production has been estimated at from zero to hundreds of thousands per hectare (e.g., Ovington and Murray 1964). This suggests that each squirrel must store in excess of several thousand nuts during mast years; however, no one, to my knowledge, has ever estimated the quantities of nuts squirrels store or what proportion of a squirrel's metabolic requirements for the winter months is derived from stored food.

The habit of storing mushrooms is well developed in tassel-eared squirrels but does not seem to have been reported in gray and fox squirrels. Tassel-eared squirrels hang *Armillariella mellea, Amanita vaginata,* and *Russula* spp. in forks of tree branches (Hall 1981) and rely heavily on them during fall. Eurasian red squirrels and Japanese squirrels also store mushrooms in clefts of branches (Moller 1983; Kato 1985; Sulkava, and Nyholm 1987). Eurasian red squirrels in northern Finland store most mushrooms at heights of 1–2 m and at mean densities of 442 mushrooms per hectare (Sulkava and Nyholm 1987). These mushrooms had a mean dry mass of 0.9 g (range 0.1–5.0 g). Fungal tissue has a caloric density similar to that of fruit and structural plant tissue but lower than that of seeds (Grönwall and Pehrson 1984).

Bones and antlers are occasionally collected and stored in cavities of trees (Allan 1935; Moller 1982, 1983).

Food caching by temperate zone tree squirrels usually occurs in late summer and fall, when nuts mature (Brown and Yeager 1945; Hall 1981; Tonkin 1983; Kato 1985). At a cemetery in Toronto, Canada, gray squirrels exhibited a bimodal pattern of hoarding; the early peak in caching occurred in August, corresponding to ripening of oak and hickory nuts, and the second peak in late September and early October, coinciding with maturation of horse chestnuts (fig. 8.16) (Thompson and Thompson 1980). Gray squirrels store food infrequently during other seasons (e.g., Cahalane 1930).

Factors affecting seasonal changes in intensity of food hoarding in tropical tree squirrels are poorly understood. Red-tailed squirrels both cache and recover nuts and fruits during the wet season (May to August)

Figure 8.16
Seasonal index of food caching (calculated as the percentage of observation periods in which caching of that species occurred) of the gray squirrel in Mount Pleasant Cemetery, Toronto, Canada. Redrawn from Thompson and Thompson 1980.

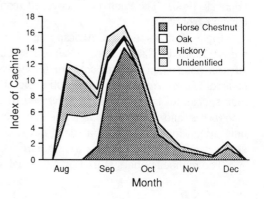

and primarily recover cached food items during the late wet season and dry season (September to January) (Heaney and Thorington 1978; Glanz et al. 1982). Most nuts are produced in the late dry and early wet seasons, and the late wet season and early dry season are periods of relative food scarcity for nut eaters. Tropical squirrels usually recover fleshy fruit within a few days (e.g., Emmons 1980).

Tree squirrels do not hibernate and so must forage frequently during winter for cached nuts and fungi. Availability of other foods at this time is at an annual low. Gray squirrels in Toronto during the winter of 1974–75 began recovering some stored nuts during October and November (see fig. 2.7), when many nuts still littered the ground and some caching was occurring. Squirrels relied on cached nuts heavily from December through February, and by March and April, squirrels' use of cached nuts had diminished greatly, but cached nuts were still included in the diet. In northern Finland, stored mushrooms comprise 11% of Eurasian red squirrels' diets in winter (Sulkava and Nyholm 1987).

Ground Squirrels. Most temperate and arctic zone ground squirrels deposit large amounts of body fat in late summer and use this fat as a metabolic reserve during winter hibernation. Some species (e.g., Belding's ground squirrel, *Spermophilus beldingi*) store little or no food (Morton, Maxwell, and Wade 1974). A number of species, however, adopt a mixed strategy, storing small quantities of food in addition to depositing fat.

Seeds, nuts, dry vegetation, fungi, or fruit may be stored, depending on squirrel species and locality (Merriam 1910; Grinnell and Dixon 1918; Shaw 1925; Howell 1938; Criddle 1939; Mayer 1953; Krog 1954). Alaska ground squirrels (*Spermophilus undulatus*), for example, store willow (*Salix* sp.) leaves, seed-laden spikes of wheatgrass (*Agropyron latiglume*), and seed capsules of the rush (*Juncus balticus*) (Krog 1954), whereas thirteen-lined (*S. tridecemlineatus*), Richardson (*S. richardsonii*), and California (*S. beecheyi*) ground squirrels store heads of wheat and oats, as well as a variety of wild grass and forb seeds such as foxtail (*Hordeum*) and filaree (*Erodium*) (Howell 1938; Criddle 1939). Howell (1938) reported that rock squirrels (*S. variegatus*) stored mostly acorns, walnuts, and the seeds of peaches, plums, and apricots.

Ground squirrels store most food within their burrows, either in small side chambers off the main tunnel or in the nest chamber (Shaw 1925; Howell 1938; Criddle 1939; Krog 1954). A storage chamber of the California ground squirrel excavated by Linsdale (1946) was 12.5×30 cm in diameter and 7.5 cm high just 10 cm below the ground surface. It contained 203 live oak acorns weighing 743 g. Stored food often is well mixed with dry sand (e.g., Grinnell and Dixon 1918). Ground squirrels also store vegetation and seed heads in tunnels just outside a nest chamber (Krog 1954), or seeds and bulbs may be incorporated into the mulch at the bottom of the nest (Shaw 1925). Antelope ground squirrels (*Ammospermophilus* sp.) store food in underground chambers (Bartholomew and Hudson 1961; Nowak and Paradiso 1983), but little seems to be known about their food-hoarding habits.

California ground squirrels (Linsdale 1946) and thirteen-lined ground squirrels (Criddle 1939) store some seeds and nuts in shallow surface caches, and Ewer (1965) reported that a captive African ground squirrel (*Xerus erythropus*) stored food primarily in surface caches. Cache preparation by this African ground squirrel was highly stereotyped. After filling its mouth with corn (the only type of food that it stored in captivity) and finding a suitable cache site, it dug a small groove 3–4 cm deep with rapid, alternating scratches of the forepaws. It then lowered its head into the hole and emptied the contents of its mouth into the groove. The squirrel pressed these seeds down by tamping them with its incisors. Next, the squirrel pushed loose earth back into the hole and tamped this down with quick patting movements of both forepaws. It sometimes camouflaged caches by raking a stone or dead leaf over the sites using scraping movements of one forepaw toward the body. Caches were widely spaced away from the burrow entrance and usually among dense vegetation. Surface caching by members of *Spermophilus* seems similar (e.g., Linsdale 1946; Saigo 1969; Griffin 1971) but has not been described in detail.

Use of stored food has not been well documented, but it is apparent that different species of ground squirrels and even sexes within a species may show strikingly different seasonal patterns of food hoarding and cache use. For species such as the arctic ground squirrel, stored food appears to be used during arousal from hibernation (McLean and Towns 1981). For most species of ground squirrels, however, stored food acts as a food reserve for a brief period at emergence from hibernation (Shaw 1925; Krog 1954; Mayer and Roche 1954). Hibernating ground squirrels are subject to variable and sometimes extreme conditions in early spring (Morton and Sherman 1978), and fresh vegetation or seeds are probably seldom available at emergence. Presence of a small food cache may permit ground squirrels to emerge earlier than they might otherwise, and thus the trait may yield important reproductive advantages. Rock squirrels, on the other hand, are active throughout the winter in much of their range, and food reserves are probably used at that time (Howell 1938).

Flying squirrels. The flying squirrels of North America (*Glaucomys*), Europe, and northern Asia (*Pteromys*) store nuts and seeds. The hoarding behavior of these squirrels, however, is poorly known compared to that of tree and red squirrels because flying squirrels are active only at night. The most persistent attempts to watch flying squirrels collect, transport, and store nuts in the wild are hampered by their ability to glide into surrounding darkness. Consequently, much of what we know about food hoarding in flying squirrels comes from observations on captive individuals in confined, unnatural settings.

Flying squirrels cut nuts, acorns, and conifer cones from trees or search for nuts among the litter on the forest floor (MacClintock 1970; Muul 1970). Hickory nuts, when available, may comprise up to 90% of nuts stored by southern flying squirrels (*Glaucomys volans*) (Muul 1968). Flying squirrels grasp a single nut in their mouth when gliding to a storage site. To achieve a very firm grip so that the nut will not be jarred loose when landing, squirrels remove the husk, gnaw notches into the shell, and insert their incisors into the notches (Sollberger 1940; Muul 1968).

Flying squirrels store nuts in trees or in shallow depressions in soil. Nuts carried to trees are deposited in cavities, wedged into cracks, inserted behind loose bark, or placed singly in branch forks (Muul 1970). They use their bared incisors to hammer nuts tightly into crevices. Flying squirrels often do not cover nuts they store in exposed sites in trees, and in captivity, they often carry nuts into their nest boxes (e.g., Hatt 1931). The contents of cavity storerooms have not been quantitatively described.

Flying squirrels prepare ground caches by scratching litter and a small amount of soil away from a site with their forepaws (Sollberger 1940; Muul 1970). Next they push the nut, held between the incisors, into the loose soil and litter between their hind feet and rap it several times in quick succession with their incisors. Flying squirrels rake litter over the half-buried nut with their forepaws, but they do not bury nuts as deeply as do tree squirrels.

Flying squirrels store nuts throughout the year, but they store most intensely in the fall. Muul (1970) found that captive southern flying squirrels stored about twenty hickory nuts each night during summer. Caching intensity increased sharply in September, peaked in November, and declined back to baseline levels by January or February (Avenoso 1968; Muul 1970). This storage season coincided with ripening and depletion of the mast crop. The onset of intensive hoarding by southern flying squirrels in fall is triggered by shortening photoperiod (Muul 1970).

Muul (1968) estimated the hoarding potential of southern flying squirrels based on seasonal changes of hoarding intensity of captive individuals. The most nuts hoarded by an individual in one night was 277. Muul assumed that seasonal changes in hoarding intensity of wild squirrels were similar to those of captive flying squirrels and thus ranged from a baseline of 20 nuts per night to a peak of 277 nuts per night. Given these assumptions, a free-ranging squirrel could store about 15,000 nuts during a harvest season. This is almost certainly an overestimation of actual hoarding performance, because search and travel times would be much greater in free-ranging squirrels. Furthermore, intensity of hoarding appears to vary with latitude. For example, Avenoso (1968) found that southern flying squirrels in Florida stored much less than did southern flying squirrels in southern Michigan (Muul 1970). Goertz, Dawson, and Mowbray (1975) could not find any evidence of food hoarding in a population of southern flying squirrels in northern Louisiana; nevertheless, Muul's calculations illustrate that flying squirrels at higher latitudes have the capacity to store many nuts.

The large quantities of nuts stored by squirrels in northern latitudes are used for winter sustenance during periods when other foods are scarce. In the south, where winters are mild and food more continuously available, stored food is apparently less important in the energetic budget (Goertz, Dawson, and Mowbray 1975). Nuts also are retrieved during spring and summer, but the role of stored food at this season is not clear.

Pocket Gophers (Geomyidae)

Pocket gophers are fossorial rodents that spend most of their lives in extensive underground burrows. Burrow systems usually consist of one long tunnel from which many short lateral tunnels extend (fig. 8.17). Food-storage chambers are small cavities 20–30 cm in diameter located at the ends of some laterals, often sealed off from the main tunnel with a plug of earth. Gophers often establish food-storage chambers within a meter or two of the nest, and they may even store some food in the nest chamber (Seton 1928; Scheffer 1940; Aldous 1945; Barnes et al. 1985). Other food-storage chambers may be located in the portion of the burrow where the gopher is actively foraging for underground plant parts (Scheffer 1940; Brown and Hickman 1973). Five burrow systems of the northern pocket gopher (*Thomomys talpoides*) excavated by Aldous (1945) in late summer contained an average of two storage chambers (range 1–3), and four burrows of this species excavated by Criddle (1930) during October and November contained an average of 3.5 storage chambers (range 1–7). Only 45% of southeastern pocket gopher (*Geomys pinetis*) burrows excavated by Brown and Hickman (1973) had food storage chambers (mean = 1.8, 1–4), and about 30% had piles of tubers scattered along runways. Burrows in sparsely vegetated areas were more likely to have stored food.

Stored food usually consists of roots and tubers (Howard and Childs 1959; Turner et al. 1973; Barnes et al. 1985). Gophers cut underground plant parts into 2- to 5-cm sections and transport them to storage chambers in their external, fur-lined cheek pouches (Scheffer 1931; Hamilton 1939; Long 1976). Segments of roots up to 20 cm long also occur in caches; gophers probably carry these in their mouths. Mean fresh weights of food caches in fall (September to November) ranged from 190–560 g, the largest cache weighing 2.9 kg (Criddle 1930, Aldous 1945; Smith 1948; Barnes et al. 1985).

Five caches of the northern pocket gopher contained mostly dandelion (*Taraxacum officinale*) roots (71%), with tubers of tuber starwart (*Stellaria jamesiana*), corms of spring beauty (*Claytonia lanceolata*), and bulbs of dogtooth violet (*Erythronium grandiflorum*) frequently represented (Aldous 1945, 1951). Yampah (*Perideridia gairdneri*) roots comprised 75% of food caches of northern pocket gophers excavated by Barnes et al. (1985). Food caches in five burrows excavated by Criddle (1930) contained twenty-five species of plant, with toad-flax (*Comandra richardsonii*), sweet clover (*Melilotus* sp.), and sweet pea (*Lathyrus venosus*) predominating. Plains pocket gophers store Jerusalem artichokes (*Helianthus tuberosa*) (Smith 1948). Some items in caches indicate that above-ground foraging must sometimes occur. For example, Vaughan (1967) found that the most important food items in the storage chambers of the northern pocket gopher in north-central Colorado were prickly pear (*Opuntia*) joints and portions of the crown of needle-and-thread grass (*Stipa comata*). Horn and Fitch (1942) reported that Botta's pocket gophers (*T. bottae*) in the San Joaquin Valley, California, stored acorns and digger pine (*Pinus sabiniana*) seeds.

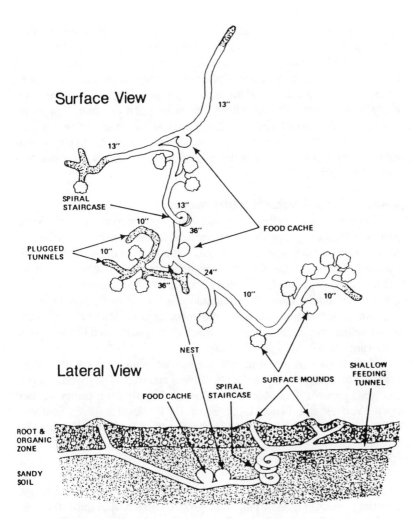

Surface View

Lateral View

Figure 8.17
Diagram of a burrow system of the southeastern pocket gopher in surface and lateral view. The food caches are in small chambers off the main burrow, often near the nest. From Brown and Hickman 1973.

During winter, pocket gophers spend considerable time burrowing under snow just above the ground surface. At this time, subnival caches may be constructed. Aldous (1945) reported two such caches of the northern pocket gopher from central Utah. One contained almost exclusively dandelion roots (320 g wet weight); the other (303 g wet weight) contained six identified food types with spring beauty (82%), tuber starwort (7%), and Indian potato (*Orogenia linearifolia;* 6%) predominating. Ingles (1952) located a subnival cache of the mountain pocket gopher (*Thomomys monticola*) composed of *Ceanothus cordulatus* leaves. Seventeen subnival caches of the northern pocket gopher in Utah contained the underground storage organs of seven species of perennial plant (Stuebe and Andersen 1985). The corms of spring beauty and Indian potato comprised 83% of the stored dry mass. Mean cache wet weight was 184 g (range 55–485 g). Some caches were within a few meters of each other and may have been made by the same pocket gopher.

Stuebe and Andersen (1985) found that the rank order of species biomass in caches was not correlated with the rank order of availability (below ground standing crops) or the rank order of caloric content. However, rank order of species biomass was negatively correlated with nitrogen content, which led Stuebe and Andersen to suggest that items encountered by pocket gophers that are high in nitrogen (i.e., protein) are more likely to be eaten, whereas those low in nitrogen are more apt to be stored. Food items stored in summer may show up in the winter diet (e.g., Bandoli 1981), but pocket gopher food caches appear to play a critical role during relatively short periods when soil conditions preclude foraging (see section 2.3).

Kangaroo Rats and Pocket Mice (Heteromyidae)

Kangaroo rats, pocket mice, and other heteromyids are New World rodents that occur primarily in arid shrublands and grasslands. In Central America and northern South America, some species inhabit tropical forests. All members of the family possess external, fur-lined cheek pouches, a trait shared with the closely related pocket gophers (Geomyidae). Two genera, *Dipodomys* (kangaroo rats) and *Microdipodops* (kangaroo mice), are named for their bipedal, saltatorial mode of locomotion. Heteromyids excavate burrows in the ground in which they spend much of their lives, venturing out only at night to forage.

Heteromyids store primarily seeds, but some species also store dried vegetation and small amounts of fungi and fruit. These rodents collect seed pods or seed-laden flower heads directly from plants as they ripen (Vorhies and Taylor 1922; Shaw 1934; Scheffer 1938; Schroder 1979), or they sift seeds from the soil using their forepaws (Hutto 1978; Lawhon and Hafner 1981). They transport seeds and seed pods in their spacious cheek pouches (Morton, Hinds, and MacMillen 1980).

Heteromyids store food three ways. First, they store seeds and occasionally other food items in special chambers in the burrow. Second, they bury seeds in shallow pits in the ground surface. And third, they pile seeds heads of grasses in haystacks on the ground surface. The third type of storage seems to be uncommon and is poorly documented (Hawbecker 1944).

The structure of heteromyid burrows and the number and arrangement of seed-storage chambers differ considerably among species (e.g., Shaw 1934; Scheffer 1938; Tappe 1941; Eisenberg 1963; Anderson and Allred 1964; Fleming and Brown 1975). The burrow systems of Santa Cruz kangaroo rats (*Diopdomys venustus*) excavated by Hawbecker (1940) and banner-tailed kangaroo rats (*D. spectabilis*) excavated by Vorhies and Taylor (1922) and Reichman et al. (1985) span the range of complexity in burrow architecture. Santa Cruz kangaroo rats excavate relatively simple burrows consisting of a tunnel with a couple of side branches (fig. 8.18) extending about 50 cm deep. Burrows harbor a nest chamber and two to five seed-storage chambers. The mound at the mouth of the burrow is small, reflecting the small amount of excavating. There are two or three entrances, which are usually plugged with soil during the day. At the other end of the spectrum, banner-tailed kangaroo rats construct a com-

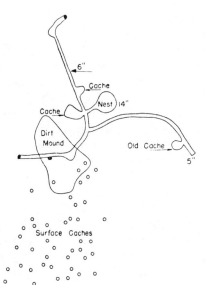

Figure 8.18
Burrow system of the Santa Cruz
kangaroo rat viewed from above.
Seeds are cached in chambers near
the nest or in scattered surface
caches (small circles). Depths of
galleries are given in inches;
plugged entrances are indicated by
filled circles. From Hawbecker 1940.

plicated labyrinth of interconnected galleries arranged in two or three
tiers capped by a large mound sometimes over 1 m high and 1.5–4.5 m in
diameter. The burrow does not extend far beyond the edge of the mound,
and there are six to twelve entrances on each mound, some of which are
left open. There is one nest chamber and numerous centrally located
seed-storage chambers measuring about 15–25 cm in diameter and 8 cm
or more in height. After banner-tailed kangaroo rats fill all the food-
storage chambers with seeds, they will use segments of tunnel as storage
chambers. Except during the breeding season, each burrow is occupied
by a single individual.

Individuals of most species of heteromyids have several subsidiary
burrows within their home ranges that they use as temporary refuges.
These are much simpler than the main burrows and, except for those of
Heermann's kangaroo rats (*D. heermanni;* Tappe 1941), do not contain
stored food (Vorhies and Taylor 1922; Hawbecker 1940; Culbertson
1946).

Grass and forb seeds comprise most of the food found in storage
chambers (Vorhies and Taylor 1922; Tappe 1941; Monson 1943; Hardy
1945; Reynolds and Glendening 1949; Hibbard and Beer 1960; Csuti
1979; Reichman, Wicklow, and Rebar 1985). Of the seeds stored by banner-tailed kangaroo rats in southeastern Arizona, 63% were grasses (primarily annual grama, *Bouteloua aristidoides;* fluffgrass, *Tridens pulchellus;*
and six-week three-awn, *Aristida adscensionis*), and 36% were forbs
(Monson 1943). Only two or three species of plant comprise most of each
store, and grass seeds often predominate. Some banner-tailed burrows
contain small amounts of fungi (puffballs), fruit (e.g., silverleaf nettle, *Solanum elaeagnifolium*), and green buds and leaves. The Great Basin kangaroo rat (*D. microps*), a specialist on the succulent leaves of shadscale

(*Atriplex confertifolia*), makes leaf caches in chambers within the burrow system in late fall (Kenagy 1973). Kenagy (1973) excavated nine burrows and found seventeen caches ranging from 3–146 g (mean = 50 g).

When more than one type of seed is stored, they are often segregated in separate chambers (Monson and Kessler 1940; Tappe 1941; Reichman, Wicklow, and Rebar 1985). Vorhies and Taylor (1922) suggested that seeds are segregated within a burrow because banner-tails typically forage for only one type of seed during a foraging excursion; different species of seed often are harvested at different seasons (e.g., McAdoo et al. 1983). The banner-tailed kangaroo rats studied by Reichman, Wicklow, and Rebar (1985), however, appeared to actively sort seeds as they stored them.

Within the burrows, heteromyids often place food-storage chambers near the ground surface. Twelve chambers of the giant kangaroo rat described by Shaw (1934), for example, were between 7.5 and 18.8 cm deep. Shaw (1934) suggested that these caches may be kept dry by the sun's warmth. Vorhies and Taylor (1922) and Hawbecker (1940), however, noted that food in chambers near the surface is sometimes wetted by rains, and the seeds may germinate or mold. Banner-tailed kangaroo rats store most seeds in chambers about 30 cm deep, with a second concentration occurring at 50 or more centimeters deep (Reichman, Wicklow, and Rebar 1985).

Reports on quantity of food stored in burrows have for the most part been sketchy. Burrows have been excavated at various times of year, and burrow contents have been reported in various units, making meaningful interspecific and intergeneric comparisons of cache sizes difficult. Even so the most avid seed storers seem to be the giant (*D. ingens*) and banner-tailed kangaroo rats. A burrow of the giant kangaroo rat excavated by Shaw (1934) in early May had nine seed-filled chambers containing the following quantities of peppergrass (*Lepidium nitidum*) and filaree (*Erodium* sp.): 2.9, 9.1, 1.9, 1.1, 5.0, 1.7, 2.2, 5.8, and 8.8 l totaling 38.5 l. Vorhies and Taylor (1922) reported nearly 5.8 kg of seeds in burrows of the banner-tailed kangaroo rat. In contrast, Heermann kangaroo rats accumulate relatively small quantities of seeds, the greatest store found by Tappe (1941) being only 157 g. The size of other kangaroo rat larders has been described by Vorhies and Taylor (1922), Shaw (1934), Hawbecker (1940), and Monson and Kessler (1940). Banner-tailed kangaroo rats actively manage their seed stores and appear to be able, somehow, to keep seeds from germinating in the humid underground chambers (Reichman, Wicklow, and Rebar 1985).

Plains pocket mice (*Perognathus flavescens*) usually store from 1.0– 4.4 l of weed seeds in their burrows in the fall (Bailey 1929; Hibbard and Beer 1960). The largest cache of the hispid pocket mouse (*P. hispidus*) found by Blair (1937) measured less than 200 ml. In foraging experiments, *Perognathus* species hoarded more seeds than did sympatric species of *Dipodomys* (Lawhon and Hafner 1981), but there have been no field studies that show this to be the general pattern.

Members of *Dipodomys, Perognathus, Heteromys,* and perhaps other genera store seeds in small surface caches. Heermann and Fresno kangaroo rats make surface caches by digging a hole in loose soil with their forepaws, depositing seeds from the cheek pouches in the depression, covering them with soil, and compacting the soil over the cache by patting it with the forepaws (Dale 1939; Culbertson 1946). They empty cheek pouches by scratching the rear of a bulging cheek with a forepaw. Cache preparation is probably similar for all species. Caches range from 1.3–10.0 cm deep (Shaw 1934; Hawbecker 1940; Reynolds and Glendening 1949; La Tourrette, Young, and Evans 1971; McAdoo et al. 1983).

Surface caches have been found clustered around entrances to the main burrow (fig. 8.18) or scattered throughout the home range (Hawbecker 1940; Reynolds and Glendening 1949; La Tourrette, Young, and Evans 1971). Shaw (1934) recorded densities of 33, 54, 55, and 64 caches on 0.37-m^2 plots placed near four giant kangaroo rat burrows on 24 March. La Tourrette, Young, and Evans (1971) reported 5.76 surface caches per square meter in sagebrush shrubland north of Reno in late fall but did not know which of several heteromyid species was responsible for the caches. Santa Cruz and Merriam's kangaroo rats transported seeds up to 32 and 50 m away from feeding stations to scatter hoard them (Hawbecker 1940; Reynolds 1958).

Each cache often contains only one type of seed. Of 875 surface caches found by Shaw (1934) around a single giant kangaroo rat burrow in early April, 873 contained peppergrass; 179, evening primrose (*Oenothera*); 169, filaree; 5, cudweed (*Gnaphalium*); 4, plantain (*Plantago*); 4, saltbush (*Atriplex*); and 1, red brome. Many of the heteromyid seed caches found by McAdoo et al. (1983) contained only Indian ricegrass (*Oryzopsis hymenoides*).

Many surface caches are only temporary storage sites. Shaw (1934) marked some seeds during late March and found the seeds incorporated in the burrow system stores in early May. Shaw concluded that the scatter-hoarded seeds were probably cured in the warm, dry soil before final storage in the cool, humid burrows. This presumably reduced the risk of fungi infecting and spoiling the seeds. Hawbecker (1940) noted that seeds stored in surface caches by Santa Cruz kangaroo rats already were ripe and cured before being placed in the surface caches and concluded that curing of seeds was not the function of scatter hoarding in this species. Further, he found no evidence that the scatter-hoarded caches were transported to the burrow larders.

Hawbecker (1944) alleged that giant kangaroo rats of San Benito County, California, constructed haystacks. He attributed construction of the haystacks, composed of the seed heads of red brome, to this kangaroo rat because they were seen foraging on and near them, but he did not observe the kangaroo rats add to or take material from the haystacks. One large haystack measured 1.2 × 1.8 m and was 10 cm thick. The haystacks were made in late spring and early summer, and Hawbecker (1944) suggested that the kangaroo rats used these sites to cure seeds. Heermann kangaroo rats make large surface caches of chess seeds near the ends of surface runways (Horn and Fitch 1942). These may measure about a liter and presumably cure the seeds before they are stored below ground.

Heteromyid rodents actively collect and store seeds as they ripen, which in much of the area inhabited by heteromyids, follows the rainy season or seasons. Seasonal changes in the quantities of seeds found in burrows have been most thoroughly documented in the banner-tailed kangaroo rat (fig. 8.19). Vorhies and Taylor (1922), Monson and Kessler (1940), and Monson (1943) reported that in southeastern Arizona, seed storage occurred primarily in two seasons: spring (April to May) and, more importantly, fall (September to November). This bimodal harvest corresponded to the ripening of seeds following the two rainy seasons (early spring and summer) typical of the region. Vorhies and Taylor (1922) pointed out that the early spring rains caused a variety of small annuals to bloom, whereas grasses, more important to the rats, produced seeds following the summer rains. The giant kangaroo rat of central California stored seeds predominantly during spring following the winter rains, the rainy season for that region (Shaw 1934). Lawhon and Hafner (1981) found that captive long-tailed (*Perognathus formosus*) and little (*P. longimembris*) pocket mice hoarded fewer seeds in winter than in spring and fall.

The seed reserves provide a food source when heteromyid rodents cannot forage. Heteromyids do not hibernate—although some of the small species may enter daily torpor—and require a supply of seeds throughout the year. At northern latitudes, the seed stores provide the only food source during the winter, when low nighttime temperatures restrict foraging. The importance of seed caches is apparent from Vorhies and Taylor's (1922) observation that, if severe drought precludes storage of seeds and food stores are depleted, banner-tailed kangaroo rats perish in large numbers. The seed stores also may be consumed during the day at any time of year and during bright, moonlit nights when foraging may be restricted (Tappe 1941). Wolff and Bateman (1978) found that captive silky pocket mice (*P. flavus*) fed on its seed stores during the afternoon after arousing from torpor, and they suggested that use of cached seeds in the afternoon improves the efficiency of nocturnal foraging. Morton and MacMillen (1982) and Frank (1988b) have suggested that heteromyid rodents also obtain significant quantities of preformed water by storing

Figure 8.19
Seasonal changes in the quantity of food stored in burrows of banner-tailed kangaroo rats. Based on data from Vorhies and Taylor 1922.

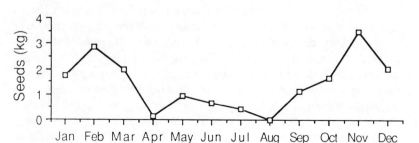

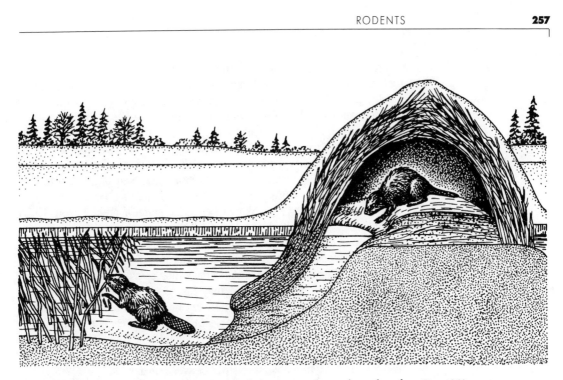

hygroscopic seeds in humid portions of their burrow, where they absorb moisture from the air, before eating them.

Figure 8.20
A winter food cache of a beaver colony under the ice. Original drawing by Marilyn Hoff Stewart.

Beavers (Castoridae)

Beavers (*Castor canadensis* and *C. fiber*) that reside in northern latitudes or high elevations, where ponds, lakes, and streams are covered with a layer of ice for several months, provide for their winter food supply by cutting live shrubs, saplings, and small trees and submerging them near their dens. These woody stems remain fresh and available to the beaver colony below the ice throughout the winter.

Stems are kept submerged several ways. Beaver jam the butts of branches and saplings into bottom sediments in deep, still water where they become fixed (Bailey 1926; Seton 1953; Conibear and Blundell 1949). When fresh, wood is nearly as dense as water and remains submerged with little holding it. Beaver construct these submerged stockpiles in about 1–2.5 m of water near the entrance to the lodge or burrow (Townsend 1953; Novakowski 1967). Beavers also form floating rafts of food (Roberts 1937; Richard 1964; Slough 1978). Slough (1978) found that the uppermost logs and branches are primarily low-preference food items such as alder (*Alnus* spp.) and peeled aspen (*Populus tremuloides*) logs. They place preferred stems, such as aspen and willow (*Salix* spp.), throughout the cache. After "freeze-up," much of the material remains available under the ice (fig. 8.20). Food caches of the European beaver are dense piles of branches extending from the bottom to the water surface and may attain a volume of 80 m^3 (Wilsson 1971). Slough (1978) reported that some beaver store stems under floating mats of vegetation.

Beavers begin constructing food caches from late August to early October (Townsend 1953; Hill 1982), with the most intensive cutting of vegetation occurring during October. Aspen, red-osier dogwood (*Cornus stolonifera*), willow, cottonwood (*Populus* spp.), birch (*Betula* spp.), maple (*Acer* spp.), viburnums (*Viburnum* spp.), and other high-quality woody plants are the most common constituents of food caches (Bailey 1926; Townsend 1953; Novakowski 1967; Northcott 1971; Slough 1978; Swenson and Knapp 1980; Echternach and Rose 1987). Cache composition often reflects woody plant availability (Novakowski 1967; Slough 1978). Novakowski (1967) and Hall (1960) found that butt diameters of most cached stems were less than 5 cm, but in other studies small trees were found in caches (e.g., Slough 1978). Dennington and Johnson (1974) found that beaver in the MacKenzie Valley and northern Yukon, Canada, cached pond lilies (*Nuphar variegatum*) in addition to woody stems. Storage of herbaceous vegetation has not been reported in other populations.

The food cache provides the primary food source for the colony during the icebound period, which may last 5–6 months (e.g., Wilsson 1971). The buds and twigs are completely consumed; only the bark is gnawed off larger stems. The usable weights of five caches examined by Novakowski (1967) in Wood Buffalo National Park, Canada, ranged from 80–512 kg. Assuming an icebound period of 150 days, this is equivalent to 0.26–0.43 kg of edible stems per beaver per day.

Hay (1958), Slough and Sadleir (1977), Payne (1981), and Swenson et al. (1983) have attempted to use the conspicuous beaver food caches as an index of population density with varying degrees of success.

Mice, Woodrats, Hamsters, and Gerbils (Cricetidae)

The family Cricetidae is a large, diverse family of rodents that vary greatly in their propensity to store food. Here I have divided the family into four groups based on similarity of food-hoarding habits: (1) New World mice, (2) woodrats, (3) hamsters, and (4) gerbils.

New World Mice. Of the approximately seventy genera of Nearctic and Neotropical cricetid rodents, only a handful of genera are known to store food. Most of these are from north-temperate regions. Future studies will likely reveal that food storage is more widespread within the group.

New World mice store nuts and seeds from a wide variety of plants including grasses, forbs, shrubs, and trees. Seed types reported in the stores of deer mice (*Peromyscus maniculatus*) in southern Michigan and Manitoba include ragweed (*Ambrosia artemisiifolia*), acorns (*Quercus velutina, Q. alba,* and *Q. rubra*), bush clover (*Lespedeza capitata*), panic grass (*Panicum* sp.), wheat, rye, sweet clover, wild buckwheat, and cherry stones (Criddle 1950; Howard and Evans 1961). Some food stores also included small quantities of insects, deer fecal pellets, and lichens. Food stores of white-footed mice (*Peromyscus leucopus*) include items such as acorns, chestnuts, beechnuts, wild cherry pits, and the seeds of basswood, conifers, raspberries, and jewelweed (*Impatiens*) (Hamilton 1942). The wide range of seeds and nuts these mice store suggests that cache composition is strongly influenced by the types of food that happen to be available. The largely carnivorous northern grasshopper mouse

(*Onychomys leucogaster*) appears to store only seeds (Ruffer 1965). Bulbs, tubers, roots, and dried vegetation have not been reported in deer mice and grasshopper mice food stores. Cotton rats (*Sigmodon* spp.), however, place small piles of grasses and sedges along their runways, and puna mice (*Punomys lemminus*) of the altiplano of Peru accumulated stems of two plants, a shrub (*Senecio adenophylloides*) and a forb (*Werneria digitata*), in chambers beneath rocks (Nowak and Paradiso 1983).

Deer and white-footed mice transport seeds in internal cheek pouches (Hamilton 1942). Large nuts are carried between the incisors. The capacity of these pouches is relatively small compared to those of Old World cricetids (e.g., hamsters). Cogshall (1928) observed that deer mice carry ten kernels of wheat in cheek pouches, and Hamilton (1942) removed thirteen jewel weed (*Impatiens biflora*) seeds from pouches of a deer mouse. Transport distances are usually less than 50 m (e.g., Abbott and Quink 1970) but may be much further (e.g., Criddle 1950).

Peromyscus store seeds in two situations: (1) in small, scattered, subsurface caches and (2) in small chambers. Mice prepare subsurface caches by digging a small hole 2.5–10 cm deep with the forepaws. Next they expel the contents of the internal cheek pouches into the hole and then cover the food with several centimeters of soil or litter, which they pat down with the forepaws (Eisenberg 1962, 1968). Summer caches sometimes are much shallower than fall caches (Abbott and Quink 1970). Cache size is limited by the capacity of the internal cheek pouches. Abbott and Quink (1970) found typical cache sizes of white-footed mice in Massachusetts to be twenty to thirty eastern white pine (*Pinus strobus*) seeds. (All the surface caches Abbott and Quink [1970] found in their study were probably made by white-footed mice, because red-backed voles (*Clethrionomys gapperi*), the only other nocturnal rodent caught on their study site, do not seem to scatter hoard food. See section on Voles, Lemmings, and Muskrats.)

Chambers used for food storage usually are located near the nest below ground, in hollow trees, or in abandoned buildings. The contents of six storage chambers provisioned by deer mice in Manitoba, Canada, weighed from 0.5–0.8 kg, and two other caches found near granaries weighed 4.4 and 20.4 kg (Criddle 1950). Deer mice stored 0.2–1.3 l of seeds in nest boxes put out by Howard and Evans (1961) in southern Michigan grassland. Some of the larger larders were the products of communal caching effort by five to eight mice. A chamber over 12 m high in a beech tree occupied by a pair of white-footed mice contained about a peck (8.8 l) of beechnuts (Hamilton 1939). Numerous laboratory and field studies have reported caching in nest boxes (Howard and Evans 1961; Dewsbury 1970; McCarty and Southwick 1975; Barry 1976; Tadlock and Klein 1979). When more than one chamber was available, the mice usually nested in one and stored food in the others.

Seed caching by *Peromyscus* occurs at all seasons but is most pronounced in the fall (Criddle 1950; Howard and Evans 1961). Abbott and Quink (1970) studied scatter hoarding and recovery of eastern white pine seeds by radio-tagging a few seeds with the isotope [46]scandium and plac-

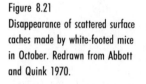

Figure 8.21

Disappearance of scattered surface caches made by white-footed mice in October. Redrawn from Abbott and Quink 1970.

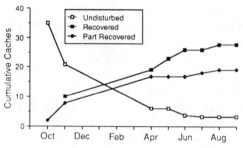

ing them in feeding trays along with untagged seeds at a ratio of one tagged to thirty-eight untagged seeds. After white-footed mice cached seeds, Abbott and Quink monitored the radio-tagged caches intermittently for a year using a gamma scintillation probe. They found that recovery of fall-cached seeds was greatest during late fall, decreased in winter, and gradually declined to zero by early summer, at which time nearly all the caches had been recovered (fig. 8.21). They found that the seeds in some caches were completely removed, whereas only a portion of the seeds in other caches had been removed. Some partially emptied caches were revisited two or more times until all seeds were recovered. When mice stored seeds in the summer, they recovered them within a week or two (Abbott and Quink 1970).

The adaptive significance of seed storage has not been demonstrated for these mice. Deer mice populations are known to have increased survival rates and to begin reproducing earlier following autumns with large seed and nut crops (e.g., Gashwiler 1979). No doubt many seeds are stored during these productive years, but the role of stored seeds in facilitating winter survival and winter breeding has not yet been determined.

Woodrats. Woodrats, or packrats (*Neotoma*), construct large, conspicuous "houses" of sticks, cactus joints, pebbles, and other materials in protected sites in rock crevices, under boulders, in dense brush, among cacti, or in trees (Vestal 1938; Finley 1958; Stone and Hayward 1968; Reichman 1988). Within the house, which often measures 2 m across and 1 m or more high, woodrats construct a system of tunnels and several chambers, one of which contains a nest of dried vegetation (fig. 8.22). When constructed near rock outcrops or hollow trees, tunnels may extend into inaccessible natural cavities. Some of the cavities and chambers are used for food storage. Except during the breeding season, each house is occupied by one adult woodrat.

Woodrats prepare two types of food stores: short-term stores for use within a day or two of collection and long-term stores for winter sustenance (Vestal 1938; Horton and Wright 1944; Finley 1958). Short-term stores usually consist of small piles of fresh vegetation. Food stored for longer periods consists of dried vegetation or various nuts, seeds, and berries. Woodrats collect vegetation on nocturnal foraging excursions and stack it near the house to dry. After it has cured, woodrats restack it in conspicuous piles on top of the house or lodged in less accessible sites

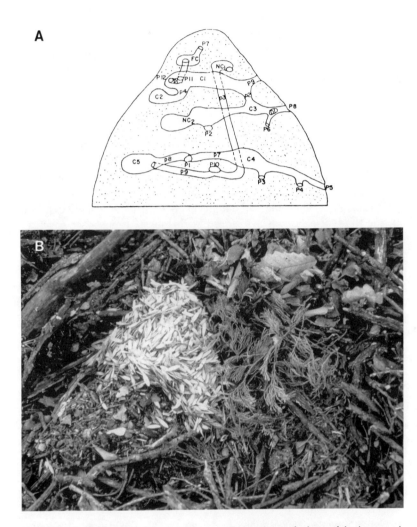

Figure 8.22   Food storage by woodrats. A, Diagrammatic cross section of a house of the desert wood-rat showing food cache chambers (FC), large interior chambers (C), nest chamber (NC), and passages (P). B, Accumulations of ash samaras (left) and buckbrush berries and leaves (right) in a dismantled nest of an eastern woodrat. A, from Stone and Hayward 1968; B, photograph courtesy Jim Reichman.

under boulders, in crevices of rocks, under shrubs, or in chambers within the house (English 1923; Linsdale and Tevis 1951; Finley 1958).

Woodrats will store the new growth of almost any plant growing within foraging range of their den sites. The following examples, taken from Finley's (1958) detailed descriptions of woodrat dens, illustrate the amounts, composition, and manner of food storage by woodrats in Colorado. A Mexican woodrat (*Neotoma mexicana*) den examined on 9 May contained about 13 l of mixed cuttings of skunkbrush sumac (*Rhus trilobata*), yucca (*Yucca glauca*), and two species of sagebrush (*Artemisia* spp.) lodged under a pile of boulders. A white-throated woodrat (*N. albigula*) had placed about 18–22 l of rat-tailed cactus (*Opuntia davisii*) joints and cuttings of gambel oak (*Q. gambelii*) and winterfat (*E. lanata*) in a vertical cleft in a cliff face near a den examined 27 Aug. desert woodrat (*N. lepida*) crammed more than 4.4 l of dried thistle (*Salsola kali*) under a rock shelter. A bushy-tailed wood *cinerea*) filled a large chamber at the rear of a horizontal shelf nearly 35 l of dry cuttings, including locoweed (*Astragalus*), lupine *nus*), owlclover (*Orthocarpus*), puccoon (*Lithospermum*), and rabbi (*Chrysothamnus nauseosus*). At another site, an unusual cache bushy-tailed woodrat contained eighty-eight dried mushrooms on a row ledge in a cave. An eastern woodrat (*N. floridana*) stored 236 bl of yucca under a block of fallen rimrock.

Nuts, seeds, and fruits, food items likely to be pilfered if left expo outside the house, are always stored in chambers, tunnels, or small re cesses within the house. Bailey (1931) found as much as 2.2 l of piñon pine (*P. edulis*) nuts in the house of a white-throated woodrat. Other houses of white-throated woodrats contained juniper berries, desert hackberries, and mesquite beans still in the pods (Bailey 1931; Vorhies and Taylor 1940; Spencer and Spencer 1941). Vestal (1938) counted 993 acorns (*Q. agrifolia*), 843 California laurel (*Umbellularia californica*) nuts, and numerous poison oak (*Rhus diversiloba*) seeds weighing a total of 4.5 kg in a storeroom of the dusky-footed woodrat (*N. fuscipes*) in Berkeley Hills, California. Woodrats are not known to scatter hoard seeds or nuts.

Stored food can be found in or near the house at any time of year, but the primary storage season is late summer and fall (Rainey 1956; Finley 1958). Seasonal comparisons of stored food in dusky-footed woodrat houses in southern California (fig. 8.23) indicate that the largest supply of stored food occurs during fall and the least during summer (Vestal 1938; Horton and Wright 1944). Types of food stored also vary with season (e.g., Fitch and Rainey 1956).

The large mass of vegetation, nuts, and seeds stored by woodrats serves as a winter food supply. For example, six dusky-footed woodrat dens that Horton and Wright (1944) had examined in November contained a total of 7.5 kg of stored food, whereas in February the same six houses contained a total of only about 0.4 kg of food. Over 94% of the food had been consumed during the winter. Short-term stores, on the other hand, are probably meant for diurnal feeding or for use on nights when nocturnal foraging is curtailed by bright moonlight or inclement weather.

Figure 8.23
Seasonal changes in the amount of food stored by dusky-footed wood-rats in southern California. Based on data from Horton and Wright 1944.

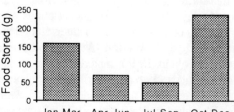

Hamsters. Hamsters of Europe, Asia Minor, and arid parts of southern Asia are prodigious food hoarders. The name hamster is derived from the German word hamstern, which means to hoard. In the Middle East, the name for the golden hamster (*Mesocricetus auratus*) means *grandfather saddlebags*, a reference to the extremely large cheek pouches used to transport food to the burrow. Near agricultural areas, farmers consider golden hamsters pests because they eat and store large quantities of cereal grains (Murphy 1971).

Golden hamsters have become favorite subjects in biomedical research since they were introduced to European and North American laboratories in 1930 (Murphy 1971). Many experiments on hoarding have been conducted on domesticated strains (see chapter 5), but unfortunately, little detailed information exists on food hoarding in the wild. Such studies are badly needed.

Hamsters are larder hoarders, storing seeds, roots, and tubers in their burrows. The burrow of the black-bellied hamster (*Cricetus cricetus*) consists of numerous interconnecting galleries and chambers about 40–80 cm below ground (Grulich 1981). Up to 90 kg of grain, peas, and potatoes have been found in burrows (Nowak and Paradiso 1983), often with each food type in a separate chamber. The greater long-tailed hamster (*Cricetulus triton*) may store more than a bushel (about 35 l) of grain in its burrow (Allen 1940).

The large cheek pouches of hamsters are specializations for collecting and storing food. The enormous pouches of the golden hamster open inside the mouth and extend back to the shoulders and, when full, more than double the width of the head. Pouches of the Djungarian hamster (*Phodopus sungorus*) are so large that when full, the shape of the body is greatly distorted (Allen 1940). Capacity of the cheek pouches has apparently not been measured, but Nowak and Paradiso (1983) reported that forty-two soybeans were taken from pouches of a long-tailed hamster (*Cricetulus* sp.), and Etienne et al. (1983) found that golden hamsters can carry a mean of sixteen hazelnuts. After the pouches are full, hamsters often carry a single large item, such as a root or tuber, between their incisors. Hamsters empty cheek pouches by pressing the rear of the pouch with a forepaw, causing seeds to pour out of their mouths. A detailed analysis of the sequential organization of hoarding has been conducted by Etienne et al. (1983).

The use of stored food by wild populations has not been established, but information from experimental studies suggests that food stores may have at least three functions. First, captive animals accumulate a food store during the dark phase of the light-dark cycle and feed from the stored food during the light phase (Silverman and Zucker 1976; Toates 1978). This suggests that in the wild, hamsters may eat from their larders during the day without leaving their burrows. Second, hamsters in captivity do not deposit body fat but store large quantities of food before hibernation. The possession of a food store in captive golden hamsters facilitated initiation of hibernation under cold temperatures (Lyman 1954). During periodic arousals, they feed on stored food, which is their only form of sustenance. Golden hamsters probably could not survive the winter without a food store. Third, the larder may play a role in the reproductive tactics of female hamsters (see section 2.4).

Gerbils. Gerbils (*Meriones, Rhombomys, Tatera,* and others) comprise a large group, widely distributed in Africa, Asia Minor, and Asia. I include with the gerbils several African cricetid rodents of uncertain relationship but that have similar food-storing habits: the tree mouse (*Beamys major*) of East Africa, the pouched mouse (*Saccostomus campestris*) of southern Africa, and the African giant rat (*Cricetomys gambianus*) of the tropical and subtropical regions of central Africa. These rodents, like gerbils, excavate fairly complicated burrow systems with one or more food-storage chambers in addition to chambers for sleeping, nesting, and defecating.

Gerbils hoard seeds, nuts, tubers, and dry vegetation. Captive Indian gerbils (*Tatera indica*) preferred to store wheat and maize over rice, ragi, legume seeds and other vegetables (Kumari and Khan 1979; Sridhara and Srihari 1980). African giant rats store large quantities of seeds (e.g., velvet beans [*Mucana aterrima*], pumpkin seeds [*Curcurbita maxima*], and ground nuts [*Arachnis hypogeae*]) and, in captivity, maize and tubers (cassavas and sweet potatoes), which they place in and around their nest (Morris 1963; Ewer 1967). A burrow of the diurnal sand rat (*Psammomys obesus*) contained 500 seed heads of barley, and another burrow contained dried.vegetation (Nowak and Paradiso 1983). The tree mouse and Namaqua gerbil (*Desmodillus auricularis*) store large numbers of seeds in short blind galleries (B. Morris 1962; Nel 1967). Captive Namaqua gerbils carry seeds (peanuts and sunflower seeds) to their nest boxes and bury them in dry sand (Pettifer and Nel 1977). When housed in a cage containing two chambers, these gerbils store most food in one chamber and sleep in the other (Christian, Enders, and Shump 1977), behavior similar to that of free-ranging individuals (Nel 1967).

The quantity of food stored can be very great, ranging up to 60 kg (Naumov and Lobachev 1975; Nowak and Paradiso 1983; Thiessen and Yahr 1977). Captive Namaqua gerbils stored 96% of their body weight in food each night (Christian et al. 1977). Naumov and Lobachev (1975) maintained that much of the food that tamarisk gerbils (*Meriones tamariscinus*) and mid-day gerbils (*M. meridianus*) store in the wild often is left unused.

Not all food is stored in subterranean storerooms. Great gerbils (*Rhombomys opimus*), in addition to storing food underground, pile dried vegetation on the ground surface near burrow entrances, behavior reminiscent of pikas and some voles. This practice is most prevalent in the northern part of the gerbil's range, especially in areas where loamy or clayey ground makes it difficult to dig underground chambers (Naumov and Lobachev 1975). Conical haystacks up to 129 cm in diameter and 75 cm in height were composed of tightly packed vegetation of shrubs, forbs, and grasses, including plant genera such as *Artemisia, Atriplex, Halocenemum,* and *Salsola* (Sabilaev and Borovskii 1970). When present, one to five haystacks (mean = 1.5) were found near the center of great

gerbil colonies. However, Sabilaev and Borovskii (1970) noted no haystacks in 9 of 12 years that they studied great gerbils. The ecological
circumstances associated with haystack construction are poorly understood. Sabilaev and Borovskii (1970) recorded haystacks only in years
characterized by high moisture and abundant food supplies. Naumov and
Lobachev (1975), however, who have summarized the Soviet literature
on gerbil food storage, found that hay pile construction was not closely
related to plant production but depended more on climatic factors. Great
gerbils constructed hay piles in the fall at the same time that they stored
food underground.

African giant rats and Namaqua gerbils scattered small surface caches
in the corners of their enclosure (Ewer 1967; Pettifer and Nel 1977), indicating that they probably scatter hoard a portion of their stores in the
wild. The giant rat made surface caches by digging a hole in dry sand 9
cm deep with its forepaws. The rat patted the bottom of the hole with its
paws and then disgorged the contents of its cheek pouches into the hole.
The rat pushed sand over the food and patted the surface of the cache
with its forepaws.

All these rodents possess cheek pouches in which they transport
food to the nest. Gerbil pouches are small compared to those of hamsters,
but the capacity of pouches seem not to have been measured. Namaqua
gerbils can carry a mean of five peanuts, but their cheek capacity was
estimated at three times this amount (Pettifer and Nel 1977). The African
giant rat can transport up to 105 ml per load (Ewer 1967).

Captive gerbils often hoard food very rapidly. Pettifer and Nel (1977),
for example, watched Namaqua gerbils make up to 175 hoarding trips in a
small enclosure in an 8-hour period following deprivation for 24 hr. Early
in the hoarding session, trips followed in rapid succession, but the hoarding rate slowed after about one-half hour. Ewer (1967) conducted a detailed study of the change in hoarding rate of a captive African giant rat.
Early in a hoarding session, the giant rat transported food as quickly as
possible. On later trips, however, hoarding excursions were often slower.
The motivation to hoard did not deteriorate in a regular fashion, but long
foraging excursions were usually followed by one or more quick trips.

The larder serves as a food reserve during the winter or dry season.
Naumov and Lobachev (1975) stated that hoarding by great gerbils en-
sures even nutrition. Large food stores prepared in the autumn are the
primary or sole food source of great gerbils in winter. Smaller stores prepared by great gerbils in spring are used during the hot, dry summer.
Sridhara and Srihari (1980) suggested that Indian gerbils, which inhabit
cultivated fields, hoard food to avoid shortages during the absence of
crops.

Mole-rats (Spalacidae)

Spalacidae consists of three species of nearly blind molelike, fossorial
rodents from eastern Europe, southern Russia, Asia Minor, and northeast Africa. In the Mediterranean region, female mole-rats (*Spalax leucodon*) construct breeding mounds of earth 160 cm across and 40 cm high

or larger (Nevo 1961; Galil 1967; Nowak and Paradiso 1983). In chambers within these mounds, they construct nests of dry grass about 20 cm in diameter. Above the nest and connected to it by a labyrinth of galleries are a set of chambers used for food storage and other chambers used for defecation (fig. 8.24). Other food-storage chambers may be located along feeding tunnels some distance from the mound and nest. Males inhabit smaller mounds with fewer galleries and storage chambers. The mound provides a site where mole-rats can establish nest and food-storage chambers above the water table to protect against flooding (Nevo 1961).

Food-storage chambers of *S. leucodon* contained bulbs of *Narcissus, Bellevalia, Gladiolus, Oxalis, Arisarum,* and *Arum* and the rhizomes and roots of *Sorghum, Prosopis,* and *Alhagi* (Nevo 1961; Galil 1967). Near agricultural areas, mole-rats often stored potatoes, carrots, onions, and ground nuts. Heth, Golenberg, and Nevo (1989) found that these mole-rats store geophytes and other food items in proportion to their availability in the soil. During lactation, female mole-rats store small quantities of fresh green foliage, including alfalfa, clover, barley, and vetch, in galleries near breeding chambers (fig. 8.24). Total food stores ranged in size from several hundred grams to 7 kilograms, which seemed to serve an especially important role during gestation and lactation for the female (Nevo 1961). In summer, nonbreeding mounds had fewer food-storage chambers, which contained roots, rhizomes, and grains. There appears to be little food storage in summer, when mole-rats subsist on food items stored the previous winter and spring.

In the Eurasian steppe, mole-rats (*Spalax microphthalmus*) construct very large food caches (Formozov 1966; Topachevskii 1976). In this region, where the soil often freezes in winter and thus prevents foraging, mole-rats must store enough food to last the winter. Four to ten storerooms can be found in each burrow; some of these are cavities up to 3.5 m long. Fourteen kilograms of rhizomes has been found in one burrow system in fall (Topachevskii 1976). Stored food includes roots, rhizomes, bulbs, tubers, and sometimes above-ground plant parts such as acorns, beans, and cucumbers. In spring, 2–3 kg may still be present in the burrows, suggesting that fall reserves are adequate to sustain the

Figure 8.24
A, Segment of mole-rat (*Spalax leucodon*) burrow with two food-storage chambers at the ends of short lateral galleries closed off with a plug of dirt. B, Portion of a mole-rat burrow in the breeding mound (viewed from above) showing the position of storage chambers and green plant material near the nest. From Galil 1967.

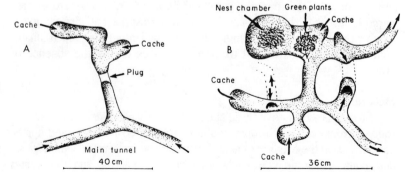

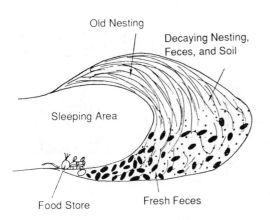

Old Nesting

Decaying Nesting, Feces, and Soil

Sleeping Area

Food Store

Fresh Feces

Figure 8.25
Lateral view of a typical mole-rat (*Tachyoryctes* sp.) burrow showing small accumulation of food to one side of the nest. From Jarvis and Sale 1971.

mole-rat through the winter. During summer, food reserves are small and last only a couple of days.

Mole-rats (Rhizomyidae)

Rhizomyidae are fossorial rodents of southeastern Asia and east Africa. At least one species, *Tachyoryctes splendens* of east Africa, is known to store food (Jarvis and Sale 1971). This mole-rat has a multipurpose nest of dry grasses and forbs used for sleeping, eliminating wastes, and storing food (fig. 8.25). The nest chamber lies near the center of the burrow system at a depth of 30–60 cm. The mole-rat deposits a small collection of food items, including bulbs (*Oxalis*), grass rhizomes (*Pennisetum*), pieces of vegetation (e.g., *Geranium, Haplosciadum,* and *Commelina*), and sometimes garden vegetables (sweet potato tubers, beans, and stalks and cobs of corn), on either side of the nest entrance. They periodically add new nest material to the bottom and back of the nest and bury an accumulation of feces, and they enlarge the front of the nest chamber. When this occurs, old food items from the food stores may be incorporated into the nest. In old nests, a gradation in age of nest material and feces can be found extending forward from the rear of the nest.

African Mole-rats (Bathyergidae)

African mole-rats are fossorial rodents and are widely distributed south of the Sahara. They excavate extensive burrow systems that include large chambers for storing food and nesting. Cape mole-rat (*Georychus capensis*), for example, store tubers, roots, and bulbs in smooth-walled, spherical chambers (Shortridge 1934; Nowak and Paradiso 1983). The food-storage chambers of the common mole-rat (*Cryptomys hottentotus*) are blind branches of shallow tunnels (Genelly 1965; Davies and Jarvis 1986). Two storage chambers, one excavated in mid-July and the other in late September, contained 1.6 and 2.6 kg of tightly-packed, bulblike bases of black seed grass (*Alloteropsis semialata*). The first of these caches occupied about 1.7 linear meters of tunnel. Small amounts of sedge (*Scleria*) rhizomes and unidentified bulbs also were found in the caches. Common mole-rats store corms of *Micranthus, Ornithogalum,*

Figure 8.26
Numbers of geophytes (e.g., bulbs and corms) in food stores of a colony of fourteen common mole-rats (*Cryptomys hottentotus*) and the expected number based on their abundance in the soil. Redrawn from Lovegrove and Jarvis 1986.

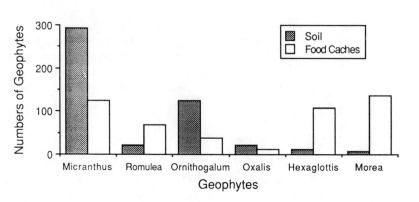

Romulea, Hexaglottis, Morea, Homeria, Oxalis, and *Cyanella* (Davies and Jarvis 1986; Lovegrove and Jarvis 1986). Both common mole-rats and cape mole-rats are selective hoarders, storing the underground parts of certain plant species out of proportion to their availability in the soil (fig. 8.26). The Cape dune mole rat (*Bathyergus suillus*) stores corms and tubers in blind, relatively deep (34–70 cm) chambers that they block with a plug of sand. In the fynbos of South Africa, these mole-rats store large quantities of *Othonna* tubers, in one case placing over 2.7 kg in a burrow (Davies and Jarvis 1986).

In low-lying areas susceptible to flooding, common mole-rats place their storage and nesting chambers in the highest ground available (Roberts 1951; Genelly 1965). In Zimbabwe, common mole-rats construct food-storage chambers near their nests in the base of termite mounds (fig. 8.27). These chambers are just a few centimeters above the rest of the burrow system but high enough to keep the nest and stored food dry when soils become saturated with water during the rainy season (Genelly 1965).

Naked mole-rats (*Heterocephalus glaber*) are not known to store large amounts of food, but they carry a few food items to their nest and eat them there (Jarvis and Sale 1971). This and a few other species of mole-rat leave tubers where they find them with their taproots undisturbed rather than storing them (Roberts 1951). The mole-rats visit the tubers and feed on them but not extensively enough to kill them. Many of the tubers Jarvis and Sale (1971) found exposed in naked mole-rat burrows were healthy and sprouting. This practice of gradually harvesting tubers in situ keeps the tubers fresh and precludes the need for caching them in chambers.

Voles, Lemmings, and Muskrats (Arvicolidae)

Voles and their relatives store lichens, mushrooms, and a wide variety of plant parts (e.g., foliage, twigs, roots, rhizomes, bulbs, corms, nuts, and fruit) in three different sorts of caches: (1) subterranean and subnival chambers and galleries, (2) piles on the ground surface, and (3) cavities and bird nests in trees. In general, each food type is stored where it will be best preserved. Roots, rhizomes, and tubers are stored

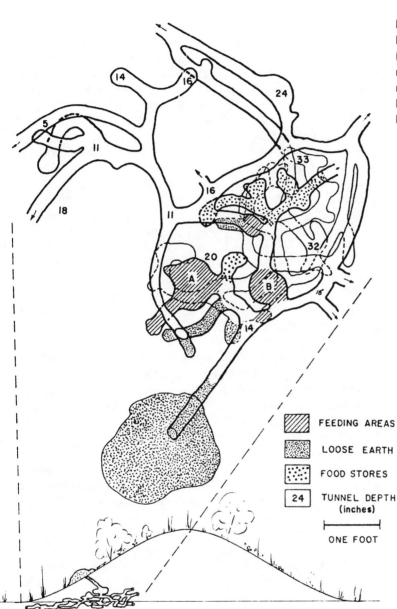

Figure 8.27
Burrow system of a mole-rat
(*Cryptomys hottentotus*) in a ter-
mite mound showing the arrange-
ment of food storage chambers.
Numbers are depths in inches.
From Genelly 1965.

FEEDING AREAS

LOOSE EARTH

FOOD STORES

24   TUNNEL DEPTH
(inches)

ONE FOOT

only underground, where they remain fresh and crisp for months. Grasses,
leafy vegetation, and twigs are stored primarily in hay piles on the ground
surface. Epiphytic beard lichens are stored in trees, where they stay dry
and well ventilated. Some voles will make two or three distinctly different
kinds of caches depending on what foods are available (e.g., Murie 1961;
Litvinov and Vasil'ev 1973; Stamatopoulos and Ondrias 1987). Voles do
not appear to scatter hoard food.

Chambers used for storage of underground plant parts are elongated
galleries of an underground system or natural cavities in stumps or logs.

Subterranean storage rooms are usually shallow, just a few centimeters beneath a layer of sod or moss. Each vole or family of voles may provision several chambers, which are connected by a network of tunnels to small, globular breeding or sleeping nests of grass a short distance away. A food-storage chamber of the Alaska vole (*Microtus miurus*) was 45 × 25 × 12.5 cm and contained 45 horsetail (*Equisetum*) tubers, 155 coltsfoot (*Petasites*) roots, 1,021 coltsfoot buds, and 2,808 grass shoots (Murie 1961). Each type of root often is stored in a separate chamber. A small underground cache of the meadow vole (*Microtus pennsylvanicus*) discovered by Lantz (1907) contained 18 g of morning-glory (*Convolvulus sepium*) roots. Underground depositories of the common vole (*Microtus arvalis*) examined by Gladkina and Chentsova (1971) contained 2.5–3.5 kg of timothy (*Phleum pratense*) bulbs. Muskrats (*Ondatra zibethica*) store large quantities of arrowhead (*Sagittaria*) tubers, bullrush roots, sedges, mints, young grasses, and reeds (Carter 1922; Seton 1928). Subterranean caches of rhizomes, roots, bulbs, and tubers stored by other voles are described by Quick and Butler (1885), Nelson (1893), Bailey (1920), Couch (1925), Hatt (1930), Hamilton (1939), Jameson (1947), Formozov (1966), Litvinov and Vasil'ev (1973), Le Louarn (1974), Meylan (1977), Wolff and Lidicker (1981), and Stamatopoulos and Ondrias (1987).

Some voles cache underground plant parts in chambers in snow. An hourglass-shaped, subnival larder of a meadow vole, exposed as snow melted in later March, contained mostly (95% dry mass) cinquefoil (*Potentilla canadensis*) rhizomes (Gates and Gates 1980). The underground parts of violets (*Viola papilionacea*), morning-glories (*C. sepium*), and buttercups (*Ranunculus bulbosa*) also were in the caches. Subnival trails radiated from the storage chamber. The mole vole (*Ellobius talpinus*) of northern Kazakhstan, USSR, also caches rhizomes in chambers in the snow (Formozov 1966). The existence of subnival caches means that collecting and storing of underground plant parts continues under the snow during winter or that subterranean stores are moved into snow chambers.

Nuts, seeds, and rarely fruits also are stored in underground chambers (Bailey 1926; Hamilton 1939). For example, Jameson (1947) found about 2,800 seeds (8.8 l) of the Kentucky coffee tree (*Gymnocladus dioica*) weighing about 5.6 kg in a storage chamber of the prairie vole (*Pitymys ochrogaster*), and a food storage chamber of the prairie vole in a hollow stump contained about 4.4 l of horse nettle (*Solanum carolinense*) fruits (Fisher 1945).

Hay storage usually occurs on the ground surface. The Alaska vole makes hay piles similar to those of pikas (*Ochotona* spp.) that range in size from a handful to a bushel basket full (Murie 1961; Youngman 1975). Common plants in hay piles of the Alaska vole examined by Murie in Mount McKinley National Park were fireweed (*Epilobium latifolium*), horsetail, coltsfoot, willow, mountain avens (*Dryas*), lupine (*Lupinus*), sage (*Artemisia hookeriana*), and lesser amounts of wintergreen (*Pyrola*), arctous (*Arctous*), alder (*Alnus*), and reed grass (*Calamogrostris canadensis*). Hay piles often contain only one species of plant. Each family of voles has several hay piles, and they add vegetation to each pile slowly enough that it cures before it is too deeply buried. Hay piles are placed in areas that stay dry or have good drainage. In shrubby and forested habitats, voles place hay piles on the exposed roots of trees, cradle them among stems of a shrub, or place them in low branches of a tree. In open habitats, Alaska voles place hay piles in protected cavities among rocks (Murie 1961). Similar haylofts are made by flat-headed voles (*Alticola strelzowi*) (Formozov 1966). This vole collects vegetation during the summer and stores it in protected sites in rock clefts. Each family usually stores 3–8 kg of hay. The voles further protect hay from being scattered by the wind and being grazed by ungulates by blocking the entrance to the hayloft with a wall constructed of hundreds of pebbles (Shubin 1959a, cited in Formozov 1966). Up to 2 kg of pebbles may be brought to the cache site in a day.

Some voles translocate cured hay into underground chambers or store vegetation directly in these chambers (Fisher 1945). The most notable example is Brandt's vole (*Microtus brandti*), a colonial vole that lives in the cold Mongolian steppes. In August, groups (perhaps families) of ten to twelve voles excavate chambers or clean existing chambers before gathering vegetation for winter and stock them with stems, leaves, and whole herbaceous plants (Formozov 1966). The burrow system may contain several chambers, the combined contents weighing up to 30 kg. At the onset of winter, the voles plug the entrance to the burrow with soil and live together in the burrow for over 4 months, using the stored hay as their only food supply. The subterranean winter existence of Brandt's vole is apparently a means of coping with the extremely cold, windy climate of the Mongolian steppes (Formozov 1966). The snow in this region is seldom thick enough to provide an insulative blanket that would permit subnival activity. The social nature of Brandt's vole and several other species of vole in winter (e.g., Alaska vole, Taiga vole [*Microtus xanthognathus*]) appears to have evolved at least in part to reduce metabolic rates and thereby extend the use of stored food (Murie 1961; Wolff and Lidicker 1981).

Fresh twigs also may be stored underground. Reddish-gray voles (*Clethrionomys rufocanus*) climb into saplings of the Siberian larch (*Larix sibiricus*) during September and gnaw off terminal branchlets up to 5 mm in diameter. They collect these on the ground, gnaw them into short segments, and carry them into underground galleries and chambers directly under the moss and litter layer (Litvinov and Vasil'ev 1973). The voles stack these cuttings to form compact bundles. Sixty-three segments were found in one small chamber. Consumption of the stored green leaves and bark of the larch begins in October, after the needles on trees turn yellow and fall. Leaves of wintergreen (*Pyrola rotundifolia*) also were stored in underground chambers, compacted in the same manner as larch twigs. Each storage chamber usually contains only one type of food. Litvinov and Vasil'ev (1973) also found that reddish-gray voles stacked freshly cut larch branches 12–20 cm long on the ground, constructing piles 20–25 cm high and 40 cm in diameter, and heather voles (*Phe-

nacomys intermedius) constructed similar twig caches (Foster 1961; Nagorsen 1987). Red tree voles (*Arborimus longicaudus*) clip three to fifteen terminal branchlets of Douglas-fir about 8–20 cm long and pile these haphazardly on top of their arboreal nest or deposit small compact bundles in one of several entrances to the nest chamber (Howell 1926). At frequent intervals during the day, voles pull branchlets one by one into the nest chamber and feed on the needles and bark.

Bank voles (*Clethrionomys glareolus*) store beard lichens (*Bryoria* spp.), which are epiphytes of conifer trees at high latitudes. The normally terrestrial bank vole collects beard lichens in the trees during summer and fall and stores them in tree cavities and in bird nests. A cache found by Pulliainen and Keränen (1979) in Finnish Lapland contained a cluster of 167 beard lichens stored in a tree cavity 1.5 m high. A second cache, located in a redwing (*Turdus iliacus*) nest, contained 538 lichens. Both caches were above the snow. *Bryoria fuscescens* comprised 95% of both caches, which weighed 32 and 46 g. Bank voles eat these lichens, and Pulliainen and Keränen suggested that the bank vole, which normally stays under a protective layer of snow during the severe, subarctic winter, also may use the large clusters of stored beard lichens as cover and shelter when foraging in trees. Other stores of beard lichens made by bank voles have been described by Formozov (1948, cited in Ognev 1964) and Nyholm (1978).

Little is known about why voles select certain plant species for storage. Gates and Gates (1980) compared the composition of roots and rhizomes found in a subnival storage chamber with plant composition of an old field in which the cache was found. They concluded that the abundance of underground plant parts in the store appeared to be proportional to abundance of those plants in the field; however, they also noted that certain plants, such as strawberry (*Fragaria virginiana*), common near the cache site, were not stored. The rank order of species dry biomass in the cache was the same as the rank order of species caloric value. The caloric values of the plant species that were not included in the cache were not determined, so it is not possible to establish whether the food value of these plants may have played a role in their not being stored. Crude protein in beard lichens (6–7%) cached by bank voles is twice that of ground lichens (*Cladonia;* 3–4%) available under the snow (Pulliainen and Keränen 1979). This may partially explain why the normally terrestrial bank vole prefers the arboreal beard lichens and risks higher chance of predation while gathering them above the snow.

Food stores usually are prepared in the fall and consumed during the winter (Couch 1925; Hamilton 1939; Formozov 1966). Murie (1961), for example, checked the sites of two known hay piles of the Alaska vole in spring and found that they had been consumed and that only a scattering of vole droppings remained. Redbacked and reddish-gray voles appeared to store food more intensely in years of high vole population density and low food availability (Litvinov and Vasil'ev 1973). Winter caches of heather voles are much larger than summer caches, the summer caches being small accumulations of twigs at secure feeding sites whereas the winter caches are substantial food reserves. Le Louarn (1974) suggested that food stores of the alpine pine vole (*Pitymys multiplex*) and common vole in France are important to winter survival, and Regnier and Pussard (1926; cited in Hatt 1930) suggested that presence of stored food is a factor in the ability of common voles to increase their populations rapidly in spring.

Old World Rats and Mice (Muridae)

The family Muridae comprises a large group of rodents closely related to the family Cricetidae indigenous to Africa, Europe, Asia, and Australia. Several species of *Rattus* and *Mus* are commensal with humans and today have a much wider distribution. Food hoarding is known in only half a dozen of the approximately 107 genera. The low frequency of food hoarding within the family may reflect the group's largely tropical center of distribution, where the food-hoarding habit may be reduced or absent, and the fact that few species have been closely studied. Food-hoarding behavior of laboratory rats (various strains of *Rattus norvegicus*) has been the object of numerous experimental studies. These studies have been treated extensively in chapter 5; here I describe only the hoarding behavior of wild populations or populations in simulated natural conditions.

Murid rodents accumulate food in underground larders, scattered surface caches, or both. The most thoroughly studied larder hoarder is the lesser bandicoot rat (*Bandicota bengalensis*), a serious agricultural pest in southern Asia from Pakistan to Indonesia. These rodents feed on and store rice, wheat, and the seeds of various legumes. During the wet season, when fields are flooded, bandicoots occupy high terrain around villages and bunds that border fields. When the paddies are drained just before harvest, bandicoots invade fields and quickly establish burrow systems (Chakraborty 1975; Poché et al. 1982). Population densities at this time may be 300 or more individuals per hectare (Poché et al. 1982). Burrows, which are usually occupied by a single adult, are simple interconnecting galleries leading to multiple entrances, with a nest chamber and one or more food-storage chambers located near the center (Chakraborty 1975; Poché et al. 1982; Sheikher and Malhi 1983; Malhi 1986). In rice fields, these galleries may extend up to 150 cm deep, but in wheat fields they are often only 6.5–24 cm deep (Poché et al. 1982).

Bandicoots begin harvesting grain within a few meters of their burrow entrances. They fell the grain by nipping the stalk near the base and then severing the stalk again just below the panicle. They transport these inflorescenses back to burrows one at a time. In some cases, panicles have been found neatly stockpiled on the ground surface next to a burrow entrance (Roy 1974). These surface stockpiles suggest that the rodents work above ground cutting and hauling food before going below ground to shuttle the panicles from the entrances to the storage chamber. In the storage chambers, the panicles again are stacked neatly with most of the inflorescenses (up to 95%) oriented in the same direction (Roy 1974; Sheikher and Malhi 1983). Bandicoots are not known to scatter hoard food.

The quantity of grain stored by lesser bandicoot rats has been well documented because of the rodents' great economic impact (Parrack 1969; Roy 1974; Chakraborty 1975; Fulk 1977; Mohana Rao 1980; Poché et al. 1982; Sheikher and Malhi 1983). For example, bandicoots in the Garhwal Himalayas of India stored an average of 390 g of wheat panicles per burrow ($N = 50$) with as much as 2 kg in one burrow (Sheikher and Malhi 1983; Malhi 1986). Twenty of the burrows contained no stored food. In Bangladesh, bandicoots cached up to 18 kg (mean $\pm$ SD = 2.1 $\pm$ 1.7 kg) of wheat in burrows (Poché et al. 1982). In west Bengal, bandicoot rats accumulated from 50 g to 4 kg (mean = 3.2 kg) of rice (Chakraborty 1975). Fulk (1977) estimated that bandicoot populations had stored 127 kg of beans (*Phaseolus mungo*) per hectare in bean fields and 93 kg of rice per hectare in rice fields in Sind Province, Pakistan. The portion of the crop stored by bandicoots has been estimated at 5.7–10.0% of the total crop (Roy 1974; Fulk 1977), all harvested with a 2-to 3-week period following draining of fields.

Other larder hoarders in this family include pest rats (*Nesokia indica*), Norway rats (*Rattus norvegicus*), Australian bush rats (*R. fuscipes*), and several species of wood mice and field mice (*Apodemus* spp.). Old World harvest mice (*Micromys minutus*) store seeds in captivity (Nowak and Paradiso 1983), but details of their hoarding behavior seem to be poorly known. Up to 1 pound (0.45 kg) of grain has been recovered from the burrow of the pest rat (Nowak and Paradiso 1983). Yabe (1981) described a Norway rat burrow with a chamber 50 cm wide and 100 cm long filled with 3 kg of peanuts in a garden of a Yokohama suburb. Steiniger (1949, cited in Wallace 1988) described special food-storage burrows separated from the more extensive nesting burrow. Calhoun (1963) described food-hoarding behavior of captive rats maintained in seminatural conditions. The hoarding behavior of Australian bush rats is similar to that of Norway rats (Wallace 1982). Many other field studies of Norway rat behavior have failed to uncover any evidence of food storing (e.g., Takahashi and Lore 1980). House mice (*Mus* sp.) also seldom store food under natural conditions (e.g., Sheikher and Malhi 1983).

Wood mice and field mice (*Apodemus*) excavate simple burrow systems of interconnecting galleries with the nest and food chambers often positioned directly under a tree (fig. 8.28). Storage chambers examined by Jennings (1975) contained English oak (*Q. robur*) acorns and hazelnuts (*Corylus avellana*) with remnants of snail shells and beetle exoskeletons. Imaizumi (1979) observed that large Japanese field mice (*A. speciosus*) and small Japanese field mice (*A. argenteus*) stored seeds in cavities under logs, in rock crevices, in underground chambers, and even in empty milk bottles and old shoes. Small Japanese field mice, but not large Japanese field mice, stored seeds in bird nest boxes positioned in trees. One of these boxes contained 1,060 maple (*Acer mono*) samaras.

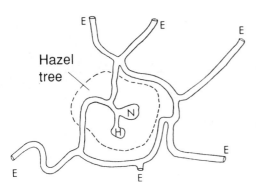

Wood mice and Japanese field mice also scatter hoard seeds (Imaizumi 1979; Jensen 1985; Jensen and Nielsen 1986). Both species studied by Imaizumi (1979) in fir forests at the foot of Mount Fuji, Japan, prepare caches by digging small holes with their forepaws, depositing one to six sunflower seeds into the hole from the mouth (they lack cheek pouches), and pushing soil or litter over the hole with the snout. Where the soil is soft, the mice push their snout directly into the soil without first digging a hole. Wood mice place many nuts and acorns in the walls of their underground runways (Jensen 1985; Jensen and Nielsen 1986). They make most caches within 2–3 m of the seed source. Japanese field mice usually make caches near conspicuous objects like tree roots, stones, and piles of dead leaves.

Under natural conditions, wood mice and Japanese field mice store maple samaras, hazelnuts, beechnuts, acorns, and pine seeds (Jennings 1975; Imaizumi 1979; Saito 1983a; Jensen 1985; Jensen and Nielsen 1986). The beechnut caches that Ashby (1967) described were probably those of the wood mouse (*A. sylvaticus*), because the voles that inhabited his study site are not known to scatter hoard.

Japanese field mice do not cache seeds uniformly during the night but hoard in distinct bouts separated by periods of no hoarding (fig. 8.29). The general hoarding activity pattern was similar for both species: storing bouts lasted for 10–70 min for small Japanese field mice and 10–90 min for large Japanese field mice. Intervals between hoarding bouts lasted 30–120 min for both species.

Wild and captive wood mice under natural conditions hoard more food during October through March than during other months (see fig. 5.15) (Dufour 1978; but see Schenk 1979). The tropical lesser bandicoot rats studied by Mohana Rao (1980) in Indian rice fields show a similar seasonal pattern in storage (see fig. 2.9), as the hoarding season coincides with availability of ripe grain in the fields.

The use of stored food by murids has not been well documented. Chakraborty (1975) stated that young bandicoot rats feed on stored grain after the grain crop has been harvested and food in fields is depleted. A large cache of maple samaras made during fall by small Japanese field

Figure 8.29
The nocturnal pattern of hoarding
trips taken by large and small
Japanese field mice. From Imaizumi 1979.

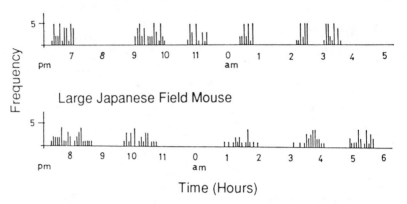

mice was repeatedly visited by these rodents until all the seeds were consumed by early April (Imaizumi 1979). Mohana Rao (1980) found a strong
relation between breeding effort (measured by mean litter size) and food
stored in burrows (see fig. 2.9), suggesting that breeding is cued to food
availability and that stored food may play a role in the breeding effort. In
Rattus and *Apodemus,* females store more than males, and they store
especially intensely just before and during lactation (Calhoun 1963; Schenk
1979). Despite their supposed strong reliance on stored food, much of
the grain stored by bandicoot rats spoils (Parrack 1969; Parrack and
Thomas 1970; Roy 1974). Imaizumi (1979) and Jensen (1985) observed
that seeds stored by individual *Apodemus* may be recovered by other individuals and suggested that stored food is used communally by the population or clan. Captive Norway rats under seminatural conditions did not
defend stored food; food that rats placed in their larders was used without strife by all members of the population (Calhoun 1963).

Dormice (Gliridae)

Dormice comprise a small, Old World family of forest-dwelling rodents. Several species accumulate nuts and fruits as a winter food supply.
The edible dormouse (*Myoxus glis*) stores acorns, hazelnuts, chestnuts,
and rose hips in tree cavities, abandoned bird houses, and underground
chambers (Koenig 1960). The mouse constructs a hibernation nest near
the food store. A captive edible dormouse stored 200 acorns (both wormy
and healthy nuts), many apple cores, bread, cookie pieces, two corks,
and a segment of wax candle in a cavity adjacent to its nest (Koenig
1960). Another individual stored sweet cherries and pieces of candy. Female edible dormice tend to store more food than do males (Koenig
1960). Two other species, the hazel dormouse (*Muscardinus avellana-
rius*) and the forest dormouse (*Dryomys nitedula*), store Swiss stone pine
(*Pinus cembra*) seeds and other foods in chambers near their hibernacula
(Oswald 1956; Nowak and Paradiso 1983).

Harvesting begins in summer and continues until early fall. The
edible dormouse enters hibernation in September or November and

emerges in late spring. During this long period of winter dormancy, the dormouse periodically arouses and feeds from the food store. Edible dormice have often been found hibernating in small groups of up to eight individuals in a single cavity; however, no information seems to be available on whether these mice construct a communal food cache.

Jumping Mice (Zapodidae)

As far as I can determine, no stored food has ever been reported in the burrows of jumping mice under natural conditions. Jumping mice hibernate and deposit body fat rather than store food. Hamilton (1941: 261), however, stated without giving any details that captive woodland jumping mice (*Napaeozapus insignis*) "practice a limited storage." This needs to be confirmed.

Jerboas (Dipodidae)

Most jerboas do not store food. However, Nowak and Paradiso (1983) stated that the greater Egyptian jerboa (*Jaculus orientalis*) and the lesser five-toed jerboa (*Alactagulus pumilio*) hoard some food in their burrows, and Eisenberg (1975) reported that *Jaculus* will store some "plant material" in its burrow. None of these authors give details concerning the type of plant material stored or where this is accumulated. On the other hand, many studies describe the burrow systems of jerboas in detail without mentioning stored food reserves (e.g., Naumov and Lobachev 1975). Of particular interest is the study of El Hilali and Veillat (1975), who excavated forty-one active burrows of the greater Egyptian jerboa during November and found hibernating jerboas but no food stores. Apparently, food storage in the greater Egyptian jerboa is an uncommon or geographically restricted practice.

Most jerboas hibernate during winter. Before dormancy, they accumulate large deposits of fat. Eisenberg (1975) suggested that these fat reserves serve as an alternative to food caching as a means of surviving periods of food scarcity.

Old World Porcupines (Hystricidae)

Ewer (1965) stated that African brush-tailed porcupines (*Atherurus africanus*) store food by thrusting it into the ground and ramming it with the incisors. No hole is dug to receive the food item, and the food is not buried by raking dirt over the cache site with the forepaws, as is done by other hystrichimorphous rodents (e.g., the green acouchi, Morris 1962). African brush-tailed porcupines are scatter hoarders (Ewer 1968). Nothing else seems to have been reported on food-hoarding behavior of this group.

Agoutis and Acouchis (Dasyproctidae)

The Dasyproctidae includes two genera of relatively large, Neotropical, forest rodents: eleven species of agoutis (*Dasyprocta*) and two species of acouchis (*Myoprocta*). Food hoarding has been described for the green acouchi (*Myoprocta acouchy* [includes *M. pratti*]; Morris 1962; Lyall-Watson 1964), the red acouchi (*Myoprocta exilis;* DuBost 1988), and the agouti (*Dasyprocta punctata;* Smythe 1970, 1978; Murie 1977;

Figure 8.30
A wild Central American agouti
(*Dasyprocta punctata*) begins to
walk away to scatter hoard the
acorn held between her incisors.
Photograph courtesy W. Hallwachs

Smythe, Glanz, and Leigh 1982; Hallwachs 1986). Some other members
of the family probably also hoard food but this has not been documented.

Agoutis and acouchis (fig. 8.30) bury individual seeds in shallow
caches on the forest floor. Morris (1962) described five steps in the food-
caching behavior of the green acouchi: (1) digging a small but deep hole
with the front feet while holding a food object in the mouth, (2) dropping
the food item into the hole, (3) pressing the food item firmly into the hole
with rapid, alternating pats of the forefeet, (4) filling the hole with soil
with forward movements of a front foot, alternating feet until the hole is
completely filled, and (5) covering the cache site with one or more objects such as leaves and pieces of bark. The food-hoarding behavior of
agoutis is similar (Hallwachs 1986), and caches are covered with 0.5–3.0
cm of soil.

In the wild, agoutis, and very likely acouchis as well, bury large
seeds and nuts (> 0.5 g) at the bases of trees and logs (Smythe 1970,
1978; Hallwachs 1986; Smythe, Glanz, and Leigh 1982). If nuts are surrounded by a fleshy pericarp, the agoutis eat this layer before they store
the nuts. Genera known to be stored by agoutis include *Anacardium*,
*Tontalea, Astrocaryum, Dipteryx, Scheelea, Quercus, Hymenaea, Gusta-
via,* and *Spondias* (Heaney and Thorington 1978; Smythe 1978; Hallwachs 1986; Smythe, Glanz, and Leigh 1982). Agoutis bury many more
nuts in the late dry and early wet season at a time when many fruits fall to
the ground where agoutis forage than during the late wet season when
fruit is scarce (Smythe 1978; Symthe, Glanz, and Leigh 1982). They retrieve the cached nuts and seeds during the late wet and early dry season
when there is a severe shortage of freshly fallen fruit.

Octodonts (Octodontidae)

Octodonts comprise a small group of rodents restricted to southern South America. Several species are known to store food, but few details are available. One Chilean species, the coruro (*Spalacopus cyanus*), a fossorial rodent that lives communally in extensive, interconnected burrow systems, stores tubers of lilies and bulbs of various other plants in underground chambers (Cabrera and Yepes 1960). Along the coastal plain of central Chile, this nomadic rodent feeds primarily on the tubers of huilli (*Leucoryne ixiodes*) during summer, rapidly excavating new galleries into unexploited patches of this lily as they deplete old feeding areas (Reig 1970). Coruros accumulate provisions in summer and use them during the harsh winter (Cabrera and Yepes 1960). The degu (*Octodon degus*) also hoards food for winter use (Woods and Boraker 1975; Nowak and Paradiso 1983), but little seems to be known about its food-storing habits.

Tuco-tucos (Ctenomyidae)

Tuco-tucos (*Ctenomys* spp.) are fossorial rodents endemic to southern South America that resemble pocket gophers (Geomyidae) but lack external cheek pouches. They feed on roots, tubers, and stems. Pearson (1959) and Nowak and Paradiso (1983) state that tuco-tucos store small amounts of food in storage chambers in their burrows, but nothing else seems to have been reported on their food-storing behavior.

8.6 LAGOMORPHS

Pikas (Ochotonidae)

Pikas are found in western North America (two species) and in Asia from the Ural Mountains to Japan and from northern Siberia to the Himalayas (sixteen species). On both continents, pikas inhabit open talus or rocky terrain in the subalpine and alpine zones but also occur at sea level in certain areas (i.e., southwestern British Columbia and northern Siberia). In Asia some species (e.g., the Daurian pika, *Ochotona daurica*) also inhabit semidesert and high steppe habitats, where they excavate burrows in soil (Loukashkin 1940; Okunev and Zonov 1980).

Most species that have been studied store vegetation in small haystacks near the entrance to their nests. Completed haystacks of the pika (*O. princeps*) of western North America are typically mounds 30–70 cm high and 60–80 cm broad at the base (fig. 8.31), although size varies greatly. Completed hay piles weighed by Millar and Zwickel (1972a) in late September ranged from 0.4–6.0 kg. Males had larger hay piles than females and adults had larger hay piles than juveniles. Productivity of foraging sites and distance between the hay pile and foraging sites also influenced the size of hay piles (Millar and Zwickle 1972a; see fig. 3.12).

Each hay pile is constructed and defended by a single territorial pika (Grinnell, Dixon, and Linsdale 1930; Kilham 1958b; Broadbooks 1965; Millar and Zwickel 1972a; Barash 1973; Conner 1983). Territories include nearly barren talus slopes, where pikas nest and store hay, and well-vegetated meadows or forests, where pikas feed and collect vegeta-

tion (Kilham 1958b; Kawamichi 1976; Huntly, Smith, and Ivins 1986). On
large talus slopes, pika territories and hay piles are located near the margins of the talus near foraging areas (Millar and Zwickel 1972a). Pikas
place hay piles in the open or under the shelter or a rock overhang or log
(Howell 1924) (fig. 8.31). Underground hay storage was reported by Kawamichi (1976), but he did not excavate or describe these stores. Locations of haystacks often are the same from year to year, even if the
occupant of the territory changes. Pikas may have more than one hay pile
(Millar and Zwickel 1972a; Kawamichi 1976). Broadbooks (1965) found
that on talus slopes with small rocks, and consequently small crevices between rocks, each pika made several small hay piles.

The bulk of the material in hay piles usually consists of grasses,
forbs, and sedges, but new growth of woody plants, conifer boughs,
moss, and lichens often are included (Beidleman and Weber 1958; Haga
1960; Broadbooks 1965; Johnson and Maxell 1966; Cahalane 1947; Elliott

Hay piles made by other species of pika may differ in size, number,
and placement. The Daurian pika, a colonial species of high steppe of
northern Manchuria, constructs small hay piles that when dry weigh
1.0–2.5 kg (Loukashkin 1940). Pallas' pika (*O. pallasi*), of the Mongolian
mountains and steppe, fills underground chambers and passageways with
dried vegetation. Each burrow system may contain from four to twelve
food chambers about 40 cm in diameter (Okunev and Zonov 1980). Food
stores may weigh from 0.2–20 kg per burrow (Shubin 1959b, cited in
Okunev and Zonov 1980). The Siberian pika (*O. alpina*) in Manchuria
hides hay in cavities of trees, among rocks, in stumps, and under wind-
felled trees (Loukashkin 1940). The same species in Japan forms hay
piles 60 cm high and 100 cm in diameter (Haga 1960). Unlike the North
American pikas, which live solitarily, Pallas' pikas and Siberian pikas work
in pairs to construct hay piles (Haga 1960; Okunev and Zonov 1980).

1980; Nikolskiy, Guricheva, and Dmitriyev 1984). Although the impressions of some observers have been that the composition of hay piles represents what is most readily available within the feeding area (Loukashkin 1940; Beidleman and Weber 1958; Haga 1960), others have found that pikas do exhibit preferences for certain plant species and that preferred plants often are harvested during their flowering time (Grinnell, Dixon, and Linsdale 1930; Broadbooks 1965; Huntly, Smith, and Ivins 1986). Millar and Zwickel (1972a) compared the use (percent by weight of hay piles) of various plants by pikas in Colorado to their availability (percent cover) and found that buffaloberry (*Shepherdia canadensis*), hedysarum (*Hedysarum* sp.), currant (*Ribes* spp.), grasses (*Elymus* spp.), and sedges (*Carex* spp..) were used more frequently than expected, whereas twin flower (*Linnaea borealis*), mountain avens (*Dryas hookeriana*), common juniper (*Juniperus communis*), and saxifrage (*Saxifraga bronchialis*) were used less frequently than expected. Preferred plants had significantly higher protein content than did nonpreferred plants, and preferred plants also tended to be clumped, which may have made them easier to harvest. Grinnell, Dixon, and Linsdale (1930) and Broadbooks (1965) also observed that pikas seemed to prefer certain species of plants. Pikas in Colorado are more selective of the types of plants they harvest and travel greater distances when collecting vegetation for hay than when grazing (Huntly, Smith, and Ivins 1986).

Fecal matter is frequently included in the hay piles. Fecal pellets of marmots have been reported in hay piles most often (Broadbooks 1965; Johnson and Maxell 1966; Elliott 1980), but coyote scats (Taylor and Shaw 1927) and elk and pika pellets (Beidleman and Weber 1958) have also been found in hay piles. Pikas, like other lagomorphs (Eden 1940; Kulwich, Struglia, and Pearson 1953), are coprophagous and routinely ingest their own feces (Haga 1960; Johnson and Maxell 1966). Broadbooks (1965) observed pikas eat marmot pellets from hay piles, and he concluded that fecal pellets were deliberately stored for their food value.

The timing of hay pile construction differs among species, and for some species it may vary considerably among localities and years. Variation in the timing of maturity and senescence of vegetation is apparently responsible for this difference. Smith (1974) observed pikas at low elevation (2,550 m) in the Sierra Nevada, California, begin collecting hay in mid-May, whereas at high elevation (3,400 m), where plant growth was delayed by the late melting of the snowpack, haying started in early July. In a year following very light winter snow, haying terminated in mid-August, when vegetation was no longer efficiently harvested; following a year with abundant winter precipitation, haying continued until early October (Smith 1974). Barash (1973) and Conner (1983), who documented seasonal changes in haying rate, found that at the peak of the harvest season, pikas may spend over half of the time they are above ground collecting hay, making ten to fifteen haying trips per hour. North American pikas begin gathering hay soon after the peak in availability of green vegetation in June or early July, but they usually do their most active haying in August and September. By October, when vegetation in meadows be-

comes sparse, the amount of time they spend collecting hay decreases
(Conner 1983). Juvenile pikas become independent of adults, establish
territories, and begin haying in late July (Broadbooks 1965; Barash 1973).
Okunev and Zonov (1980) found that Pallas' pika began making small hay
piles in early June and that collection of hay for winter began in late July
and early August. The Siberian pika begins collecting hay in October on
Hokkaido, Japan, and haying continues until snow falls (Haga 1960). The
same species begins collecting hay in July in Manchuria (Loukashkin
1940).

There is disagreement as to whether pikas actively dry vegetation.
Dalquest (1948) states that pikas cure hay by spreading it out on rocks in
the sun; if rain threatens, the hay is collected and stored under rocks.
Loukashkin (1940) observed Daurian pikas turn over hay piles, a behavior that may facilitate drying, and he stated that Siberian pikas place small
packets of vegetation on warm rocks in the sun to dry and later move the
cured hay to protected places. Grinnell, Dixon, and Linsdale (1930) suggested that the many dead twigs found in some hay piles may facilitate
hay ventilation. Others, however, have not observed curing behavior
(e.g., Barash 1973). Hayward (1952) and Millar and Zwickel (1972a)
found the contents of many hay piles to be black and moldy, and Hayward
believed the curing of hay to be purely coincidental with good weather
and the slow building of the haystack.

Pikas use stored hay to supplement their winter diets of lichens,
bark, roots, and other available plant materials (Johnson and Maxell
1966; Okunev and Zonov 1980; Conner 1983). Conner (1983) concluded
that hay piles may be especially important when variable environments
prevent surface foraging during winter or delay the production of new
green vegetation in the spring (see chapter 2).

9

Food-Hoarding Birds

Food hoarding is well developed in certain raptors and in omnivorous birds that include seeds as a major component of their diets. Many jays, crows and nutcrackers (Corvidae), tits and chickadees (Paridae), falcons and hawks (Falconidae), true owls (Strigidae), shrikes (Laniidae), nuthatches (Sittidae), and woodpeckers (Picidae) are conspicuous and well-known food hoarders, but eight other families each include only one or a few food-hoarding species (table 9.1). Except for several species of falcons, woodpeckers, and owls, birds are scatter hoarders that hide food items in soil, among foliage, in forks of tree branches, and in bark crevices. Appendix I lists food-hoarding birds mentioned in the text.

9.1 RAPTORS

Hawks (Falconidae, Accipitridae, and Sagittariidae)

Of 287 species of hawks, falcons, eagles, vultures, and related raptors (Brown and Amadon 1968), I have found mention of food storage for only 25 species (table 9.2). Table 9.3 summarizes the taxonomic distribution of food hoarding within the Falconiformes. Even though it is unlikely that all reported occurrences of food hoarding have been included in these tables and even though future studies will no doubt identify other food-storing raptors, additions to this list are not likely to change the general pattern already evident. That is, food hoarding is widespread in true falcons (*Falco*) and absent or weakly developed in most other hawks and eagles. Why this is so is not clear. It is not surprising that some groups do not store food. New World vultures (Cathartidae), Old World vultures (Aegypiinae), and other scavengers, for example, would seem to gain little benefit by storing carrion. Tropical hawks may store less food than temperate zone species because of the likelihood of flesh decomposing before they retrieve it. Ospreys (Pandionidae) may not store fish, their primary prey, because fish decompose quickly. But members of a number of other groups, such as harriers, kites, and accipiters, would seem to have as much to gain from food caching as do falcons.

Hawks (used here as a general term to refer to all food-hoarding members of the order) store excess prey by furtively depositing them at a storage site, usually near hunting perches, nests, or roost sites. They do not cover cached prey, but they may conceal prey in foliage. Hawks handle prey with their beaks or feet. American kestrels (*Falco spar-*

Table 9.1 Bird taxa known to store food with a summary of their food-storing behavior

Order Family	Dispersion[a]	Food Type[b]	Substrate/ Location[c]	Storage Duration[d]
Falcons and hawks				
Falcons (Falconidae)	S,L	SM,Bi,Re,A	F,T	S
Hawks et al. (Accipitridae)	S	MM,SM,Bi	F,T	S
Secretary bird (Sagittariidae)	S	SM	G	S
Owls				
Barn owls (Tytonidae)	L	SM	N	S,L
True owls (Strigidae)	L,S	SM,Bi,I	N,C,T,F	S,L
Woodpeckers				
Woodpeckers (Picidae)	L,S	Nu,I	T,C	L
Perching birds				
Crows, jays, and nutcrackers (Corvidae)				
Jays	S	Nu,I,Ca	S,F,L	L
Nutcrackers, pinyon jays	S	Nu	S,L	L
Crows, raven, magpies	S	I,Nu,E,Mi	S,F,L	S,L
Australian butcherbirds (Cracticidae)	S	Re,I,Ca	Sp,F	S,L
White-winged chough (Grallinidae)	S	Mi	S	S
Bowerbirds (Ptilonorhychidae)	S	Fr	T	S
Chickadees and tits (Paridae)	S	S,Nu,I	T,F,S,L	S,L
Nuthatches (Sittidae)	S	S,I	T,S	S,L
Thrushes et al. (Muscicapidae)	S	I	T	S
Shrikes (Laniidae)	S,L	SM,Bi,Re,I	Sp,T	S,L
Mynas (Sturnidae)[e]	?	S	T	?

Note: See text for references and exceptions to the general patterns within taxa.

[a] Dispersion patterns: L, larder; S, scattered.

[b] Food types: A, amphibians; Bi, birds; Ca, carrion; E, eggs; Fr, fruit; I, invertebrates; Mi, miscellaneous; MM, medium mammals; Nu, nuts; Re, reptiles; S, seeds; SM, small mammals.

[c] Substrates and locations: C, cavity or chamber; F, foliage; G, ground surface; L, litter; N, nest; S, soil; Sp, spines; T, tree trunk and branches.

[d] Storage duration: S, short-term (generally <10 days); L, long-term (generally >10 days).

[e] Needs confirmation; see text.

verius), the species whose storing behavior has been studied most extensively, wedge, tamp, and push prey with their beak to lodge them securely in cache sites (Mueller 1974; Balgooyen 1976; Collopy 1977; Toland 1984; Nunn et al. 1976). A merlin concealed a small bird by holding it in its talons and hovering while lowering the prey into a shrub (Greaves 1968). Hawks often inspect storage sites from a short distance after caching and may reposition or remove and recache prey if it is not well hidden (Mueller 1974; Fox 1979).

Hawks usually hide prey in shrubs, trees, and clumps of grass. For example, Collopy (1977) found that sparrow hawks near Arcata, California, stored 88% of prey in grass clumps, 9% on posts, and 3% in shrubs. Toland (1984), working in Missouri, found that male American kestrels used tree limbs and holes as cache sites most frequently (58%), whereas

Table 9.2 Hawks (Falconiformes) known to store food with a list of prey types stored

Family Species	Prey Type[a]	References
Accipitridae		
Accipiter gentilis	R, P	Schnell 1958
Accipiter nisus	R, P	Owen 1931; Ashmole 1987
Kaupifalco monogrammicus	NP	Brown and Amadon 1968
Buteo platypterus	NP	Lyons and Mosher 1982
Buteo buteo	R	Melchior 1975
Buteo regalis[b]	L	Angell 1969
Stephanoaetus coronatus[b]	MM	Brown 1966, 1971
Sagittariidae		
Sagittarius serpentarius	R	Brown and Amadon 1968
Falconidae		
Falco rupicoloides[c]	?	Cade 1982
Falco sparverius	R, S, P, L, Sn, F	Stendall and Waian 1968; Mueller 1974; Balgooyen 1976; Collopy 1977; Toland 1984
Falco tinnunculus	R	Clegg 1971; Rijnsdorp et al. 1981
Falco punctatus[c]	?	Cade 1982
Falco chicquera[c]	?	Cade 1982
Falco columbarius	P, R	Greaves 1968; Oliphant and Thompson 1976; Warkentin and Oliphant 1985
Falco novaeseelandiae	P	Fox 1979
Falco cuvieri[c]	?	Brown and Amadon 1968
Falco eleanorae	P	Vaughan 1961; Walter 1979
Falco rufigularis (= *albigularis*)	B, P	Beebe 1950
Falco femoralis[c]	?	Cade 1982
Falco biarmicus[c]	?	Cade 1982
Falco mexicanus	P, NP	Oliphant and Thompson 1976
Falco cherrug[c]	?	Cade 1982
Falco rusticola[c]	?	Cade 1982
Falco deiroleucus	Ba	Jenny and Cade 1986
Falco peregrinus	P, NP	Beebe 1960, Trealeaven 1980

[a] Ba, bat; F, frog; La, lagomorph (rabbit); L, lizard; MM, medium-sized mammal;
NP, non-passerine bird; P, passerine bird; R, rodent; S, shrew; Sn, snake.
[b] Evidence for food storing is circumstantial.
[c] Apparently no details of caching behavior have been published.

females used grass clumps most often (68%). As a consequence of this cache site selection, males stored prey in more elevated sites than did females. European kestrels conceal prey within shrubs or clumps of grass (Parker 1977), and bat falcons (*Falco rufigularis*) in Venezuela cache bats and small birds in bromeliads and other epiphytes (Beebe 1950). Eleonora's falcons (*Falco eleanorae*) tuck small birds head first into niches of rocks and small shrubs (Vaughan 1961). Merlins (*Falco columbarius*) de-

Table 9.3 Taxonomic distribution of food-hoarding species in the Falconiformes

Family Subfamily	Common Name	Species in Taxon	Species Known to Store Food in Taxon
Cathartidae	New World vultures	7	0
Sagittariidae	Secretary birds	1	1
Accipitridae			
Elaninae	White-tailed kites	8	0
Perninae	Honey buzzards	11	0
Milvinae	Kites	12	0
Accipitrinae	Accipiters	53	2
Buteoninae	Eagles and round-winged hawks	93	5
Aegypiinae	Old World vultures	15	0
Circinae	Harriers	13	0
Circaetinae	Serpent eagles	12	0
Pandionidae	Ospreys	1	0
Falconidae			
Herpetotherinae	Laughing and forest falcons	6	0
Polyborinae	Caracaras	10	0
Polihieracinae	Pigmy falcons	8	0
Falconinae	True falcons	37	17
Total		287	25

posit small passerines inside shrubs (Greaves 1968; Dickson 1979) and on branches near the trunk of deciduous and coniferous trees (Oliphant and Thompson 1976).

Prey stored by hawks are always vertebrates, most frequently rodents, shrews, small birds, lizards, snakes, and frogs (e.g., Schnell 1958; Balgooyen 1976; Collopy 1977; Toland 1984). Lyons and Mosher (1982) reported the caching and consumption of a nestling broad-winged hawk (*Buteo platypterus*) by its parent. Crowned eagles (*Stephanoaetus coronatus*) apparently stored a suni (*Neotragus moschatus;* a small antelope), a tree hyrax (*Dendrohyrax arboreus*), and a portion of a bushbuck (*Traelaphus scriptus*) (Brown 1966, 1971). The prey stored by American kestrels does not differ significantly in species composition from vertebrate prey captured (Collopy 1977). Hawks usually cache freshly caught prey (often minus the head) or remains of prey that are too large to be consumed by an adult or group of nestlings in one meal (e.g., Schnell 1958; Collopy 1977; Toland 1984; Bildstein 1987).

Hawks are usually scatter hoarders, placing one prey item at each cache site. Hawks may use the same cache sites repeatedly, or they may use a site only once and then abandon it. A transmitter-equipped merlin tracked for 225 hours near Saskatoon, Saskatchewan, Canada, repeatedly cached prey in two spruce trees (Warkentin and Oliphant 1985), whereas another merlin in the same vicinity seldom cached prey in the same site more than once (Oliphant and Thompson 1976). Female American kestrels in the Sierra Nevada, California, repeatedly used from two to six cache sites, which they provisioned with food brought by the male (Balgooyen 1976). Males maintained fewer cache sites than did females. No joint cache sites were used; cached prey were the exclusive property of the cacher (Balgooyen 1976). In Missouri, however, male American kestrels often cached food in view of females, and after males departed, females retrieved the prey (Toland 1984). During courtship, female New Zealand falcons (*Falco novaeseelandiae*) take prey from the male's caches, a form of remote food passing (Fox 1979).

Some falcons store prey in larders. An American kestrel studied by Stendell and Waian (1968) in southern California repeatedly used a small pine tree as a cache site, storing seventeen prey items in the tree over a 40-day period in winter. As many as five prey were found in the tree simultaneously. Toland (1984) described how an American kestrel killed and stored seven white mice in two grass clumps in close proximity. Larders of Eleonora's falcon often contain fifteen to twenty birds or remains of birds placed together within several meters of a nest (Vaughan 1961).

Hawks store food throughout the year, but the intensity of hoarding may change with the season. American kestrels in the Sierra Nevada, where the species is a summer resident, store only during the breeding season, initiating caches shortly after arrival in spring and stopping 1–2 weeks after young hatched (Balgooyen 1976). In southern and coastal California, where American kestrels are permanent residents, they also store food in winter (Stendell and Waian 1968; Collopy 1977). In Missouri, American kestrels stored 58% of prey handled in fall and winter but only 7% of prey handled in spring and summer (Toland 1984).

Prey caching may occur at any time of day, but retrieval of prey often is most prevalent in the evening (see fig. 2.3) (Mueller 1974; Oliphant and Thompson 1976; Collopy 1977; Rijnsdorp, Daan, and Dijkstra 1981). Evening consumption of stored prey may be a mechanism for delaying meals that could have a detrimental influence on flight dynamics (Rijnsdorp, Daan, and Dijkstra 1981) and for ensuring adequate food for maintenance during the period of nocturnal inactivity (Collopy 1977). Food items retrieved late in the day often were cached earlier the same day. Thus caching is usually brief—a few hours to 2–3 days (e.g., Schnell 1958; Oliphant and Thompson 1976; Ashmole 1987)—but sometimes longer (Stendell and Waian 1968; Balgooyen 1976; Fox 1979).

Hawks appear to retrieve a high proportion of the items they store (Collopy 1977; Toland 1984; Bildstein 1987). Toland (1984) found that American kestrels in central Missouri recovered 78% of all prey (N = 116) they cached. Collopy (1977), working on the assumption that the probability of observing prey caching and retrieving were equal, estimated that American kestrels recovered 70% of all prey they stored. Unrecovered prey may be removed by scavengers, abandoned, or lost. High retrieval rates also have been noted under experimental conditions (Mueller 1974). Although no quantitative data on cache recovery rates appear to exist for other hawks, anecdotal observations suggest that they are similarly high. A high incidence of prey retrieval suggests that cached food is important in energetic budgets of hawks.

Hawks cache food for a variety of reasons. First, some hawks store prey during the breeding season to be eaten by the attending female or to be used by the female to feed nestlings (Schnell 1958; Sperber and Sperber 1963; Balgooyen 1976; Oliphant and Thompson 1976; Lyons and Mosher 1982). Second, hawks store prey during winter. A cache of one to several prey would lessen the impact of sudden inclement weather or rapid changes in availability of prey, for example, caused by a thick blanket of snow (Tordoff 1955; Collopy 1977; Fox 1979; Rijnsdorp, Daan, and Dijkstra 1981; Warkentin and Oliphant 1985). Third, goshawks, New Zealand falcons, and American kestrels have been observed storing prey in response to an intruder (Schnell 1958; Fox 1979; Bildstein 1982). In the case of the goshawk, the female cached prey she was feeding to young and then returned to attack the intruder. The adaptive advantages of prey caching were discussed in chapter 2.

Owls (Strigidae and Tytonidae)

Food storing has been reported frequently in owls from north-temperate and arctic regions but has not been reported, to my knowledge, in species inhabiting tropical climates. Unlike hawks, food hoarding is widespread within the order, being represented in barn owls (Tytonidae) and both subfamilies of true owls (Striginae and Buboninae) (table 9.4).

Owls usually store small mammals (rodents and shrews) or small birds (e.g., Reese 1972; Phelan 1977; Cope and Barber 1978; Ritchie 1980; Solheim 1984; Korpimäki 1987). Screech owls in northern Ohio, however, store frogs, fish, and invertebrates (e.g., crayfish and leeches) as well as rodents and birds (table 9.5; Van Camp and Henny 1975); elf owls (*Micrathene whitneyi*), which are insectivorous, cache moths and crickets in nest cavities (Ligon 1968). Eurasian pygmy owls (*Glaucidium passerinum*) store small mammals and birds but apparently prefer to store and feed on rodents (Solheim 1984). Only when rodents were scarce did pygmy owls store large numbers of birds and shrews.

Caching behavior of free-ranging owls has seldom been described, probably because most owls forage at night when observation is difficult. A captive hawk owl deposited mice dorsal side up next to a wall, nudged them against the baseboard several times with its beak, and then retreated and visually scrutinized the cached prey (Collins 1976). Its behavior was furtive, similar to that displayed by hawks. A captive great horned owl aggressively defended a cache from a person that entered the room (Collins 1976). After the person left, the owl retrieved and recached the prey elsewhere. Prey cached by owls in the wild are frequently decapitated, eviscerated, or partially consumed.

Most owls, in marked contrast to hawks, accumulate prey items in larders. Kaufman (1973), Collins (1976), and others referred to these larders as stockpiles. Stockpiles occur in several situations. During the breeding season, stockpiles may be made on the nest rim, (e.g., snowy owls, *Nyctea scandiaca;* Pitelka, Tomich, and Treichel 1955). Cavity-

Table 9.4 Owls (Strigiformes) known to store food with a summary of type of prey stored, cache dispersion patterns, and nest type

Family Subfamily *Species*	Prey Type[a]	Prey Dispersion[b]	Nest Type[c]	References
Tytonidae				
Tyto alba	R	L	C	Hawbecker 1945; Wallace 1948; Reese 1972; Kaufman 1973
Strigidae				
Buboninae				
Otus asio	R, P, F, Fi, I	L	C	VanCamp and Henny 1975; Phelan 1977; Cope and Barber 1978
Bubo virginianus	R, L	S	O	Collins 1976
Bubo bubo	NP	S	O	Willgohs 1974; Cugnasse 1977
Nyctea scandiaca	R	L	O	Pitelka et al. 1955; Tulloch 1968
Surnia ulula	R	S	O	Collins 1976; Ritchie 1980; Fridzen 1984; Lane and Duncan 1987
Glaucidium passerinum	R, P	L, S	C	Solheim 1984
Glaucidium gnoma	R	?	C	Bent 1938
Micrathene whitneyi	I	L	C	Ligon 1968
Athene cunicularia	R, I	L	C	Rich and Trentlage 1983
Athene noctua	NP	L	C	Lockley 1938
Striginae				
Strix aluco	?	?	C	Räber 1950
Strix varia	?	?	C, O	Fox 1979
Strix nebulosa	R	L	O	Högland and Lansgren 1968
Asio otus	?	?	O	Räber 1950
Asio flammeus	?	?	O	Lyons and Mosher 1982
Aegolius funereus	R	S	C	Catling 1972; Korpimäki 1987
Aegolius acadicus	R	S	C	Mumford and Zusi 1958

[a]Prey type codes: Fi, fish; F, frog; I, invertebrate; L, lagomorph (rabbit); NP, non-passerine bird; P, passerine bird; R, rodent.
[b]Prey dispersion codes: L, larder; S, scattered prey.
[c]Nest type codes: C, cavity nest; O, nest in the open.

nesting species like barn owls (*Tyto alba* [Hawbecker 1945; Wallace 1948; Reese 1972]), Eurasian pygmy owls (Solheim 1984), screech owls (*Otus asio;* Van Camp and Henny 1975), and boreal (Tengmalm's) owls (*Aegolius funereus;* Korpimäki 1987) stockpile prey in cavities, and, in general, cavity-nesting owls tend to be larder hoarders (table 9.4). Eurasian pygmy owls may maintain more than one larder (Solheim 1984). A screech owl that made its way into a barn stored twenty-two 1-day old chicks on a shelf (Cope and Barber 1978). Although stockpiles seem to be the predominant mode of storage, some owls scatter hoard prey (Rich and Trentlage 1983; Solheim 1984). Hawk owls (*Surnia ulula*), for example, cache individual rodents on tree branches near nest sites (Ritchie

Table 9.5 Types of prey found in nest boxes of screech owls in northern Ohio

Season[a]	Number of Prey Items	Percent of Prey Items				
		Mammals	Birds	Frogs	Fishes	Invertebrates
Breeding	477	30.4	64.8	1.0	2.9	0.8
Nonbreeding	121	60.3	26.4	5.0	5.8	2.5

Source: After Van Camp and Henny 1975.

[a]Breeding, 26 March–7 June; nonbreeding, 15 October–23 February.

1980; Lane and Duncan 1987), and wintering saw-whet owls (*Aegolius acadicus*) and boreal owls store rodents on tree branches near roost sites (fig. 9.1; Mumford and Zusi 1958; Catling 1972; Bondrup-Nielsen 1977).

Few data seem to be available concerning daily patterns of prey storage. Barred owls (*Strix varia*) cached prey in the early morning, at the end of their foraging period, and consumed this stored food in the early evening before other prey became active (Fox 1979).

Most reports of food storing in owls are from the breeding season, and for many owls, food caching plays an integral role in the breeding effort (e.g., Pitelka, Tomich, and Treichel 1955; Ligon 1968; Tulloch 1968; Ritchie 1980). Korpimäki (1987) found larders of the boreal owl to be largest during the egg-laying and hatching periods. Barn owls consume hundreds of mice during the nesting season, many of which are temporarily stored in the nest in contact with eggs (Wallace 1948). Turnover (removal and replenishment) of rodents comprising a barn owl stockpile suggested that the larder was actively managed as a food reserve (Wallace 1948). Van Camp and Henny (1975) stated that screech owls in northern Ohio stored prey in the nest mostly during the first 2 weeks of nestling life but did not say whether the young consumed stored prey.

Food storage also may occur during winter (e.g., Frazar 1877; Van Camp and Henny 1975; Korpimäki 1987). Eurasian pygmy owls studied by Solheim (1984) placed numerous prey in nest boxes they used as winter roosts. Storage occurred from late October through April, with the largest caches (mean ranged from eighteen to twenty-six items) occurring from November to February (fig. 9.2). Caches were made most frequently when daily maximum temperatures were below 0°C. Boreal owls often deplete their supply of stored rodents during early spring storms, and Korpimäki (1987) suggested that food storage in this owl buffers the breeding female and small young from food shortages caused by adverse weather. Cold temperatures that prevail in northern latitudes during winter cause stored prey to freeze. Boreal, saw-whet, great horned owls, and possibly other owls are able to thaw frozen prey by assuming a posture on the carcass similar to that adopted when incubating to apply heat from the lower abdomen (Bondrup-Nielsen 1977; George

Figure 9.1
A saw-whet owl perching at a diurnal roost near a vole it has cached. Original drawing by Marilyn Hoff Stewart.

and Sulski 1984; Solheim 1984). This behavioral adaptation permits storage of prey in cold regions and may be an essential behavior if these owls are to inhabit northern latitudes in winter.

There are several important differences in the way hawks and owls store prey (table 9.6). The most obvious is cache dispersion; most food-storing owls construct larders at nest or roost sites, whereas hawks are much more inclined to scatter prey. Hawks occasionally leave prey remains in or near the nest, but this does not qualify as food storage because the food is either eaten in short order or ignored. Owls often establish larders in cavities, sites apparently never used by hawks. Use of cavities for storage apparently helps owls protect larders from diurnal predators (Solheim 1984). Fox (1979) suggested two reasons why prey storage may be more important for owls than for hawks. First, owls lack a crop (typically found in hawks), in which a small amount of excess food can be carried. Second, mammalian predators, which can be important scavengers of prey stored by raptors, are less active during the day when owl prey are in the cache.

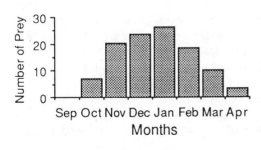

Figure 9.2
Mean number of prey stored in nest boxes during winter months by Eurasian pygmy owls in southeastern Norway. Based on data from Solheim 1984.

Table 9.6 Comparison of caching behavior of hawks and owls

Hoarding Traits	Hawks	Owls
Dispersion of stored prey	Usually scattered	Often in larder
Storage sites	Tree branches, shrubs, ground	Nest cavities, roost cavities, tree branches
Number of prey stored	One to several	Several to many
Type of prey reportedly stored		
Mammals	Frequent	Frequent
Birds	Frequent	Frequent
Reptiles	Occasional	Never
Amphibians	Occasional	Occasional
Fish	Never	Occasional
Invertebrates	Never	Occasional

9.2 WOODPECKERS

Woodpeckers (Picidae)

The family Picidae contains 198 species of woodpeckers (Short 1982), but only 10 species have been observed storing food. Six species—hairy woodpecker, *Picoides villosus;* great spotted woodpecker, *P. major;* downy woodpecker, *P. pubescens;* golden-fronted woodpecker, *Melanerpes aurifrons;* gila woodpecker, *M. uropygialis;* yellow-bellied sapsucker, *Sphyrapicus varius*—store food only infrequently (Alderson 1890; Coward 1928; Martin and Kroll 1975; Conner and Kroll 1979; Andersson 1985; MacRoberts and MacRoberts 1985; Burchsted 1987). The other four species (acorn woodpecker, *M. formicivorus;* Lewis' woodpecker, *M. lewis;* red-headed woodpecker, *M. erythrocephalus;* and red-bellied woodpecker, *M. carolinus*) regularly store food, and their habits have been studied in some detail. Storage of acorns by the common flicker (*Colaptes auratus cafer*) has been reported by de Saussure (1858, cited by Baird, Brewer, and Ridgeway 1905), but it seems certain that de Saussure inaccurately attributed the work of the acorn woodpecker to the flicker. Further study will probably show that other woodpeckers store food, but it is significant—given that woodpeckers are a large, diverse, nearly cosmopolitan group—that seven of the ten food-hoarding species belong to the tribe Melanerpini, and nine occur in temperate North America (the breeding ranges of acorn and golden-fronted woodpeckers extend southward well into the tropics). The tribe Melanerpini is a New World—primarily Neotropical—lineage of twenty-six species of omnivorous, unspecialized woodpeckers that seldom drill into wood for grubs (Short 1982). Most members of the tribe subsist on mast, fruit, sap, and glean insects from the surface of bark. Bock (1970) pointed out that dependence on free-living rather than wood-boring insects is apparently related to the aquisition of mast-storing behavior in woodpeckers.

Woodpeckers store primarily mast. Acorns (*Quercus* and *Lithocarpus*) are a staple food of most food-storing woodpeckers, and whenever a good crop is produced, woodpeckers store them extensively. Other types of nuts are regionally important. For example, red-headed woodpeckers store beechnuts in Indiana (Hay 1887), Lewis' woodpeckers store almonds in California (Bock 1970), yellow-bellied sapsuckers store pecans in Texas (Conner and Kroll 1979), and acorn woodpeckers store piñon pine (*P. edulis*) nuts in New Mexico (Stacey and Jansma 1977) and digger pine (*P. sabiniana*) nuts and almonds in California (Gignoux 1921; MacRoberts and MacRoberts 1976).

Woodpeckers harvest nuts directly from trees on or within a short distance of their territory; those nuts that fall to the ground are generally ignored. All woodpeckers except acorn woodpeckers usually prepare nuts for storage by removing the hull and breaking the endosperm into two or more fragments. This they usually do on an anvil, a favored perch where they can wedge a nut into a crack or depression to facilitate breaking its tough hull. Anvils are important in this process because woodpeckers do not use their feet to hold food while feeding as do tits and corvids. Woodpeckers transport pieces of mast, or occasionally whole nuts, to scattered sites within their territory and hammer them into crevices in bark or desiccation cracks of dead trees, fence posts, and utility poles. They also may insert food behind exfoliating bark or into small natural cavities (Dorsey 1926; Kilham 1958c; Howard 1980). Lewis' and red-headed woodpeckers sometimes widen cracks and crevices to better receive a morsel or whole nut (Bock 1970; Moskovits 1978), but only acorn woodpeckers routinely prepare holes for caching food. Red-headed woodpeckers sometimes block the entrance to cavities containing stored food with bits of wood and bark (Kilham 1958a; MacRoberts 1975; Moskovits 1978); other woodpeckers do not cover caches.

Mast is not the only food stored. Red-headed and Lewis' woodpeckers occasionally store relatively large insects, including wasps, ants, beetles, grasshoppers, and crickets (Bent 1939; Kilham 1958a; Bock 1970). Some woodpeckers store small quantities of fruit (Hay 1887; Bent 1939; Kilham 1963; Conner and Kroll 1979; Kattan 1988). Inedible objects such as pebbles also may be stored (Kilham 1974b; Ritter 1921).

The nut-storing behavior of the social acorn woodpecker differs from that described previously in several important respects. Acorn woodpeckers excavate thousands of holes in special "granary" trees in which they store acorns and other nuts. Construction of storage holes occurs mainly during the latter half of the acorn storage season and continues through the winter. Each hole takes about an hour to excavate, but a hole is rarely completely excavated by a single woodpecker at one sitting. Typically, several woodpeckers work on a particular hole, completing it over a period of several days. Initially a woodpecker chisels a V-shaped notch in the surface of a tree trunk or branch. Later it or another woodpecker expands the excavation to form a cavity 25–40 mm deep and 12–20 mm in diameter with a constricted opening (fig. 9.3) (MacRoberts and MacRoberts 1976; Gutierrez and Koenig 1978). Group members

Figure 9.3
Three stages in the excavation of a storage hole by acorn woodpeckers showing shape of the hole in cross-section. A, Chiseling a V-shaped notch. B, Expanding the interior of the hole. C, Finishing the hole leaving a constriction around the entrance. From MacRoberts and Mac-Roberts 1976. Used by permission, American Ornithologists' Union.

spend a minimum of 1.4–2.7% of their time excavating storage holes (Mumme and de Queiroz 1985). Acorn woodpeckers drill holes in dead wood or bark of living trees, but they rarely penetrate to the living cambium beneath the bark. Holes are spaced several centimeters apart (fig. 9.4). In dead trees and utility poles, acorn woodpeckers often excavate holes in columns following desiccation cracks. The nuts and acorns that acorn woodpeckers store differ considerably in size and shape in different parts of their geographic range. At each locality, the woodpeckers excavate holes of appropriate size to fit the nuts available (Gignoux 1921; Ritter 1922; MacRoberts and MacRoberts 1976).

During fall, acorn woodpeckers harvest acorns and, after removing the caps, insert them into these cavities. They may try several holes before finding one of the appropriate size. Then the woodpecker hammers the acorn in tightly. The acorn is usually slightly recessed in the hole, with the apex of the acorn almost invariably pointing inward. When acorn

Figure 9.4
Acorns stored in a dead tree by acorn woodpeckers in California. Photograph courtesy Walter Koenig.

crops are large and holes in the granary are filled to capacity, these woodpeckers store acorns and fragments of acorn mast in any available crevices or cavities near the granary trees (MacRoberts and MacRoberts 1976). Jehl (1979), for example, described use of Coulter pine (*P. coulteri*) cones for acorn storage.

Establishment of a large granary may take many years. At Hastings Natural History Reservation, Monterey County, California, where the behavior of the acorn woodpecker has been studied annually since 1971, the mean number of storage holes at twenty-six granaries was 4,100 (MacRoberts and MacRoberts 1976). The largest granary in the sample at Hastings contained 11,000 holes. Ritter (1921, 1938) estimated 31,800 and 50,000 holes in two granaries in the San Jacinto Mountains, California. Except when the acorn crop is small, the quantity of acorns that woodpeckers can store is limited by the number of storage holes (Koenig and Mumme 1987). Each year woodpeckers add several hundred holes to granaries, many of these only replacing holes lost when branches fall or wood decays. Thus the capacity of a storage facility is a balance between construction and loss of holes, and large granaries can only be established through the work of woodpeckers over many generations. The granary is a focal point of the woodpeckers' communal territory throughout the year (MacRobert and MacRobert 1976; Mumme and de Queiroz 1985; Koenig and Mumme 1987).

Granaries comprise one primary storage tree and often one or more supplemental storage trees nearby. At Hastings, MacRoberts and MacRoberts (1976) found a mean of 2.1 granary trees (the maximum was seven trees) in fifty-three woodpecker groups sampled. Almost any dead tree or tree with thick dry bark can be used as a granary, but acorn woodpeckers prefer certain types of trees. In pine-oak woodlands studied by Gutierrez and Koenig (1978) and Koenig and Williams (1979), primary storage trees often were dead or dying trees that emerged above the canopy of surrounding woodland. Most primary trees were pines. When first selected, many granary trees were probably living, but because they were overmature, they soon died. At Hastings, MacRoberts and MacRoberts (1976) found that 78% of granary trees were valley oak (*Q. lobata*), the largest species of oak in the woodland. Acorn woodpeckers sometimes excavate holes and store acorns in the eaves of buildings, causing considerable damage (e.g., Henshaw 1921).

Food-hoarding behavior of acorn woodpeckers in part of their geographic range differs from that practiced in western North America. Acorn woodpeckers in some tropical localities do not store nuts in specially excavated receptacles but instead tuck acorns, fruits, and insects into cracks, under lichens, or among the leaves of bromeliads (Skutch 1969; Kattan 1988). However, acorn woodpeckers do prepare cavities for acorn storage in some tropical localities, for example, Chiapas, Mexico (Goldman 1951), Honduras (Skutch 1969), Colombia (Miller 1963), Panama (Eisenmann 1946; Wong 1989), Belize (Peck 1921), and Guatemala (Salvin 1876). The pattern of geographic variation in this behavioral trait is not well documented and is probably complex.

The degree to which woodpeckers scatter the items they cache differs among woodpecker species. The stores of the acorn woodpecker are among the most highly aggregated stores of any bird. Red-bellied woodpeckers, on the other hand, hide food in widely scattered sites (Blincoe 1923; Kilham 1963). Lewis' and sometimes red-headed woodpeckers crowd numerous nut fragments into cracks and cavities of special storage trees (Brewster 1898; Law 1929; Kilham 1958a), and a hairy woodpecker reportedly stored two bushes of potato beetles in a hollow stump (Alderson 1890), but it is uncertain from this single observation how these woodpeckers typically disperse stored food.

The amount of food stored by woodpeckers other than the acorn woodpecker has seldom been estimated. Bock (1970) estimated that two Lewis' woodpeckers harvested about 800 and 2,400 almonds in one fall, but not all of these almonds were stored. Koenig (1980) estimated that a population of 120–130 acorn woodpeckers at Hastings stored a mean of 43,607 acorns per year (41,927–45,470 acorns per year over 3 years), or 344 acorns per bird per year. In fall, acorn woodpeckers store acorns at a mean rate of about 12 acorns per group per hour, with breeders storing at rates of 2.5 acorns per hour and helpers at rates of 1.5 acorns per hour (Mumme and de Queiroz 1985).

Woodpeckers frequently recache and manipulate food items they have stored. Acorns stored by acorn woodpeckers dry out and shrink after they have been stored for several weeks. When acorn woodpeckers encounter these acorns, they extract and recache them in smaller holes. Kilham (1958a) and Moskovits (1978) noted that red-headed woodpeckers initially stored whole or large sections of acorns in any available crevice and later, after the acorn crop was depleted, these acorns were recovered, broken up, and recached throughout the territory. They suggested that this behavior allows woodpeckers to increase their share of the harvest when acorns are abundant but still protect the stores when acorns are scarce. Michael (1926) and Bock (1970) suggested that Lewis' woodpeckers managed their food stores to inhibit fungal growth. The inner surfaces of nut fragments become moist and infected with mold. Turning the nuts allows them to dry out, suppressing fungal infestation.

Woodpeckers store food at any time of year but primarily during fall and early winter, when nuts and acorns are ripe. Moskovits (1978), for example, found that red-headed woodpeckers move to winter territories during late summer and early fall and harvest acorns until mid-December. Lewis' woodpeckers harvest almonds extensively from September through December, when large almond crops are available (Bock 1970). Acorn woodpeckers begin storing acorns as they ripen in September and continue storing until their storage facilities are full, usually by December or January (fig. 9.5).

Lewis' and red-headed woodpeckers aggressively defend their private food stores from conspecifics and heterospecifics alike (Bock 1970; Hadow 1973; Moskovits 1978; Kilham 1958c). For acorn woodpeckers, on the other hand, granaries serve as a communal food supply, freely par-

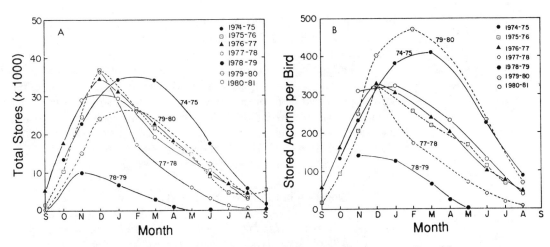

Figure 9.5
Seasonal changes in the number of acorns in acorn woodpecker granaries. A, Total acorns in stores. B, Number of stored acorns per bird. From Koenig and Mumme 1987.

taken of by all members of social groups but cooperatively defended from interlopers (MacRoberts 1970). Red-bellied woodpeckers do not defend their scattered, nearly inaccessible food stores (Kilham 1963).

The food stores of woodpeckers are an important supplement to winter diets when little other food is available (Beal 1911; Bock 1970; MacRoberts and MacRoberts 1976; Moskovits 1978; Koenig 1980; Koenig and Mumme 1987; Stacey and Ligon 1987). In Florida, red-headed woodpeckers consumed most stored mast in late winter (Moskovits 1978). Lewis' woodpeckers may exhaust their food stores by the breeding season in some situations (Bock 1970) and hardly feed on their stores at all in others (Law 1929). Similar observations have been made on the acorn woodpecker by Ritter (1938), but at Hasting, all groups of acorn woodpeckers had exhausted their stores by the time storage commenced anew in the fall (MacRoberts and MacRoberts 1976). The extent of use of stores probably depends on availability of other foods. Acorn woodpeckers also feed on stored mast during other seasons (MacRoberts and MacRoberts 1976). Acorn woodpeckers feed on the kernel of the acorn and not grubs that sometimes infect them (MacRoberts 1974), as some early accounts claimed.

9.3 PERCHING BIRDS

Jays, Crows, and Nutcrackers (Corvidae)

Many of the 100 or so species of corvids are known to store food. I divide this diverse family into three groups that store different types of food in distinctly different ways: (1) jays, (2) nutcrackers and pinyon jays, and (3) crows, ravens, and magpies. Turcek and Kelso (1968) provide an excellent review of hoarding behavior in the Corvidae, emphasizing the European and Asian literature.

Jays. The habit of storing food is well developed in jays inhabiting the forests and woodlands of North America, Europe, and northern Asia. Tropical and desert species of jays store small amounts of food (e.g., Dementév and Gladkov 1954; Goodwin 1976), but it is unclear whether food storage is as important in the biology of these species as it is to their northern relatives.

Jays store single food items or small collections of items in scattered, concealed caches. Many species store acorns and soft-shelled nuts, but virtually any other food, including invertebrates, small vertebrates, scraps of meat, small quantities of fruit, and various items supplied by humans, such as bread, fat, and grains, are stored (Laskey 1942; Blomgren 1971; Bevanger 1974; Haemig 1989). Inedible materials such as deer hair also may be stored, perhaps to be used later as nest material (e.g., Lawrence 1968).

Jays hide food in three types of sites: (1) loose soil, (2) cracks and crevices of tree trunks, and (3) foliage of coniferous trees. When Eurasian jays (*Garrulus glandarius*), blue jays (*Cyanocitta cristata*), Steller's jays (*C. stelleri*), scrub jays (*Aphelocoma coerulescens*), and Mexican jays (*A. ultramarina*) store nuts in soil, they thrust their bills into the substrate at an angle of about 45° to the ground surface. This they repeat several times as if to "test" potential cache sites and loosen the soil (Brown 1963b; Bossema 1979; DeGange et al. 1989). They also may select natural holes and crevices. When a suitable site is found, the jay grasps a nut in the tip of its bill and inserts it into the loosened soil or depression. The jay hammers the nut with its bill to drive it further into the ground and then draws litter and soil over the cache with sideswipes of the bill and places a twig, pebble, or clod of earth on top of the site. Cache depth ranges from a few millimeters (e.g., DeGange et al. 1989) to about 3 cm, the maximum depth being constrained by length of the bill. Alternatively, if the ground is too hard, an acorn or nut is simply thrust under litter and other leaves are raked over it (Michener and Michener 1945; Darley-Hill and Johnson 1981). Jays seem to prefer caching nuts in forest clearings and fallow meadows (Goodwin 1976; Johnson and Adkisson 1985).

Food storing by gray jays (*Perisoreus canadensis*) is markedly different from that of any other corvid. Gray jays have enlarged salivary glands that secrete copious, sticky mucus (Bock 1961), which they use to cement together food items, including invertebrates and scraps of meat from carrion, to form a sticky bolus (Dow 1965). Jays form boluses by repeatedly rotating food items in the oral cavity with the highly manipulative tongue. Typical boluses measure about 17 mm long by 8 mm wide. Gray jays lodge boluses in foliage and bark of coniferous trees, where they quickly dry and become cemented in place. They do not conceal boluses by placing debris on them but space caches widely to minimize cache loss to robbers (Waite 1988). Dried boluses resist weathering and persist intact for many months (Dow 1965; Rutter 1972). Dow (1965) observed that captive gray jays stored some boluses in soil, but this has not been observed in free-ranging gray jays. The adaptive value of bolus formation and firm attachment to above-ground substrates seems to be to ensure persistence of stored food in sites accessible during winter,

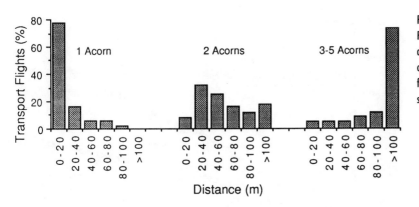

Figure 9.6
Frequency distribution of transport distances of Eurasian jays carrying one acorn, two acorns, or three to five acorns. Redrawn from Bossema 1979.

when the ground in boreal forests is covered with a thick blanket of snow. When gray jays recover a bolus, they consume the food items along with the dried mucus. This method of food storage has not been reported, to my knowledge, in the closely related Siberian jay (*P. infaustus*) (Bevanger 1974; Pravosudov 1984).

Jays are well known for their ability to transport large loads of food (Novikov 1948; Eigelis and Nekrasov 1967; Bock, Balda, and Vander Wall 1973; Vander Wall and Balda 1981) (see section 3.3). When collecting acorns, jays appear to swallow whole two to five acorns, but they actually hold these acorns in the upper portion of the esophagus. They also may carry an additional acorn in the bill. When Eurasian jays transport acorns long distances (i.e., >100 m) they carry larger loads (fig. 9.6).

Jays from throughout a region will converge on productive nut-bearing trees to collect seeds and carry them back to their territories (Chettleburgh 1952; Goodwin 1956; Swanberg 1969; Johnson and Adkisson 1985). Each jay harvests and transports nuts independently of other jays. During the peak nut harvest, a nearly constant stream of nut-laden jays can be observed dispersing from woodlots, often along fencerows and hedges just above the treetops. Eurasian jays and blue jays regularly perch in treetops along the route (Swanberg 1969; Johnson and Adkisson 1985).

Food storage is most prevalent in the fall, coincident with ripening of nuts and acorns. Chettleburgh (1952) determined that the acorn harvest season for Eurasian jays in Essex, England, began in early September and ended in mid-November, when the crop was depleted. During peak activity, a 10-day period in mid-October, thirty-five to forty jays transported and stored acorns for about 10 hours each day. In contrast, at the beginning and end of the harvest, jays actively stored acorns for only about 3 hours each morning and only a few (no more than eight) jays participated. Johnson and Adkisson (1985) illustrated a similar pattern in seasonal change in hoarding behavior by counting the number of blue jays leaving a Wisconsin woodlot with beechnuts (fig. 9.7). The slightly concave lines in figure 9.7 indicate that hoarding intensity was greater during

Figure 9.7
Cumulative number of trips by blue
jays leaving a woodlot with beechnuts on 4 sample days during September 1981. Redrawn from Johnson and Adkisson 1985.

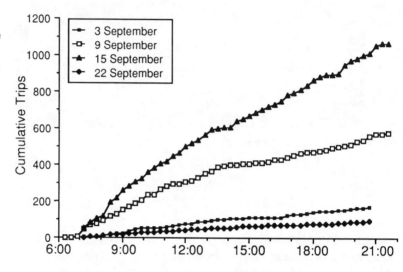

Figure 9.7
Cumulative number of trips by blue jays leaving a woodlot with beechnuts on 4 sample days during September 1981. Redrawn from Johnson and Adkisson 1985.

the first half of each day. Also, hoarding rates in early and late September were much lower than those during mid-September, the peak of the beechnut harvest. Jays store food at other seasons, including the breeding season (e.g., Pravosudov 1984).

Jays store thousands of nuts (see section 7.2), and these supplies figure importantly in the diet of jays in winter (Novikov 1948; Harrison 1954; Chettleburgh 1955; Swanberg 1969). For example, Eurasian jays in the Netherlands derived much of their sustenance in the winter of 1963–64 from acorns they stored the previous fall (see fig. 2.6) (Bossema 1968). By early spring 1964, the occurrence of acorns in the diet had decreased markedly, but it increased again in June. Bossema (1968) asserted that this early summer increase of acorns in jay diets was a result of their removing acorns from the bases of recently germinated seedlings. Jays rarely feed acorns to nestlings, but acorns form an important part of fledgling diets. Gray jays rely heavily on stored food for winter sustenance (Dow 1965; Rutter 1972), and some of the food gray jays feed their young is stored before the breeding season (Rutter 1969). Siberian jays stored food during the breeding season (Pravosudov 1984), but whether these stores played a role in the breeding effort was not determined. The importance of stored food in the energetic budgets of jays remains one of the least understood aspects of their hoarding ecology.

Nutcrackers and Pinyon Jays. The two species of nutcrackers are the most specialized food-hoarding birds. Both species inhabit coniferous forests, Clark's nutcracker (*Nucifraga columbiana*) in the mountainous regions of western North America and the Eurasian nutcracker (*N. caryocatactes*) disjunctly distributed in mountains of central Europe, foothills and low mountains of southern Sweden and Norway, slopes of the Himalayas, taiga forests of Siberia to eastern Asia, and the islands of Japan and Taiwan. Pinyon jays (*Gymnorhinus cyanocephalus*) inhabit piñon-juniper woodlands of the western United States and northern Mexico. Although pinyon jays are not closely related to nutcrackers, I have included them here because when harvesting and storing seeds they behave like little blue nutcrackers. Other aspects of their behavior, especially their colonial nesting behavior (Balda and Bateman 1971), are unlike that of nutcrackers. Their similarities are the result of convergent adaptations for exploiting conifer seeds, on which all three species are highly dependent.

Nutcrackers and pinyon jays store primarily large, wingless pine seeds (see table 7.3) but they also harvest winged pine seeds (e.g., ponderosa pine [*P. ponderosa*], Jeffrey pine [*P. jeffreyi*], bristlecone pine [*P. longaeva*]) (Tomback 1978; Lanner, Hutchins, and Lanner 1984). Nutcrackers eat and, in some areas, store spruce (*Picea*) seeds (Kirikov 1936; Korelov 1948), but they neither eat nor store the seeds of true firs (*Abies*). Clark's nutcrackers store Douglas-fir (*Pseudotsuga menziesii*) seeds when wingless pine seeds are unavailable (Vander Wall, Hoffman, and Potts 1981). In Sweden and portions of Germany, where no wingless-seeded pines occur, Eurasian nutcrackers store hazelnuts (*Corylus avellana*) (Swanberg 1951, 1956; Volker and Rudat 1978). Captive Clark's nutcrackers store insects and the carcasses of small mammals.

Nutcrackers and pinyon jays begin harvesting pine seeds in late July, but they do not commence storing seeds until late August or early September, when seeds are mature (Balda and Bateman 1971; Tomback 1978, 1982; Hutchins and Lanner 1982; Mattes 1982; Vander Wall 1988). Early in the season, nutcrackers and pinyon jays pry seeds from green, unopened cones by jabbing their bills between the overlapping scales. They often detach cones and carry them to sturdy branches in the interior of trees to remove seeds. Later, birds pick seeds from open cones or from the ground beneath trees. They sort and discard inedible (e.g., unfertilized, diseased) seeds. Seed discrimination is accomplished by several mechanisms, including bill clicking, the rapid manipulating of seeds between the mandibles, and bill weighing, the brief holding of seeds in the bill (Ligon and Martin 1974; Vander Wall and Balda 1977; Johnson, Marzluff, and Balda 1987).

Nutcrackers transport seeds to caching areas alone, in pairs, or in small, loose flocks, but the gregarious pinyon jays may travel in cohesive flocks of 300 or more. Caching behaviors of the three species are similar (Bibikov 1948; Vander Wall and Balda 1977). A bird first loosens the soil by probing its bill into the substrate to a depth of 2–3 cm. No hole is excavated. A seed is next ejected into the bill from the sublingual pouch (nutcrackers) or esophagus (pinyon jay) by a quick upward and forward jerk of the head. The bird grasps the seed in the tip of the bill and then jabs the bill into the previously loosened soil (fig. 9.8). Several seeds may be added to a cache by repeating this process. When finished, the bird rakes soil or litter over the area with its bill. Finally, the bird may deposit a cone, twig, pebble, or other object on the cache site. Additional caches may be made within a few meters of the first cache.

In the Tien Shan Mountains of central Asia, where no pines occur, Eurasian nutcrackers (*N. c. rothschildi*) collect and store intact spruce (*Picea schrenkiana*) cones under moss (Kirikov 1936; Korelov 1948).

Figure 9.8
A Clark's nutcracker burying pine
seeds carried in its bulging sublingual pouch. The clumping of
adult trees results from several
seedlings establishing from a cache
and is typical of bird-dispersed
pines such as whitebark pine and
limber pine. Original drawing by
Marilyn Hoff Stewart.

The systematic caching of cones has not been reported elsewhere.

Pinyon jays make smaller caches than do nutcrackers. Eighty-three pinyon jay caches excavated by Vander Wall and Balda (1981) contained between one and seven piñon pine (*P. edulis*) seeds with a mean of 1.4 seeds per cache; sixty-eight caches contained only one seed. In the same study, Clark's nutcracker cache size ranged from one to fourteen seeds, with a mean of four seeds per cache ($N = 94$). Similar cache sizes have been reported by Tomback (1978), Hutchins and Lanner (1982), and Balda and Kamil (1989). Eurasian nutcrackers in southern Sweden placed three to four, but sometimes as many as eighteen, hazelnuts in each cache (Swanberg 1951). Mattes (1982) found three to four, but sometimes as many as twenty-four, Swiss stone pine (*P. cembra*) seeds in caches in Switzerland. Kuznetsov (1959) found three to twenty-eight Siberian stone pine (*P. sibirica*) seeds in Eurasian nutcracker caches in the Ural Mountains, and Egorov (1961) recorded a mean of 11.7 (range 3 to 48, $N = 39$) Japanese stone pine (*P. pumila*) seeds per cache in Yukat, eastern USSR.

Pinyon jays cache food within their traditional colony home range in piñon-juniper woodland and ponderosa pine forests. They often make caches on the south sides of trees, where snow melts first after winter storms, permitting seed recovery (Balda and Bateman 1971). Nutcrackers use a wide variety of cache areas. At high elevations, where deep snow precludes cache recovery in winter, nutcrackers cache many seeds on sites likely to have only sparse snow cover. These sites include windswept ridges (Oswald 1956; Lanner and Vander Wall 1980), cliff faces (French 1955; Dulkeit 1960; Kishchinskii 1968), and south-facing slopes (Vander Wall and Balda 1977; Hutchins and Lanner 1982). A south-facing slope used for caching by Clark's nutcrackers in the San Francisco Peaks, Arizona, was free of snow 1–2 months earlier in the spring than a nearby north-facing slope (Vander Wall and Balda 1977). However, Clark's nutcrackers also cache many seeds at sites that will receive more than 3 m of snow, preventing seed recovery until spring or even midsummer (Vander

Wall and Hutchins 1983). On both sparse snow and deep snow sites, many Clark's nutcrackers working independently intermix caches in the same area without aggressive interactions (Vander Wall and Balda 1977; Hutchins and Lanner 1982). Although cache areas are used communally, the caches themselves are not shared (see section 6.3). These communal sites have been referred to as convergent storage sites by Tomback (1978). Similar communal sites are used by Eurasian nutcrackers (Bibikov 1948; Reimers 1953; Kuznetsov 1959; Mattes 1982), but in some areas members of this species store seeds on breeding territories (Swanberg 1951, 1956; Mattes 1982). Caching of seeds on breeding territories has not been reported for Clark's nutcracker (e.g., Mewaldt 1956).

Within cache areas, nutcrackers cache seeds in diverse sites. Both species cache in meadows, open woodlands, and closed-canopied forest. Clark's nutcrackers, for example, cache seeds in alpine tundra, coniferous forests, piñon-juniper woodland, mountain mahogany (*Cercocarpus ledifolius*) forests, and sagebrush (*Artemisia*) shrublands (Vander Wall and Balda 1977, 1981; Tomback 1978; Lanner and Vander Wall 1980; Hutchins and Lanner 1982). Eurasian nutcrackers cache in larch (*Larix decidua*), pine forests, spruce forests, burned areas, meadows, and alpine tundra (Bibikov 1948; Kondratov 1953; Kuznetsov 1959; Reimers 1959b; Mattes 1982). Substrates for both species are equally diverse: loamy soil, rocky soil, cinders, rock outcrops, and mats of conifer needles. Eurasian nutcrackers cache extensively beneath dense moss and lichens (Bibikov 1948; Kondratov 1953; Kishchinskii 1968), and these nutcrackers place a small portion of their caches in cracks of fallen logs or in trees under lichens or behind exfoliating bark (Dulkeit 1960; Kishchinskii 1968; Mattes 1982). Kuznetsov (1959) and Hutchins and Lanner (1982) found that nutcrackers will cache in very moist sites, but Tomback (1982) stated that Clark's nutcrackers prefer dry, well-drained sites. The only generalizations that seem to pertain to nutcracker cache site selection are that they often cache near large objects (e.g., logs, rocks, tree trunks, roots) and that they cache where herbaceous vegetation is sparse (Pivnik 1960; Mezhennyi 1964; Vander Wall and Balda 1977; Tomback 1978; Mattes 1982).

Morphological specializations of these corvids enhance their ability to transport seeds from widely scattered cone-bearing trees to territories or communal caching grounds. Nutcrackers collecting seeds for storage deposit them in their sublingual pouches, the distensible floor of the mouth opening in front of the tongue (Bock et al. 1973). The sublingual pouch may appear greatly distended when full of seeds (see fig. 3.11) or collapse and disappear from view when empty. Pinyon jays lack a sublingual pouch but hold numerous seeds in their distensible esophagi. The capacity of these structures is summarized in table 3.1. The number of seeds reportedly found in nutcracker pouches and pinyon jay esophagi varies greatly because of differences in seed size and because not all birds were carrying full loads when collected.

The spacious food-transporting structures of these corvids enables them to transport seeds long distances. Pinyon jays typically transport seed about 1 km to caching areas (Balda and Bateman 1971), but when pine seeds are scarce, these jays have been seen carrying seeds 8–10 km (Vander Wall and Balda 1981, 1983). Clark's nutcrackers near Flagstaff, Arizona, routinely transport piñon pine seeds 7.5–15 km to cache sites in the San Francisco Peaks, and some birds have been reported carrying seeds as far as 22 km (Vander Wall and Balda 1977). Eurasian nutcrackers are known to carry seeds 15 km (Reimers 1958; Mattes 1982). Although establishing the maximum distance seeds are carried by these corvids gives important insights to their seed-harvesting behavior, in many (if not most) situations nutcrackers transport seeds less than 10 km (e.g., Swanberg 1951; Tomback 1978; Lanner and Vander Wall 1980; Hutchins and Lanner 1982; Mattes 1982; Vander Wall 1988).

Seed harvesting by nutcrackers may in some situations proceed in two stages: initial caching near the harvest trees and, when seeds in cones are depleted, recovery and transport of those seeds greater distances to final cache sites (Kuznetsov 1959; Mezhennyi 1964). Hutchins and Lanner (1982), for example, found that in mid-October Clark's nutcrackers recovered whitebark pine seeds previously cached near harvest trees in Squaw Basin meadows, Wyoming, and transported them 3.5 km to cache sites on the steep, south-facing slopes of the Breccia Cliffs, where they recached the seeds. Caching of seeds near harvest trees increases seed harvest efficiency, but restorage of these seeds may be necessary if rodent predation on caches in the area is likely to be high or if accessibility is likely to be low because of deep winter snows.

Individual nutcrackers have been estimated to store from 20,000 to over 100,000 seeds in the fall (Vander Wall and Balda 1977; Tomback 1982; Hutchins and Lanner 1982; Mattes 1982; Vander Wall 1988), and pinyon jays can store from 18,000–21,500 seeds (Ligon 1978, Balda 1980a) (see section 7.2). Variability in the estimates for nutcrackers is caused by several factors, the three most important being (1) seed size, an important determinant of number of seeds transported each harvest trip; (2) distance seeds are transported, which limits the number of daily foraging excursions between seed collecting and caching areas; and (3) seed crop size, which determines length of harvest period. The largest quantities of seeds stored have been when the seeds are relatively small, the transport distance only a few kilometers, and the cone crops very large.

Nutcrackers and pinyon jays begin to recover cached seeds in early winter as soon as the fall seed crop is depleted. During winter, Eurasian nutcrackers may dig through snow to obtain stored seeds (Bibikov 1948; Kuznetsov 1959; Crocq 1977). Dulkeit (1960), for example, found that Eurasian nutcrackers in the taiga of northeastern Altai, USSR, had burrowed from 26–85 cm through the snow to reach cached seeds. Swanberg (1951) stated that nutcrackers in Sweden can find stored seeds in 18 inches (45 cm) of snow. At 351 winter excavations tallied by Swanberg (1951), 86% had hazelnut shells nearby, indicating that at least that proportion of the caches had been accurately relocated. Mattes (1982) found that 82% of the excavations made by Eurasian nutcrackers in the snow of the Engadine Valley, Switzerland, had pine seed shells near the hole. Clark's nutcrackers also dig through snow to reach food caches, but the behavior is poorly documented. Nutcrackers in Squaw Basin, Wyoming, dug through 10 cm of snow to reach caches in November (Hutchins and Lanner 1982), and I saw a bird peck through 1 cm of ice at the edge of a snowbank in the Teton Mountains to reach a cache (Vander Wall and Hutchins 1983). Cahalane (1944) observed a Clark's nutcracker digging through about 8 inches (20 cm) of snow to reach a Douglas-fir cone frozen to the ground litter, but the nutcracker probably had not cached this cone because it was frozen to the surface of the ground.

For all three species, stored food forms an important component of the winter diets. Ligon (1978) found that 70–90% of pinyon jay diets between November and February came from seed stores. Giuntoli and Mewaldt (1978) reported that 80–100% of Clark's nutcracker diets during winter (November to February) were comprised of conifer seeds, no doubt taken from seed caches.

During the breeding season, stored pine seeds continue to be important in the diets of both adult and young nutcrackers, but pine seeds form only about 10% of diets of nestling pinyon jays (Bateman and Balda 1973). However, stored piñon pine seeds may form a large and very important part of nestling pinyon jay diets when adverse conditions preclude foraging for insects (Ligon 1978). Nestling nutcrackers are fed almost exclusively pine seed (Bendire 1889; Johnson 1900; Bradbury 1917; Mewaldt 1956; Mezhennyi 1964; Crocq 1974; Volker and Rudat 1978; Mattes 1982). Kishchinskii (1968), for example, found that 62% of nestling diets in the Kolyma Highlands of eastern USSR were comprised of Japanese stone pine seeds, and Reimers (1959a) reported that Eurasian nutcrackers in the Ural Mountains, USSR, brought 110–206 Siberian stone pine seeds to the nest each day. Swanberg (1951) found that hazelnut kernels are a principal part of nestling Eurasian nutcracker diets in Sweden. After fledging, young of both Clark's and Eurasian nutcrackers follow adults to communal cache areas and continue to be fed largely on stored pine seeds during June and July (Kuznetsov 1959; Vander Wall and Hutchins 1983).

Crows, Ravens, and Magpies. Crows, ravens, magpies, and their relatives occupy a broad range of habitats from alpine tundra to intertidal mudflats to tropical forests on all continents except South America. Their great adaptability is reflected in the broad range of foods stored: walnuts, acorns, and whole pine cones by rooks (*Corvus frugilegus*) (Forster 1967; Andrew 1969; Purchas 1975, 1980; Källander 1978); mice, fish, frogs, salamanders, pecans, and dung by common crows (*C. brachyrhynchos*) (George and Kimmell 1977; Phillips 1978; Conner and Williamson 1984; Kilham 1984a, 1984b); carrion by carrion crows (*C. corone*) and common ravens (*C. corax*) (Magoun 1979; Hewson 1981; Heinrich 1988); clams, crabs, fish, and marine worms by northwestern crows (*C. caurinus*) (James and Verbeek 1983, 1984, 1985); and bread, cheese, peanuts, corn, fat, and similar foods by various species (e.g., Goodwin 1955; Chisholm 1972; Robson 1974; Chapman 1978; Leonard 1978; Lewis 1978; Walters 1979; King 1988).

Crows and magpies transport food items to storage sites in the bill or antelingual pouch, the distensible floor of the oral cavity between the rami of the lower mandible. Large items, such as walnuts, eggs, and clams, are usually carried singly in the bill tip (Källander 1978; James and Verbeek 1983; Sonerud and Fjeld 1985); several small, soft food items, such as bread and carrion, may be wadded into a bolus and carried in the mouth or antelingual pouch (Goodwin 1955; Magoun 1979). The rook has the largest antelingual pouch of any member of the genus *Corvus* with a capacity of 22 g, or 5% of its body weight (see table 3.1). Most crows, ravens, and magpies transport food less than a few hundred meters and often less than 50 m from the food source (Rowley 1973; Phillips 1978; Walters 1979; Hewson 1981; James and Verbeek 1983, 1985; Kilham 1984a, 1984b), but rooks routinely carry walnuts up to 2.5 km (Källander 1978). Among the highly social rooks, areas used for caching by individuals broadly overlap (Purchas 1980; Källander 1978). Northwestern crows use both exclusive (defended) and nonexclusive (shared) caching areas (James and Verbeek 1983).

Most members of this group exhibit two types of caching behavior: burial in soil and hiding in dense vegetation. Crows bury food by simply thrusting the bill holding the food into loose soil. They may then place a clod of dirt or clump of dry grass on the cache site. Crows may cache in natural depressions, but only the rook routinely digs a small hole in which to bury food (Richards 1958; Simmons 1970). The single food item to be buried, which rooks lay on the ground near the site while they prepare the hole, is picked up in the bill, inserted into the hole, and hammered with the bill until it is well seated in the hole. Crows and magpies also hide food in dense grass or foliage when these are available. Cache sites are always covered by grass or debris. Caching occurs primarily on the ground, but Kilham (1984b) observed a common crow regularly storing food items in trees in the foliage of epiphytic air plants (*Tillandsia setacea*), and fish crows (*C. ossifragus*) stored bread in trees (McNair 1985). Black-billed magpies (*Pica pica*) bury some food in soil (Henty 1975) but also store food on the ground among dense grass and foliage (Hayman 1958; Butlin 1971); They place leaves or other debris on food to conceal it (Linsdale 1937). Choughs (*Pyrrhocorax graculus* and *P. pyrrhocorax*), inhabitants of alpine and coastal rocky cliffs in Great Britain and Europe, hide food in cracks and fissures of rock ledges and place three to ten pebbles on each cache site (Turner 1959; Fitzpatrick 1978).

The types of cache sites chosen appear to be independent of food type and crow species. For example, both burial in soil and hiding in dense vegetation have been exhibited by rooks storing walnuts (Källander 1978; Purchas 1980), northwestern crows storing clams (James and Verbeek 1983), and common crows storing small vertebrates (Phillips 1978; Kilham 1984b).

The caching behavior of jackdaws (*C. monedula*) differs from that of other crows and ravens. These small, cavity-nesting crows cache food inside their nest cavities and other darkened chambers (Lorenz 1970).

Novikov (1948) described a cavity in a hollow linden tree containing numerous acorns and a nest of jackdaws.

Most crows and ravens store food whenever it is available in excess of immediate needs (e.g., George and Kimmel 1977; Sonerud and Fjeld 1985). Superimposed on this background of opportunistic caching, several species exhibit marked seasonal and diurnal patterns in caching intensity. Rooks cache walnuts and acorns extensively in autumn and early winter (Källander 1978; Purchas 1980; Waite 1985). Northwestern crows store clams and other intertidal prey most intensively during their early spring (May and June) breeding season, when other food is scarce (James and Verbeek 1984). The pattern of caching and recovery of intertidal prey by northwestern crows is synchronized with the tidal cycle (see fig. 2.4). Caching occurs most frequently as the tide is falling and prey are exposed, peaking 4 hours before low tide. Cache recovery, on the other hand, is greatest when the tide is in and prey are scarce, peaking 4–6 hours after low tide.

Most foods stored by crows and ravens are perishable. Because of this perishability, stored food represents an ephemeral food supply that must be retrieved and consumed within hours or, at most, a day or two. This short tenure of caches is well illustrated by northwestern crows (see fig. 2.4), which retrieve and eat most cached prey within 1 day (James and Verbeek 1984). Black-billed magpies also recover items within hours or days of storing them (Hayman 1958; Clarkson et al. 1986). Burial of walnuts and acorns by rooks (Källander 1978; Purchas 1980; Waite 1985) and pecans by common crows (Conner and Williamson 1984) are major exceptions to this pattern. For rooks, cached nuts provide a significant proportion of the winter food source for several months.

Cached food plays an integral role in nesting behavior of some crows. During the breeding season, male common crows and fish crows store food near the nest for the females (Kilham 1984b; McNair 1985). Male northwestern crows store clams and later, when the tide is in, retrieve these clams and deliver some of them to incubating females, allowing the females to be more attentive of their nests. These impacts of stored food on reproduction have been considered further in section 2.4. Crows and ravens also use stored food during inclement weather (e.g., Waite 1986).

Australian Butcherbirds (Cracticidae)

Australian butcherbirds (*Cracticus*) and their relatives, currawongs (*Strepera*) and Australian magpies (*Gymnorhina*), form a small group of Australian and Papuan birds that are probably related to crows and jays. Food storing has been reported for several species, but no thorough studies of the behavior have been made.

Butcherbirds derive their name from their habit of impaling large insects and small vertebrates on thorns or wedging them in crevices when feeding, as shrikes do (Laniidae). Like shrikes, Australian butcherbirds leave some prey hanging to be eaten later (Amadon 1951; Serventy and Whittell 1951; van Tyne and Berger 1959; Austin 1961; Pizzey 1980).

Walters (1980) described how grey butcherbirds (*Cracticus torquatus*) took meat from a feeder and wedged it in crevices of a powerline insulator, a television antenna, and among foliage of a pine tree. Pizzey (1980) noted that butcherbirds cached beetles and remains of lizards in cavities of limbs and bark. Reynolds (1969) watched a pied currawong (*Strepera graculina*) cache meat in the forks of branches; the currawong returned as many as six times to the same site to add chunks of meat to this small larder. Bell (1983) observed pied currawongs dig up food in friable soil and thought that this food may have been cached by the birds.

White-winged Chough (Grallinidae)

White-winged choughs (*Corcorax melanorhamphos*), named for their superficial resemblance to the alpine chough (Corvidae, *Pyrrhocorax*) of the Mediterranean region, belong to a small family of Australian birds, the Grallinidae (Amadon 1944, 1950). Klapste (1981) observed a small flock of white-winged choughs bury bread in soft sand at a picnic ground in northwestern Victoria. The birds scatter hoarded the bread, carrying pieces in their beaks about 100 m before burying them. Both adults and juveniles stored food. Nothing else seems to be known about food storing in this group.

Bowerbirds (Ptilonorhynchidae)

Of the eighteen species of bowerbirds native to New Guinea and northern Australia, only MacGregor's bowerbird (*Amblyornis macgregoriae*) is known to store food (Pruett-Jones and Pruett-Jones 1985). The absence of food storing in other bowerbirds is surprising, as the decorating of a bower, often with edible fruits, is similar in some respects to food-storing behavior and would seem to predispose bowerbirds to store food. Future studies may show that other forest-dwelling bowerbirds also store food.

MacGregor's bowerbirds inhabit humid tropical forests, where males construct elaborate maypole bowers of sticks, which act as courtship arenas. Males store fresh, ripe fruit in shrubs and understory trees in the vicinity of their bowers (Pruett-Jones and Pruett-Jones 1985). They wedge or place fruit in forks of branches, in small cavities, on top of stumps, and in gaps between vines and tree trunks (fig. 9.9). Bowerbirds do not cover cached fruit. Cache sites range from 0.2–9.0 m in height and may range up to 12.5 m horizontal distance from the bower (fig. 9.10). The number of stored fruits near bowers ranged from 0–82, with a mean of 17.6. There was a mean of 1.3 fruits per cache, and only one fruit was cached at 84% of the sites. On average, the bowerbird replaced the fruit at each cache site every 6–7 days. Pruett-Jones and Pruett-Jones recorded forty species of fruit from 421 bowerbird caches.

Only male MacGregor's bowerbirds store fruit, and adult males store much more than do juveniles. Furthermore, there is a significant correlation between the completeness of a male's bower and number of fruits it stores (Pruett-Jones and Pruett-Jones 1985). Thus males that actively attend and maintain their bowers cache more fruit than do less attentive males. Males usually attend bowers from May to February, but fruit stor-

Figure 9.9
Fruit stored by MacGregor's bowerbird. Left, a *Timonius* sp. fruit
cached in a fork of a tree. Right,
two *Elaeocarpus* sp. fruits. The
fruits are about 20 mm in diameter. Photographs courtesy Stephen
Pruett-Jones.

ing occurs only during the breeding season (September to February), when males court visiting females.

Food-storing behavior of MacGregor's bowerbirds is unusual in several respects. First, MacGregor's bowerbirds are one of the few animals known to regularly store fleshy fruit in the humid tropics. Fruit storage is practiced by only a few species in temperate climates (e.g., Parrott 1980), and it is not a common practice in these species. The paucity of fruit storage, especially in the tropics, probably occurs because ripe fruit cannot be stored for more than a few days before it is attacked by microbes. Second, MacGregor's bowerbird is the only species in which food storing is limited to males. Male northwestern crows store clams and feed them to their mates during the incubation and nestling period, but females store food at other seasons (James and Verbeek 1983, 1984). In many Hymenoptera, food storing is exclusively the domain of females. Third, this is the only species of vertebrate in which the only function of food caching is to facilitate or promote courtship behavior. Storing fruit permits males to spend more time at the bower and court females if they visit the bower.

Chickadees, Tits, and Titmice (Paridae)

The food-hoarding behavior of chickadees, tits, and titmice (*Parus* spp.), common inhabitants of forests and woodlands in north-temperate regions, has been reviewed by Sherry (1989). Parids store arthropods,

Figure 9.10 Spatial distribution
of cache sites (solid circles) around
a MacGregor's bowerbird bower
(star). Most of the sites were in
small trees. The scales are in
meters. From Pruett-Jones and
Pruett-Jones 1985.

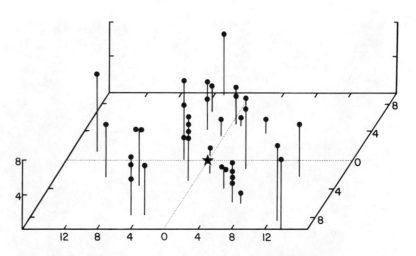

seeds, and nuts (e.g., Haftorn 1954, 1956a, 1956b, 1956c, 1974; Gibb
1960). The importance of various seeds and nuts depends on tit species
and habitat. For example, crested tits (*Parus cristatus*) in Norway and
boreal chickadees (*P. hudsonicus*) in Alaska store many spruce seeds
(Haftorn 1954, 1974); varied tits (*P. varius*) in Japan hide *Castanopsis*
nuts (Higuchi 1977); tufted titmice (*P. bicolor*) cache acorns in Mississippi (McNair 1982); and willow tits (*P. montanus*) store hempnettle
(*Galeopsis*) seeds in Norway and Britain (Haftorn 1956b; Gibb 1960). At
feeding stations, parids collect and hide items supplied by humans, such
as sunflower seeds, peanuts, and pieces of fat (Hart 1958a, 1958b; Haftorn 1974; Cowie, Krebs, and Sherry 1981; Moreno, Lundberg, and
Carlson 1981; Eyre 1984).

Chickadees and their relatives are scatter hoarders and store prey
throughout their home ranges. The wide spacing of cached food reduces
cache loss to cache robbers (Sherry, Avery, and Stevens 1982). They
transport single food items or small collections of prey in their bill, and
most species store each load of food at one site. Coal tits (*P. ater*), for
example, store twenty to fifty aphids in a compact mass (Haftorn 1956a;
Gibb 1960). Willow tits, however, may carry up to six *Galeopsis* seeds
per foraging excursion, and each seed is stored in a different site (Haftorn
1956b). Tits never cache items carried on different transport flights together. Transport distances are usually less than 100 m (fig. 9.11).

Figure 9.11
Distribution of transport distances
for food items stored by marsh tits
and willow tits. Redrawn from
Cowie et al. 1981 and Haftorn
1956b.

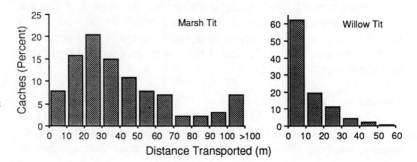

Tits and chickadees hide food in crevices of bark, under lichens, among foliage, inside hollow stems, under moss, and buried in soil (Lewis 1923; Richards 1949; Haftorn 1954, 1956c, 1974; Gibb 1960; Higuchi 1977; Moreno, Lundberg, and Carlson 1981; Alatalo and Carlson 1987; Nakamura and Wako 1988; Petit, Petit, and Petit 1989). At storage sites, tits simply insert items into recesses or tuck them under loose substrates (fig. 9.12). Next they hammer seeds and nuts with their bills, forcing the objects securely into small crevices or cavities. In addition to this mechanical fixture, crested tits also secrete a copious saliva, which when dry, causes food items to adhere to the caching substrate (Haftorn 1954). Invertebrate prey are "cemented" in place with their own coagulated body fluids. Haftorn (1974) observed that these fixtures were so effective that he could vigorously shake branches and rap them against his writing desk without dislodging stored food items. Crested and coal tits also cover stored food with tufts of lichen or debris, but other parids, notably boreal chickadees (Haftorn 1974), leave stored prey partially exposed.

There is considerable interspecific variation in types of sites parids select to cache food (Pravosudov 1986; Alatalo and Carlson 1987). One important difference among Norwegian species of tits studied by Haftorn (1954, 1956a, 1956b, 1956c) was the distribution of cache sites along branches from trunk to branch tip in conifer trees (fig. 9.13). Crested tits distributed spruce seeds and insect prey fairly uniformly among dead main branches, the inner portion of live main branches (lacking foliage), and the outer portion of live branches (with green foliage). These tits placed over 75% of their caches under or contacting lichens and left most

Figure 9.12
Coal tit in typical storing posture at the end of a spruce twig, just before inserting a conifer seed (in bill) into a bud capsule. Original drawing by Marilyn Hoff Stewart; after Haftorn 1956a.

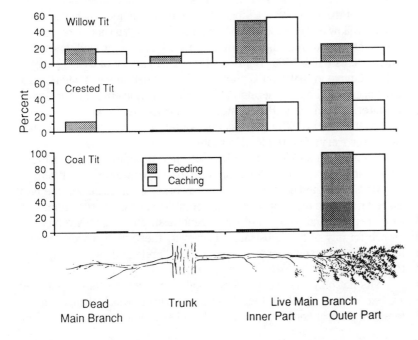

Figure 9.13
Horizontal distribution of feeding sites and storage sites from trunk to branch tip of spruce boughs for willow, crested, and coal tits in Norway. Redrawn from Haftorn 1956c.

Figure 9.14
Vertical distribution of storage sites used by willow, crested, and coal tits in pine and spruce in Norway. Relative height refers to four height zones of equal width. Redrawn from Haftorn 1956c.

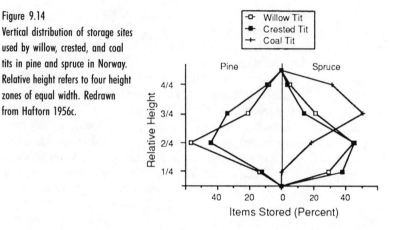

items partially exposed (Haftorn 1954). The coal tit—a specialist in the storage of *Galeopsis* seeds—strongly favored branch tips for cache sites where it stored large numbers of these seeds in bud capsules (fig. 9.12) (Haftorn 1956a). They thus cached seeds in small-diameter, foliated portions of branches. The willow tits stored more food on the trunk and inner branches—needle-free sections of the tree—than any other species. Like crested tits, willow tits used lichens most frequently as storage points, but unlike crested tits, they inserted seeds and other prey much more deeply under lichens so that they were seldom visible. For all three species illustrated in figure 9.13, cache site distributions are similar to distributions of foraging sites.

Another axis of interspecific variation in cache placement is height above ground. Many tits and chickadees preferentially store food in midsections of trees (fig. 9.14). Coal tits generally hide food higher in trees than do willow and crested tits. Marsh tits (*P. palustris*) and boreal chickadees carry food from the canopy downward and the ground upward to storage sites (Gibb 1960; Haftorn 1974). This pattern of caching seems to be adaptive in that food is stored in that section of trees where least snow accumulates and, therefore, where it is most likely to be accessible in winter. Marsh tits in Britain store much closer to the ground (Cowie, Krebs, and Sherry 1981). In this area, winter snow is scant and does not hinder recovery of stored food in winter.

Even within species, cache site selection is highly variable (Cowie, Krebs, and Sherry 1981; Pravosudov 1986; Stevens and Krebs 1986). In the Marsh tits studied by Cowie and coworkers, for example, one bird cached mostly in moss, a second cached most seeds in nettle stems and bark crevices, and a third bird was a generalist, showing no strong preference for any of five substrates. However, these preferences seemed to be transient, some birds changing their "preferred" site from week to week.

Tits not only have very diverse food-hoarding repertoires, but they are exceedingly plastic in their hoarding behavior, which changes in different ecological contexts. Moreno, Lundberg, and Carlson (1981), for example, found that marsh tits cached sunflower seeds in the soil when they were within a territory of a pair of European nuthatches (*Sitta europaea*) but stored seeds in tree bark in areas where nuthatches were absent. In areas of overlap, nuthatches would doubtless find many of the marsh tit caches stored in tree bark, and thus the change in cache site selection may have served to protect caches from nuthatches. Willow tits in Sweden store most food in the upper portion of conifer trees, but in sympatry with Siberian tits (*P. cinctus*), which also store most food high in conifer trees and which dominate willow tits in interspecific foraging flocks, willow tits shift storage sites to the lower portion of trees and thereby reduce cache site overlap with Siberian tits (see fig. 4.5) (Alatalo and Carlson 1987). Storage sites also may vary with food type. Boreal chickadees store most seeds under lichen but hide most insect larvae in bark crevices (Haftorn 1974). Coal tits store *Galeopsis* seeds in bud capsules of spruce and insects in foliage (Haftorn 1956a).

Tits store food at all times of year. Ludescher (1980) demonstrated a strong seasonal trend in the hoarding behavior of captive willow tits in West Germany. The storing of seeds was most pronounced from September to early March and much reduced from late April to early August (see fig. 5.14). Crested tits in Norway had a spring and early summer (April or July) peak in hoarding, when they stored mostly pine seeds, and a fall (September or October) peak, when they stored mostly insects and spruce seeds (Haftorn 1954). Spring storage of pine seeds by coal, willow, crested, and marsh tits in Norway and Britain (Haftorn 1954, 1956a, 1956b; Gibb 1960) is related to the opening of scotch pine (*Pinus sylvestris*) cones in early spring. Crested tits cache most food items at midday (fig. 9.15).

Data summarized by Haftorn (1956c) suggest that the quantity of food stored by tits may be very large. Mean storing intensity (storing frequency) within foraging flocks of crested, willow, and coal tits ranged from twenty-eight to thirty-eight items per hour during the spring, summer, and fall. Tits stored a mean of 30–37% of all food items they handled. These rates of storing result in each individual storing many thousands of food items within their home range each year. Pravosudov (1985, cited in Pravosudov 1986), for example, estimated that tits can store over 1,000 items in a day, and Haftorn (1959) found that individual tits in a Norwegian spruce forest can potentially store between 50,000 and 80,000 spruce seeds each fall. Tits and chickadees may recover and eat stored food within a few hours or days of caching it or they may re-

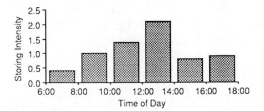

Figure 9.15
Daily pattern of storing in crested tits. Storing intensity refers to number of individuals engaged in hoarding per observer hour. Redrawn from Haftorn 1954.

trieve food several months later (e.g., Nakamura and Wako 1988). Marsh tits in England, where winters are mild, recover most seeds within 2–3 days of storage (Cowie, Krebs, and Sherry 1981; Stevens and Krebs 1986). Stevens and Krebs (1986), using minute magnetic detectors placed near cached seeds, determined that marsh tits (which carried a small magnet on one leg) retrieved about 24% of the seeds they stored, most within 12 hours of caching them. Only a small portion of the winter diets of coal and crested tits in England appear to come from caches made in the fall (Gibb 1960). Tits of the northern boreal forests, on the other hand, rely on food they store in the fall for a large share of their winter diet. Without stored food, crested and willow tits probably could not exist in northern boreal forests in winter (Haftorn 1956b). Stored food also may be used to supplement the nestling diet (Higuchi 1977).

Not all tits store food as actively as the species mentioned previously. Great tits (*P. major*) and blue tits (*P. caeruleus*), for example, store little or no food (Southern 1946; Hinde 1952). These tits tend to wander during the nonbreeding season and do not establish winter territories (Gibb 1960). In some areas, these tits depend more on feeding stations established by humans as sources of winter food. Food storage is most highly developed in sedentary species at high latitudes.

Nuthatches (Sittidae)

Nuthatches (*Sitta* spp.) store conifer seeds, fragments of nuts, and small invertebrates in crevices of tree bark. Items such as bread and suet obtained from bird feeders also are cached. Nuthatches store each seed or bill full of food separately. They prepare caches by inserting prey into crevices and then hammering it with the bill until it is wedged firmly in the hole. Nuthatches use bits of bark or lichen gathered near the site to conceal cached food (see fig. 3.1) (Richards 1949, 1958; Kilham 1974a; Dorka 1980).

Moreno, Lundberg, and Carlson (1981) studied the cache site selection of a pair of European nuthatches (*S. europaea*) in central Sweden by establishing a feeding station near the center of a nuthatch territory and observing their caching of sunflower seeds from mid-November to early February. The pair of nuthatches defended the territory and feeding station from other nuthatches and several marsh tits. They cached all seeds within 40 m of the feeding station ($N = 211$), and they hid most of these from 6–14 m above the ground. The male and female nuthatches stored seeds similarly, except that the male carried seeds further, caching them near the periphery of the territory. Nuthatches cached 71% of the seeds in bark crevices and most of the remaining seeds under lichens; they used trunks and large branches for over 96% of their caching, concentrating the seeds in the inner portion of the tree crown. The pair cached nearly 94% of the seeds in English oaks (*Quercus robur*), even though oaks comprised only 56% of the trees available. Trunks and large branches of oaks were probably preferred for caching because these substrates provide deeply furrowed bark. The observations of Richards (1958) and Ball (1978) on food storage by European nuthatches are qualitatively similar to those of Moreno, Lundberg, and Carlson (1981). Amur nuthatches (*S. e. amurensis*) store numerous Korean stone pine (*Pinus koraiensis*) seeds in soil and serve as a vehicle of dispersal for this tree (Bromley, Kostenko, and Okhotina 1974).

Caching behavior of North American nuthatches is similar to that European nuthatches but less well documented. White-breasted (*S. carolinensis*) and red-breasted nuthatches (*S. canadensis*) often cache seeds in trees with furrowed bark, such as elm, oak, and hemlock (Kilham 1974a, 1975; Petit, Petit, and Petit 1989). Brown-headed (*S. pusilla*) and pygmy nuthatches (*S. pygmaea*), which typically forage among the needles and twigs of pines, scatter hoard pine seeds and invertebrates under bark flakes of trunks and large branches of pine trees (Norris 1958; Stallcup 1968; McNair 1983; Sealy 1984).

Richards (1958) states that Eurasian nuthatches cache throughout the year, except possibly during the breeding season. Löhrl (1958) and Enoksson (1987) observed that Eurasian nuthatches cached most intensely during autumn. Most descriptions of North American nuthatches caching food pertain to the fall (Kilham 1974a, 1975; McNair 1983), but Norris (1958) observed pygmy nuthatches cache seeds in mid-July, and Stallcup (1968) stated that pygmy nuthatches store food in reproductive and nonreproductive seasons. Grubb and Waite (1987) provide circumstantial evidence that red-breasted nuthatches cache food during winter at sites outside their breeding range (i.e., outside their usual territories).

The amount of time between the caching and recovery of a food item has not been well documented in nuthatches. Waite and Grubb (1988) described a diurnal caching rhythm in captive white-breasted nuthatches in winter. Pairs cached nearly half of the food energy they handled in the morning and decreased the proportion cached gradually during the day (fig. 9.16). Food consumed and food recovered from caches by the nuthatches did not change significantly during the day, the latter result perhaps being an artifact of ad libitum feeding in the captive situation. Waite and Grubb suggested that this caching rhythm may be adaptive in two ways. First, caching early in the day may minimize the risk of finding inadequate food in the late afternoon, just before going to roost for the night. The energy demands of long winter nights are great, and caching food early in the day facilitates foraging late in the day when the nuthatches must obtain energy to pay the expense of nocturnal thermoregulation. Second, nuthatches may minimize their risk of predation by maintaining low body mass and consequently greater maneuverability during the day. Nuthatches also derive some long-term benefits from hoarding food. For example, fall food storage by Eurasian nuthatches permits individuals to maintain territories through the winter (Enoksson 1987).

Thrushes and Old World Flycatchers (Muscicapidae)

Thrushes and their relatives (Muscicapidae: subfamily Turdinae) constitute a large, well-studied, nearly cosmopolitan group containing over 300 species. Food hoarding has been reported in only two species. A blackbird (*Turdus merula*), a European thrush, was observed by Hamp-

Figure 9.16
Daily changes in the proportion of
food energy handled that was cached
by white-breasted nuthatches. Redrawn from Waite and Grubb 1988.

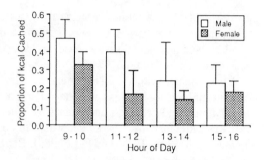

ton (1983) briefly leaving earthworms lying on the ground while it extracted other worms from their burrows. Hampton (1983) called this
hoarding, however, this behavior does not constitute food hoarding, as
defined here, because the blackbird did not handle the worms so as to
conserve them for future use. Wallace (1963) reported food hoarding in a
captive Neotropical thrush of the genus *Catharus,* probably one of the
nightingale thrushes. Wallace gives no details of the hoarding behavior;
he reported only that the thrush left "little stockpiles hidden behind fur-
niture." Further observations are needed to establish the extent and details of food-hoarding behavior in this species.

The only other known food-hoarding species in this family is the
South Island robin (*Petroica australis australis*), an Old World flycatcher
(Muscicapidae: subfamily Muscicapinae) endemic to New Zealand. These
robins store invertebrates in crotches of tree branches, in stump ends,
and in cracks and crevices of bark (Soper 1976; Powlesland 1980). South
Island robins store prey singly, usually within 10 m of capture sites, and
from 30 cm to 8 m above the ground (mean = 2.9 m high), even though
they conduct 61% of their foraging on the ground (Powlesland 1980,
1981). They do not cover or camouflage cached prey. Robins protect
caches through territorial defense and cache spacing.

Robins store food during most months, but they cache most intensely from April through July (austral fall and winter) (Powlesland
1980). Earthworms comprise 70% of the prey items cached, but the
robins also cached slugs, stick insects, cicadas, beetle larvae, and snails.

Stored invertebrates serve as food reserves only for a short time.
Even during winter, stored prey decompose quickly in the mild climate of
southern New Zealand, rendering them useless after about 3 days. Of
forty caches monitored by Powlesland (1980), 58% disappeared the same
day they were made, and Powlesland presumed that the individuals that
cached them also had retrieved them.

South Island robins hoard and retrieve food on a daily cycle. During
the morning and early afternoon, hoarding occurs more frequently than
retrieval of caches, and in the late afternoon and evening, retrieval of
caches occurs more frequently than caching (see fig. 2.2). Powlesland
suggested, based on these observations, that prey stored early in the day
serve as a readily available source of energy during the evening, when
South Island robins require more energy to sustain themselves through

long, cool winter nights. In addition, stored food could aid survival during inclement weather. During the breeding season, male South Island robins feed stored prey to their mates, a behavior that may allow females to be more attentive to nests.

Shrikes (Laniidae)

The impaling of prey on sharp objects is a well-studied aspect of shrike predatory behavior (Miller 1931, 1937; Lorenz and von Saint Paul 1968; Beven and England 1969; Wemmer 1969; Smith 1972; Olsson 1985). Where there are no spines, shrikes wedge prey into narrow V-shaped forks. All members of the genus *Lanius* that have been studied exhibit these habits, but impaling seems to be much less prevalent in other groups of shrikes (e.g., boubou shrikes, *Laniarius ferrugineus;* Sonnenschein and Reyer 1984).

Securing prey by impaling or wedging serves two functions: feeding and storing. Shrikes have weak feet (Miller 1931; Beven and England 1969; Smith 1972), which makes it difficult for them to feed while holding prey in their feet as hawks and owls do. Impaling and wedging of large prey anchors them, permitting the shrike to pull strips of meat from the carcass. Impaling and wedging of prey for storage has apparently evolved from this unusual mode of feeding.

Shrikes store a wide variety of prey types. Mice, small birds, reptiles, and large arthropods are the prey most frequently found impaled, but frogs, shrews, fish, earthworms, crustaceans, snails, and fruit also have been reported (Owen 1929, 1948; Miller 1931; Durango 1951; Cade 1967; Beven and England 1969; Karasawa 1976; Reid and Fulbright 1981). Shrikes also impale nonfood items such as fecal sacs, egg shells, leaves, flowers, and nest material (Owen 1948; Durango 1951; Lorenz and von Saint Paul 1968). Shrikes are opportunistic and will impale virtually any type of food. Northern shrikes (*Lanius excubitor*) in North America, for example, impale mostly rodents and small birds (Miller 1931; Cade 1967), but in Algeria, Africa, the same species impales large quantities of dates (*Phoenis dactylifera*) on spines at the base of palm fronds (Beven and England 1969; Parrott 1980).

Shrikes impale prey on virtually any type of sharp object, but certain types of impaling sites seem to be preferred. They often use thorns, spines, and barbed wire as impaling sites (fig. 9.17). Loggerhead (*L. ludovicianus*) and northern shrikes prefer exposed sites for impaling (Miller 1937; Yosef and Pinshow 1989), whereas red-backed shrikes (*L. collurio*) conceal prey inside bushes (Durango 1951). Wemmer (1969) found that captive loggerhead shrikes used long (5 cm) nails as impaling sites more often than medium (3 cm) and short (1 cm) nails. Preferences for certain types of impalements are at least partially learned (see section 5.1). Impalements also occur more frequently on low sites within 2–3 m of the ground (Miller 1931; Cade 1967). Shrikes impale prey at widely spaced sites or group them in larders with ten or more items impaled in the same thorn bush (Owen 1948; Durango 1956; Beven and England 1969; Ash 1970). Red-backed shrikes often construct larders within a

Figure 9.17
A, A Kalahari barking gecko
(*Ptenopus garrulus*) impaled (still
alive) on a thorn by a shrike in the
Kalahari Gemsbok National Park,
Republic of South Africa. B, A juvenile rufous-sided towhee impaled
on barbed wire by a loggerhead
shrike in eastern Utah. A, photograph courtesy Eric Pianka; B,
photograph by author.

few meters of their nests (Owen 1948; Durango 1956), but northern
shrikes in Alaska establish larders about 50–200 m from nest sites (Cade
1967).

The duration of storage ranges from a few hours to several months.
Short-term storage seems to be especially important during inclement
weather, when foraging is impaired (Owen 1948; Durango 1956), and
during the breeding season, when males impale prey for consumption by
females or nestlings (Beven and England 1969; Applegate 1977; Carlson
1985; Yosef and Pinshow 1989). Among northern shrikes in Israel, the
incidence of food caching increases during the reproductive season and
appears to strongly support the reproductive effort (Yosef and Pinshow
1989). Long-term storage is more often associated with winter or other
season of food scarcity. The bull-headed shrike (*L. bucephalus*) in Japan
impales large numbers of prey during fall and early winter (November to
January) and retrieves these prey throughout the winter (December to
March) (Karasawa 1976). Bull-headed shrikes ate 65% of prey within
1 month of impaling them and 94% within 2 months of impalement. Parrott
(1980) suggested that dates impaled by northern shrikes in Algeria during spring aided the shrikes during the lean summer months. Shrikes do
not use some impaled prey, and these eventually dry out or spoil (Miller
1931; Owen 1948; Olsson 1985). The ecological and energetic aspects of
prey storage (e.g., Carlson 1985; Yosef and Pinshow 1989) have not been

studied as thoroughly as has the ontogeny of impaling behavior (Lorenz and von Saint Paul 1968; Wemmer 1969; Smith 1972).

Mynas (Sturnidae)

The family Sturnidae is a large (111 species) Old World group of birds that includes the starlings, mynas, and oxpeckers. Malhi (1987) reported common mynas (*Acridotheres tristis*) harvesting the ripened seed heads of wheat and carrying them to crotches of tree branches and to their nests. Malhi interpreted the activities of the mynas as food-hoarding behavior, but it is uncertain, from his short description, whether the mynas were actually accumulating the wheat seed heads and conserving them for future use. No other member of this family has been reported to store food. More detailed studies of this interesting behavior are needed before food hoarding can be unequivocally attributed to this species.

10

Food-Hoarding Arthropods

Food storing has been reported in three orders of arthropods: spiders (Araneae), beetles (Coleoptera), and ants, wasps, and bees (Hymenoptera), including a total of twenty-two families (table 10.1). Belostomatid bugs (Victor and Wigwe 1989) do not hoard food as defined here. Most species store invertebrates, seeds, pollen, nectar, carcasses of small vertebrates, or vertebrate feces in inconspicuous chambers in the soil, galleries in wood, plant stems, or natural chambers. Some species store food in special receptacles constructed of mud, plant fibers, or wax. Food storage takes one of two general forms: (1) a reserve food supply for the individual or individuals that stored the food and (2) an exclusive provision to nurture a developing larva.

10.1 SPIDERS

Spiders (Araneidae, Diguetidae, Lycosidae)

Spiders store prey by wrapping them in silk and suspending them from their webs or from some other substrate. The wrapping of prey is a component of the prey handling and feeding behavior of many spiders that construct aerial webs (e.g., araneid and diguetid spiders), as well as some spiders that do not build webs but forage on elevated substrates (e.g., some lycosid spiders). In many species, wrapping is the initial step in prey handling. Some araneid spiders, such as *Argiope argentata,* use silk as an attack weapon to immobilize and further ensnare prey so that it cannot escape (Robinson, Mirick, and Turner 1969). *Argiope* may also bite the prey to immobilize it. If the spider is not ready to feed, it may leave the prey stored in situ, or it may cut or pull the prey from the web, carry it to the hub, rewrap it in silk, and suspend it from a thread (Robinson 1969; Vollrath 1979). *Nephila clavipes* always bites prey before wrapping it and always stores prey near the hub of its orb web (Vollrath 1979). *Argiope,* by storing most prey in situ in the web, is able to respond more quickly to a second prey caught in the web than *Nephila,* which invests time in cutting, transporting, rewrapping, and restoring prey (Robinson, Mirick, and Turner 1969). This quicker response results in fewer prey escaping. *Diguetia* sp. first bite a prey to immobilize it, cut it from the web, carry it to its retreat, and there feed on the prey. If another prey gets caught in the web, *Diguetia* wraps the first prey in silk and attaches it to the web near the retreat before rushing off to attack the second prey

(Eberhard 1967). Some *Lycosa* that hunt and feed in herbaceous vegetation use small amounts of silk to attach prey to the vegetation as they feed on it. If they are disturbed during feeding, they flee but return to the suspended prey later (Rovner and Knost 1974; Greenquist and Rovner 1976).

Storing prey is advantageous to the spider in several ways. Web-building spiders may catch prey more quickly than they can handle or consume them. Storing the excess prey in the web or at the hub frees the

Table 10.1 Arthropod taxa known to store food with a summary of their food-storing behavior

Group Family	Dispersion[a]	Food Type[b]	Substrate/ Location[c]	Storage Duration/Use[d]
Spiders				
Orb-weavers (Araneidae)	S,L	I	W	S
Diguetid spiders (Diguetidae)	S	I	W	S
Wolf spiders (Lycosidae)	S	I	F	S
Beetles				
Dung beetles (Scarabaeidae)	L	D,V,Ca	B,S	M,S
Ground beetles (Carabidae)	L	S	B	S
Burying beetles (Silphidae)	S	SM,Bi	S,L	M
Rove beetles (Staphylinidae)	L	A	N	S
Ants (Formicidae)				
Harvester ants	L	S	N	L
Honey ants	L	N,W	N	L
Carnivorous ants	L	I	N	L
Wasps				
Spider wasps (Pompilidae)	S	Sp	N,F	M,S
Sphecid wasps (Sphecidae)	L	I,Sp	N,F	M,S
Potter, mason wasps (Eumenidae)	L	I	N	M
Paper wasps (Vespidae)	L	N	N	L
Masarid wasps (Masaridae)	L	P,N,I	N	M
Bees				
Membrane bees (Colletidae)	L	P,N	N	M
Digger bees (Andrenidae)	L	P,N	N	M
Sweat bees (Halictidae)	L	P,N	N	M
Melittid bees (Melittidae)	L	P,N,O	N	M
Leafcutter bees (Megachilidae)	L	P,N	N	M
Carpenter bees (Anthophoridae)	L	P,N,O	N	M
Apid bees (Apidae)				
Stingless bees	L	P,N	N	L,M
Bumblebees and orchid bees	L	P,N	N	L,M
Honey bees	L	P,N	N	L

Note: See text for references and exceptions to the general patterns within taxa.

[a] Dispersion patterns: L, larder; S, scattered.

[b] Food types: A, algae; Bi, birds; Ca, carrion; D, dung; I, insects; N, nectar; O, floral oil; P, pollen; S, seeds; SM, small mammals; Sp, spiders; V, vegetation; W, water.

[c] Substrates and locations: B, burrow; F, foliage; L, litter; N, nest; S, soil; W, web.

[d] Storage duration or use: S, short term (generally <10 days); L, long-term (generally >10 days); M, mass provisions.

spider to engage in other activities such as monitoring the web or attacking and handling other prey that may get caught (Eberhard 1967; Robinson, Mirick, and Turner 1969; Vollrath 1979). Being unencumbered by prey also may permit spiders to avoid their own predators more effectively (Rovner and Knost 1974; Greenquist and Rovner 1976) or interact with other spiders. When preparing to feed, spiders can bite stored prey, injecting them with enzymes that liquify their soft tissues, and return later to ingest these liquids (Vollrath 1979).

Some species have become specialized in kleptoparasitizing the prey stored in the orb webs of other spiders (Robinson, Mirick, and Turner 1969; Vollrath 1979). *Argyrodes elevatus* (Theridiidae) waits at the edge of *Argiope* and *Nephila* orb webs; when it senses the host spider attacking and wrapping a prey item, it slowly advances onto the web. When the host spider stores the prey and retreats to the hub to monitor the web, *Argyrodes* moves to the prey, cuts it out of the web, and carries it away. *Argyrodes* is much smaller than its host spiders but can steal prey up to ten times its own mass.

Eberhard (1967) and Robinson, Mirick, and Turner (1969) described possible pathways for the evolution of prey wrapping behavior and food storage.

10.2 BEETLES

Of the several groups of beetles known to store food, by far the most important are the dung beetles, a group that comprises three subfamilies of Scarabaeidae: Aphodiinae, Geotrupinae, and Scarabaeinae. Members of all three subfamilies use dung and other sources of organic detritus as adult and larval food. In the Geotrupinae and Scarabaeinae, a nest is constructed for the purpose of reproduction (Halffter and Edmonds 1982), but aphodiine dung beetles lay eggs directly in the food source; no provisioning or food storage takes place. Elsewhere in the Coleoptera, several species of ground beetles (Carabidae), burying beetles (Silphidae), and a few species of rove beetles (Staphylinidae) also store food.

Dung Beetles (Scarabaeidae)

Dung beetle food hoarding takes two forms: (1) large reservoirs of food accumulated in feeding burrows and (2) larval provisions stockpiled in breeding nests. Feeding burrows are usually oblique or vertical shafts into which beetles carry dung or other foods and consume it apart from competitors and predators. *Coproecus hemisphaericus* of Western Australia digs a tunnel that ends in a small chamber 5–12 cm below ground, which they stock with three to eight fecal pellets (Matthews 1974). A single beetle resides in the storage chamber or in a vertical feeding shaft below the chamber (fig. 10.1). Most dung beetles require fresh, moist dung for feeding, but the humid storage chambers of *C. hemisphaericus* allow it to use dry fecal pellets, an abundant resource in arid Western Australia. These buried pellets absorb moisture during the wet season, reviving bacterial and fungal activity, and the beetles feed on these microorganisms. Several other Australian dung beetles have similar habits (e.g., Matthews 1974; Halffter and Edmonds 1982).

Most nest-building dung beetles are coprophagous, feeding on and provisioning nests with the excrement of vertebrates. Host specificity of most dung beetles is weak (Fincher, Stewart, and Davis 1970; Gordon and Cartwright 1974), and for many species, the habitat and soil type at the site of the dung deposit have a greater influence on whether a species of dung beetle will be present than does type of dung (e.g., Mohr 1943). Some species of dung beetles use other organic substances as both adult and larval food. For example, *Phanaeus halffterorum* uses decomposing fungi to construct brood balls in the same way that coprophagous species use dung (Halffter and Edmonds 1982). The provisions of *Canthidium granivorum* consist of seeds (*Pithecellobium dulce;* Fabaceae), which the female beetle mixes with her own feces. *Canthidium puncticolle* is necrophagous, provisioning its nest with carrion (Halffter and Edmonds 1982). Female *Cephalodesmius armiger* process flowers, seeds, small fruits, and partially decayed leaf litter to produce a dunglike substance (Monteith and Storey 1981).

Halffter and Edmonds (1982) recognized seven nesting patterns of dung beetles that differ in placement of the nest, manner in which provisions are prepared, degree of female-male cooperation, and extent of parental care. Three of these illustrate the diversity of food-storage behavior in dung beetles. The discussion emphasizes handling of provisions; consult Halffter (1977) or Halffter and Edmonds (1982) for further examples and details of other aspects of nesting behavior.

The most common form of nesting behavior in dung beetles is the formation of a brood mass in a chamber below a dung pat. A female *Onthophagus taurus,* sometimes with the aid of a male, excavates a gallery beneath a fresh dung pat (fig. 10.2). The female carries dung into a slightly enlarged cavity at the end of the gallery and compacts it into a cylindrical or spherical brood mass. At the proximal end of the brood mass, she lays an egg in a small chamber and caps it with a thin layer of dung. Finally, she fills the gallery above the brood mass with loose soil. Each female constructs from one to many brood nests, depending on the species. Compound nests may be linear or racemously branching with nest chambers separated by a plug of soil.

An interesting variation of this nesting pattern occurs in *O. parvus* and several other Australian dung beetles that use macropod pellets for feeding and nesting (Matthews 1972). These beetles ride on wallabies, congregating around the anal opening, and when the wallabies defecate, they fall to the ground with the fecal pellets. They immediately bury a pellet about 4 cm deep and use the unmodified pellet as a brood mass. These unusual resource exploiting and nesting behaviors are apparently adaptations for using fecal pellets before they desiccate in the arid north Australian outback.

Some dung beetles construct a large "cake" in an underground chamber that they later use to form brood balls. *Copris armatus* excavates a tunnel terminating in a large chamber beneath or beside a dung pat.

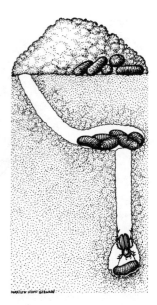

Figure 10.1
Diagrammatic section of a feeding burrow of an Australian dung beetle *Coproecus hemisphaericus,* which it has stocked with rabbit pellets. Original drawing by Marilyn Hoff Stewart; after Matthews 1974.

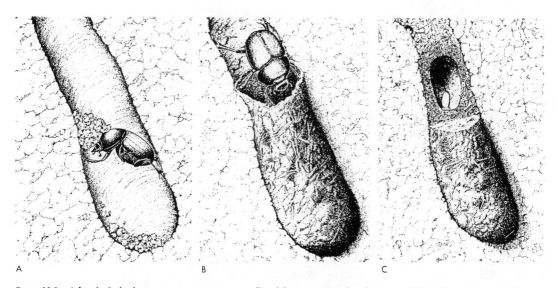

A   B   C

Figure 10.2   A female *Onthophagus taurus* excavating a nest gallery (A), constructing a brood mass out of dung (B), and an egg positioned in a small chamber at the top of the brood mass (C). From Goidanich and Malan 1964.

A female beetle, often with the help of a male, works to fill this chamber with excrement (fig. 10.3). Next, the adults compress the dung cake by interposing themselves between the dung mass and the chamber ceiling. This evidently promotes fermentation of the dung mass and makes the cake more homogeneous and plastic. At this point, the male usually leaves the nest chamber, and the female carves several elliptical brood balls from the dung cake. She lays an egg in a small chamber formed in excrement at the upper end of each brood ball. The number of brood balls a female forms is determined by the size of the dung cake, but usually three to four brood balls form a cluster in the nest chamber. The female attends the nest during development, and some time during larval development she adds a soil layer to the brood balls.

Some species with similar nesting habits, such as *Heliocopris japetus*, temporarily store dung in a large shallow chamber and later transport it to a deeper nest chamber (fig. 10.4). In *Heliocopris,* the temporarily stockpiled dung may weigh 2–3 kg (Klemperer and Boulton 1976).

Some of the most interesting dung beetles are the ball-rollers, which carve a dung ball from the margin of a dung pat and roll it away and bury it at a distance. Male-female cooperation is considerable, but the male usually assumes the more active role during ball formation, rolling (see fig. 3.6), and burial. The beetles gradually bury the dung ball several centimeters deep by excavating a crater directly under it. Ball-rolling dung beetles do not excavate tunnels or galleries. Once the ball is buried, the male departs while the female reworks the dung ball into a pear-shaped brood ball. When the ball is finished, the female deposits an egg within a small cavity at the top of the ball and then abandons the nest. Many Australian ball-rolling dung beetles do not form balls by carving them from the margin of a dung pat but use fecal pellets of kangaroos, other mar-

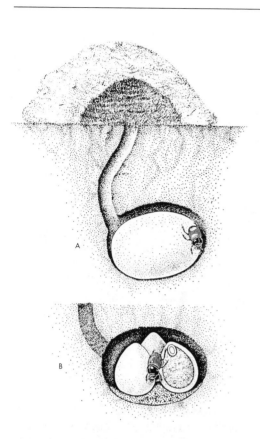

Figure 10.3
A, A female *Copris armatus* has transported dung from a dung pat into an underground chamber and is working the dung "cake," which is surrounded by an air layer created as she formed the cake. B, The female has reformed the dung cake into three brood balls and has laid an egg at the upper end of each ball. From Huerta et al. 1981.

supials, and rabbits, which are already the appropriate size and shape (Matthews 1974).

For all dung beetles, regardless of nesting behavior, food for the developing larva is limited to that supplied to the brood mass by the parent. In most species, the size of the brood mass or brood ball is fixed before or shortly after the egg is laid, but in some species (e.g., *Cephalodesmius armiger*), the female parent tends the nest and adds to the brood ball as the larva grows (Monteith and Storey 1981).

Ground Beetles (Carabidae)

Most ground beetles are carnivorous, but the larvae of a few species collect and store grass seeds. The larvae of an unidentified carabid beetle studied by Alcock (1976) in Washington stored the seeds of *Festuca myuros* in burrows. The burrows, probably abandoned nests of halictid bees, were nearly vertical shafts about 10 cm deep. At the bottom of each shaft was a small chamber where the larva resided. At intervals of approximately 1–3 days, larvae leave their burrows in search of seeds. Foraging trips are usually short, covering a distance of about 20 cm and lasting only a couple of minutes. When they encounter a *Festuca* seed, larvae transport it back to the nest in their jaws so that the long awn projects posteriorly over their back. Back in the burrow, the larvae deposit the seed just inside the burrow entrance and then leave the burrow

Figure 10.4

Heliocopris japetus brood burrow construction. The dung is indicated by shading, superficial soil by fine stippling, subsoil by heavy stippling. The beetles quickly move the dung below ground to a storage chamber (A–C) and later dig a deeper nest chamber (D and E), where brood balls will be formed. From Klemperer and Boulton 1976.

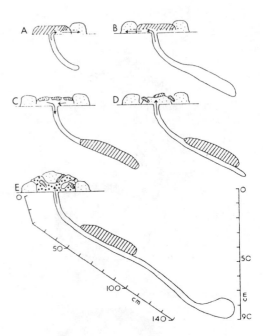

to search for more seeds. Fourteen foraging bouts observed by Alcock (1976) yielded from two to thirty-seven seeds (mean = thirteen).

After a foraging bout, larvae rearrange the seeds, pulling them deeper into the burrow. The larvae then move into the chamber at the bottom of the burrow and consume the seeds one at a time. A burrow excavated by Alcock contained seventeen closely packed *Festuca myuros* seeds just above the terminal chamber and a beetle larva and partially eaten seed in the chamber (fig. 10.5).

Larvae of two species of ground beetle (*Harpalus pensylvanicus* and *H. erraticus*), common in the croplands of South Dakota, inhabit burrows as deep as 70 cm (Kirk 1972). Shortly after the seeds of yellow and green foxtail (*Setaria lutescens* and *S. viridis*) mature and fall in early October, seeds can be found pressed into the walls of these burrows. Although Kirk did not see larvae collect and store seeds, the arrangement of the seeds suggests that they had been placed there by the beetle larvae. One burrow, for example, contained 177 seeds pressed into the walls of the burrow between 10 and 25 cm below ground so that the flat sides of the seeds looked like tiles on a wall. A larva resides at the bottom of each burrow. Kirk did not observe the larvae feeding on the seeds.

The adult ground beetle (*Synuchus impunctatus*) is said to hoard the seeds of *Malampyrum lineare* (Cantlon et al. 1963), but this is not true food storage. The beetles collect the carunculate seeds and carry them one at a time to secluded feeding sites (e.g., under logs). There they eat the caruncle, a food body attached to the seed coat, and discard the seed. The seed piles that were first thought to be caches are actually refuse piles (Manley 1971).

Burying Beetles (Silphidae)

Burying beetles (*Nicrophorus* spp.) provision nests with vertebrate carrion, such as small mammals and birds, a behavior unique in several respects (Leech 1935; Milne and Milne 1976; Wilson and Fudge 1984, Scott and Traniello 1989). These beetles are the only insects that store vertebrates and are among the few species that hoard food items considerably larger than themselves. Adding to the uniqueness of these insects, food hoarding is associated with a high degree of parental cooperation and care of the brood.

When burying beetles discover a carcass, they immediately dig under the carcass, roll over on their backs, and attempt to lift the corpse with their legs, a behavior that may serve to assess the size of the carcass (Milne and Milne 1944). Small carcasses, about 100 g or less, that are not infested with fly larvae are quickly buried by beetles (Wilson 1985). Beetles bury a carcass by burrowing just under it, plowing soil from beneath the carcass with their pronotums. When they emerge at the edge of the carcass, they push up a small pile of soil and quickly turn around, submerge, and repeat the process. Gradually the body sinks as earth piles up around it (fig. 10.6). Once buried, a dominant pair of beetles often drives other beetles away (Wilson and Fudge 1984). Burying beetles move a carcass if the soil under it is unsuitable for digging (see fig. 3.7).

Burying beetles may bury carcasses incompletely, with part of the corpse level with the soil surface but concealed under plant litter, or 8 cm or more deep (Wilson 1985). As burial occurs, the corpse's extremities are drawn in until the bloating carcass forms a nearly spherical mass. The beetles remove all feathers or hair from the carcass and form a chamber around it by compacting the soil. The carcass gradually becomes an amorphous mass of putrified tissue. After the pair copulates and the female lays eggs in a small receptacle at the top of the chamber, the female further prepares the carcass by eating and clawing a depression in its upper surface. Male and female regurgitate droplets of partly digested

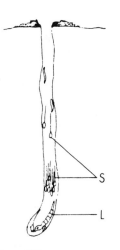

Figure 10.5
A burrow of a carabid beetle larva showing the larder of seeds (S) part way down the burrow and the larva (L) consuming a seed in the chamber at the bottom of the burrow. From Alcock 1976.

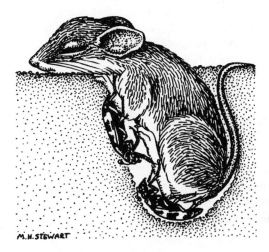

M.H. STEWART

Figure 10.6
A pair of burying beetles burying a mouse by plowing soil from under the carcass. Original drawing by Marilyn Hoff Stewart.

tissue into this depression (Milne and Milne 1976; Halffter 1982). Later the 1- to 2-mm long larvae migrate to the top of the cup-shaped carrion mass in response to stridulation by the female (Milne and Milne 1976). There an adult (usually the female) broods and feeds the larvae fluid and tissue from the carrion mass in much the way that birds feed nestlings. Parental care often continues until pupation, at which time the parents dig their way out of the chamber and begin the process anew.

A host of insects, microbes, and vertebrates compete for carcasses, so beetles must work industriously to get carcasses buried as quickly as possible (Wilson and Fudge 1984; Wilson 1985). Numerous burying beetles of several species may cooperate in burying a carcass. After the carcass is safely concealed, a dominant pair of beetles drives the others away. Flies are one of the most troublesome competitors of burying beetles because they lay their eggs before the beetles arrive at a carcass or as the beetles bury the carcass. Fly larvae can quickly consume a carcass, making it unfit for the beetle brood. Burying beetles counteract the effect of the flies in several ways. They remove some of the fly eggs by cleaning the carcass of all of its hair. In addition, burying beetles transport phoretic mites to carcasses. These mites feed on fly eggs and thereby indirectly benefit the burying beetles (Wilson 1985). If, in spite of these defenses, fly larvae infest a corpse, the adult *Nicrophorus* may abandon it.

Rove Beetles (Staphylinidae)

Three species of rove beetles (Staphylinidae: *Bledius* spp.) studied by Bro Larsen (1952) store algae. Most of the 350 known species of *Bledius* inhabit littoral and supralittoral zones in freshwater and marine ecosystems. The species studied by Bro Larsen excavate burrows in sediments of saltwater marshes of the European coast—*B. spectabilis* in the intertidal zone, where burrows are submerged daily, and *B. diota* and *B. furcatus* higher up on the shore. For *B. spectabilis,* the species most intensively studied, the burrow provides refuge for a pair of beetles and a nest site for rearing young (Wyatt 1986). The female constructs and maintains the burrow. About half way down the 4- to 6-cm deep burrow, the female constructs a set of small chambers and lays one egg in each. Ventilation of the burrow and nest chambers is promoted by the constant digging of the beetles, an activity made necessary by the low oxygen content of the anaerobic sediments. During high tides, surface-tension effects trap air in the burrow, and the nest is not flooded (Wyatt 1986). After the eggs hatch, the larvae remain in the nest and feed on algae stored by the female. The female periodically leaves the burrow at low tide to collect more algae. Bro Larsen (1952) did not report the type and quality of algae stored.

Female *Bledius* avoid high salt concentrations in the algae they store for larvae. Adult *B. spectabilis* tolerate 5–6% NaCl concentrations in the algae they eat but will not store algae with concentrations over 4% NaCl (Bro Larsen 1952). *B. diota* and *B. furcata* often collect algae after rains, when the concentration of salt in stagnant tidal pools and in algae is lowest.

10.3   ANTS

Ants (Formicidae)

Ants exhibit three types of food storage. Harvester ants store seeds, honey ants store nectar and other liquids, and carnivorous ants store invertebrate prey. Within each group, the evolution of food storage has apparently had several independent origins.

Harvester Ants. Seed-harvesting ants have been the objects of natural history investigations since antiquity (the pre-1900 literature has been reviewed by Moggridge [1873], McCook [1879], and Wheeler [1910]), but despite this interest, the below-ground dynamics of seed storage and use are still poorly understood. Our ignorance is due largely to the laborious and difficult task of carefully excavating ant nests that may extend 3 m or more deep into sandy, unstable soils, where storage rooms and tunnels collapse during excavation or may lie beneath a layer of hard caliche that cannot be easily penetrated. Descriptions of harvester ant nests are merely fuzzy snapshots of the state of nests at a point in time. Further, there seems to be great inter- and intraspecific variation in the status of colonies in a region, making it necessary to excavate many nests to gain accurate insights into the dynamics of seed stores.

Seed-harvesting ants are important components of arid and semiarid grasslands, deserts, and shrub-steppe ecosystems throughout the world. Most seed-storing species belong to *Messor* and *Monomorium* in the Old World and *Pogonomyrmex* and *Veromessor* in the New World. The genus *Pheidole* is cosmopolitan (Sudd 1967). One or more seed-storing species have been identified in *Solenopsis, Pheidologeton, Tetramorium, Meranoplus, Ischnomyrmex,* and *Oxyopomyrmex* (Wheeler 1910). All these genera belong to the ant subfamily Myrmicinae.

Harvester ants establish subterranean nests in open, sunny sites with sparse vegetation. Nests of some harvester ants are very conspicuous, so much so that they can be easily counted on aerial photographs. Western harvester ants (*Pogonomyrmex occidentalis*) of the United States, for example, construct conical mounds of soil 10–30 cm high covered with fine gravel. One or more entrances are located near the base of the mound. *P. occidentalis* and related ants, such as *P. owyheei* of the northwestern United States, remove all the vegetation within a radius of 2 m or more of the cone (Headlee and Dean 1908; Cole 1968), a habit that may serve to keep the nest warm and dry. Not all harvester ant nests are this conspicuous; e.g., *P. imberbiculus* constructs a small crater of coarse soil a few centimeters high and 10–20 cm in diameter (Wheeler 1910). They often establish nests under rocks (Creighton 1956), and vegetation may grow up to the base of the crater.

Inside harvester ant nests are numerous, small, flattened or domed chambers connected by a network of galleries (fig. 10.7). In *P. occiden-*

Figure 10.7
A portion of the interior of a harvester ant nest showing granaries, galleries, and ants managing the stored seeds. Original drawing by Marilyn Hoff Stewart.

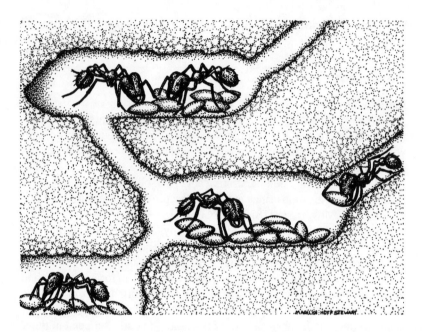

talis nests, chambers can be found from a few centimeters below the apex of the mound to 3.5 m or more below ground (McCook 1882; Lavigne 1969; Clark and Comanor 1973). The ants brood eggs, rear larvae, deposit refuse, or accumulate seeds in these chambers. *P. occidentalis* nests excavated by Lavigne (1969) near Casper, Wyoming, contained between thirteen and sixty-four chambers. Storage chambers are typically 2–10 cm across and only 0.5–3 cm high (McCook 1882; Headlee and Dean 1908; Cole 1932; Wray 1938; Costello 1947; Tevis 1958; Hutchins 1966; Lavigne 1969). The nests of some harvester ants are very simple. *Messor arenarius* nests, for example, are a relatively simple system of radiating galleries joining a half-dozen granaries (fig. 10.8) (Delye 1971). The largest granaries are 40 cm long, 10 cm wide, and 2 cm high.

Foraging harvester ants leave their nests during favorable weather to search for food (see Carroll and Janzen [1973] for a review of harvester ant foraging behavior). Workers may travel over 20 m from the nest looking for seeds and invertebrates. Harvester ants carry seeds one at a time between their mandibles or psammophores. Some specialized species of *Messor* have a small notch on the underside of the head that facilitates transport of seeds (Wheeler 1910). The ants carry unthreshed seeds into the nest and deposit them in chambers near the surface (Tevis 1958), where workers remove the chaff and other debris (Lavigne 1969). Cleaned seeds are probably redistributed to other chambers deeper in the nest. Ants usually dismember and consume invertebrate prey shortly after they are delivered to the nest, but in some cases the ants may store insects and other foods, such as pollen robbed from bee nests (Bohart and Knowlton 1953; Creighton 1956).

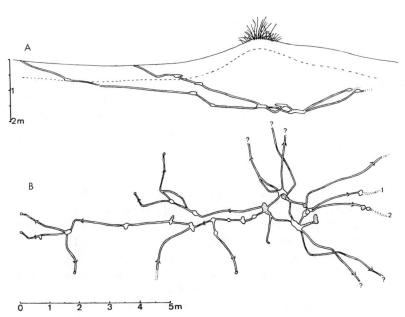

Figure 10.8
The burrow system of the harvest
ant *Messor arenarius* in sand dunes
of the Sahara Desert. A, Sectional
view of the nest from the side. B,
view of chambers and galleries
from above. The stippled chambers
are granaries. From Délye 1971.

Harvester ants collect seeds throughout the warmer months of the year. Seasonal changes in amounts of seeds stored in nests show great intra- and interspecific variation. Lavigne (1969) found that nests of western harvester ants near Casper, Wyoming, contained up to 23 g (mean = 8.4 g) of seeds in the fall and winter (November to May) of 1966–67 and 3–67 g (mean = 37.8 g) of seeds in summer (June to November) of 1967. MacKay (1981) studied the behavior of three species of *Pogonomyrmex* harvester ants in southern California. The amount of seeds stored by these ants was inversely related to elevation. *P. montanus,* which occurred at highest elevations, stored only trace amounts of seeds. Seed reserves of *P. subnitidus,* a midelevation harvester ant, were between 2 and 6 mg per ant from winter to early summer and then increased sharply in some nests from August to December to about 10–30 mg (in one case about 75 mg) per ant (fig. 10.9). Seed reserves of *P. rugosus,* a lowland species, ranged from about 10–75 mg per ant throughout the year with no strong seasonal patterns evident. From about 170–600 g of seeds remained in nests of *P. rugosus* colonies in early spring before foraging began, and the quantity of seeds in a nest was highly correlated with number of adult workers (MacKay and MacKay 1984). Nests of *P. badius* have been found with nearly 2 l of ragweed (*Ambrosia elatior*), crabgrass (*Digitaria sanguinalis* and *D. ischaemum*), and other seeds stored in granaries (Wray 1938); fire ants, *Solenopsis germinata,* store up to 100 ml of the annual grass *Paspalum* in their nests (Carroll and Risch 1984).

Seeds of grasses and forbs seem to be preferred by harvester ants. Plant genera commonly stored by *Pogonomyrmex* include the grasses

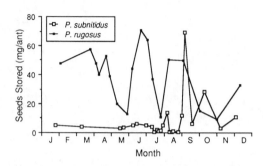

Figure 10.9
Seasonal changes in the quantity of seed stored by two species of harvester ants (*Pogonomyrmex rugosus* and *P. subnitidus*). Redrawn from MacKay 1981.

Bromus, Oryzopsis, Stipa, and *Agropyron,* the forbs *Lepidium, Capsella, Chenopodium, Polygonum, Carduus, Salsola, Hordeum,* and *Malva,* and the shrub *Atriplex* (McCook 1882; Headlee and Dean 1908; Cole 1932; Costello 1947; Lavigne 1969). *Messor arenarius,* studied by Delye (1971) in the Sahara Desert, stored almost exclusively seed of drin (*Aristida pungens*). The nests of *Veromessor pergandei,* examined by Tevis (1958) in the southern California desert, contained the seeds of *Amaranthus.* The seeds that ants deliver to the nest are not necessarily representative of those they store (Lavigne 1969).

The distributions of granaries in some harvester ant nests appear to change seasonally. Nests of *P. occidentalis* excavated during winter have fewer seeds in chambers near the surface than in similar nests during summer (fig. 10.10). Cole (1934) reported fewer stored seeds within mounds of *P. owyheei* from fall to spring. This change has two possible causes. First, ants may have differentially consumed seeds near the surface before the onset of winter. Second, ants may have redistributed the seeds, moving them from the upper to the lower chambers before winter. It is unlikely that the seeds in the upper chambers were eaten or moved during winter because the ants become dormant at this season (MacKay and MacKay 1984).

Harvester ants consume stored seeds (MacKay and MacKay 1984) and feed seeds to larvae. The large soldier castes of species such as

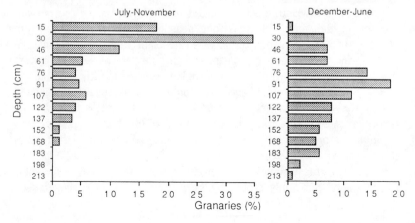

Figure 10.10
Vertical distribution of seed storage chambers in nests of the western harvester ant near Riverton, Wyoming. Seeds occur closer to the surface during the summer and fall. Based on data from Lavigne 1969.

Pheidole serve the function of seed crushers, permitting the smaller castes easier access to contents of hard-shelled seeds (Wheeler 1910). The harvester ant *Messor capitatus* masticates seeds and then licks the paste until it is thoroughly impregnated with saliva from the labial salivary glands. This saliva contains the enzymes maltase and amylase, which hydrolyze the starch first into maltose and then glucose (Delage 1962). Wheeler (1910) noted that *Pogonomyrmex barbatus* workers in artificial formicaria coated seed fragments with saliva before feeding the fragments to larvae.

MacKay and MacKay (1984) presented four hypotheses to explain the functional significance of seed storage by several species of *Pogonomyrmex* harvester ants: (1) ants store seeds for winter use, (2) ants store seeds as a food resource to rapidly produce new individuals, (3) ants store seeds as insurance against "bad years" in environments with unpredictable primary production, and (4) ants store seeds as protection against predation on foragers. MacKay and MacKay (1984) considered the evidence for each of these hypotheses and rejected the first three for lack of supporting information. They favored the fourth hypothesis, citing cases where ants have been known to close nest entrances in the presence of high forager mortality. In my opinion, however, the case for hypothesis 4 seems to be no stronger than those for some of the other hypotheses. Further, MacKay and MacKay (1984) failed to consider other plausible hypotheses. One such hypothesis is that harvester ants store seeds because they inhabit predictably harsh physical environments where foraging is frequently interrupted or precluded for short periods (hours or days). Such interruptions include cool summer nights, hot afternoons, rainy weather, and periods during spring and fall when soil surface temperatures are low, but deep ground temperatures still permit activity. A store of seeds may permit colonies to remain active below ground for a much greater proportion of nonforaging time than if individuals relied solely on body fat reserves. It is, of course, possible that two or more of these hypotheses may be correct. Further observations and experiments are needed to resolve this issue.

Honey Ants. Many ants collect nectar and honeydew and transport these sweet exudates in their distensible crops to nests. Their abdomens often become so enlarged that movement is slow and cumbersome. In most species, however, this swollen condition persists only until the foragers return to the nest and distribute the liquid to nestmates. But, in certain species of ants belonging to the subfamilies Formicinae and Dolichoderinae, the habit of storing liquid in the crop has been carried to an extreme; certain individuals act as living receptacles for nectar and honeydew brought to the nest by foragers (Wheeler 1908). In these species, collectively known as honey ants, individuals called repletes imbibe so much liquid that their gasters become greatly distended; the sclerites of the abdomen are forced apart by the volume of liquid, exposing the intersegmental membranes. The gaster becomes spherical and so large that locomotion is extremely difficult. These repletes cling to the ceilings of underground chambers (fig. 10.11). Thus honey ants, which lack the

Figure 10.11
A replete of the honey ant *Myrmecocystus mexicanus* hanging from the ceiling of an underground chamber. Original drawing by Marilyn Hoff Stewart.

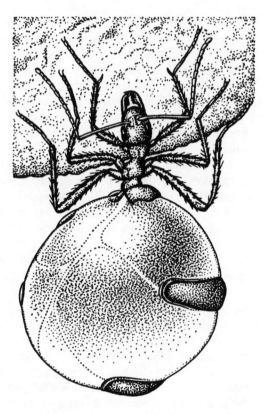

ability to produce wax, use modified workers as the functional equivalent of bee honeypots (Eisner and Brown 1958).

Honey ants occur in arid and semiarid areas of North America, eastern Europe, Asia Minor, south Africa, and Australia (Wheeler 1910; Creighton 1950; Hutchins 1967). One of the most specialized representatives of this group is *Myrmecocystus mexicanus,* which occurs from Idaho to central Mexico (McCook 1882; Wheeler 1908; Snelling 1976). These honey ants forage nocturnally for liquids exuding from plant galls, floral and extrafloral nectaries, and secretions of aphids and coccids (Wheeler 1908; Snelling 1976). They also feed on the body fluids of insects (Conway 1977).

M. mexicanus establishes nests in well-drained, rocky soils, often on sparsely vegetated ridge tops (McCook 1882; Wheeler 1910), but moist soils of valley bottoms also harbor nests (Snelling 1976). From the inconspicuous nest entrance, a single passage descends about 10–30 cm before branching into a network of galleries that lead deeper to chambers 15–180 cm below ground (Wheeler 1908; Burgett and Young 1974; Conway 1977). These honey ants excavate most chambers within the zone of permanent soil moisture (Snelling 1976). Chambers are used as refuse piles for dead insects, for brood rearing, or as pantries, where repletes are gradually transformed into living bags of nectar. The storage chambers are small, only about 7–15 cm across and 1.5–4 cm high (McCook

1882). Three nests excavated by Conway (1977) contained 7–21 replete chambers; one nest contained 1,030 repletes. A nest excavated by McCook (1882) contained over 300 repletes segregated in ten chambers. Abdomens of large repletes may be nearly a centimeter in diameter and contain from 0.3–0.4 g of fluid, about eight times the mass of the ant's body (Wetherill 1852; Conway 1977).

Repletes of honey ants from other parts of the world are similar to those of *M. mexicanus* without, however, attaining such dimensions. The black honey ant (*Camponotus inflatus*) constructs nests to a depth of nearly 200 cm in arid regions of central and western Australia (Wheeler 1908, 1910). Although the repletes are not quite as large as those of *M. mexicanus,* they are nonetheless unable to move about freely and must be cared for by workers. Repletes of *Melophorus bagoti* of Australia (Wheeler 1908) and *Prenolepis imparis,* widely distributed in North America, eastern Europe, and Asia Minor (Wheeler 1930; Talbot 1943), on the other hand, appear as workers with enlarged abdomens and retain considerable mobility (see fig. 3.2). *Plagiolepis trimeni* is a small honey ant of southern Africa whose abdomen, when fully distended, is only 4.5 mm in diameter. These species represent a level of repletion between that of *Prenolepis* and *Myrmecocystus* (Wheeler 1910).

The question of which ants in a colony become repletes is only partially understood. Workers that become repletes are anatomically identical to other workers except that major workers often become repletes, probably because their greater size permits a greater reservoir for nectar storage (McCook 1882). Wheeler (1908) suggested that the transition from worker to replete must begin at the callow stage, while the gaster is still soft and distensible. The number of repletes formed by a colony, which varies greatly within and among species, is largely influenced by the amount of food and callow workers available and has proved to be of little value as a taxonomic character (Snelling 1976). Under laboratory conditions, it required 4–6 weeks for some ants to achieve the replete condition.

The color of *M. mexicanus* repletes in a nest may vary because of differences in the types of liquids they contain (Conway 1977). Dark amber repletes, which often predominate in a nest, contain fluid rich in dissolved solids, primarily glucose (61%) and fructose (39%). Light amber and clear repletes contain weak solutions of sucrose and may serve primarily as water-storage vesicles (Conway 1977). Other individuals may contain lipids (glyercol and cholesterol esters) derived from insect prey (Burgett and Young 1974). Honey in Australian honey ants (*Melophorus inflatus*) is 59% sugar with a fructose:glucose ratio of 0.67:1 (Badger and Korytryk 1956).

The number of repletes in a colony increases when sources of nectar and honeydew are available in excess, which usually occurs during spring or the wet season. During hot, dry seasons, honey ant workers curtail above-ground foraging and workers take liquids from the repletes to nourish the colony (McCook 1882; Wheeler 1910; Talbot 1943). Repletes comprised about 80% of colonies of *Prenolepis imparis* during winter and

spring and 67% during summer, when no food was being brought to the
nest and the brood was developing (Talbot 1943). Stumper (1961), using
radioactively labeled nectar, demonstrated that ambient temperature is
an important influence on nectar allocation in a colony. At 20–21°C there
was a net flow of nectar from workers to repletes in a laboratory colony
of *Proformis nasuta,* an inhabitant of the arid steppes of eastern Europe
and southern Russia. Above 30°C, the flow of nectar was reversed, moving from repletes to the foragers. Well-stocked colonies can persist for
over 4 months with no outside source of food or water (McCook 1882).

   Carnivorous Ants. Several species of ants in the subfamily Ponerinae
store invertebrates. These are carnivorous ants that raid the nests of
other ants or termites or forage for other small insects on the ground
surface. Foraging ants capture prey by stinging them and then carry them
back to the nest. There the ants pile the prey in chambers, where they
remain alive but quiescent for weeks or months (Maschwitz, Hahn, and
Schönegge 1979; Hölldobler 1982). *Leptogenys neutralis* of southwestern
Australia store dealate female ants, termites, and various beetles in
chambers apart from the brood (Wheeler 1933). *Leptogenys chinensis* and
Harpegnathus saltator are solitary foragers that search for solitary prey
(Maschwitz et al. 1979), but *Cerapachys turneri* of northern Queensland,
Australia, form raiding parties and ravage the nests of *Pheidole* ants,
carrying away adults and brood. Raiding parties take an extreme form in
some of the army ants of the Neotropics. The army ants *Eciton burchelli*
and *E. hamatum* on Barro Colorado Island, Panama, stage massive raids
in which thousands of ants swarm over invertebrates and small vertebrates on the forest floor. During the nomadic phase of these colonies,
booty is concentrated in caches along the columns (fig. 10.12) (Rettenmeyer 1963).

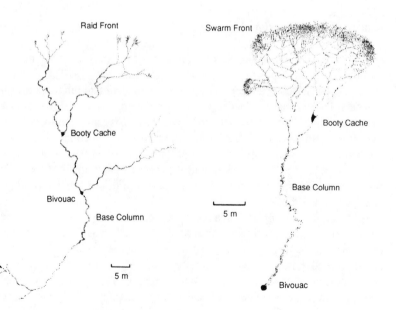

Figure 10.12
Carnivorous army ants (e.g., *Eciton
hamatus* [left] and *E. burchelli*
[right]) form booty caches along
columns leading away from the
raid front. From Rettenmeyer 1963.

Maschwitz, Hahn, and Schönegge (1979) have suggested that food storage may be widespread in carnivorous ponerine ants because these ants lack a crop in which they can store internally a small amount of food, and because reciprocal exchange of food among colony members is lacking. Paralysis of prey may have evolved because it enables ants to store excess food, which nourishes the colony when other food is scarce. Hölldobler (1982) noticed that a captive colony of *Cerapachys* began producing eggs and raising a brood following a lucrative raid that supplied the colony with food for over 2 months.

10.4 WASPS

Nearly all wasps that construct nests either mass or progressively provision those nests with invertebrate prey. The nesting and provisioning behavior of wasps is extremely diverse, and the literature on this aspect of wasp biology is very large. Wasps that provision nests are thought to have evolved from a parasitoid ancestor through intermediate stages in which prey were partially concealed in natural crevices or, when the prey were burrowing forms, in the prey's own retreat. Such primitive forms of provisioning behavior are represented by certain species of Bethylidae, Tiphiidae, and Scoliidae—small families composed largely of parasitoids that by and large do not construct nests (e.g., Burdick and Wasbauer 1959; Malyshev 1969). Nest-building and provisioning behavior are well developed in five families of wasps: Pompilidae, Sphecidae, Eumenidae, Vespidae, and Masaridae. I illustrate the range of provisioning behavior that exists by describing the nesting of several representative species in each wasp family. I emphasize but do not limit myself to the literature published since 1970. Excellent summaries of earlier literature are provided by Evans (1966a, 1966b), Krombein (1967), Spradbery (1973), Iwata (1976), Krombein et al. (1979), and others.

Spider Wasps (Pompilidae)

Spider wasps, which get their name from provisioning their nests with spiders (e.g., Kurczewski and Kurczewski 1972), are among the most primitive of the wasps that provision nests. Nesting behavior varies considerably, but the behavior of *Anoplius tenebrosus* described by Alm and Kurczewski (1984) is representative of the more generalized spider wasps. On warm spring days, female *A. tenebrosus* search out and attack a large spider, usually of the family Lycosidae (wolf spiders) or Thomisidae (crab spiders). The spider must be of sufficient size to permit complete development of the spider wasp larva. Spider wasps begin hunting before they construct a nest. The wasp paralyzes the prey by stinging, drags it backward by the base of the hindlegs to a clump of grass or other vegetation, and caches it among the foliage. Or the wasp may simply leave the prey lying on the sand while it prepares the nest. Female *A. tenebrosus* dig nests in sandy soil exposed to the sun. The burrow is usually a simple, nearly vertical, corridor extending 2–7 cm into the soil and ending in a single, oval-shaped cell averaging 9×7.5 mm. When completed, the wasp retrieves the spider, drags it to the burrow, deposits it dorsum upward in the cell, and lays a single egg laterally on its abdomen (see fig. 4.3). The wasp then uses soil to fill the short corridor leading from the cell, tamping loose soil with the tip of its abdomen. Paralysis is usually temporary; the spider recovers in several hours, but it cannot escape from the sealed cell. The egg hatches in 2–3 days, and the tiny larva immediately begins to feed on the living spider. After it has consumed the contents of the cell, the larva pupates. An adult emerges from the pupa case in midsummer and digs out of the nest. Most species of spider wasps provision about one cell per day under favorable conditions (Evans and Yoshimoto 1962).

The nesting behavior of most spider wasps follows this pattern, differing most notably in the types of sites selected for nesting and the types of spiders captured for provisions. Some species of spider wasps construct nests in galleries in wood, hollow stems, and even in snail shells (Evans and Yoshimoto 1962; Fye 1965; Kurczewski and Spofford 1986). Those that dig their own nests often do so in the wall of a natural crevice or small mammal burrow. The means by which wasps transport prey, arrange the prey in the cell, and close the nest varies among species. A number of pompilids, especially those in the tribe Auplopodini, amputate some or all of the legs of their spider prey, an act that is thought to facilitate prey transport and storage (Evans and Yoshimoto 1962; Kurczewski and Kurczewski 1972; Kurczewski and Spofford 1986). Some spider wasps hunt for trapdoor or burrowing spiders, which the wasps sting and leave in the spider's own burrow. *Anoplius marginalis* attacks and paralyzes burrowing wolf spiders (*Geolycosa*), constructs an 8- to 12-cm-long burrow off one side of the spider's burrow, deposits the paralyzed spider in a cell at the end of the side burrow, lays an egg, and then backfills the side burrow and much of the spider's original burrow (Gwynne 1979). The diversity of provisioning habits of the Pompilidae is described by Evans and Yoshimoto (1962).

Most spider wasps capture prey before preparing a nest and consequently it is usually necessary for wasps to temporarily store the spider while they prepare a cell. Spider wasps often hide prey under debris or in low vegetation (fig. 10.13) (e.g., Olberg 1959; Kurczewski and Snyder 1964; Kurczewski and Kurczewski 1973; Gwynne 1979; Alm and Kurczewski 1984; Kurczewski and Spofford 1986). *Calicurgus hyalinatus* temporarily stores spider prey by suspending it from a low twig using silk thread from the spider's spinnerets (Kurczewski and Spofford 1985). During nest construction, wasps repeatedly visit the temporarily stored spiders, apparently to protect the prey from scavengers and to monitor the state of paralysis so the spider does not escape. *Anoplius apiculatus* drags prey to the nest entrance, where it covers it with excavated soil, a habit that reduces parasitism and kleptoparasitism (Evans, Lin, and Yoshimoto 1953). When the wasp completes the cell, it exhumes the spider and drags it into the nest.

The mass of the accumulated provisions is usually greater than the mass of the provisioning female (see table 2.3). Alm and Kurczewski (1984) found that the mean fresh mass of spiders paralyzed by *A. tene-*

Figure 10.13
Spider wasps often cache spiders in foliage while they excavate a nest chamber. This spider was stored by the spider wasp *Anoplius marginatus*. Later, they retrieve the spiders and restore them in nest chambers. Photograph courtesy Frank Kurczewski.

brosus was 92 mg (range 16–210 mg), whereas the mean mass of the wasps was 40 mg (range 11–72 mg), a ratio of 2.3:1. Two prey of the wasp *C. hyalinatus* weighed five and six times more than the 8 mg wasp (Kurczewski and Spofford 1985). The large size of the prey is, of course, to ensure sufficient food for the larva to develop completely.

Sphecid Wasps (Sphecidae)

The family Sphecidae is a large group with diverse nesting habits. Most species provision nests with arthropod prey or are parasitic on other wasps that do so. Virtually all species are solitary (i.e., each female constructs and provisions her own nest), although in many species numerous females may nest in close proximity. Nesting behavior in the Sphecidae is much more diverse than that exhibited by spider wasps, and nest architecture is more complex. Furthermore, sphecid wasps usually construct nests before searching for provisions and usually provision each cell with two or more prey. There are three major types of nests: (1) nests excavated in soil, (2) nests constructed with mud, and (3) nests built in stems and existing galleries in wood.

Numerous genera, including *Philanthus, Sphex, Ammophila, Cerceris,* and *Astata,* excavate nests in soil. A common and well-studied member of this group is the great golden digger wasp (*Sphex ichneumoneus*) (e.g., Ristich 1953; Brockmann 1985). Great golden digger wasps excavate burrows in soft substrates, terminating in oval, smooth-walled cells 3 cm long (fig. 10.14). The great golden digger wasp provisions this cell with katydids, crickets, and related Orthoptera, which it paralyzes by stinging. The wasp lays an egg on one of the last prey placed in the cell, seals the chamber with a plug of soil, and then begins to excavate another chamber off the same burrow a few centimeters away from the first chamber. After the wasp has completed from one to seven cells, it fills

Figure 10.14
The complete nest of a golden digger wasp. A, Sometimes brood cells are closed without being provisioned. B, The fully provisioned brood cell with a larva. C, The first brood cell dug and provisioned in this nest contains a prepupa. D, A detailed view of a provisioned brood cell with three katydids. From Brockmann 1985.

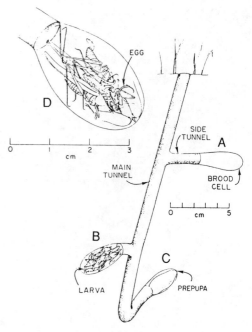

the main burrow with soil and begins to excavate a new nest elsewhere. The eggs hatch in about 2 days, and the larvae consume all the provisions in about a week before spinning cocoons.

The nest-building and provisioning behavior of other digger wasps follows this same general pattern but often differs in many important details (cf. Evans 1962b, 1966a; Kurczewski 1968, 1987; Evans and Hook 1982; Kurczewski and Miller 1984; Kurczewski and Kurczewski 1987). Most digger wasps provision each cell with from several to a dozen prey, but some species use only one large prey item (e.g., Evans 1959; O'Brien and Kurczewski 1982). Paralysis may be permanent and even fatal, as in *Astata* (Evans 1957a), or temporary, as in *Chlorion* (Peckham and Kurczewski 1978). Wasps transport prey in their mandibles, held in their legs, impaled on the sting, or attached to a uniquely modified "clamp" on the terminal abdominal segment (Evans 1962b). When closing the nest, a wasp compacts the soil with its abdomen, its head, or a small pebble held in its mandibles. The wasp also may excavate short, blind accessory burrows near the mouth of the original burrow that apparently confuse parasitic wasps and flies that try to excavate the nest. Several species of *Cerceris* construct nests and provision cells communally (Alcock 1980; Evans and Hook 1982). Most species of *Bembix* provision larvae progressively, occasionally checking cells and adding prey to meet the larva's needs (e.g., Evans 1957b).

Several groups of sphecid wasps, generally known as mud-daubers, construct complex nests of mud and clay. They place their nests in sheltered sites under rocks, on cliffs, in tree cavities, under bridges, and in abandoned buildings protected from the weather. Organ-pipe mud-

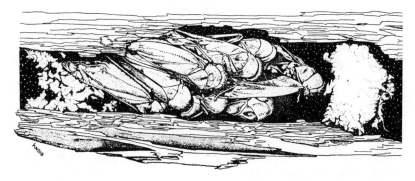

Figure 10.15
Leafhoppers stored in a cavity of a rotting log by the wasp *Crabro davidsoni*. From Davidson and Landis 1938.

daubers (*Trypoxylon politum*), for example, construct inverted tubes of mud under rock overhangs and other protected sites, which they provision with seven to ten medium-sized spiders (e.g., Dorris 1970; Cross, Stith, and Bauman 1975; Brockmann 1988). When the first cell is fully provisioned they construct a partition of mud below the prey and begin stocking the next cell. Other sphecid wasps construct nests in cavities that they excavate in pithy stems or rotten wood or in galleries that have been previously excavated by other insects (fig. 10.15). After the wasp lays an egg on one of the prey, it constructs a partition several millimeters from the store of food and begins provisioning the next cell. The wasp repeats the provisioning and partitioning cycle, gradually working closer to the mouth of the cavity. Near the end of the cavity, the wasp seals the nest with a closing plug (Krombein 1967).

Most sphecid wasps exhibit marked specificity in prey selection, often capturing prey of one species or genus of arthropod. The digger wasp *Clypeadon laticinctus,* for example, preys exclusively on workers of the harvester ant *Pogonomyrmex occidentalis* (Alexander 1986), whereas *Aphilanthops* spp. preys on winged queen ants of the genus *Formica* (Evans 1962a). *Chlorion aerarium* in central New York preys on one species of cricket, *Gryllus pennsylvanicus* (Peckham and Kurczewski 1978). *Podium* spp. provisions tube nests with cockroach nymphs and adults (Krombein 1970); *Passaloecus* spp. provisions tube nests with aphids (Fye 1965). *Astata unicolor* preys on immature stinkbugs (Pentatomidae), whereas *Astata occidentalis* preys only on medium-sized adult stinkbugs (Evans 1957a). Other wasp species have slightly broader tastes in prey. The great golden digger wasp preys on three families of Orthoptera; species of prey may change seasonally, apparently as availability changes (Brockmann 1985). *Trypoxylon,* which includes some mud-daubers and tube-nesters, preys on spiders from a wide variety of families (Dorris 1970). Some species that appear to have very narrow preferences in prey at one site vary significantly in prey choice geographically. It is tempting to conclude that prey specificity is a means of reducing interspecific competition among wasps, but it seems more likely that specialization on a prey type is simply an adaptation of foraging behavior to increase efficiency of prey capture, transport, and handling. These hypotheses are not mutually exclusive.

The prey stored by sphecid wasps are relatively small compared to those stored by spider wasps. Mean mass of individual prey is often less than half the mass of the wasp (e.g., Kurczewski 1968). Combined mass of provisions, however, may greatly exceed the mass of the wasp (table 2.3). The digger wasp *Chlorion aerarium,* for example, provisions cells with two to nine crickets with a combined mean mass of 1,400 mg (range 827–1,938 mg) (Peckham and Kurczewski 1978). The adult wasp has a mass of less than 100 mg. The number of prey and the mass of individual prey that wasps store often vary inversely (e.g., Kurczewski 1968; Kurczewski and Miller 1984). The total mass of provisions stored by the digger wasp *Plenoculus davisi,* for example, increases with the number of prey stored, but the average size of individual prey decreases (fig. 10.16), suggesting that the wasp partially compensates for small prey by storing more prey per cell. The organ-pipe mud-dauber provisions cells with many small spiders early in the season and fewer but larger prey late in the season (Cross et al. 1975, 1978). The biomass of provisions in early and late cells is statistically indistinguishable.

Temporary storage of prey is common among wasps in the subfamilies Philanthinae and Astatinae (Evans 1962a, 1966a). *Astata unicolor,* for example, stores stinkbugs in loose soil at the bottom of nests. Only after they store numerous bugs do the wasps construct cells and provision each with two to four bugs from the stockpile, depositing an egg on the first bug in each cell (Evans 1957a). *Cerceris watlingensis* stores weevils in the main burrow during the day and then constructs and provisions cells at night (Salbert and Elliott 1979). The digger wasp *Clypeadon laticinctus* accumulates ants in the upper portion of the burrow to use in provisioning cells (Alexander 1986).

Potter and Mason Wasps (Eumenidae)

As the names suggest, these wasps usually use mud or clay in nest construction. Although potter and mason wasps comprise a smaller group than the sphecid wasps, their nesting habits are even more diverse. Some members of the family construct nests similar to those of mud-dauber wasps (e.g., certain *Ancistrocerus*) (Spradbery 1973); others nest in hollow stems or old burrows of other insects (e.g., some *Ancistrocerus, Leptochilus, Symmorphus, Euodynerus*) (Cooper 1953, Fye

Figure 10.16
The combined mass of prey and the mean mass of individual prey stored in cells by the sphecid wasp *Plenoculus davisi* vary as functions of the number of prey stored in each cell. Redrawn from Kurczewski 1968.

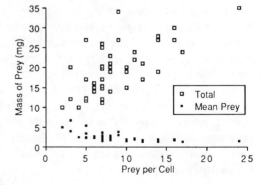

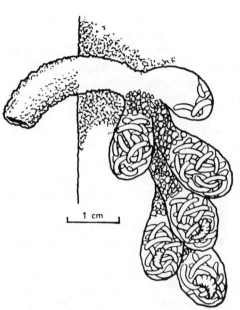

Figure 10.17
The nest of the chimney-making wasp *Odynerus spinipes*. Cells dug in a mud bank are filled with caterpillars. The female has suspended an egg from a filament, still visible in the upper cells; in the lower cells, wasp larvae are consuming the provisions. From Bristowe 1948.

1 cm

1965, Parker 1966); and still others excavate nests in the ground like digger wasps (e.g., certain *Pseudepipona, Euodynerus*) (Spradbery 1973; Freeman and Taffe 1974; Cowan 1981). To illustrate some general patterns of nest construction and provisioning in this family, I describe two other types of nests that seem to have no parallels among the Sphecidae.

Chimney-making wasps excavate shallow nests in heavy soils and construct temporary "chimneys" around the entrances. *Odynerus spinipes*, for example, excavates burrows into vertical banks, depositing the excavated material on the rim of the chimney until it extends out from the wall as much as 3 cm, curving sharply downward and opening from below (fig. 10.17) (Bristowe 1948). The chimney is thought to protect the nest from the weather and possibly from kleptoparasites (Spradbery 1973). At the end of the burrow, the wasp excavates an oval cell. Before provisioning this cell with five to ten Lepidoptera larvae, the wasp lays an egg, suspending it from the roof of the cell by a fine filament. The caterpillars, packed in a tangled mass, completely fill the cell. When provisioning is complete, the wasp seals the chamber with a thin layer of mud and constructs another cell above and to the side of the first cell. Five to six cells constitute a completed nest. When the nest is finished, the wasp breaks down the chimney, using it to seal off the main entrance to the nest.

Potter wasps (*Eumenes* spp.) construct miniature spherical pots in shrubs or on roots of plants under overhangs. *Eumenes colona* in Jamaica constructs pots with a maximum inside diameter of 14 mm and walls about 0.5 mm thick (Freeman and Taffe 1974). After suspending an egg from the roof of the pot, the female potter wasp hunts Lepidoptera larvae, paralyzes them temporarily, transports them back to the nest, and inserts them into the pot. *Eumenes colona* places four to fourteen caterpillars in each pot, depending on their size; the largest prey weigh about 75 mg (Freeman and Taffe 1974). When the pot is full, the female potter wasp collects mud and seals the entrance to the nest. Similar nesting behavior has been described for *Zeta abdominale* (Taffe 1983).

Most eumenid wasps provision nests with Lepidoptera larvae, which they paralyze by stinging them in the thoracic region (Spradbery 1973). Eumenid wasps carry prey ventral side up, holding the larva with their mandibles and often all three pairs of legs. Many species are host specific, feeding on only a single species of larvae. From three to thirty-five tightly packed prey constitute complete provisions, depending the species of wasp, size of prey, and sex of the wasp larva. Cells that produce females are larger than cells that produce males, and female-producing cells contain more provisions (see fig. 2.14). *Ancistrocerus adiabatus,* for example, provision female-producing cells with a mean of 82 mg of prey and male-producing cells with a mean of 51 mg of prey (Cowan 1981). Temporary storage of prey does not seem to occur in eumenids. The practice of suspending the egg from the top of the cell before provisioning is widespread in the family and presumably functions to place the larva at eclosion in a favorable position near the top of the mass of partially paralyzed caterpillars.

Paper Wasps (Vespidae)

The adults of most paper wasps (Vespidae) do not mass provision but feed larvae directly with well-masticated insect flesh. A few species of polybine and polistine wasps, however, store honey in their nests. Paper wasps construct intricate nests of well-masticated plant fibers, sometimes mixed with soil to produce a paperlike substance called carton. The basis of all nests is one or more combs composed of several to several thousands cells, often arranged in clusters on horizontal planes and usually opening from below. The combs of some nests are enclosed in a protective carton envelope. The diversity of nest architecture is described by Spradbery (1973). Wasps of the family Vespidae are mostly subsocial or eusocial wasps living in kin groups ranging from several to several thousand individuals. Some colonies may contain numerous queens. In warm climates, nests may be inhabited throughout the year.

Paper wasps store honey in cells identical to those used for brood rearing. Neither the quantity nor the composition of honey stored by paper wasps appear to have been described. Honey stored by paper wasps (*Polistes annularis*) in southern Texas is similar in taste to that produced by honey bees but much more viscous (Strassman 1979). The greater viscosity is probably necessary because paper wasps store honey in vertically oriented cells that open from below, whereas honey bees construct nearly horizontal cells that open from the side.

Small droplets of honey can be found in the nests of some vespids at any time of year, but they store large amounts of honey only at certain seasons. *P. annularis* begins to store honey in September; by November, all occupied colonies contain some honey (Strassman 1979). Queens of *Protopolybia pumila,* a Neotropical wasp, do not forage but rely in part on honey stored by workers as an energy source (Akre 1982). The populous colonies of *Brachygastra mellifica* use honey to survive short periods of inclement weather during the winter (Schwarz 1929), storing nectar in such large quantities that some Mexican Indians use it as a source of honey. *P. annularis,* which lacks a protective envelope surrounding the comb, leaves the nest in late fall to seek shelter in protective cracks and crevices of rocky cliff faces not far from their nest. On warm winter days they return to their own nest to feed on honey (Strassman 1979).

The ecological context in which paper wasps store honey is not well understood. Wasps store honey in the humid tropics (*Protopolybia*) and in the temperate zone with long, cold winters (e.g., *Polistes aurifer*). Members of the neotropical genus *Polybia* that store honey usually inhabit arid regions. An analysis of how colony size, length of productive season, and environmental conditions influence the quantity of honey stored is needed.

Masarid Wasps (Masaridae)

The family Masaridae contains the only wasps that have the unusual habit (for wasps) of provisioning nests with pollen and nectar. Members of the genus *Pseudomasaris* construct cells of clay (like some mud-dauber wasps) and provision them with an admixture of pollen and nectar (Hicks 1927; Parker 1967; Torchio 1970). Other members of the family, e.g., *Paragia tricolor,* an Australian chimney-building wasp, provision cells with loaves of pollen mixed with nectar that they regurgitate from their crops (Houston 1984). One genus, *Euparagia,* provisions cells with weevil larvae (e.g., Clements and Grissell 1968).

10.5 BEES

The nine families of bees (about 20,000 species) include a diverse array of solitary and eusocial forms that are thought to share a common ancestor with the sphecid wasps. A major difference in the provisioning behavior of bees and wasps is the type of food they use to provision brood cells. Whereas nearly all wasps use food of animal origin, nearly all bees collect pollen and nectar as larval food. Most solitary bees mass provision pollen and nectar for offspring in individual cells. Eusocial bees either mass provision (e.g., stingless bees) or directly feed larvae a food derived from pollen and nectar stored elsewhere in the nest (e.g., honey bees). The nest and provisions of several hundred species of bees have been described. In the synopsis that follows, I emphasize the literature published since 1970 (for a summary of earlier literature see Stephen, Bohart, and Torchio 1969; Plateaux-Quénu 1972; and Michener 1974).

Membrane Bees (Colletidae)

Membrane bees are thought to be the most primitive of all living bees. The common name of the group refers to their use of glandular secretions to construct a thin, transparent membrane that lines each cell. Some other families of bees produce similar linings.

Membrane bees excavate nests in soil or in pithy stems, or use existing burrows in wood. Soil-nesting is the most common type. Female

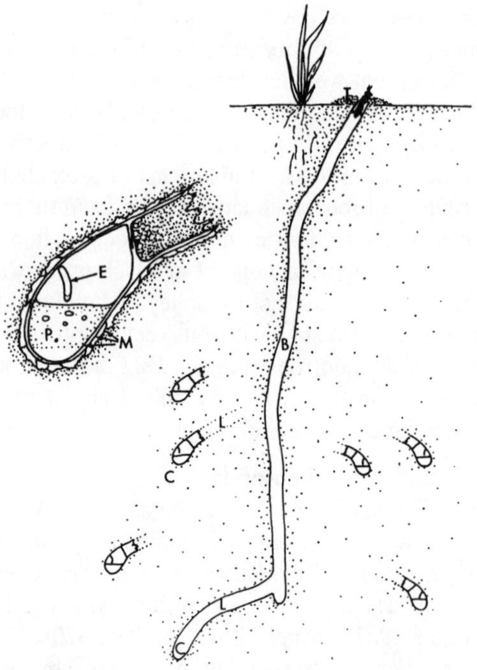

Figure 10.18
Nest of the colletid bee *Colletes thoracicus* showing main burrow (B) and cells (C) at the end of short lateral galleries (L). The detailed cell shows the liquid provisions (P) inside the polyester membrane (M) with an egg (E) suspended from the ceiling of the cell. From Batra 1980.

Colletes thoracicus, for example, excavate burrows with small cells at the ends of short lateral galleries (fig. 10.18). The female lines this cell with secretions from Dufour's gland, a gland associated with the sting apparatus, opening near the posterior tip of the abdomen. The bee imbibes these oily secretions and then applies them to the cell wall and a short segment of burrow adjacent to the cell using the broad, brushlike glossa or tongue. The secretions, composed primarily of macrocyclic lactones, rapidly polymerize when added to salivary enzymes to form a polyester membrane, a cellophanelike sheet that is impermeable to water (Hefetz, Fales, and Batra 1979; Albans et al. 1980; Batra 1980; Cane 1981; Torchio, Trostle, and Burdick 1988). Once the membrane is complete, the female bee gathers pollen and nectar for larval provisions. The bee disgorges the contents of its honey stomach into the bottom of the cell, turns around, and scrapes pollen from the hind leg scopae. Other species, such as *Hylaeus* spp., transport nectar and pollen mixed together in their crops. Several trips are required to complete the provisions. Finally, before closing the cell and backfilling the lateral burrow, the bee lays an egg attached to the upper surface of the cell directly over the pool of provisions. When the egg hatches, the larva falls onto the provisions where it feeds, floating on its side partially submerged.

Under favorable conditions, membrane bees prepare one cell each day. Female *C. thoracicus* construct cells at night, provision cells during the day, and seal cells in the evening (Batra 1980).

The composition of membrane bee provisions has not, to my knowledge, been quantitatively analyzed. In *Colletes,* the provisions are a more or less uniform, semiliquid suspension of pollen in nectar. *Ptiloglossa,* another genus of ground-nesting bees, produces stratified provisions in which the pollen settles to the bottom to form an opaque fluid grading into a clear nectar on top (Rozen 1984b). Oily droplets of unknown origin cover the top of the provisions in *P. arizonensis.* In this genus, the egg floats on top of the provisions. The consistency of provisions of various membrane bees has been described as soupy, ropey, or similar to that of egg yolks (e.g., Roubik and Michener 1984; Rozen 1984b). In *Hylaeus,* bees that nest in pithy stems, the provisions are much more viscous (e.g., Torchio 1984). For most membrane bees, the provisions occupy about half of the cell volume, but in *Colletes ciliatoides* the provisions nearly fill the cell (Torchio 1965).

The insoluble membrane that lines each cell helps to maintain the proper moisture content of the provisions by preventing leaching, drying, and flooding of the cell. Further, the membrane is an effective barrier to contamination of provisions by soil microorganisms. The fungal spores and bacteria that inhabit the provisions are primarily from contaminated nectar. In both *Colletes* and *Ptiloglossa,* the provisions frequently ferment before they are completely consumed by the larvae. Gases released during fermentation cause small bubbles to form in the provisions (fig. 10.18).

The family Oxaeidae comprises a small group of bees whose nests and provisions resemble those of the membrane bees. Single, vertically oriented cells positioned at the ends of horizontal lateral galleries contain semiliquid provisions of nectar and pollen (Linsley and Michener 1962; Hurd and Linsley 1976). Completely provisioned cells are usually one-fourth to one-third full.

Digger Bees (Andrenidae)

Digger bees, as the name implies, nest almost exclusively in the ground. Although there are many species, their nesting habits are poorly understood (Parker and Bohart 1982). Those species that have been studied excavate relatively simple branched or unbranched burrows and construct single cells at the ends of short lateral burrows (e.g., Rozen 1973; Schrader and LaBerge 1978) or opening directly into the main burrow (e.g., Johnson 1981). Cells are oval chambers with maximum diameters slightly greater than the burrow. Walls of cells are smoothed and impregnated with secretions, which penetrate into the surrounding soil and harden, forming a durable, waterproof soil lining 0.5 mm or more thick (Stephen 1966; Rozen 1967; Johnson 1981). The inner surface of the cell also may be coated with a thin, transparent secretion. Some species do not line cells but coat the provisions with a thin waxy secretion, which probably maintains appropriate moisture conditions within the provisions (Rozen 1967; Torchio 1975; Eickwort, Matthews, and Carpenter 1981).

Provisions consist of a doughy mixture of pollen held together with nectar. Many species form provisions into a moist spherical or subspherical ball (Rozen 1965, 1973; Torchio 1975; Norden and Scarbrough 1979). The presence of dry pollen at the center of balls prepared by some

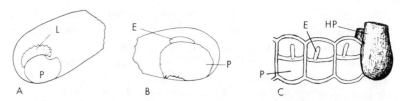

Figure 10.19   Provisioned cells of various bees. A, A larva (L) of *Melitturga clavicornis* (Andrenidae) consuming a provision ball (P). B, The provisions of *Macropis nuda* (Melittidae) with an egg (E) on the upper surface. C, A portion of the comb of the stingless bee *Trigona carbonaria* (Apidae) showing three sealed cells with eggs projecting above the provisions and two honey pots (HP). A, from Rozen 1965; B, from Rozen and Jacobson 1980; C, from Michener 1961.

species suggests that bees add nectar to the surface of the provisions after the ball is formed. The pollen balls of *Melitturga clavicornis* are 6.5–8.5 mm long and 5–6 mm high (fig. 10.19A). Other digger bees, including *Andrena regularis* (Schrader and LaBerge 1978) and *A. dunningi* (Johnson 1981), make pollen masses that are vertically compressed with a small depression in the upper surface. The latter provisions are very moist, with liquid settling in the upper depression. The female lays an egg directly on the upper surface of the finished provisions and then seals the cell. Development is slow; the egg hatches in 4–12 days, and the larva consumes the provisions within a couple of weeks.

Sweat Bees (Halictidae)

Sweat bees are relatively small bees that get their name from their habit of lapping perspiration from the skin of humans, although members of other groups (notably Colletidae and Megachilidae) have similar habits. Nearly all species excavate nests in soil. Nest structure varies in complexity, ranging from short burrows terminating in single cells to branched nests with dozens of cells. In the latter nests, cells may occur directly off the main burrow, at the ends of short lateral galleries, in linear arrays, or in clusters forming primitive combs in small chambers (Sakagami and Michener 1962). Cell structure and the shape and position of provisions within cells, on the other hand, vary relatively little within the family. Cells are small, urn-shaped excavations with constricted entrances and are lined, as in the Colletidae, with cellophanelike secretions from Dufour's gland and salivary glands (Batra 1972; Cane 1981). The provisions are composed of pollen and nectar formed into a spherical or subspherical mass that rests at the bottom of the cell.

Sweat bees vary in social structure from solitary to primitively eusocial. Among the more social species, provisioning of cells is often accomplished by two or more females working cooperatively. In nests of *Lasioglossum zephyrum*, for example, some females remain in the nest with their mothers and assist with nest construction (Batra 1964, 1966). Some of these females mate, resulting in weakly defined castes of fertile egg layers and sterile workers. Workers construct cells perpendicular to the main burrow and then forage for pollen and nectar. A returning forager first inspects the cell and then turns around and scrapes pollen from its scopae, pollen-carrying setae on the femor and tibia of the hind legs (fig. 10.20). The bee disgorges a small quantity of nectar onto the pollen and then mixes the pollen and nectar into a pasty ball by biting with its mandibles and patting with its tarsi. Alternatively, bees may dump several loads of pollen in a powdery mass in the cell before they add any nectar. From one to six workers cooperate in provisioning each cell, and foragers tend to return repeatedly to the same cell until it is complete. Bees seem to recognize whether pollen or nectar is needed to produce provisions of the proper consistency. From six to eight trips are needed to complete a pollen ball. Once a pollen ball nears full size (3–4 mm long and 2–2.5 mm high), egg-laying bees in the colony begin to compete among themselves to lay an egg on the provisions. The winner of this contest kneads the pollen ball again, oviposits directly on the provisions, and then seals the cell.

Some members of the colony rarely leave the nest to forage. These bees feed from pollen balls at all stages of the provisioning process (Batra 1964).

The proportions of pollen and nectar in provisions have not been determined. The mixture is much more viscous than the provisions of membrane bees, being described as mealy-moist in *Conanthalictus* (Rozen and McGinley 1976); soft buttery in *Lasioglossum zephyrum* (Batra 1964); and doughy in the subfamily Halictinae (Sakagami and Michener 1962). Female *Pseudapis diversipes* make pollen balls 4–5 mm in diameter (Rozen 1986). The production of egg layers or queens does not require special provisions. Sweat bees construct cells and form provision balls on a daily

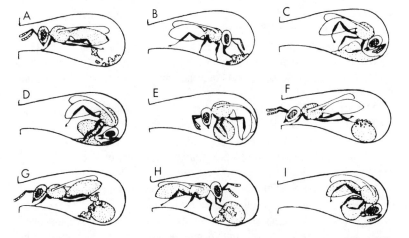

Figure 10.20   Cell provisioning in the social halictid bee *Lasioglossum zephyrum*. Several bees (different individuals are indicated by different patterns of shading) participate in forming a pollen ball. A, Unloading pollen from the scopal hairs on the hind legs; B, Regurgitating nectar on the pollen; C–I, Kneading the pollen and nectar mixture and forming a pollen ball. Six to eight pollen loads are needed to form the provisions in a cell. From Batra 1964.

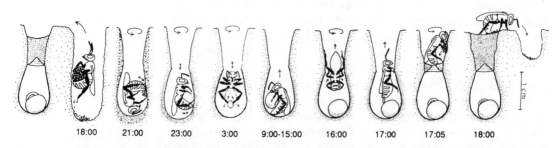

Figure 10.21   Usual temporal sequence of provisioning in *Nomia melanderi* (Halictidae): 18:00, cell closure and initiation of a new cell; 21:00, addition of a soil lining to a new cell; 23:00, tamping the cell; 3:00, application of waterproof lining by glossa and penicilli of the hind basitarsi; 9:00–15:00, pollen ball manipulation, grooming, and removal of debris from cell; 16:00, rubbing cell neck; 17:00, oviposition; 17:05, construction of a cell cap; 18:00, cell closure and initiation of a new cell. From Batra 1970.

cycle; they excavate and line cells at night, provision cells during midday, and seal cells in the evening (Batra 1970) (fig. 10.21).

Melittid Bees (Melittidae)

The nests of melittid bees are simple affairs, similar to those described for digger bees (Andrenidae); however, the provisions are very different. Pollen masses resemble loaves of bread in the bottoms of slightly inclined cells. In some species, such as *Macropis nuda,* the pollen mass is partially hollowed underneath so that the mass rests on a broad base at the rear of the cell and a narrow "foot" at the front (fig. 10.19B). Provisions of *M. nuda* are 5–5.5 mm long and 3.5–4 mm in diameter (Rozen and Jacobson 1980). The unusual nature of the provisions, however, is not shape but composition. *M. nuda* forms provisions from pollen mixed with oils secreted by flowers of fringed loosestrife (*Lysimachia ciliata*) (Cane et al. 1983), resulting in soft, moist pollen loaves that lack a sweet flavor (Rozen 1977a; Rozen and Jacobson 1980). Furthermore, pollen masses are water-repellent (Rozen and Jacobson 1980). The bees also use floral oils to coat the inside of cells. Other melittid bees, such as *Hesperapis* spp., make provisions of pollen and nectar (e.g., Burdick and Torchio 1959).

Many melittid bees are oligolectic, obtaining pollen from a few species of plant (e.g., Rozen 1977a). *M. nuda* in New York, for example, collected pollen and oil from only fringed loosestrife (Rozen and Jacobson 1980; Cane et al. 1983). The plant produces oil in tiny droplets at the tips of trichomous elaiophores located on the base of petals and filaments. Nectaries are absent. The bees collect oil on brushlike pads on the fore- and midtarsi, to which the oil adheres by capillary forces. They transfer this oil to the pollen carried on the scopae of the hind leg (Cane et al. 1983). Thus, bees carry oil, unlike nectar, externally. The energetics and nutritional benefits to bees of using oil rather than nectar as a constituent of provisions has not been investigated.

Leafcutter and Mason Bees (Megachilidae)

The nesting and provisioning behavior of megachilid bees has at-

tracted a great deal of attention in recent years (e.g., Rust 1980; Eickwort, Matthews, and Carpenter 1981; Parker and Tepedino 1982;
Frohlich 1983; Parker 1984, 1985; Frohlich and Parker 1985; Williams et
al. 1986; Rozen 1987). These bees differ markedly from most other solitary bees in that most construct cells of materials (leaf fragments, mud,
pebbles, resin) brought from outside the nest. Many megachilids construct nests in cylindrical cavities in wood or pithy stems; others construct nests in soil or on the surfaces of rocks or twigs. The sunflower
leafcutter bee (*Eumegachile pugnata*), for example, cleans out pithy
stems and forms a partition of oval leaf pieces, masticated leaves, and soil
particles deep within the cavity (Frohlich and Parker 1983). Provisions
consist of pollen and nectar. The bee builds the provisions gradually after
numerous foraging trips. Upon returning from a foraging trip, the bee enters the nest head first, regurgitates a droplet of nectar onto the pollen
mass, and mixes the nectar and pollen with the forelegs and mandibles.
Then the bee leaves the nest, reentering backward to unload pollen. The
sunflower leafcutter bee transports pollen on scopal hairs on the ventral
surface of the abdomen and removes the pollen by brushing the surface of
the abdomen with the hind tarsi. Subsequent provisioning trips continue
in like manner until the pollen mass, a pasty, oval loaf, is completed. The
bee oviposits directly onto the provisions. Finally, it builds a second partition of leaf pieces and masticated leaves several millimeters from the pollen mass.

The nectar concentration of provisions varies. Provisions range from
Not all megachilids use leaves for cell construction. Mason bees construct cells from pebbles and mortar. The mason bee, *Hoplitis anthocopoides,* cements pebbles to rocks to form roughly hemispherical,
mortar-lined cells 7–10 mm long and 5–8.5 mm wide (Eickwort 1975).
When a cell is completed, mason bees mix pollen and nectar in the bottom
of the cell to form a viscous, semiliquid mass that fills about one-half of
the cell. Mason bees then lay an egg on the upper surface of the provisions, seal the cell with mortar, and apply a layer of mortar over the outer
surface of the cell.

The nectar concentration of provisions varies. Provisions range from
dry in *Lithurgus apicalis* (Parker and Potter 1973) to semiliquid in *Dianthidium ulkei* (Frohlich and Parker 1985), but as in nearly all other solitary bees, the actual proportion of nectar rarely has been measured. The
attending bee usually forms the pollen-nectar mass into a subspherical to
oval loaf or packs provisions into the bottom of the cell. Certain *Megachile* and *Osmia* often deposit a final load of nectar just before oviposition
so that the egg lies in a soft, moist depression in the provisions (Gerber
and Klostermeyer 1972; Phillips and Klostermeyer 1978). Some megachilid bees add small amounts of secretions from Dufour's gland (Frohlich
1983; Frohlich and Parker 1985; Williams et al. 1986). The pungent, oily
secretions of *Megachile integra* consist largely of triglycerides, which
give the provisions an unwholesome odor (Williams et al. 1986). The
function of these secretions is unknown.

Many megachilid bees are oligoleges, collecting only a few species
of pollen for provisions. *Osmia latisulcata,* for example, collects almost

exclusively locoweed (*Astragalus*) pollen (Parker 1984), whereas *Hoplitis biscutellae* uses exclusively creosote bush (*Larrea tridentata*) pollen (Rust 1980). Parker (1978) found that the pollen-collecting preferences of *Proteriades bullifacies* for scorpionweed (*Phacelia*) was consistent over a large area of Nevada and California. Small quantities of pollen from various species of plants frequently are found mixed with a large quantity of pollen from one plant species, a pattern likely due to "pollen contamination" as bees forage widely for nectar.

Most published descriptions of solitary bee provisions concerns only qualitative characteristics important in taxonomy such as shape, position in the cell, and consistency. Rarely is enough information provided to draw conclusions regarding energetics and nutrition. The quantity of provisions usually is reported in linear dimensions or proportion of the cell containing stored food. The proportion of provisions comprised of nectar and the sugar concentration of nectar are seldom measured, although some recent studies on megachilid bees are exceptions to this pattern. *Osmia lignaria* provision male-producing cells with about 80–215 mg and female-producing cells with about 165–330 mg of food (estimated from data in Phillips and Klostermeyer 1978). The size of bees emerging from cells is significantly correlated with the size of the pollen mass (see fig. 2.12), and consequently females are larger than males. The provisions of *Megachile rotundata* consist of 64% nectar and 36% pollen (Klostermeyer, Mech, and Rasmussen 1973), but these proportions vary widely within and between species. Females of this species initially bring primarily pollen to the cell but gradually bring mostly nectar; on the last couple of provisioning trips they deliver only nectar to the cell (fig. 10.22). The effect of this change in provisioning behavior is that young larvae initially have a liquid diet, but as they grow and consume the provisions, their diet progressively increases in pollen content (Klostermeyer, Mech, and Rasmussen 1973). The nutritional consequences of this dietary change are unknown.

The family Fideliidae is a small group of desert-adapted bees closely related to the Megachilidae. All species excavate nests in soil and provision unlined cells with pollen slightly moistened with nectar. The mealy-

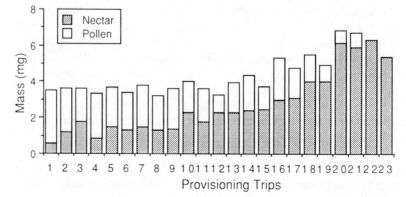

Figure 10.22
Record of the masses of pollen and nectar loads per provisioning trip in a nest of *Megachile rotundata* (Megachilidae). Initially the bee brings primarily pollen to form the provision mass, but as provisioning progresses, she brings primarily or only nectar. Redrawn from Klostermeyer et al. 1973.

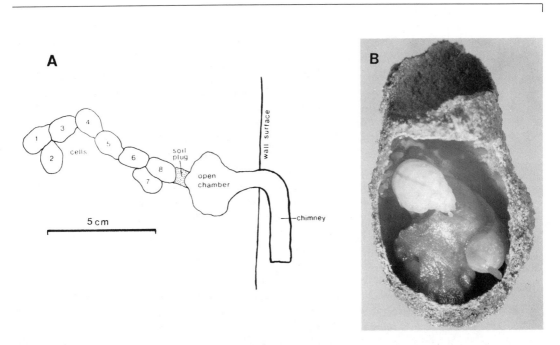

Figure 10.23   The nest burrow and provisions of two *Anthophora* bees (Anthophoridae). A, Typical nest of *Anthophora abrupta* in a wall. Numbers indicate the order in which cells were provisioned. B, The larva of *A. peritomae* in its soil-lined cell as it consumes the basal portion of the pollen and nectar provisions. A, from Norden 1984; B, from Torchio 1971.

moist pollen mass of *Fidelia villosa* is roughly hemispherical (Rozen 1970). Some female fideliid bees oviposit into a small central chamber within the provisions. Female *Parafidelia pallidula* of southern Africa are unusual in that they construct very large pollen masses (11–15 mm long and 10–12 mm in diameter) with two or three chambers, each of which contains an egg (Rozen 1977b).

Carpenter Bees and Miner Bees (Anthophoridae)

Carpenter and miner bees comprise a large group with diverse nesting and provisioning habits. Many species construct nests in loose soil similar to those described for andrenid and melittid bees (e.g., Rozen 1984a, 1986; Vinson, Frankie, and Coville 1987). Many others excavate short galleries in a variety of hard substrates such as compacted soil (Norden 1984; Ordway 1984), pithy stems (Johnson 1988), and dead wood (Anzenberger 1977; Gerling, Hurd, and Hefetz 1983). Such sites are often difficult to excavate; consequently, these bees often crowd numerous cells into confined areas, usually by constructing cells in series, end to end, along a gallery. Most *Anthophora* spp., for example, excavate galleries in hard soil, soft sandstone cliffs and even adobe walls, in which they provision a series of wax-lined cells with a pasty or soupy mixture of pollen and nectar (e.g., Torchio 1971; Norden 1984) (fig. 10.23).

Carpenter bees (*Xylocopa*) are large bees that excavate or enlarge existing galleries in dead, partially decayed wood using their heavily sclerotized mandibles (Anzenberger 1977; Gerling, Hurd, and Hefetz 1983).

The provisions of carpenter bees (*Xylocopa*) are very large. The fresh provisions of *X. imitator* weigh 1.2–1.3 g, those of *X. flavorufa* weigh 2.0–2.1 g, and those of *X. nigrita* weigh 2.4–2.5 g (Anzenberger 1977). Fresh provisions of *X. sulcatipes* and *X. pubescens* weigh a mean of 1.2 and 1.1 g, respectively, and contained 20–22% water and 4.2–4.6 kcal per cell. The low variance in provision mass indicates that bees stock cells with similar-sized pollen loaves. How female bees gauge the size of provisions is not clear, but Anzenberger (1977) determined that if the provision mass is repeatedly reduced in size when the female bee is away collecting pollen, she will continue to add to the provisions for 5 days— two to three times the normal time. Although the size of provisions largely determines the size of the adult bee, there seem to be no data available in this group that male-producing and female-producing cells contain different amounts of provisions.

Some members of the Anthophoridae construct pollen masses that contain other liquids in addition to or instead of nectar. Some bees add Dufour's gland secretions to provisions, which cause the pollen mass to acquire a pungent, disagreeable odor. In addition to mixing these secretions with the provisions, *Anthophora abrupta* paints the Dufour's gland secretion onto the cell wall (Norden 1984). These secretions contain oily lipids, which are rapidly converted (probably by salivary enzymes) to an opaque, waxy, waterproof lining (Norden et al. 1980). After the larva consumes the provisions, it eats the cell lining, an easily digested source of lipids. Other anthophorid bees, like some melittids, add floral oils to provisions. The provisions of *Centris flavofasciata,* for example, consist of a pollen mass mixed with floral oils topped by a pool of clear oil about 2.4 mm deep (Vinson, Frankie, and Coville 1987). Over 1,000 plant species in the Malpighiaceae, Krameriaceae, Scrophulariaceae, Iridaceae, and Orchidaceae are known to produce oils instead of nectar. These oils are produced by trichomous or epithelial elaiophores usually located at the base of the corolla, caylx, or filaments (Cane et al. 1983; Simpson, Neff, and Seigler 1977). Female bees transport floral oils on specialized setae located on the tarsi and ventral surface of the abdomen and also possess special "combs" for manipulating oil (Roberts and Vallespir 1978; Neff and Simpson 1981). The role of the oil in provisioning and the advantage of oil over nectar as a constituent of provisions has not been determined, but it is likely that the oils provide a rich source of energy for the larvae.

Stingless Bees, Bumblebees, and Honey Bees (Apidae)

The family Apidae contains most of the eusocial bees, species characterized by differentiated castes, division of labor within social units, overlapping generations, and presence of a queen that produces all new individuals for the colony (Michener 1974). The family Apidae includes three subfamilies: Meliponinae—the stingless bees; Bombinae—the bumblebees and parasocial orchid or euglossine bees; and Apinae—the honey bees. Food storage is more extensive and more complex in these bees than in any other insects, due largely, it seems, to their highly organized social structure.

Stingless Bees. Stingless bees inhabit areas with tropical climates, reaching greatest species richness in the Neotropical and Oriental regions. Colonies composed of several hundred to many thousands of individuals establish nests in preexisting cavities or in the open. Cavities suitable for nests occur in hollow trees, underground chambers, within termite nests, and in an assortment of human-made structures (Schwarz 1948; Michener 1961, 1974; Roubik 1983; Wille 1983). Nests are architecturally complex structures (fig. 10.24). The core of a nest is a series of brood cells, usually arranged in several single-layered combs oriented in horizontal planes. A multi-layered envelope, the involucrum, often surrounds the brood combs, separating them from the rest of the nest. Storage pots occur in dense clusters around or to one side of the involucrum. Stingless bees construct these structures of cerumen, a mixture of wax and plant resins. They also use cerumen to coat the walls and entrance of the cavity. The storage pots and brood combs are suspended in the cavity by numerous tiny pillars and struts. Unused portions of large cavities may be partitioned off with plant resins (propolis), cerumen, mud, and organic matter.

Stingless bees accumulate honey, pollen, and other substances in nearly spherical "pots." Large nests often contain 200 or more pots, which range in size from 3.5 mm in diameter in *Trigona schrottkyi* (Michener 1974) to 55 mm high and 48 mm wide in *Melipona nigra* (von Ihering 1903, cited by Schwarz 1948). Stingless bees usually segregate nectar and pollen but often place them in similar-shaped vessels, interspersing the two types of food randomly within a cluster of pots. *Trigona flavicornis* is an exception to this pattern. This species constructs special pollen pots in the form of inverted cones (fig. 10.25). *Trigona silvertrii* con-

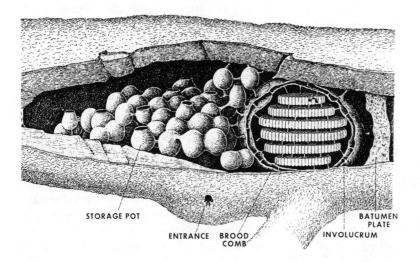

STORAGE POT
ENTRANCE   BROOD
COMB
BATUMEN
PLATE
INVOLUCRUM

Figure 10.24
Nest of the stingless bee *Melipona interrupta grandis* (Apidae) in a hollow branch. Pollen and nectar are stored in the large storage pots to the left of the brood nest. Original drawing by J. M. F. de Camargo; from Michener 1974.

Figure 10.25
Pollen pots (P), honey pots (H), brood cells (B), and gyne (reproductive) cell (G) of the stingless bee *Trigona flavicornis*. Original drawing by J. M. F. de Camargo; from Michener 1974.

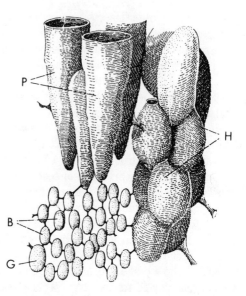

structs pollen storage tubes 10–20 cm tall (Michener 1974).

Stingless bees store relatively small quantities of honey compared to honey bees (*Apis*). Roubik (1983) inspected nests of thirty-eight species of stingless bees in Panama and found only seven species that regularly stored more than 1 l of honey. Taste and consistency of stingless bee honey vary from sweet and viscous to bitter, sour, and watery (Wille 1983). Concentration of dissolved sugar in ripened honey of twenty-seven species of *Melipona* and *Trigona* ranged from 55–86%; most honeys had concentrations of less than 75% (Roubik 1983). This relatively dilute honey (compared to about 75–85% dissolved sugar in honey bee honey) is probably due to the slow evaporation of water from honey in the humid tropics.

Total pollen stores of stingless bees vary greatly. Roubik (1983) found that most stingless bee colonies stored less than 0.5 l of pollen, but one colony of *Trigona corvina* had accumulated 2.2 l. In some nests, some pollen pots contained old, blackened pollen, unsuitable as food.

Occasionally, stingless bees store substances other than honey and pollen. The Panamanian bee *Trigona hypogea* derives protein for larval growth and worker maturation from carrion rather than pollen, as do other stingless bees (Roubik 1982a). This bee stores no pollen, but Roubik found storage pots containing a proteinaceous, glandular secretion similar to that produced by the hypopharyngeal glands of workers. These stored glandular secretions are about 20% protein and apparently take the place of pollen as food for larvae and recently emerged adult workers.

Stored food is important in several aspects of bee life. Brood production by *Melipona fulva* and *M. favosa* depends in part on an adequate supply of stored pollen (Roubik 1982b). Stingless bees are mass provisioners, stocking cells with a glandular secretion mixed with about 16% pollen and 8% honey (Kerr and Laidlaw 1956). Provisions often consist of two distinct layers, a thick basal layer comprised mostly of pollen, and a clear, fluid upper layer (see fig. 10.19C) (Michener 1974). The queen lays an egg on top of the provisions, and then the queen or workers quickly cap the cell with soft cerumen. When pollen supplies are low, stingless bees may slow or stop brood production (see fig. 2.5). Stored honey serves as an important food reserve for the queen and workers during periods when workers cannot forage or when floral nectar is in short supply. During more productive times, stingless bees transport some stored honey and pollen to future nest sites before swarming (Kerr and Laidlaw 1956). This transfer of stored food to daughter nests probably plays an important role in colony reproduction.

Bumblebees and Orchid Bees. Bumblebees (Bombinae: Tribe Bombini) store nectar and pollen in storage pots interspersed among similar-shaped brood cells. At first sight, bumblebee nests appear to be haphazard conglomerates of oval cells (fig. 10.26), much less organized than the precisely arranged combs of honey bees. If one follows the development of bumblebee nests over the summer, however, patterns in placement of brood cells and stored food become apparent.

The queen is the only member of the colony that survives the winter. After she emerges from her hibernaculum in early spring, she forages for nectar and pollen and eventually establishes a nest in a small cavity such as an abandoned mouse nest. The queen forages for pollen and constructs a pollen cake about 10 mm in diameter in the center of this nest, on which she lays eight to sixteen eggs (Free and Butler 1959; Alford 1975). The queen next encases this mass in a thin layer of wax. After this is finished, she constructs a wax "honey pot" about 15–20 mm tall and 15 mm wide near the entrance to the nest. During fair weather, the queen forages for nectar and fills this pot. After the eggs hatch, much of the queen's time is spent incubating the brood mass by lying on top of it; nectar stored in the honey pot serves as a food reserve during inclement

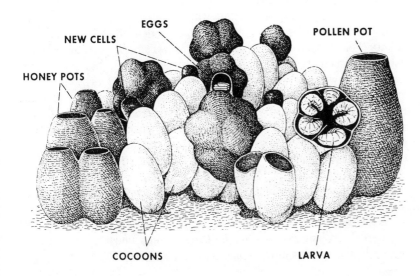

EGGS
NEW CELLS
POLLEN POT
HONEY POTS
COCOONS
LARVA

Figure 10.26
Nest of a bumblebee. Honey pots and pollen pots are either remodeled cocoons or are constructed specially for the purpose of food storage. Original drawing by J. M. F. de Camargo; from Michener 1974.

weather when she cannot forage. The queen periodically opens the wax cell containing the brood and regurgitates nectar and sometimes pollen into the mass. The larvae spin individual silk cocoons when about 5 days old and pupate at about 14 days of age. After they pupate, the queen lays another batch of eggs inside a small pollen mass at the top and near one side of the pupal chamber. The queen removes wax from the top of the cocoons and uses it to encase the new brood mass. When the first cohort of workers emerges, they assist the queen in the nest and assume all the foraging duties. One of their first chores is to clean and enlarge the empty cocoons and begin storing nectar in them. These cells are capped with wax when filled. As summer progresses, several batches of brood are reared, each initiated on top of existing cocoons, thus creating a nest that grows irregularly upward and outward (Free and Butler 1959; Alford 1975).

As the colony grows, the workers convert the old brood cells into honey pots. Workers emerge from pupal cells by biting a jagged hole in the top of the wax cocoon. Workers trim off these jagged edges with their mandibles, forming a circular opening, and then extend the sides of the wax chamber upward. They construct other honey pots solely of wax. A returning forager inserts her head into a honey pot and, with rapid contractions of her abdomen, regurgitates nectar from her crop into the pot. Most honey pots contain a thin nectar, which is consumed quickly by nest workers if foraging is interrupted. Some old honey pots, however, contain thick and viscous honey (Free and Butler 1959). Knee and Medler (1965) determined that thick honey contained 70–87% dissolved sugars and thin honey, less than 42% to as much as 54% dissolved sugars. The transformation of nectar into honey is facilitated by the enzyme invertase (added to the nectar in the forager's saliva), which converts sucrose to fructose and glucose (Alford 1975). After foragers fill the honey pots, workers cap them with wax. As new cocoons become available, workers may empty old honey pots that are being buried under the growing nest (Alford 1975).

Bumblebees store pollen in two ways. Sladen (1912, cited by Free and Butler 1959) classified bumblebees as pollen storers and pocket makers. Pollen storers such as *Bombus lucorum* and *B. terrestris* accumulate pollen in empty cocoons. As they fill these cocoons, they extend the margins with wax so that they may eventually form cylinders as tall as 7.5 cm (fig. 10.26). Workers take pollen from the pollen pots, mix it with nectar, and regurgitate it into the cells of developing larvae. Pocket makers like *B. agrorum* and *B. hortorum* place small quantities of pollen in openings of the wax envelope surrounding the larvae. As the larvae grow, they eventually come to rest directly on this bed of pollen, from which they feed directly.

Unlike honey stored by honey bees, nectar hoarded by bumblebees is not used during winter. The nectar nourishes the brood and is a reserve for the workers and queen during the summer. Near the end of summer, the colony raises primarily queens and drones. Stored food is especially important in production of queens, because the primary determinant in caste differentiation in females is that larvae that become queens are fed more pollen and nectar. Shortly after maturity, the drones and queens leave the nest and mate. All workers and drones die in the fall, and the nest is abandoned; the fertilized queens disperse and seek hibernacula, where they overwinter in a dormant state.

The Neotropical orchid bees (Bombinae: Tribe Euglossini), close relatives of bumblebees, construct nests that superficially resemble those of bumblebees. However, the aggregations of up to twenty-five to thirty female orchid bees that share a nest lack caste differentiation and division of labor and thus are not true societies (Michener 1974; Dressler 1982). Some species nest solitarily, and all females that nest in colonies reproduce independently. Associated with this primitive social system (compared to other Apidae), food storage in orchid bees is simple, consisting of small amounts of pollen and honey prepared as larval provisions.

Orchid bees construct small nests in cavities of trees or in the ground; some solitary species construct exposed nests. Females construct brood cells of mud, resins, organic matter, and feces (Dodson 1966). Groups of cells form tight, irregular clusters similar to those of *Bombus* (Roberts and Dodson 1967), but each female in the colony constructs her own brood cells. Females never use old cells for food storage.

Provisions are a soft, pasty mixture of pollen and sweet honey (Dodson 1966). Foraging females mix this substance in the field and carry it to the nest in their corbiculae (Zucchi, Sakagami, and de Camargo 1969). Provisioning of a cell requires several days. Each female usually provisions her own cell, but other females may sometimes assist provisioning or may steal the provisions (Zucchi, Sakagami, and de Camargo 1969). When the pollen and honey paste fills the cell about one-third full, the female lays a single egg on the provisions and caps the cell. Provisions in two cells of *Eulaema nigrita* weighed 1.05–1.34 g (Zucchi, Sakagami, and de Camargo 1969). Pollen and honey are not accumulated in cells or storage pots, as is done by nearly all other members of the Apidae.

Honey Bees. The four to six species of honey bees (*Apis*) are endemic to temperate and tropical regions of Europe, Asia, and Africa. The honey bee (*Apis mellifera*), which is without question the most thoroughly studied of all food-storing animals, has been domesticated and spread throughout much of the world; its honey-storing habits forms the basis of a major food industry. The voluminous literature on honey bee social behavior and management has been summarized by numerous authors (e.g., Wilson 1971; Dadant et al. 1975; Gojmerac 1980; Free 1982; Seeley 1985).

All honey bees store honey and pollen in specially constructed wax combs that hang vertically from supports. Unmanaged colonies of honey bees (*Apis mellifera*) and the Asian honey bee (*A. cerana*) usually establish nests in cavities of trees or rocky cliffs, where they attach the combs to the ceiling and walls of the cavity with propolis (Seeley 1983; Seeley and Morse 1976; Seeley, Seeley, and Akratanakul 1982). The dwarf honey bee of southern Asia (*A. florea*) suspends its small nest on exposed branches of shrubs near the ground; the giant honey bee (*A. dor-*

sata) of southern Asia hangs its nest from large trunks, branches, and rock overhangs high above the ground (Seeley, Seeley, and Akratanakul 1982). Combs consist of a double layer of hexagonal cells joined along a medial plane, with the walls of the cells sloping upward at an angle of 9°–14° from horizontal. The number of combs in a nest ranges from one to eight, depending on bee species and nest site configuration (Seeley and Morse 1976). The combs are the core of the nest and are used for brood rearing and food storage (fig. 10.27).

Foraging *A. mellifera* collect nectar and pollen from flowers up to 10 km from the nest, although most foraging occurs within 3 km (Beutler 1951; Visscher and Seeley 1982). Bees extract nectar from nectaries with their specialized mouth parts and temporarily hold the nectar in a honey crop. Bees gather pollen in their mouth parts by biting and licking anthers. They collect other pollen by grooming the body hairs where pollen has adhered. This pollen is passed to the mouth, where it is thoroughly moistened. As small cakes of pollen form, the bee passes these back to the middle legs and eventually the hind legs, where they are deposited in pollen baskets or corbiculae (Michener, Winston, and Jander 1978).

Upon returning to the hive, the pollen is deposited in certain cells. Hive workers lick the pollen, adding a small amount of nectar or honey, and pack it firmly into the cells with their heads. This modified pollen is called bee bread. Bees do not cap pollen storage cells with wax.

The formation of honey from nectar entails evaporation of water from nectar and a number of changes in chemical composition. Nectar brought back to the hive is regurgitated to workers that carry it to combs for storage. Bees also secrete the enzyme invertase in their saliva, which

Figure 10.27
Honey bee comb with honey-filled cells, capped cells, and worker bees. Photograph courtesy E. S. Ross.

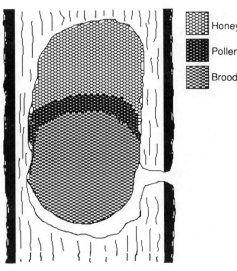

Honey
Pollen
Brood

Figure 10.28
The position of the brood nest and the distribution of stored honey and pollen in a comb of a honey bee nest in a hollow tree. Redrawn from Seeley and Morse 1976.

they add to the nectar. The invertase converts the sucrose in nectar to dextrose and levulose (White 1975). Hive bees place small droplets of the modified nectar in numerous cells, often spreading it on the walls of cells to obtain a large surface area for evaporation. Bees near the hive entrance fan their wings, an activity that creates convection currents that help ventilate the hive, speeding evaporation of moisture from the nectar. After the cell is filled and water content has been reduced to 15–25%, worker bees cap the unripened honey with wax. Conversion of the sugars in honey (curing) continues after the honey has been capped.

Honey bees store honey and pollen in specific regions of the combs (Seeley and Morse 1976; Budathoki and Madge 1987). Cells used to raise brood occupy the lower portion of the nest. Above the brood, bees store pollen in a narrow band of cells; the region above the pollen is used for honey storage (fig. 10.28).

Unmanaged honey bees in temperate areas of North America produce about 60 kg of honey per colony each year (Seeley 1985). This storage occurs during a relatively short period of about 13–17 weeks (Seeley and Visscher 1985). Approximately 35 kg of this honey is consumed during summer to raise brood, produce wax for comb construction, and maintain the colony, but there is a net increase in stored honey during summer. During especially favorable periods, when flowers are abundant, honey may be stored at a rate of about 1 kg/day. About 25 kg of honey is stored as an energy source to get the colony through the winter. Honey production varies considerably, depending on type and number of flowers available, weather, and colony size. Unmanaged honey bees produce less honey than that produced by colonies managed for honey production, which may yield 50 kg or more of surplus honey. Honey production also is important in the process of colony reproduction or swarming, which usually occurs in spring and early summer, when stored food reserves in hives are increasing rapidly (see fig. 2.11). Swarming honey bees do not transfer stored honey and pollen to daughter colonies as some stingless bees do.

Honey bees use stored pollen as a protein source for the brood, but workers do not feed it to them directly. Workers eat the pollen and secrete a proteinaceous substance from their hypopharyngeal glands, which they mix with honey and feed to the larvae. It takes about 130 mg of pollen to produce one bee (Haydak 1935). Most of the brood rearing occurs in spring and summer. A healthy hive in temperate climates may consume 20 kg of pollen each year (Seeley 1985). Some pollen is stored over winter as a supplemental food for the colony and to provide for early spring brood rearing before pollen becomes available (Seeley and Visscher 1985).

11

Food Hoarding and Community Structure

Much of the discussion of the previous ten chapters has centered on the behavior of food-hoarding animals and on the immediate advantages that accrue to individuals as a result of this behavior. In this final chapter, I examine hoarding from a broader perspective. How have the food-hoarding activities of animals influenced the biotic communities to which they belong? Three types of effects are especially evident. First, by becoming better able to cope with the rigors of varying environments, food-hoarding animals have contributed to the diversity and complexity of the animal communities in these environments. Second, storing food potentially increases the competitive abilities of hoarders relative to nonhoarders, and this may affect the outcome of competitive interactions within communities, altering the way these communities are organized. And third, through their mutualistic interactions with certain plants whose propagules they disperse, food hoarders have had far-reaching effects on other animals as well as on plants. None of these "community effects" has received much attention.

11.1 SPECIES DIVERSITY IN VARIABLE ENVIRONMENTS

The buffering effect of having a reserve food supply has some important consequences for how animals adapt to variable environments. Hoarding food, in a sense, changes the conditions under which an animal must live, and hoarding, by changing these conditions, may in many cases allow animals to persist in environments where they could not otherwise survive. This is especially obvious in the case of some long-term food hoarders that live at high altitudes or latitudes where severe winters demand substantial energy reserves. Gray jays (*Perisoreus canadensis*), crested tits (*Parus cristatus*), arctic ground squirrels (*Spermophilus parryii*), Clark's nutcrackers (*Nucifraga columbiana*), honey bees (*Apis mellifera*), red squirrels (*Tamiasciurus hudsonicus*), beavers (*Castor canadensis*), and many other species probably could not maintain viable populations for more than a few years throughout much of their distributional ranges if it were not for the food they store in the fall (Haftorn 1954; Novakowski 1967; M. C. Smith 1968; Rutter 1969; McLean and Towns 1981; Vander Wall, Hoffman, and Potts 1981; Seeley and Visscher 1985). This also may be true of some short-term food hoarders, although the importance of stored food in the long-term persistence of these species at a site is often less apparent. But if the presence of a small supply of stored food is critical at infrequent intervals, say, during several days of inclement weather or following a late spring freeze, the existence of that stored food may influence the future population size and, perhaps, even the persistence of that species in that environment.

At issue here is how animals achieve permanent residency in variable environments. Roberts (1979) pointed out that many food-hoarding birds are territorial, arguing that if food storing is to be adaptive, the hoarder must accumulate food in a territory where it can restrict the access of potential competitors. And Andersson and Krebs (1978) argued that hoarding can evolve more readily in species in which individuals occupy exclusive feeding areas. There seems to be little doubt that territoriality favors hoarding, although not all food-hoarding animals store food on their territories (e.g., Clark's nutcracker; Vander Wall and Balda 1977). But a related issue seems not to be widely appreciated. That is, storing food permits some animals to be permanent residents in an area (e.g., Enoksson 1987), with or without a territory. In other words, territoriality not only facilitates the management of stored food, but also the permanent occupation of a territory or home range may often be a consequence of storing a sufficiently large quantity of food.

Permanent residency in an area confers many advantages. For example, a permanent resident in a territory almost always is able to dominant conspecific vagrants and neighbors. Because territory holders are familiar with food sources and refuges within the territory, they can forage more efficiently and avoid predators more adeptly than can wandering individuals. Permanent residents avoid the costs of migration, including the depletion of metabolic reserves and increased risk of predation. By maintaining a territory throughout the period of food scarcity, animals do not have to compete for limited territories when migrants return. Further, permanent residency allows an animal to begin breeding as soon as conditions become favorable, and consequently permanent residents that hoard food often initiate breeding much earlier than nonhoarding migrants (e.g., Ligon 1978). Food hoarding is one way of acquiring many of the benefits that permanent residency confers.

Food hoarding, of course, is not the only way animals can adapt to variable environments. In fact, some alternative strategies have been even more widely adopted. Many animals, especially birds, migrate out of variable environments to avoid periods of food scarcity. Other species, particularly mammals with more limited powers of mobility, accumulate large internal fat reserves and become dormant during such periods. Other species, especially invertebrates, complete their life cycles during the short period when conditions are favorable and persist during the lean season as eggs or pupae. And yet other species (e.g., mule deer, jackrabbits, grouse) switch to available but nutritionally marginal foods and remain active throughout the winter. Many animals combine two or more of these strategies, often in conjunction with food hoarding.

A consequence of permanent residency resulting from food hoarding is the increased diversity and complexity of animal communities in highly variable environments, however, the extent of this effect is poorly known. It is widely believed that food-hoarding animals are much more prevalent at higher latitudes, but we have no quantitative surveys of animal communities at different latitudes or regions of the world to test this assumption. It is also generally accepted that hoarding is more prevalent in highly variable environments; again, no studies have tallied the number of food-hoarding species across communities as a function of environmental variability. A number of studies have compared the hoarding behavior of related species or races in areas that differ in environmental variability (Barry 1976; Rinderer, Collins, and Tucker 1985; Korpimäki 1987; Tannenbaum and Pivorum 1987; see also section 3.3), but these studies do not get at the issue of the prevalence of food hoarding in communities in areas of differing environmental variability. As mentioned previously, hoarding is only one of several strategies that animals exhibit to cope with variable environments, but we do not know how the importance of hoarding relative to these other strategies changes with the degree and types of environmental variability. Although we vaguely realize that there are some broad ecogeographical patterns in the distribution and numbers of food-hoarding species, these patterns have not been described with any degree of precision.

11.2 FOOD HOARDING AND COMPETITIVE ABILITY

The adaptive value of food hoarding has usually been examined in the context of an organism meeting the demands of a varying, abiotic environment. But to what extent does hoarding improve an individual's ability to compete with other animals (both conspecifics and heterospecifics) that forage for the same pool of resources? When food is a limiting factor, food hoarders may often have an advantage over nonhoarding animals in exploiting that resource. The competitive asymmetry that often exists between hoarders and nonhoarders is due to the fact that the rate at which nonhoarders can gather food is often constrained by the rate at which they can consume and digest that food. There is no advantage for animals that do not store food to collect food any faster than they can consume it (or feed it to their young). But for food hoarders (and some other species that accumulate food internally in a crop), foraging and feeding can be optimized with regard to different criteria. Food-hoarding animals can collect or capture prey at very high rates, often much faster than nonhoarders, and hide it to be used later. The result of these differences in food acquisition rates is that a food hoarder can often appropriate a disproportionately large share of available food.

The potential benefits to hoarders is well illustrated by Hart's (1958b) observation of a willow tit (*Parus montanus*), blue tits (*P. caeruleus*), and great tits (*P. major*) foraging at a feeding station near a garden in Kent, England. Willow tits are active food hoarders, whereas the blue and great tits rarely store food. Each morning forty to fifty peanut halves were placed in a feeder to which all birds had access. Rather than eat the peanut halves, the willow tit carried them into the garden one at a time, hid them in a fence and in bark crevices, and quickly returned for more. By the time all the peanut halves had been removed from the feeder, the willow tit had stored fifteen to twenty, which it then began to retrieve and consume. Each of the ten or so blue and great tits seldom obtained more than two peanut halves apiece. The rest of the peanuts were eaten by house sparrows (*Passer domesticus*). The willow tit, which is behaviorally dominated by the much larger great tit, sequestered nearly half of the peanuts and obtained nearly ten times more peanut halves than other individual tits. Obviously, hoarding in this situation gave the willow tit an enormous advantage in competing for a limited, ephemeral food supply.

Similar benefits are thought to accrue to hoarders in more natural settings. One of the principal advantages that dung beetles (Scarabaeidae) and burying beetles (Silphidae) derive from quickly burying dung and carcasses of small vertebrates is that they prevent the use of these resources by flies, other beetles, and scavengers (e.g., Heinrich and Bartholomew 1979; Wilson and Fudge 1984). Hutchins and Lanner (1982) estimated that red squirrels and Clark's nutcrackers, the two most efficient storers of whitebark pine (*Pinus albicaulis*) cones and seeds, consumed or stored over 99% of all seeds produced (not including developing seeds damaged by insects) at their western Wyoming study site.

It seems reasonable to conclude that some animals gain a tremendous competitive advantage as a result of storing food, but the extent to which this advantage applies to food-hoarding animals in general is not clear. How large is the competitive edge, if any, gained by hoarding species such as hamsters, pikas, kangaroo rats, chipmunks, nuthatches, bowerbirds, owls, and woodpeckers in their interaction with other animals? Can food hoarding change the outcomes of competitive contests? And to what extent have the competitive advantages of food hoarders altered the structures of communities and guilds?

Relatively little attention has been brought to bear on these questions, despite the fact that they may be crucial to understanding the structure and functioning of some animal communities. One community in which the role of food hoarding in mediating community structure seems especially important is that of desert granivores. These communities are usually dominated by various rodents (primarily kangaroo rats and pocket mice of the family Heteromyidae and mice and woodrats of the family Cricetidae); harvester ants (*Pogonomyrmex* spp.); and various birds (usually sparrows, finches, quails, and doves). The harvester ants and nearly all of the rodents avidly store seeds. The structure of these diverse communities, and in particular the organization of the rodent assemblages, has been the object of much research (e.g., Reichman and Oberstein 1977; Hutto 1978; Brown, Reichman, and Davidson 1979; Brown 1989). Many hypotheses have been advanced to explain how so many species can coexist on seeds: perhaps animals partition resources by foraging in different microhabitats, animals may select different types and sizes of seeds within the spectrum of seeds available, or temporal and spatial variability in seed production may prevent competitive exclusion. Regardless of which of these hypothesized causes most strongly shapes the structure of these communities, the fact that most of these animals store seeds has an enormous influence on the structure of the communities to which they belong. The rodents and ants remove seeds from the resource pool by gathering them and moving them to underground chambers. Although it is widely recognized that the collection and storage of seeds is important in the adaptive strategies of these desert-dwelling animals, the role of seed storing in their competitive relationships has not been widely studied.

11.3 INDIRECT EFFECTS OF HOARDER-PLANT MUTUALISMS

In chapters 3 and 7, I described how certain food-hoarding animals have entered into mutualistic relationships with certain plants by dispersing seeds, nuts, or other propagules to sites where new plants eventually establish. These interactions appear to have shaped the behavior and morphology of food-hoarding animals (making them more efficient harvesters of propagules), the morphology of seeds, nuts, and supporting structures, and the fruiting phenology of the plants (making the propagules more attractive and rewarding food items for food-hoarding animals). Both the plants and the food-hoarding animals benefit from this interaction, but they are not the only ones to benefit. Many animals that do not store food or that store food but do not contribute to the establishment of the plants are also attracted to nut and seed crops. These seed and nut predators can often claim over half a crop (see section 7.3). The number of species of nondispersing consumers of seeds and nuts at a food source may often exceed the number of species that disperse the propagule. Formozov (1933), for example, reported that numerous animals, including chipmunks, squirrels, sable, brown bear, deer, boar, numerous mice and voles, and many species of birds, in addition to Eurasian nutcrackers (*Nucifraga caryocatactes*), the primary disperser of the seeds, forage for the seeds of Siberian stone pine (*Pinus sibirica*). Similar observations have been made for the wingless-seeded pines of western North America (Smith and Balda 1979; Hutchins and Lanner 1982). Acorns, beechnuts, and hickory nuts are valuable resources for quail, turkey, pigeons, woodpeckers, deer, peccaries, and dozens of other wildlife species in many north-temperate regions (e.g., Christisen and Korschgen 1955; Gysel 1971; Nixon, McClain, and Hansen 1980). Even humans have profited greatly from nutritious and tasty nuts, the evolution of which has been strongly affected by nut-burying animals. In mast years, these seeds and nuts may remain available well into winter and continue to be important food resources for animals. Even following dispersal of the seeds and nuts, animals continue to raid the caches of food-storing animals or dig up and consume seedlings establishing from these caches (e.g., Bossema 1968; Vander Wall and Hutchins 1983).

To what extent have the characteristics of animal communities in habitats dominated by hoarder-dispersed plants (e.g., oak-hickory forests, piñon-juniper woodlands, Siberian stone pine forests) been influenced by the production, often at infrequent intervals, of large, attractive nuts and seeds? And, if food-hoarding animals act, through their selective foraging and caching behavior, to a large extent as the selective agents for these large attractive nuts and seeds, then how, over evolutionary time, have these animals indirectly shaped the composition and structure of the animal communities of which they are a part? Do the seasonal pulses in the availability of nuts and seeds, often at intervals of several years, have any lasting effects on community structure? We need long-term studies of animal communities, especially the seed-eating guilds, in habitats dominated by hoarder-dispersed plants to answer these and other questions about the influence of food hoarding on community structure.

APPENDIX I

Food-Hoarding Mammals and Birds Mentioned in the Text

Accipiter gentilis	Goshawk	*Canis latrans*	Coyote
Accipiter nisus	Sparrowhawk	*Canis lupus*	Wolf
Aegolius acadicus	Saw-whet owl	*Canis mesomelus*	Black-backed jackal
Aegolius funereus	Boreal (Tengmalm's) owl	*Castor canadensis*	American beaver
Alactagulus pumilio	Lesser five-toed jerboa	*Castor fiber*	Eurasian beaver
Alopex lagopus	Arctic fox	*Catharus* sp.	Nightingale thrush
Alticola strelzowi	Flat-headed vole	*Cercopithecus aethiops*	Barbados green monkey
Amblyornis macgregoriae	MacGregor's bowerbird	*Clethrionomys gapperi*	Red-backed vole
Ammospermophilus sp.	Antelope ground squirrel	*Clethrionomys glareolus*	Bank vole
		Clethrionomys rufocanus	Reddish-gray vole
Aphelocoma coerulescens	Scrub jay	*Corcorax melanorhamphos*	White-winged chough
Aphelocoma ultramarina	Mexican jay	*Corvus brachyrhynchos*	Common crow
Aplodontia rufa	Mountain beaver	*Corvus caurinus*	Northwestern crow
Apodemus argenteus	Small Japanese field mouse	*Corvus corax*	Common raven
Apodemus flavicollis	Yellow-necked mouse	*Corvus corone*	Carrion crow
Apodemus speciosus	Large Japanese field mouse	*Corvus frugilegus*	Rook
		Corvus monedula	Jackdaw
Apodemus sylvaticus	Wood mouse	*Corvus ossifragus*	Fish crow
Arborimus longicaudus	Red tree vole	*Cracticus torquatus*	Grey butcherbird
Asio flammeus	Short-eared owl	*Cricetomys gambianus*	African giant rat
Asio otus	Long-eared owl	*Cricetulus triton*	Greater long-tailed hamster
Athene cunicularia	Burrowing owl		
Athene noctua	Little owl	*Cricetus cricetus*	Black-bellied hamster
Atherurus africanus	African brush-tailed porcupine	*Crocuta crocuta*	Spotted hyena
		Cryptomys hottentotus	Common mole-rat
		Cryptotis parva	Least shrew
Bandicota bengalensis	Lesser bandicoot rat	*Ctenomys* sp.	Tuco-tuco
Bathyergus suillus	Cape dune mole-rat	*Cyanocitta cristata*	Blue jay
Beamys major	Tree mouse	*Cyanocitta stelleri*	Steller's jay
Blarina brevicauda	Short-tailed shrew	*Cyanopica cyana*	Azure-winged magpie
Bubo bubo	Eagle owl		
Bubo virginianus	Great horned owl	*Dasyprocta punctata*	Agouti
Burramys parvus	Mountain pygmy possum	*Desmodillus auricularis*	Namaqua gerbil
		Dipodomys deserti	Desert kangaroo rat
Buteo buteo	Buzzard	*Dipodomys heermanni*	Heermann's kangaroo rat
Buteo platypterus	Broad-winged hawk		
Buteo regalis	Ferruginous hawk	*Dipodomys ingens*	Giant kangaroo rat
		Dipodomys merriami	Merriam's kangaroo rat
Callosciurus prevosti	Prevost's squirrel	*Dipodomys microps*	Great Basin kangaroo rat
Canis aureus	Golden jackal		

Dipodomys spectabilis	Banner-tailed kangaroo rat	*Heteromys desmarestianus*	Spiny pocket mouse
Dipodomys venustus	Santa Cruz kangaroo rat	*Homo sapiens*	Human
Dryomys nitedula	Forest dormouse		
Dusicyon thous	Crab-eating fox	*Ictonyx striatus*	Striped polecat
Ellobius talpinus	Mole vole	*Jaculus orientalis*	Greater Egyptian jerboa
Epixerus ebii	African palm squirrel		
		Laniarius ferrugineus	Boubou shrike
Falco biarmicus	Lanner	*Lanius bucephalus*	Bull-headed shrike
Falco cherrug	Saker	*Lanius collurio*	Red-backed shrike
Falco chicquera	Red-headed falcon	*Lanius excubitor*	Northern shrike
Falco columbarius	Merlin	*Lanius ludovicianus*	Loggerhead shrike
Falco cuvieri	African hobby	*Liomys irroratus*	Mexican spiny pocket mouse
Falco deiroleucus	Orange-breasted falcon		
Falco eleonorae	Eleonora's falcon	*Liomys salvini*	Spiny pocket mouse
Falco femoralis	Aplomado falcon	*Lutra lutra*	River otter
Falco mexicanus	Prairie falcon	*Lycaon pictus*	African wild dog
Falco novaeseelandiae	New Zealand falcon	*Lynx canadensis*	Canadian lynx
Falco peregrinus	Peregrine falcon	*Lynx lynx*	European lynx
Falco punctatus	Mauritius kestrel	*Lynx rufus*	Bobcat
Falco rufigularis	Bat falcon		
Falco rupicoloides	Greater kestrel	*Martes foina*	Stone martin
Falco rusticola	Gyrfalcon	*Martes martes*	Pine martin
Falco sparverius	American kestrel	*Martes pennanti*	Fisher
Falco tinnunculus	European kestrel	*Melanerpes aurifrons*	Golden-fronted woodpecker
Felis concolor	Mountain lion		
Funisciurus anerythrus	African striped squirrel	*Melanerpes carolinus*	Red-bellied woodpecker
Funisciurus isabella	Gray's four-striped squirrel	*Melanerpes erythrocephalus*	Red-headed woodpecker
Funisciurus lemniscatus	Leconte's four-striped squirrel	*Melanerpes formicivorus*	Acorn woodpecker
		Melanerpes lewis	Lewis' woodpecker
Funisciurus pyrrhopus	Cuvier's tree squirrel	*Melanerpes uropygialis*	Gila woodpecker
		Meles meles	European badger
Garrulus glandarius	Eurasian jay	*Meriones meridianus*	Mid-day gerbil
Geomys bursarius	Plains pocket gopher	*Meriones tamariscinus*	Tamarisk gerbil
Geomys pinetis	Southeastern pocket gopher	*Meriones unguiculatus*	Mongolian gerbil
		Mesocricetus auratus	Syrian golden hamster
Georychus capensis	Cape mole-rat	*Micrathene whitneyi*	Elf owl
Glaucidium gnoma	Northern pygmy owl	*Microdipodops pallidus*	Pale kangaroo mouse
Glaucidium passerinum	Eurasian pygmy owl	*Micromys minutus*	Old World harvest mouse
Glaucomys volans	Southern flying squirrel		
Gulo gulo	Wolverine	*Microsorex hoyi*	Pygmy shrew
Gymnorhinus cyanocephalus	Pinyon jay	*Microtus arvalis*	Common vole
		Microtus brandti	Brandt's vole
		Microtus pennsylvanicus	Meadow vole
Heliosciurus rufobrachium	Sun squirrel	*Microtus miurus*	Alaska vole
		Microtus socialis	Social vole
Heterocephalus glaber	Naked mole-rat	*Microtus xanthognathus*	Taiga vole

Mus musculus	House mouse	*Parus montanus*	Willow tit
Muscardinus		*Parus palustris*	Marsh tit
avellanarius	Hazel dormouse	*Parus varius*	Varied tit
Mustela erminea	Short-tailed weasel	*Perisoreus canadensis*	Gray jay
Mustela frenata	Long-tailed weasel	*Perisoreus infaustus*	Siberian jay
Mustela nivalis	Least weasel	*Perognathus flavescens*	Plains pocket mouse
Mustela putorius	Polecat	*Perognathus flavus*	Silky pocket mouse
Mustela vison	Mink	*Perognathus formosus*	Long-tailed pocket
Myoprocta acouchy	Green acouchi		mouse
Myoprocta exilis	Red acouchi	*Perognathus hispidus*	Hispid pocket mouse
Myoxus glis	Edible dormouse	*Perognathus intermedius*	Rock pocket mouse
		Perognathus	
Neotoma albigula	White-throated woodrat	*longimembris*	Little pocket mouse
Neotoma cinerea	Bushy-tailed woodrat	*Perognathus parvus*	Great Basin pocket
Neotoma floridana	Eastern woodrat		mouse
Neotoma fuscipes	Dusky-footed woodrat	*Peromyscus leucopus*	White-footed mouse
Neotoma lepida	Desert woodrat	*Peromyscus maniculatus*	Deer mouse
Neotoma mexicana	Mexican woodrat	*Petroica australis*	South Island robin
Nesokia indica	Pest rat	*Phenacomys intermedius*	Heather vole
Nucifraga caryocatactes	Eurasian nutcracker	*Phodopus sungorus*	Djungarian hamster
Nucifraga columbiana	Clark's nutcracker	*Pica pica*	Black-billed magpie
Nyctea scandiaca	Snowy owl	*Picoides major*	Great spotted
			woodpecker
Ochotona alpina	Siberian pika	*Picoides pubescens*	Downy woodpecker
Ochotona daurica	Daurian pika	*Picoides villosus*	Hairy woodpecker
Ochotona pallasi	Pallas' pika	*Pitymys multiplex*	Alpine pine vole
Ochotona princeps	North American pika	*Pitymys ochrogaster*	Prairie vole
Octodon degus	Degu	*Pitymys pinetorum*	Pine vole
Ondatra zibethica	Muskrat	*Poecilogale albinucha*	African striped weasel
Onychomys leucogaster	Northern grasshopper	*Protoxerus stangeri*	Oil-palm squirrel
	mouse	*Psammomys obesus*	Diurnal sand rat
Onychomys torridus	Southern grasshopper	*Pteromys* spp.	Eurasian flying squirrels
	mouse	*Punomys lemminus*	Puna mouse
Oryzomys palustris	Rice rat	*Pyrrhocorax graculus*	Alpine chough
Otus asio	Screech owl	*Pyrrhocorax pyrrhocorax*	Red-billed chough
Pan troglodytes	Chimpanzee	*Rattus fuscipes*	Australian bush rat
Panthera pardus	Leopard	*Rattus norvegicus*	Norway rats
Panthera tigris	Tiger	*Rhombomys opimus*	Great gerbil
Paraxerus poensis	Bush squirrel		
Parus ater	Coal tit	*Saccostomus campestris*	Pouched mouse
Parus atricapillus	Black-capped	*Sagittarius serpentarius*	Secretary bird
	chickadee	*Saimiri sciureus*	Squirrel monkey
Parus bicolor	Tufted titmouse	*Sciurus aberti*	Tassel-eared squirrel
Parus caeruleus	Blue tit	*Sciurus carolinensis*	Eastern gray squirrel
Parus cinctus	Siberian tit	*Sciurus granatensis*	Red-tailed squirrel
Parus cristatus	Crested tit	*Sciurus lis*	Japanese squirrel
Parus hudsonicus	Boreal chickadee	*Sciurus niger*	Fox squirrel
Parus major	Great tit	*Sciurus vulgaris*	Eurasian red squirrel

Sigmodon sp.	Cotton rat	*Tachyoryctes splendens*	East African mole-rat
Sitta canadensis	Red-breasted nuthatch	*Talpa altaica*	Siberian mole
Sitta carolinensis	White-breasted nuthatch	*Talpa europaea*	European mole
Sitta europaea	Eurasian nuthatch	*Tamias alpinus*	Alpine chipmunk
Sitta pusilla	Brown-headed nuthatch	*Tamias amoenus*	Yellow pine chipmunk
Sitta pygmaea	Pygmy nuthatch	*Tamias dorsalis*	Cliff chipmunk
Sorex arcticus	Arctic shrew	*Tamias minimus*	Least chipmunk
Sorex cinereus	Masked shrew	*Tamias ruficaudus*	Red-tailed chipmunk
Sorex palustris	Water shrew	*Tamias sibiricus*	Siberian chipmunk
Spalacopus cyanus	Coruro	*Tamias speciosus*	Lodgepole pine
Spalax leucodon	Mole-rat		chipmunk
Spalax microphthalmus	Mole-rat	*Tamias striatus*	Eastern chipmunk
Spermophilus beecheyi	California ground	*Tamiasciurus douglasii*	Douglas' squirrel
	squirrel	*Tamiasciurus*	
Spermophilus lateralis	Golden-mantled ground	*hudsonicus*	Red squirrel
	squirrel	*Tatera indica*	Indian gerbil
Spermophilus parryii	Arctic ground squirrel	*Taxidea taxus*	Badger
Spermophilus	Richardson ground	*Thomomys bottae*	Botta's pocket gopher
richardsonii	squirrel	*Thomomys monticola*	Mountain pocket gopher
Spermophilus	Thirteen-lined ground	*Thomomys talpoides*	Northern pocket gopher
tridecemlineatus	squirrel	*Tyto alba*	Barn Owl
Spermophilus undulatus	Alaska ground squirrel		
Spermophilus variegatus	Rock squirrel	*Ursus americanus*	Black bear
Sphyrapicus varius	Yellow-bellied sapsucker	*Ursus arctos*	Brown bear
Stephanoaetus coronatus	Crowned eagle	*Ursus maritimus*	Polar bear
Strepera graculina	Pied currawong		
Strix aluco	Tawny owl	*Vulpes vulpes*	Red fox
Strix nebulosa	Great gray owl	*Vulpes zerda*	Fennec fox
Strix varia	Barred owl		
Surnia ulula	Northern hawk owl	*Xerus erythropus*	African ground squirrel

*Plants, Plant parts, Fungi, and Lichens Mentioned in the Text
that are Stored by Animals*

Abies concolor	White fir	Cones
Acer mono	Japanese maple	Samaras
Acer negundo	Boxelder	Samaras
Acer pensylvanicum	Striped maple	Samaras
Acer rubrum	Red maple	Samaras
Acer saccharum	Sugar maple	Samaras
Acer spp.	Maples	Woody stems
Aesculus hippocastanum	Horse chestnut	Nuts
Agropyron latiglume	Wheatgrass	Seed heads
Alhagi sp.		Rhizomes
Alloteropsis semialata	Black seed grass	Bulbs
Alnus sp.	Alder	Woody stems, foliage
Amanita vaginata	Amanita fungi	Fruiting bodies
Amaranthus sp.	Amaranth	Seeds
Ambrosia artemisiifolia	Ragweed	Seeds
Anacardium spp.	Cashew	Nuts
Arachnis hypogeae	Ground nut	Tubers
Arctostaphylos patula	Manzanita	Seeds
Arctostaphylos sp.	Kinnikinnick	Fruit
Arctous sp.	Arctous	Foliage
Arisarum sp.		Bulbs
Aristida adscensionis	Six-week three-awn grass	Seeds
Armillariella mellea	Honey mushroom	Fruiting bodies
Artemisia hookeriana	Sagebrush	Foliage
Arum sp.		Bulbs
Astragalus sp.	Locoweed	Foliage
Astrocaryum sp.	Black palm	Fruits, nuts
Atriplex confertifolia	Shadscale	Foliage

Bellevalia sp.		Bulbs
Betula spp.	Birch	Woody stems
Bouteloua aristidoides	Annual grama	Seeds
Bromus tectorum	Cheatgrass	Seeds
Bryoria fuscescens	Beard lichen	Lichens
Calamogrostris canadensis	Reed grass	Foliage
Capsella sp.	Shepherd's purse	Seeds
Carduus sp.	Thistle	Seeds
Carex spp.	Sedges	Foliage
Carya cordiformis	Bitternut hickory	Nuts
Carya illinoensis	Pecan	Nuts
Castanea dentata	American chestnut	Nuts
Castanopsis sp.	Chinquapin	Nuts
Ceanothus cordulatus	Mountain whitethorn	Foliage
Ceanothus prostratus	Squawcarpet	Seeds
Ceanothus velutinus	Snowbrush ceanothus	Seeds
Chenopodium	Goosefoot	Seeds
Chrysothamnus nauseosus	Rubber rabbitbrush	Foliage
Chrysothamnus viscidiflorus	Green rabbitbrush	Seeds
Cirsium sp.	Canadian thistle	Roots
Cladonia spp.	Lichen	Lichens
Claytonia lanceolata	Spring beauty	Corms
Comandra richardsonii	Toad-flax	Roots
Commelina sp.		Foliage
Convolvulus sepium	Morning-glory	Roots

Cornus stolonifera	Red-osier dogwood	Woody stems
Corylus avellana	European hazel	Nuts
Corylus rostrata	Beaked hazel	Nuts
Crataegus sp.	Hawthorn	Fruits
Curcurbita maxima	Pumpkin	Seeds
Cyanella sp.		Corms
Dipteryx spp.		Fruits, seeds
Dryas sp.	Mountain avens	Foliage
Elaeocarpus sp.		Fruits
Elymus spp.	Rye grass	Foliage
Epilobium latifolium	Fireweed	Foliage
Equisetum spp.	Horsetail	Tubers
Eriogonum sp.	Buckwheat	Seed heads
Erodium sp.	Filaree	Seeds
Erythronium americanum	Trout lily	Bulbs
Erythronium grandiflorum	Dogtooth violet	Bulbs
Eurotia lanata	Winterfat	Foliage
Fagus grandifolia	American beech	Nuts
Fagus sylvatica	European beech	Nuts
Festuca idahoensis	Idaho fescue	Seeds
Galeopsis sp.	Hempnettle	Seeds
Geranium sp.	Geranium	Foliage
Gladiolus sp.	Gladiola	Bulbs
Gnaphalium sp.	Cudweed	Seeds
Gustavia superba	Membrillo	Fruits, seeds
Gymnocladus dioica	Kentucky coffee tree	Seeds
Halocenemum sp.		Foliage
Haplosciadum sp.		Foliage
Hedysarum sp.	Hedysarum	Foliage
Helianthus tuberosa	Jerusalem artichoke	Tubers
Hexaglottis sp.		Corms
Homeria sp.		Corms

Hordeum sp.	Foxtail grass	Seeds
Hymenaea courbaril	Guapinol	Fruits, seeds
Hymenogaster sp.	Fungi	Fruiting bodies
Impatiens biflora	Jewelweed	Seeds
Inga sp.		Fruits, nuts
Juglans ailantifolia	Japanese walnut	Nuts
Juglans cinerea	Bitternut	Nuts
Juglans nigra	Black walnut	Nuts
Juglans regia	English walnut	Nuts
Juncus balticus	Rush	Capsules, roots
Juniperus communis	Common juniper	Foliage
Juniperus osteosperma	Utah juniper	Seeds
Larix sibiricus	Siberian larch	Foliage
Lathyrus venosus	Sweet pea	Roots
Lepidium nitidum	Peppergrass	Seeds, capules
Lespedeza capitata	Bush clover	Seeds
Leucoryne ixiodes	Huilli	Tubers
Linnaea borealis	Twinflower	Foliage
Liquidambar styraciflua	Sweet gum	Capsules
Lithocarpus spp.	Tan oak	Acorns
Lithospermum sp.	Puccoon	Foliage
Lupinus spp.	Lupines	Foliage
Malus sp.	Golden apple	Fruits
Malva sp.	Mallow	Seeds
Melilotus sp.	Sweet clover	Roots
Micranthus junceus		Corms
Morea sp.		Corms
Mucana aterrima	Velvet bean	Seeds
Muscari sp.		Bulbs
Narcissus sp.	Narcissus	Bulbs
Nuphar variegatum	Pond lily	Foliage
Oenothera sp.	Evening primrose	Seeds

Opuntia davisii	Rat-tailed cactus	Joints
Opuntia spp.	Prickly pear cactus	Joints
Ornithogalum sp.		Corms
Orogenia linearifolia	Indian potato	Tubers
Orthocarpus sp.	Owlclover	Foliage
Oryzopsis hymenoides	Indian ricegrass	Seeds
Othonna sp.		Tubers
Oxalis cernua	Oxalis	Bulbs
Panda oleosa		Nuts
Panicum sp.	Panic grass	Seeds
Pennisetum sp.	Grass	Rhizomes
Perideridia gairdneri	Yampah	Roots
Persea sp.		Fruits, nuts
Petasites sp.	Coltsfoot	Roots
Phaseolus mungo	Bean	Seeds
Phleum pratense	Timothy	Bulbs
Picea engelmanii	Engelmann spruce	Seeds
Picea glauca	White spruce	Cones
Picea pungens	Black spruce	Cones
Picea schrenkiana	Asian spruce	Cones, seeds
Picea spp.	Spruce	Seeds
Pinus albicaulis	Whitebark pine	Cones, seeds
Pinus armandii	Armand pine	Seeds
Pinus ayacahuite	Mexican white pine	Seeds
Pinus bungeana	Lacebark pine	Seeds
Pinus cembra	Swiss stone pine	Seeds
Pinus contorta	Lodgepole pine	Cones
Pinus coulteri	Coulter pine	Seeds
Pinus edulis	Colorado piñon pine	Cones, seeds
Pinus flexilis	Limber pine	Cones, seeds
Pinus gerardiana	Chilgoza pine	Seeds
Pinus jeffreyi	Jeffrey pine	Seeds
Pinus koraiensis	Korean stone pine	Seeds
Pinus lambertiana	Sugar pine	Seeds
Pinus longaeva	Bristlecone pine	Seeds
Pinus monophylla	Singleleaf piñon pine	Seeds
Pinus parviflora	Japanese white pine	Seeds
Pinus pinea	Italian stone pine	Seeds
Pinus ponderosa	Ponderosa pine	Cones, seeds
Pinus pumila	Japanese stone pine	Seeds
Pinus sabiniana	Digger pine	Seeds
Pinus sibirica	Siberian stone pine	Seeds
Pinus strobiformis	Southwestern white pine	Seeds
Pinus strobus	Eastern white pine	Seeds
Pinus sylvestris	Scotch pine	Seeds
Plantago sp.	Plantain	Seeds
Polygonum sp.	Knotweed	Seeds
Polystichum munitum	Western sword fern	Foliage
Populus tremuloides	Aspen	Woody stems
Populus spp.	Cottonwood	Woody stems
Potentilla canadensis	Cinquefoil	Rhizomes
Prosopis juliflora	Velvet mesquite	Seeds
Prosopis sp.	Mesquite	Roots
Prunus armeniaca	Apricot	Stones
Prunus dulcis	Almond	Nuts
Prunus pensylvanica	Wild cherry	Stones
Prunus persica	Peach	Stones
Pseudotsuga menziesii	Douglas-fir	Cones, seeds, foliage
Pteridium aquilinum	Bracken fern	Foliage
Purshia glandulosa	Desert bitterbrush	Seeds
Purshia tridentata	Antelope bitterbrush	Seeds
Pyrola rotundifolia	Wintergreen	Foliage

Quercus agrifolia	Coast live oak	Acorns
Quercus alba	White oak	Acorns
Quercus dentata	Oak	Acorns
Quercus gambelii	Gambel oak	Foliage
Quercus lobata	Valley oak	Acorns
Quercus macrocarpa	Bur oak	Acorns
Quercus mongolica	Mongolian oak	Acorns
Quercus montana	Chestnut oak	Acorns
Quercus muehlenbergii	Chinquapin oak	Acorns
Quercus nigra	Water oak	Acorns
Quercus palustris	Pin oak	Acorns
Quercus petraea	Sessile oak	Acorns
Quercus robur	English oak	Acorns
Quercus rubra	Red oak	Acorns
Quercus velutina	Black oak	Acorns
Quercus virginiana	Live oak	Acorns
Ranunculus bulbosa	Buttercup	Bulbs
Raphia sp.	Palm	Fruits
Rhus diversiloba	Poison oak	Seeds
Rhus trilobata	Skunkbrush sumac	Foliage
Ribes cereum	Wax currant	Fruits, seeds
Romulea sp.		Corms
Russula sp.	Russula fungi	Fruiting bodies
Sagittaria sp.	Arrowhead	Tubers
Salix sp.	Willow	Foliage, woody stems
Salsola kali	Russian thistle	Foliage
Saxifraga bronchialis	Saxifrage	Foliage
Scheelea zonensis	Palm	Fruits, nuts
Scleria sp.	Sedge	Rhizomes
Senecio adenophylloides	Ragwort	Stems

Sequoiadendron giganteum	Sequoia	Cones
Setaria lutescens	Yellow foxtail	Seeds
Setaria viridus	Green foxtail	Seeds
Shepherdia canadensis	Buffaloberry	Foliage
Sitanion hystrix	Squirreltail	Seeds
Solanum carolinense	Horse nettle	Fruits
Solanum elaeagnifolium	Silverleaf nettle	Fruits
Sorghum sp.	Sorghum	Rhizomes
Spondias spp.	Hog plum	Fruits, seeds
Stellaria jamesiana	Tuber starwart	Tubers
Stipa comata	Needle-and-thread grass	Crowns
Stipa occidentalis	Needlegrass	Seeds
Symphoricarpos albus	Snowberry	Seeds
Taraxacum officinale	Dandelion	Roots
Tilia sp.	Basswood	Seeds
Timonius sp.		Fruits
Tontalea spp.		Fruits, seeds
Tridens pulchellus	Fluffgrass	Seeds
Umbellularia californica	California laurel	Nuts
Viburnum pauciflorum	Cranberry	Fruits
Viburnum spp.	Viburnums	Woody stems
Viola papilionacea	Blue violet	Roots
Werneria digitata		Stems
Yucca glauca	Yucca	Foliage

The numbers in brackets following each work are the pages on which that work is cited.

Abbott, H. G. 1962. Tree seed preferences of mice and voles in the Northeast. *J. Forest.* 60:97–99. [194]

Abbott, H. G., and T. F. Quink. 1970. Ecology of eastern white pine seed caches made by small forest mammals. *Ecology* 51:271–78. [4, 62, 106, 141, 158, 181, 183–84, 191–94, 197, 259, 260]

Adovasio, J. M., J. Donahue, K. Cushman, R. C. Carlisle, R. Stuckenrath, J. D. Gunn, and W. C. Johnson. 1983. Evidence from Meadowcroft Rockshelter. In *Early man in the New World,* edited by R. Shutler, Jr., 163–89. Beverly Hills, Calif.: Sage Publishers. [214–15]

Ågren, G., Q. Zhou, and W. Zhong. 1989. Territoriality, cooperation and resource priority: Hoarding in the Mongolian gerbil, *Meriones unguiculatus. Anim. Behav.* 37:28–32. [67]

Akre, R. D. 1982. Social wasps. In *Social insects,* Vol. IV, edited by H. R. Hermann, 1–105. New York: Academic Press. [345]

Alatalo, R. V., and A. Carlson. 1987. Hoarding-site selection of the willow tit *Parus montanus* in the presence of the Siberian tit *Parus cinctus. Ornis Fenn.* 64:1–9. [92, 162, 170, 311, 313]

Albans, K. R., R. T. Aplin, J. Brehcist, J. F. Moore, and C. O'Toole. 1980. Dufour's gland and its role in secretion of nest cell lining in bees of the genus *Colletes* (Hymenoptera: Colletidae). *J. Chem. Ecol.* 6:549–64. [346]

Albino, R. C., and M. Long. 1951. The effect of infant food-deprivation upon adult hoarding in the white rat. *Br. J. Psychol.* 42:146–54. [117]

Alcock, J. 1976. The behaviour of the seed-collecting larvae of a carabid beetle (Coleoptera). *J. Nat. Hist.* 10:367–75. [2, 12, 48, 179, 325–27]

———. 1980. Communal nesting in an Australian solitary wasp, *Cerceris antipodes* Smith (Hymenoptera, Sphecidae). *J. Aust. Entomol. Soc.* 19:223–28. [68, 340]

Alderson, V. A. 1890. Hairy woodpecker and potato bugs. *Oologist* 7:147. [292, 296]

Aldous, C. M. 1945. Pocket gopher caches in central Utah. *J. Wildl. Mgmt.* 9:327–28. [250–51]

———. 1951. The feeding habits of pocket gophers (*Thomomys talpoides moorei*) in the high mountain ranges of central Utah. *J. Mamm.* 32:84–87. [250]

Aleksiuk, M. 1969. The function of the tail as a fat storage depot in the beaver (*Castor canadensis*). *J. Mamm.* 51:145–48. [23]

Alexander, A., and R. R. Ewer. 1959. Observations on the biology and behavior of the smaller African polecat (*Poecilogale albinucha*). *Afr. Wild Life* 13:313–20. [226, 232]

Alexander, B. 1986. Alternative methods of nest provisioning in the digger wasp *Clypeadon laticinctus* (Hymenoptera: Sphecidae). *J. Kans. Entomol. Soc.* 59:59–63. [89, 341–42]

Alford, D. V. 1975. *Bumblebees.* London: Davis-Poynter. [357–58]

Allan, P. F. 1935. Bone cache of a grey squirrel. *J. Mamm.* 16:326. [246]

Allen, D. L. 1943. *Michigan fox squirrel management.* Michigan Department of Conservation, Game Division Publication 100:1–404. [244, 246]

Allen, E. G. 1938. The habits and life history of the eastern chipmunk, *Tamias striatus lysteri. N.Y. State Mus. Bull.* 314:1–122. [238]

Allen, G. M. 1940. *The mammals of China and Mongolia.* New York: American Museum of Natural History. [263]

Alm, S. R., and F. E. Kurczewski. 1984. Ethology of *Anoplius tenebrosus* (Cresson) (Hymenoptera: Pompilidae). *Proc. Entomol. Soc. Wash.* 86:110–19. [87, 89, 337]

Amadon, D. 1944. The genera of Corvidae and their relationships. *Am. Mus. Novit.* 1251: 1–21. [72, 308]

———. 1950. Australian mud nest builders. *Emu* 50: 123–27. [308].

———. 1951. Taxonomic notes on the Australian butcher-birds (Family Cracticidae). *Am. Mus. Novit.* 1504: 1–33. [47, 307]

Andersen-Harild, P., C. A. Blume, E. Kramshøj, and O. Schelde. 1966. Nogle invasioner af Nøddekrige (*Nucifraga caryocatactes*). *Dansk, Ornithol. Foren. Tidiskr.* 60: 1–13. [65]

Anderson, A. O., and D. M. Allred. 1964. Kangaroo rat burrows at the Nevada test site. *Great Basin Nat.* 24: 93–101. [252]

Andersson, M. 1985. Food storing. In *A dictionary of birds,* edited by B. Campbell and E. Lack, 235–37. Vermillion, S. Dak.: Buteo Books. [292]

Andersson, M., and J. Krebs. 1978. On the evolution of hoarding behaviour. *Anim. Behav.* 26: 707–11. [44–46, 67, 79, 81, 149, 176, 230, 364]

Andrew, D. G. 1969. Food-hiding by rooks. *Br. Birds* 62: 334–36. [305]

Angell, T. 1969. A study of the ferruginous hawk: Adult and brood behavior. *Living Bird* 8: 225–41. [285]

Ansell, W. F. H. 1960. The African striped weasel, *Poecilogale albinucha* (Gray). *Proc. Zool. Soc. Lond.* 134: 59–64. [226, 232]

Anzenberger, G. 1977. Ethological study of African carpenter bees of the genus *Xylocopa* (Hymenoptera, Anthophoridae). *Z. Tierpsychol.* 44: 337–74. [353–54]

———. 1986. How do carpenter bees recognize the entrance of their nests? An experimental investigation in a natural habitat. *Ethology* 71: 54–62. [150, 152]

Applegate, R. D. 1977. Possible ecological role of food caches of loggerhead shrike. *Auk* 94: 391–92. [12, 30, 89, 126, 317]

Ash, J. S. 1970. Observations on the decreasing population of red-backed shrikes. *Br. Birds* 63: 185–205, 225–39. [317]

Ashby, K. R. 1967. Studies on the ecology of field mice and voles (*Apodemus sylvaticus, Clethrionomys glareolus* and *Microtus agrestis*) in Houghall Wood, Durham. *J. Zool.* 152: 389–513. [28, 199, 275]

Ashmole, N. P. 1987. Sparrow hawk caching and returning repeatedly to prey. *Scott. Birds* 14: 182–83. [12, 285, 287]

Audubon, J. J., and J. Bachman. 1846. *The viviparous quadrupeds of North America,* vol. 1. New York: Audubon Society. [243]

Austin, O. L., Jr. 1961. *Song birds of the world.* New York: Golden Press. [307]

Avenoso, A. C. 1968. Selection and processing of nuts by the flying squirrel, *Glaucomys volans. Diss. Abstr.* 30B: 437–38. [64, 172, 249]

Bachman, J., and J. J. Audubon. 1849. *The quadrupeds of North America,* vol. 2. New York: Audubon Society. [238]

Badger, G. M., and W. Korytnyk. 1956. Examination of honey in Australian honey ants. *Nature* 178: 320–21. [335]

Bailey, B. 1929. Mammals of Sherburne County, Minnesota. *J. Mamm.* 10: 153–64. [25, 254]

Bailey, V. 1920. Identity of the bean mouse of Lewis and Clark. *J. Mamm.* 1: 70–72. [270]

———. 1926. A biological survey of North Dakota. II. The mammals of North Dakota. *North Am. Fauna* 49: 17–226. [226, 257–58, 270]

———. 1931. Mammals of New Mexico. *North Am. Fauna* 53: 1–412. [116, 262]

Baird, S. F., T. M. Brewer, and R. Ridgeway. 1905. *A history of North American birds,* vol. II. Boston: Little, Brown. [292]

Baker, M. C., E. Stone, A. E. Baker, R. J. Shelden, P. Skillicorn, and M. D. Mantych. 1988. Evidence against observational learning in storage and recovery of seeds by black-capped chickadees. *Auk* 105: 492–97. [89, 94, 160, 165, 167]

Balda, R. P. 1980a. Are seed caching systems coevolved? *Acta Congr. Int. Ornithol.* 2: 1185–91. [71, 181, 304]

———. 1980b. Recovery of cached seeds by a captive *Nucifraga caryocatactes. Z. Tierpsychol.* 52: 331–46. [102, 155–57, 162, 167–69, 176]

Balda, R. P., and G. C. Bateman. 1971. Flocking and annual cycle of the piñon jay, *Gymnorhinus cyanocephalus. Condor* 73: 287–302. [67, 72, 79, 109, 183, 301–2, 304]

———. 1973. The breeding biology of the piñon jay. *Living Bird* 11: 5–42. [27–28, 30, 32]

Balda, R. P., K. G. Bunch, A. C. Kamil, D. F.

Sherry, and D. F. Tomback. 1987. Cache site memory in birds. In *Foraging behavior,* edited by A. C. Kamil, J. R. Krebs, and H. R. Pulliam, 645–66. New York: Plenum Press. [1, 148]

Balda, R. P., and A. C. Kamil. 1988. The spatial memory of Clark's nutcrackers (*Nucifraga columbiana*) in an analogue of the radial arm maze. *Anim. Learn. Behav.* 16:116–22. [170]

———. 1989. A comparative study of cache recovery by three corvid species. *Anim. Behav.* 38:486–95. [63, 167, 170, 302]

Balda, R. P., A. C. Kamil, and K. Grim. 1986. Revisits to emptied cache sites by Clark's nutcrackers (*Nucifraga columbiana*). *Anim. Behav.* 34:1289–98. [169]

Balda, R. P., and R. J. Turek. 1984. The cache-recovery system as an example of memory capabilities in Clark's nutcracker. In *Animal cognition,* edited by H. L. Roitblat, T. G. Beaver, and H. S. Terrace, 513–32. London: Erlbaum Associates. [162, 167–68]

Baldwin, H. I. 1942. *Forest tree seed of the north temperate region with special reference to North America.* Waltham, Mass.: Chronica Botanica. [241–42]

Balgooyen, T. G. 1976. Behavior and ecology of the American kestral (*Falco sparverius*) in the Sierra Nevada of California. *Univ. Calif. Publ. Zool.* 103:1–83. [12, 31, 88, 95, 284–88]

Ball, A. R. 1978. Nuthatch caching insect larvae. *Br. Birds* 71:539–40. [314]

Bandoli, J. H. 1981. Factors influencing seasonal burrowing activity in the pocket gopher, *Thomomys bottae. J. Mamm.* 62:293–303. [252]

Barash, D. P. 1973. Territorial and foraging behavior of pika (*Ochotona princeps*) in Montana. *Am. Midl. Nat.* 89:202–7. [279, 281–82]

Bard, G. E. 1952. Secondary succession on the piedmont of New Jersey. *Ecol. Monogr.* 22:195–215. [210]

Barnes, V. G., R. M. Anthony, K. A. Fagerstone, and J. Evans. 1985. Hazards to grizzly bears of strychnine baiting for pocket gopher control. *Wildl. Soc. Bull.* 13:552–58. [250]

Barnett, R. J. 1977. The effect of burial by squirrels on germination and survival of oak and hickory nuts. *Am. Midl. Nat.* 98:319–30. [75–76, 191, 202]

Barrett, L. I. 1931. Influence of forest litter on the germination and early survival of chestnut oak, *Quercus montana,* Willd. *Ecology* 12:476–84. [187–88]

Barry, W. J. 1976. Environmental effects on food hoarding in deermice (*Peromyscus*). *J. Mamm.* 57:731–46. [64, 132, 134–36, 138, 259, 365]

Bartholomew, G. A., and J. W. Hudson. 1961. Desert ground squirrels. *Sci. Am.* 205(5):107–16. [247]

Basile, J. V., and R. C. Holmgren. 1957. Seeding depth trials with bitterbrush (*Purshia tridentata*) in Idaho. Intermountain Forest and Range Experiment Station, Research Paper, U.S. Department of Agriculture, Forest Service 54:1–10. [188]

Bateman, G. C., and R. P. Balda. 1973. Growth, development, and food habits of young piñon jays. *Auk* 90:39–61. [305]

Batra, S. W. T. 1964. Behavior of the social bee, *Lasioglossum zephyrum,* within the nest (Hymenoptera: Halictidae). *Insect. Soc.* 11:159–85. [68, 348–49]

———. 1966. The life cycle and behavior of the primitively social bee, *Lasioglossum zephyrum* (Halictidae). *Univ. Kans. Sci. Bull.* 46:359–423. [348]

———. 1970. Behavior of the alkali bee, *Nomia melanderi,* within the nest (Hymenoptera: Halictidae). *Ann Entomol. Soc. Am.* 63:400–406. [88, 350]

———. 1972. Some properties of the nest-building secretions of *Nomia, Anthophora, Hylaeus* and other bees. *J. Kans. Entomol. Soc.* 45:208–18. [348]

———. 1980. Ecology, behavior, pheromones, parasites and management of sympatric vernal bees *Colletes inaequalis, C. thoracicus* and *C. validus. J. Kans. Entomol. Soc.* 53:509–38. [88, 346]

Baudy, R. E. 1976. Breeding techniques for felines destined for release in the wild: Recommendations for the Florida panther. In *Proc. Fla. Panther Conf.,* edited by P. C. H. Pritchard, Florida Audubon Society and Florida Game and Fresh Water Fish Commission. [233]

Baulu, J., J. Rossi, and J. A. Horrocks. 1980. Food hoarding by a Barbados green monkey (*Cercopithecus aethiops sabaeus*). *Lab. Primate Newsletter* 19(3):13–14. [1, 49, 119, 141, 224]

Beach, F. A. 1950. The snark was a boojum. *Am. Psychol.* 5:115–24. [111, 144]

Beal, F. E. L. 1911. Food of the woodpeckers of the United States. *U.S. Biol. Surv. Bull.* 37:1–64. [297]

Bean, W. J. 1981. *Trees and shrubs hardy in the British Isles.* 8th ed. New York: St. Martin's Press. [75]

Beatley, J. C. 1969. Dependence of desert rodents on winter annuals and precipitation. *Ecology* 50:721–24. [106]

Beattie, A. J. 1985. *The evolutionary ecology of ant-plant mutualisms.* New York: Cambridge University Press. [180]

Becker, P., M. Leighton, and J. B. Payne. 1985. Why tropical squirrels carry seeds out of source crowns. *J. Trop. Ecol.* 1:183–86. [184, 245]

Beddard, F. E. 1902. *Mammalia.* London: Macmillan. [230]

Beebe, C. W. 1950. Home life of the bat falcon, *Falco albigularis albigularis* Daudin. *Zoologica* 35:69–86. [285]

Beebe, F. L. 1960. The marine peregrines of the northwest Pacific coast. *Condor* 62:145–89. [285]

Beidleman, R. G., and W. A. Weber. 1958. Analysis of a pika hay pile. *J. Mamm.* 39:599–600. [280–81]

Beigneux, F., J. M. Lassalle, G. Le Pape. 1980. Hoarding behavior of AKR, C57BL/6 mice and their F1 in their home living space: An automatic recording technique. *Physiol. Behav.* 24:1191–93. [114, 130–31]

Bell, H. L. 1983. Possible burying of food by the pied currawong *Strepera graculina. Sunbird* 13:59. [308]

Bendire, C. E. 1889. *Picicorvus columbiana* (Wils.), Clarke's nutcracker. Its nest and eggs, etc. *Auk* 6:226–36. [305]

Benkman, C. W., R. P. Balda, and C. C. Smith. 1984. Adaptations for seed dispersal and the compromises due to seed predation in limber pine. *Ecology* 65:632–42. [73, 197, 242]

Bennett, K. D. 1985. The spread of *Fagus grandifolia* across eastern North America during the last 18,000 years. *J. Biogeog.* 12:147–64. [213–14]

Bent, A. C. 1938. Life histories of North American birds of prey, Part 2. *U.S. Natl. Mus. Bull.* 170:1–482. [289]

———. 1939. Life histories of North American woodpeckers. *U.S. Natl. Mus. Bull.* 174:1–334. [293]

Beusekom, G. van. 1948. Some experiments on the optical orientation in *Philanthus triangulum* Fabr. *Behaviour* 1:195–226. [151]

Beutler, R. 1951. Time and distance in the life of the foraging bee. *Bee World* 32:25–27. [360]

Bevan, W., and M. A. Grodsky. 1958. Hoarding in hamsters with systematically controlled pretest experience. *J. Comp. Physiol. Psychol.* 51:342–45. [117]

Bevanger, K. 1974. Klatremus som hamstringsobjekt fur lavskrike. *Sterna* 13:51–52. [298–99]

Beven, G., and M. D. England. 1969. The impaling of prey by shrikes. *Br. Birds* 62:192–99. [12, 30, 86, 89, 126, 317–18]

Bibikov, D. I. 1948. On the ecology of the nutcraker. Translated from the Russian by Leon Kelso. *Trudy Pechorskogo-llychskogo Gosudarstvennogo Zapovednika* 4:89–112. [181, 192, 209, 301, 304]

Bildstein, K. L. 1982. Prey concealment by American kestrels. *Raptor Res.* 16:83–88. [96, 288]

———. 1987. Behavioral ecology of red-tailed hawks (*Buteo jamaicensis*), rough-legged hawks (*Buteo lagopus*), northern harriers (*Circus cyaneus*), and American kestrels (*Falco sparverius*) in south central Ohio. *Ohio Biol. Surv. Biol. Notes* 18:1–53. [286–87]

Bindra, D. 1947. Water-hoarding in rats. *J. Comp. Physiol. Psychol.* 40:149–56. [122]

———. 1948a. The nature of motivation for hoarding food. *J. Comp. Physiol. Psychol.* 41:211–18. [120, 122, 140]

———. 1948b. What makes rats hoard? *J. Comp. Physiol. Psychol.* 41:397–402. [47, 121, 138]

Blair, W. F. 1937. The burrows and food of the prairie pocket mouse. *J. Mamm.* 18:188–91. [254]

Blincoe, B. J. 1923. Random notes on the feeding habits of some Kentucky birds. *Wilson Bull.* 35:63–71. [296]

Blomgren, A. 1971. Studies of less familiar birds. 162. Siberian jay. *Br. Birds* 64:25–28. [298]

Blundell, J. E., and L. J. Herberg. 1973. Effectiveness of lateral hypothalamic stimulation, arousal, and food deprivation in the initiation of hoarding behaviour in naive rats. *Physiol. Behav.* 10:763–67. [124]

Bock, C. E. 1970. The ecology and behavior of the Lewis' woodpecker (*Asyndesmus lewis*). *Univ. Calif. Publ. Zool.* 92:1–91. [47, 62–63, 83, 93, 97, 105, 292–93, 296–97]

Bock, C. E., and L. W. Lepthien. 1976. Synchronous eruptions of boreal seed-eating birds. *Am. Nat.* 110:559–71. [65]

Bock, W. J. 1961. Salivary glands in the gray jay (*Perisoreus*). *Auk* 78:355–65. [107, 298]

Bock, W. J., R. P. Balda, and S. B. Vander Wall. 1973. Morphology of the sublingual pouch and tongue musculature in Clark's nutcracker. *Auk* 90:491–519. [55, 299, 303]

Bogue, G., and M. Ferrari. 1976. The predatory "training" of captive puma. *World's Cats* 3:36–45. [233]

Bohart, G. B., and G. F. Knowlton. 1953. Notes on food habits of the western harvester ant (Hymenoptera, Formicidae). *Proc. Entomol. Soc. Wash.* 55:151–53. [330]

Bond, W., and P. Slingsby. 1984. Collapse of an ant-plant mutualism: The Argentine ant (*Iridomyrmex humilis*) and myrmecochorous Proteaceae. *Ecology* 65:1031–37. [180]

Bondrup-Nielsen, S. 1977. Thawing of frozen prey by boreal and saw-whet owls. *Can. J. Zool.* 55:595–601. [88, 290]

Borchert, M. I., F. W. Davis, J. Michaelsen, and L. D. Oyler. 1989. Interaction of factors affecting seedling recruitment of blue oak (*Quercus douglasii*) in California. *Ecology* 70:389–404. [185–86]

Borker, A. S., R. A. Dhume, and M. G. Gogate. 1985. Measure of motivational drive for hoarding in laboratory rats. *Ind. J. Physiol. Pharmac.* 29:126–28. [128]

Borker, A. S., and M. G. Gogate. 1981. Hunger versus hoarding and body weight in rats. *Indian J. Physiol. Pharmacol.* 25:365–68. [120, 123]

———. 1984a. Role of ovarian hormones on hoarding in rats—Feedback mechanisms. *Indian J. Physiol. Pharmacol.* 28:253–58. [128]

———. 1984b. Role of ovarian hormones in hoarding pattern. *Ind. J. Physiol. Pharmacol.* 28:115–20. [128]

Bossema, I. 1968. Recovery of acorns in the European jay (*Garrulus g. glandarius* L.). *Koninklijke Nederlandse Akademie van Wetenschappen, Series C* 71:1–5. [18, 106, 158, 164, 192, 198, 200, 300, 367]

———. 1979. Jays and oaks: An eco-ethological study of a symbiosis. *Behaviour* 70:1–117. [83, 94–95, 100, 106, 155–56, 158, 162, 164–67, 179, 182–83, 186, 192, 198, 200, 209, 298–99]

Bossema, I., and W. Pot. 1974. Het terugvinden van verstopt voedsel door de Vlaamse gaai (*Garrulus g. glandarius* L.). *Levende Natuur.* 77:265–79. [158, 164–65]

Botkin, C. W., and L. B. Shires. 1948. The composition and value of piñon nuts. *New Mexico Exp. Stn. Bull.* 344:3–14. [32, 201]

Bourliere, F. 1954. *The natural history of mammals.* New York: Alfred Knopf. [107]

Bradbury, W. C. 1917. Notes on the nesting habits of the Clarke nutcracker in Colorado. *Condor* 19:149–55. [305]

Bradford, D. F., and C. C. Smith. 1977. Seed predation and seed number in *Scheelea* palm fruits. *Ecology* 58:667–73. [75]

Braestrup, F. W. 1941. A study on the arctic fox in Greenland: Immigration, fluctuations in numbers based mainly on trading statistics. Kommissionen for Videnskabelige Undersøgelser I Grønland. *Meddelelser om Grønland.* 131(4):1–101. [226, 228]

Brander, A. 1923. *Wild animals in central India.* London: Edward Arnold. [226, 233]

Breed, M. D., K. B. Rogers, J. A. Hunley, and A. J. Moore. 1989. A correlation between guard behaviour and defensive response in the honey bee, *Apis mellifera. Anim. Behav.* 37:515–16. [105]

Brenner, F. J., and P. D. Lyle. 1975. Effect of previous photoperiodic conditions and visual stimulation on food storage and hibernation in the eastern chipmunk (*Tamias striatus*). *Am. Midl. Nat.* 93:227–34. [24, 126, 133, 240]

Brewster, W. 1898. Lewis' woodpecker storing acorns. *Auk.* 15:188. [296]

Bristowe, W. S. 1948. Notes on the habits and prey

of twenty species of British hunting wasp. *Proc. Linn. Soc. London* 160: 12–37. [343]

Broadbooks, H. E. 1958. Life history and ecology of the chipmunk, *Eutamias amoenus,* in eastern Washington. Miscellaneous Publications, The Museum of Zoology, University of Michigan 103: 1–48. [24, 50, 237–39]

———. 1965. Ecology and distribution of the pikas of Washington and Alaska. *Am. Midl. Nat.* 73: 299–335. [83, 279–82]

Brockmann, H. J. 1980. House sparrows kleptoparasitize digger wasps. *Wilson Bull.* 92: 394–98. [94]

———. 1985. Provisioning behavior of the great golden digger wasp, *Sphex ichneumoneus* (L.) (Sphecidae). *J. Kans. Entomol. Soc.* 58: 631–55. [135, 339–40]

———. 1988. Father of the brood. *Nat. Hist.* 97(7): 32–37. [104, 126, 341]

Brockmann, H. J., and C. J. Barnard. 1979. Kletoparasitism in birds. *Anim. Behav.* 27: 487–514. [94]

Bro Larsen, E. 1952. On subsocial beetles from the salt-marsh, their care of progeny and adaptation to salt and tide. In *Transactions of the 9th International Congress on Entomology,* 502–6. Amsterdam, The Netherlands. [328]

Bromley, G. F., V. A. Kostenko, and M. V. Okhotina. 1974. The role of the Amur nuthatch (*Sitta europaea amurensis* Swinh.) in the revival of the Korean cedar. *Trudy Biol.-pochvenn. Inst. Vladivostok* 17: 162–66. [179,315]

Brower, J. E. 1970. The relationship of food and temperature to torpidity in pocket mice. *Proc. Penn. Acad. Sci.* 44: 160–71. [25]

Brown, B. W., and G. O. Batzli. 1985. Foraging ability, dominance relations and competition for food by fox and gray squirrels. *Trans. Ill. Acad. Sci.* 78: 61–66. [185]

Brown, J. H., and G. A. Bartholomew. 1969. Periodicity and energetics of torpor in the kangaroo mouse, *Microdipodops pallidus. Ecology* 50: 705–9. [25, 42]

Brown, J. H., O. J. Reichman, and D. W. Davidson. 1979. Granivory in desert ecosystems. *Ann. Rev. Ecol. Syst.* 10: 210–27. [366]

Brown, J. L. 1963a. Aggressiveness, dominance and social organization in the Steller's jay. *Condor* 65: 460–84. [108]

———. 1963b. Social organization and behavior of the Mexican jay. *Condor.* 65: 126–53. [298]

Brown, J. S. 1989. Desert rodent community structure: A test of four mechanisms of coexistence. *Ecol. Monogr.* 59: 1–20. [366]

Brown, L. G., and L. E. Yeager. 1945. Fox squirrel and gray squirrels in Illinois. *Bull. Ill. Nat. Hist. Surv.* 23: 449–536. [199, 244–46]

Brown, L. H. 1966. Observations on some Kenya eagles. *Ibis* 108: 531–72. [285–86]

———. 1971. The relations of the crowned eagle *Stephanoaetus coronatus* and some of its prey animals. *Ibis* 113: 240–43. [285–86]

Brown, L. H., and D. Amadon. 1968. *Eagles, hawks, and falcons of the world.* New York: McGraw-Hill. [283, 285]

Brown, L. N., and G. C. Hickman. 1973. Tunnel system structure of the southeastern pocket gopher. *Fla. Sci.* 36: 97–103. [205–51]

Bruce, H. M., and E. Hindle. 1934. The golden hamster, *Cricetus (Mesocricetus) auratus* Waterhouse. Notes on its breeding and growth. *Proc. Zool. Soc. London.* 1934: 361–66. [33]

Bruce, K., and D. Estep. 1987. Body weight regulation and gonadal hormone manipulation in female eastern chipmunks. *Bull. Psychonom. Soc.* 25: 20–22. [127, 129–30]

Bruckner, D. 1980. Hoarding behavior and life span of inbred, noninbred and hybrid honeybees (*Apis mellifera carnica*). *J. Apic. Res.* 19: 35–41. [113]

Brylski, P., and B. K. Hall. 1988a. Epithelial behaviors and threshold effects in the development and evolution of internal and external cheek pouches in rodents. *Z. Zool. Syst. Evol.-forsch.* 26: 144–54. [54]

———. 1988b. Ontogeny of a macroevolutionary phenotype: The external cheek pouches of geomyoid rodents. *Evolution* 42: 391–95. [54]

Buckner, C. H. 1964. Metabolism, food capacity, and feeding behavior in four species of shrews. *Can. J. Zool.* 42: 259–79. [221]

Budathoki, L. K., and D. S. Madge. 1987. Distribution of brood and food stores in combs of the honeybee, *Apis mellifera* L. *Apidologie* 18: 43–52. [361]

Buitron, D., and G. L. Nuechterlein. 1985. Experiments on olfactory detection of food caches by black-billed magpies. *Condor* 87: 92–95. [155]

Buller, A. H. R. 1920. The red squirrel of North America as a mycophagist. *Trans. Br. Mycol. Soc.* 6:355–62. [83, 139, 243]

Bullock, S. H. 1981. Aggregation of *Prunus ilicifolia* (Rosaceae) during dispersal and its effect on survival and growth. *Madroño* 28:94–95. [208]

Bunch, K. G., and D. F. Tomback. 1986. Bolus recovery by gray jays: An experimental analysis. *Anim. Behav.* 34:754–62. [155–57]

Burchsted, A. E. 1987. Downy woodpecker caches food. *Wilson Bull.* 99:136–37. [292]

Burdick, D. J., and P. F. Torchio. 1959. Notes on the biology of *Hesperapis regularis* (Cresson) (Hymenoptera: Melittidae). *J. Kans. Entomol. Soc.* 32:83–87. [350]

Burdick, D. J., and M. S. Wasbauer. 1959. Biology of *Methocha californica* Westwood (Hymenoptera: Tiphidae). *Wasmann J. Biol.* 17:75–88. [337]

Burgett, D. M., and R. Young. 1974. Lipid storage by honey ant repletes. *Ann. Entomol. Soc. Am.* 67:743–44. [89, 334–35]

Burnell, K. L., and D. F. Tomback. 1985. Steller's jays steal gray jay caches: Field and laboratory observations. *Auk* 102:417–19. [94, 141]

Burton, S. S. 1930. A new diet for the red squirrel. *J. Forest.* 28:233. [243]

Butler, C. G., and J. B. Free. 1952. The behaviour of worker honeybees at the hive entrance. *Behaviour* 4:262–92. [89, 105, 110]

Butlin, S. M. 1971. Food-hiding by magpie. *Br. Birds* 64:422. [306].

Cabrera, A., and J. Yepes. 1960. *Mamiferos Sud Americanos*, Vol. 2. Buenos Aires: Ediar. [279]

Cade, T. J. 1963. Observations on torpidity in captive chipmunks of the genus *Eutamias*. *Ecology* 44:255–61. [24, 240]

————. 1967. Ecological and behavioral aspects of predation by the northern shrike. *Living Bird* 6:43–86. [317–18]

————. 1982. *Falcons of the world.* London: Collins. [285]

Cahalane, V. H. 1930. Out-of-season caching by fox squirrel. *J. Mamm.* 11:78. [246]

————. 1942. Caching and recovery of food by the western fox squirrel. *J. Wildl. Mgmt.* 6:338–52. [19, 116, 171–72, 181, 185, 190–92, 199, 244]

————. 1944. A nutcracker's search for buried food. *Auk* 61:643. [305]

————. 1947. *Mammals of North America.* New York: Macmillan. [236, 280]

Calaby, J. H., H. Dimpel, and I. McTaggart Cown. 1971. The mountain pigmy possum, *Burramys parvus* Broom (Marsupialia), in the Kosciusko National Park, New South Wales. Melbourne, Australia: Commonwealth Scientific and Industrial Research Organization, Division of Wildlife Research, Technical Paper 23. [217]

Cale, G. H., and J. W. Gowen. 1956. Heterosis in the honeybee (*Apis mellifera* L.). *Genetics* 41:292–303. [113]

Calhoun, J. B. 1963. *The ecology and sociology of the Norway rat.* Washington, D.C.: U.S. Department of Health, Education, and Welfare. [33, 111, 144, 274, 276]

Camp, C. L. 1918. Excavations of burrows of the rodent *Aplodontia*, with observations on the habits of the animal. *Univ. Calif. Publ. Zool.* 17:517–36. [236]

Cane, J. H. 1981. Dufour's gland secretion in the cell linings of bees (Hymenoptera: Apoidea). *J. Chem. Ecol.* 7:403–10. [346, 348]

Cane, J. H., G. C. Eickwort, F. R. Wesley, and J. Spielholz. 1983. Foraging, grooming and mate-seeking behavior of *Macropis nuda* (Hymenoptera, Melittidae) and use of *Lysimachia ciliata* (Primulaceae) oils in larval provisions and cell linings. *Am. Midl. Nat.* 110:257–64. [59, 350, 354]

Cantlon, J. E., E. J. C. Curtis, and W. M. Malcolm. 1963. Studies of *Melampyrum lineare*. *Ecology* 44:466–74. [326]

Carlson, A. 1985. Central place food caching: A field experiment with red-backed shrikes (*Lanius collurio* L.). *Behav. Ecol. Sociobiol.* 16:317–22. [13, 32, 53, 317]

Carroll, C. R., and D. H. Janzen. 1973. Ecology of foraging by ants. *Ann. Rev. Ecol. Syst.* 4:231–58. [330]

Carroll, C. R., and S. J. Risch. 1984. The dynamics of seed harvesting in early successional communities by a tropical ant, *Solenopsis geminata*. *Oecologia* 61:388–92. [105, 179, 331]

Carter, T. D. 1922. Notes on a Saskatchewan muskrat colony. *Can. Field-Nat.* 36:176. [270]

Cartwright, B. A., and T. S. Collett. 1979. How honey-bees know their distance from a near-by visual landmark. *J. Exp. Biol.* 82:367–72. [152]

———. 1982. How honey bees use landmarks to guide their return to a food source. *Nature* 295: 560–4. [152–53]

Catling, P. M. 1972. Food and pellet analysis studies of the saw-whet owl (*Aegolius acadicus*). *Ont. Field Biol.* 26:1–15. [47, 289–90]

Chakraborty, S. 1975. Field observations on the biology and ecology of the lesser bandicoot rat, *Bandicota bengalensis* (Gray) in West Bengal. In *Proceedings of All Indian Rodent Seminar,* pp. 102–12. Ahmedabad, India. [273–75]

Chapman, G. 1978. Caching of food by the Australian crow. *Emu* 78:98. [305]

Charlton, S. G. 1984. Hoarding-induced lever pressing in golden hamsters (*Mesocricetus auratus*): Illumination, time of day, and shock as motivating operations. *J. Comp. Psychol.* 98:327–32. [132–33]

Chesemore, D. L. 1975. Ecology of the arctic fox (*Alopex lagopus*) in North America: A review. In *The wild canids: Their systematics, behavioral ecology, and evolution,* edited by M. W. Fox, 143–63. New York: Van Nostrand Reinhold. [228, 230]

Chettleburgh, M. R. 1952. Observations on the collection and burial of acorns by jays in Hainault Forest. *Br. Birds* 45:359–64. [181, 183, 200, 299]

———. 1955. Further notes on recovery of acorns by jays. *Br. Birds* 48:183–84. [300]

Chiasson, R. B. 1954. The phylogenetic significance of rodent cheek pouches. *J. Mamm.* 35:425–27. [54]

Chisholm, A. 1972. Concerning birds that store food. *Vict. Nat.* 89:20. [305]

Chmurzynski, J. A. 1964. Studies on the stages of spatial orientation in female *Bembex rostrata* (Linné 1758) returning to their nests (Hymenoptera, Sphecidae). *Acta Biol. Exper. Warsaw* 24:103–32. [175]

Christian, D. P., J. E. Enders, and K. A. Shump, Jr. 1977. A laboratory study of caching in *Desmodillus auricularis. Zool Afr.* 12:505–7. [126, 264]

Christisen, D. M., and L. J. Korschgen. 1955. Acorn yields and wildlife usage in Missouri. *Trans. North Am. Wildl. Conf.* 20:337–56. [367]

Clark, L. D., and P. E. Gay. 1976. Some observations of the operant behavior of desert pack rats (*Neotoma lepida*). *Bull. Psychonom. Soc.* 8:309–11. [140]

Clark, W. H., and P. L. Comanor. 1973. The use of western harvester ant, *Pogonomyrmex occidentalis* (Cresson), seed stores by heteromyid rodents. *Occas. Pap. Biol. Soc. Nev.* 34:1–6. [89, 330]

Clarke, C. H. D. 1939. Some notes on hoarding and territorial behaviour of the red squirrel, *Sciurus hudsonicus. Can. Field-Nat.* 53:42–43. [242]

Clarkson, K., S. F. Eden, W. J. Sutherland, and A. I. Houston. 1986. Density dependence and magpie food hoarding. *J. Anim. Ecol.* 55:111–21. [89, 98–101, 103–4, 184, 307]

Clausen, J. 1965. Population studies of alpine and subalpine races of conifers and willows in the California high Sierra Nevada. *Evolution* 19: 56–68. [211]

Clegg, T. M. 1971. Kestrel hiding prey. *Scott. Birds* 6:276–77. [285]

Clements, S. L., and E. E. Grissell. 1968. Observations on the nesting habits of *Euparagia scutellaris* Cresson. *Pan-Pac. Entomol.* 44:34–37. [345]

Cochrane, C. T. 1968. Effects of visual group and isolation conditions upon hoarding behavior in female albino rats. *Diss. Abst.* 28(10B):4308. [141]

Cogshall, A. S. 1928. Food habits of deer mice of the genus *Peromyscus* in captivity. *J. Mamm.* 9:217–21. [259]

Cole, A. C., Jr. 1932. The relation of the ant, *Pogonomyrmex occidentalis* Cr., to its habitat. *Ohio J. Sci.* 32:133–46. [330, 332]

———. 1934. A brief account of aestivation and overwintering of the occident ant, *Pogonomyrmex occidentalis* Cresson, in Idaho. *Can. Entomol.* 66:193–98. [20, 179, 332]

———. 1968. *Pogonomyrmex harvester ants, a study of the genus in North America.* Knoxville: University Tennessee Press. [329]

Coling, J. G., and L. J. Herberg. 1982. Effect of ovarian and exogenous hormones on defended body weight, actual body weight, and the paradoxical hoarding of food by female rats. *Physiol. Behav.* 29:687–91. [129–30, 134]

Collins, A. M., T. E. Rinderer, J. R. Harbo, and M. A. Brown. 1984. Heritabilities and correlations for several characteristics in the honey bee. *J. Hered.* 75:135–40. [133]

Collins, C. T. 1976. Food-caching behavior in owls. *Raptor Res.* 10:74–76. [87, 93, 97, 116, 288–89]

Collopy, M. W. 1977. Food caching by female American kestrels in winter. *Condor* 79:63–68. [13–14, 94, 96, 149, 284–88]

Conibear, F., and J. L. Blundell. 1949. *The wise one.* New York: William Sloan Associates. [257]

Conner, D. A. 1983. Seasonal changes in activity patterns and the adaptive value of haying in pikas (*Ochotona princeps*). *Can. J. Zool.* 61:411–16. [22, 83, 97, 279, 281–82]

Conner, R. N., and J. C. Kroll. 1979. Food-storing by yellow-bellied sapsuckers. *Auk* 96:195. [292–93]

Conner, R. N., and J. H. Williamson. 1984. Food storing by American crows. *Bull. Tex. Ornithol. Soc.* 17:13–14. [199, 305, 307]

Conrads, K., and R. P. Balda. 1979. Überwinterungschancen Sibirischer Tannenhäher (*Nucifraga caryocatactes macrorhynchos*) im Invasionsgebiet. *Naturwissenschaftlichen Vereins Bielefeld* 24:115–37. [65]

Conway, J. R. 1977. Analysis of clear and dark amber repletes of the honey ant, *Myrmecocystus mexicanus hortideorum*. *Ann. Entomol. Soc. Am.* 70:367–69. [334, 335]

Cook, J. B. 1939. Pocket gophers spread Canada thistle. *Calif. Dept. Agric. Bull.* 28:142–43. [179, 205–6]

Cooper, K. W. 1953. Biology of eumenine wasps. I. The ecology, predation, nesting and competition of *Ancistrocerus antilope* (Panzer). *Trans. Am. Entomol. Soc.* 79:13–35. [342]

Cope, J. B., and J. C. Barber. 1978. Caching behavior of screech owls in Indiana. *Wilson Bull.* 90:450. [288–89]

Costello, D. F. 1947. Harvesters of the plains. *Nature Mag.* 40:146–49, 164. [83, 89, 330, 332]

Couch, L. K. 1925. Storing habits of *Microtus townsendii*. *J. Mamm.* 6:200–201. [20, 270, 272]

Covich, A. P. 1987. Optimal use of space by neighboring central place foragers: When and where to store surplus resources. In *Adv. Behav. Econ.* 1:249–94. [1, 53]

Coville, R., and C. Griswold. 1983. Nesting biology of *Trypoxylon xanthandrum* in Costa Rica with observations on its spider prey (Hymenoptera: Sphecidae; Araneae: Senoculidae). *J. Kans. Entomol. Soc.* 56:205–16. [90]

Cowan, D. P. 1981. Parental investment in two solitary wasps *Ancistrocerus adiabatus* and *Euodynerus foraminatus* (Eumenidae: Hymenoptera). *Behav. Ecol. Sociobiol.* 9:95–102. [36–39, 343–44]

———. 1983. Hypotheses on cell provisioning in eumenid wasps. *Biol. J. Linn. Soc.* 20:245–47. [40]

Cowan, I. M. 1947. The timber wolf in the Rocky Mountain national parks of Canada. *Can. J. Res.* 25:139–74. [228]

Coward, T. A. 1928. *The birds of the British Isles and their eggs,* 3d ed. London: Frederick Warne. [292]

Cowie, R. J., J. R. Krebs, and D. F. Sherry. 1981. Food storing in marsh tits. *Anim. Behav.* 29:1252–59. [3, 12, 18, 65, 94, 100–101, 105, 119, 155–56, 158, 160, 164, 167, 179, 310, 312, 314]

Cox, W. T. 1911. Reforestation on the national forests. U.S. Department of Agriculture, Forest Service, *Bulletin* 98:1–57. [242]

Craighead, F. C., and J. J. Craighead. 1972. Data on grizzly bear denning activities and behavior by using wildlife telemetry. In *Bears—Their biology and management,* edited by S. Herrero, 84–106. International Union for Conservation of Nature and National Resources, Publisher's News Service. [231]

Cram, W. E. 1924. The red squirrel. *J. Mamm.* 5:37–41. [243]

Creighton, W. S. 1950. The ants of North America. *Bull. Mus. Comp. Zool.* 104:1–585. [334]

———. 1956. Studies of the North American representatives of *Ephebomyrmex* (Hymenoptera: Formicidae). *Psyche* 63:54–66. [329–30]

Criddle, S. 1930. The prairie pocket gopher, *Thomomys talpoides rufescens. J. Mamm.* 11:265–80. [250]

————. 1939. The thirteen-striped ground squirrel in Manitoba. *Can. Field-Nat.* 53:1–6 [247–48]

————. 1943. The little northern chipmunk in southern Manitoba. *Can. Field-Nat.* 57:81–86. [24, 237–38]

————. 1947. A nest of the least weasel. *Can. Field-Nat.* 61:69. [88, 226, 232]

————. 1950. The *Peromyscus maniculatus bairdii* complex in Manitoba. *Can. Field-Nat.* 64: 169–77. [258–59]

————. 1973. The granivorous habts of shrews. *Can. Field-Nat.* 87:69–70. [220]

Critchfield, W. B., and E. L. Little, Jr. 1966. Geographic distribution of the pines of the world. Washington, D.C.: U.S. Department of Agriculture, Forest Service 991:1–97. [194]

Crocq, C. 1974. Notes complémentaires sur la nidification du casse-noix *Nucifraga caryocatactes* dans les Alpes Francaises. *Alauda* 42:39–50. [305]

————. 1977. Biologie de l'alimentation du Casse-noix *Nucifraga caryocatactes caryocatactes* (L.) dans les Alpes: Etude des caches. *L'Oiseau et la Revue Francaise D'Ornithologie.* 47:319–34. [166, 304]

————. 1978. Ecologie du Casse-noix (*Nucifraga caryocatactes* L.) dans les Alpes Francaises du sud: Ses relations avec L'Arolle (*Pinus cembra* L.). Thése, l'Univ. d'Aix-Marseille, France. [207]

Cross, E. A., A. E. S. Mostafa, T. R. Bauman, and I. J. Lancaster. 1978. Some aspects of energy transfer between the organ-pipe mud-dauber *Trypoxylon politum* and its araneid spider prey. *Envir. Entomol.* 7:647–52. [35, 37–38, 42, 342]

Cross, E. A., M. G. Stith, and T. R. Bauman. 1975. Bionomics of the organ-pipe mud-dauber, *Trypoxylon politum* (Hymenoptera: Sphecidae). *Ann. Entomol. Soc. Am.* 68:901–16. [104, 341–42]

Croze, H. 1970. Searching image in carrion crows. *Z. Tierpsychol.* Beiheft 5. [98]

Csuti, B. A. 1979. Patterns of adaptation and variation in the Great Basin kangaroo rat (*Dipodomys microps*). *Univ. Calif. Publ. Zool.* 111:1–69. [253]

Cugnasse, J. M. 1977. Mise en réserve de nourriture chez les rapaces. *Alauda* 45:241–42. [289]

Culbertson, A. E. 1946. Observations on the natural history of the Fresno kangaroo rat. *J. Mamm.* 27:189–203. [253, 255]

Daan, S. 1981. Adaptive daily strategies in behavior. *Handb. Behav. Neurobiol.* 4:275–98. [14]

Dadant, C. C. 1975. *The hive and the honey bee.* Hamilton, Ill.: Dadant and Sons. [359]

Dahl, Fr. 1891. Die Nahrungsvorrate des Maulwarfs. *Zool. Anz. Bel.* 14:9–11. [223]

Dale, F. H. 1939. Variability and environmental response of the kangaroo rat, *Dipodomys heermanni saxatilis. Am. Midl. Nat.* 22:703–31. [255]

Dalquest, W. W. 1948. Mammals of Washington. *Univ. Kans., Publ. Mus. Nat. Hist.* 2:1–444. [282]

Daly, C., and D. Shankman. 1985. Seedling establishment by conifers above tree limit on Niwot Ridge, Front Range, Colorado, U.S.A. *Arctic Alpine Res.* 17:389–400. [211]

Danka, R. G., R. L. Hellmich II, T. E. Rinderer, and A. M. Collins. 1987. Diet selection ecology of tropically and temperately adapted honey bees. *Anim. Behav.* 35:1858–63. [16, 64]

Darley-Hill, S., and W. C. Johnson. 1981. Acorn dispersal by blue jays (*Cyanocitta cristata*). *Oecologia* 50:231–32. [91, 181–82, 189, 200, 211, 298]

Davidson, R. H., and B. J. Landis. 1938. *Crabro davidsoni* Sandh., a wasp predaceous on adult leafhoppers. *Ann. Entomol. Soc. Am.* 31:5–8. [341]

Davies, K. C., and J. U. M. Jarvis. 1986. The burrow systems and burrowing dynamics of the mole-rats *Bathyergus suillus* and *Cryptomys hottentotus* in the fynbos of the south-western Cape, South Africa. *J. Zool. London* 209:125–47. [107, 205, 267–68]

Davis, J., and L. Williams. 1957. Irruptions of the Clark nutcracker in California. *Condor* 59: 297–307. [65]

————. 1964. The 1961 irruption of the Clark's nutcracker in California. *Wilson Bull.* 76:10–18. [65]

Davis, M. B. 1981. Quaternary history and the stability of forest communities. In *Forest succession: Concepts and application,* edited by D. C. West,

H. H. Shugart, and D. B. Botkin, 132–53. New York: Springer-Verlag. [212–14]

De Bruin, J. P. C. 1988. Sex differences in food hoarding behavior of Long Evans rats. *Behav. Proc.* 17:191–98. [126]

DeGange, A. R., J. W. Fitzpatrick, J. N. Layne, and G. E. Woolfenden. 1989. Acorn harvesting by Florida scrub jays. *Ecology* 70:348–56. [18, 31, 59, 67, 92–94, 191, 200, 298]

Degerbøl, M. 1927. Do moles (*Talpa europaea* L.) store up worms? *Vidensk. Medd. Damsk. Naturh. Foren. Dbh.* 84:195–202. [222–23]

Degerbøl, M., and P. Freuchen. 1935. Report of the mammals collected by the fifth Thule Expedition to artic North America. In *Report of the Fifth Thule Expedition 1921–24*, Vol. 2, 1–278. Copenhagen: Nordisk Forlag. [62, 227–28]

Delage, B. 1962. Recherches sur l'alimentation des fourmis granivores *Messor capitatus* Latr. *Insect. Soc.* 9:137–43. [33, 333]

Delcourt, H. R., and P. A. Delcourt. 1984. Ice age haven for hardwoods. *Nat. Hist.* 93:22–28. [212]

Delye, G. 1971. Observations on the nest and construction behavior of *Messor arenarius* (Hymenoptera, Formicidae). *Insect. Soc.* 18:15–20. [330–32]

Dementév, G. P., and N. A. Gladkov. 1954. *Birds of the Soviet Union*. Translated and published by Israel Program for Scientific Translations, Jerusalem. [298]

DeMeules, D. H. 1954. Possible anti-adrenalin action of shrew venom. *J. Mamm.* 34:425. [219]

de Monte, M., and J.-J. Roeder. 1987. Prey storing and washing by *Martes martes*. *Mammalia* 52:621–22. [226]

Denenberg, V. H. 1952. Hoarding in the white rat under isolation and group conditions. *J. Comp. Physiol. Psychol.* 45:497–503. [138, 141]

Dennington, M., and B. Johnson. 1974. Studies of beaver habitat in the MacKenzie Valley and northern Yukon. *Can. Wildl. Serv. Rep.* 39:1–169. [258]

Dennis, W. 1930. Rejection of wormy nuts by squirrels. *J. Mamm.* 11:195–201. [83, 182, 244]

Dewsbury, D. A. 1970. Food hoarding in rice rats and cotton rats. *Psychol. Rep.* 26:174. [126, 140–41, 259]

Dice, L. R. 1921. Notes on the mammals of interior Alaska. *J. Mamm.* 2:20–28. [242]

———. 1927. How do squirrels find buried nuts? *J. Mamm.* 8:55. [171]

Dickson, R. C. 1979. Food concealment by merlins. *Br. Birds* 72:118–19 [286]

Dimpel, H., and J. H. Calaby. 1972. Further observations on the mountain pigmy possum (*Burramys parvus*). *Vict. Nat.* 89:101–6. [217]

Dixon, J. M. 1971. *Burramys parvus* Broom (Marsupialia) from the Falls Creek area of the Bogong High Plains, Victoria. *Vict. Nat.* 88:133–38. [217]

Dodson, C. H. 1966. Ethology of some bees of the tribe Euglossini (Hymenoptera: Apidae). *J. Kans. Entomol. Soc.* 39:607–29. [359]

Dorka, V. 1980. Nuthatches *Sitta europaea* storing insects. A differentiation between longterm- and shortterm-food storing behavior. *Ökol. Vögel* 2:145–50. [10, 314]

Dorris, P. R. 1970. Spiders collected from mud-dauber nests in Mississippi. *J. Kans. Entomol. Soc.* 43:10–11. [341]

Dorsey, G. A. 1926. A red-headed woodpecker storing acorns. Bird-Lore 28:333–34. [293]

Dow, D. D. 1965. The role of saliva in food storage by the gray jay. *Auk* 82:139–54. [18, 94, 107–8, 298, 300]

Dow, R. 1942. The relation of the prey of *Sphecius speciosus* to the size and sex of the adult wasp (Hym.: Sphecidae). *Ann. Entomol. Soc. Am.* 35:310–17. [37–38]

Dressler, R. L. 1982. Biology of the orchid bees (Euglossini). *Ann. Rev. Ecol. Syst.* 13:373–94. [359]

DuBost, G. 1988. Ecology and social life of the red acouchy, *Myoprocta exilis;* comparison with the orange-rumped agouti, *Dasyprocta leporina*. *J. Zool. London* 214:107–23. [277]

Dufour, B. A. 1978. Le terrier d'*Apodemus sylvaticus* L.: Sa construction en terrarium et son adaptation a des facteurs extremes et internes. *Behav. Proc.* 3:57–76. [132–33, 135, 275]

Dulkeit, G. D. 1960. The winter life of birds in the taiga of northeastern Altai. *Trudy Problemnykh i Tematicheskikh soveshchanii, Zool. Inst. Akad. Nauk* 9:175–90. Leningrad, U.S.S.R. [302–4]

Durango, S. 1951. The impaling habits of the red-backed shrike. *Vår Fågelvärld* 10:49–65. [86, 317]

———. 1956. Territory in the red-backed shrike *Lanius collurio*. *Ibis* 98:476–84. [12–13, 317–18]

Eberhard, W. 1967. Attack behavior of diguetid spiders and the origin of prey wrapping in spiders. *Psyche* 74:173–81. [321–22]

Echternach, J. L., and R. K. Rose. 1987. Use of woody vegetation by beavers in southeastern Virginia. *Va. J. Sci.* 38:226–32. [258]

Eden, A. 1940. Coprophagy in the rabbit: origin of the 'night' faeces. *Nature* 145:628–29. [281]

Egorov, O. V. 1961. *Ecology and economics of the Yakut squirrel*. Publ. Akademiya Nauk, Moscow, U.S.S.R. [195, 302]

Eibl-Eibesfeldt, I. 1951. Beobachtungen zur Fortpflanzungsbiologie und Jugendentwicklung des Eichhörnchens (*Sciurus vulgaris* L.). *Z. Tierpsychol.* 8:370–400. [244]

Eickwort, G. C. 1975. Nest-building behavior of the mason bee *Hoplitis anthocopoides* (Hymenoptera: Megachilidae). *Z. Tierpsychol.* 37:237–54. [351]

Eickwort, G. C., and K. R. Eickwort. 1972. Aspects of the biology of Costa Rican halictine bees, III. *Sphecodes kathleenae*, a social cleptoparasite of *Dialictus umbripennis*. *J. Kans. Entomol. Soc.* 45:529–41. [89, 96]

Eickwort, G. C., R. W. Matthews, and J. Carpenter. 1981. Observations on the nesting behavior of *Megachile rubi* and *M. texana* with a discussion of the significance of soil nesting in the evolution of megachilid bees (Hymenoptera: Megachilidae). *J. Kans. Entomol. Soc.* 54:557–70. [347, 351]

Eigelis, Y. V., and B. V. Nekrasov. 1967. The morphological peculiarities of the buccal cavity of Corvidae as related to food transportation. Translated by Leon Kelso. *Zool. Zh.* 46:258–63. [55, 57, 299]

Eisenberg, J. F. 1962. Studies on the behavior of *Peromyscus maniculatus gambelii* and *Peromyscus californicus parasiticus*. *Behaviour* 19:177–207. [259]

———. 1963. The behavior of heteromyid rodents. *Univ. Calif. Publ. Zool.* 69:1–100. [252]

———. 1968. Behavior patterns. In *Biology of Peromyscus (Rodentia)*, edited by J. A. King, 451–

95. Lawrence, Kans.: American Society of Mammalogists. [259]

———. 1975. The behavior patterns of desert rodents. In *Rodents in desert environments*, edited by I. Prakash and P. K. Ghosh, 189–224. The Hague: Dr. W. Junk Publishers. [277]

Eisenberg, J. F., and D. G. Kleiman. 1972. Olfactory communication in mammals. *Ann. Rev. Ecol. Syst.* 3:1–32. [173]

Eisenberg, J. F., and M. Lockhard. 1972. An ecological reconnaissance of the Wilpattu National Park, Ceylon. *Smithson. Contr. Zool.* 101:1–118. [235]

Eisenmann, E. 1946. Acorn storing by *Balanosphyra formicivora* in Panama. *Auk* 63:250. [295]

Eisner, T., and W. L. Brown. 1958. The evolution and social significance of the ant proventriculus. *Proc. 10th Int. Congr. Entomol.* 2:503–8. [334]

Ekman, J. 1979. Coherence, composition and territories of winter social groups of the willow tit *Parus montanus* and the crested tit *P. cristatus*. *Ornis Scand.* 10:56–68. [17]

Elgmork, K. 1982. Caching behavior of brown bears (*Ursus arctos*). *J. Mamm.* 63:607–12. [12, 48, 87, 104, 225–26, 231]

———. 1983. Caching and covering activity of the brown bear. *Fauna, Oslo* 36:41–53. [226, 231–32]

El Hilali, M., and J. P. Veillat. 1975. *Jaculus orientalis:* A true hibernator. *Mammalia* 39:401–4. [277]

Elliott, C. L. 1980. Quantitative analysis of pika (*Ochotona princeps*) hay piles in central Idaho. *Northwest Sci.* 54:207–9. [280–81]

Elliott, L. 1978. Social behavior and foraging ecology of the eastern chipmunk (*Tamias striatus*) in the Adirondack Mountains. *Smithson. Contr. Zool.* 265:1–107. [24, 62, 89, 97, 104–6, 109, 149, 179, 181, 183, 199, 237–40]

Elliott, P. F. 1974. Evolutionary responses of plants to seed-eaters: Pine squirrel predation on lodgepole pine. *Evolution* 28:221–31. [70]

———. 1988. Foraging behavior of a central-place forager: Field tests of theoretical predictions. *Am. Nat.* 131:159–74. [53]

Eltringham, S. K. 1979. *The ecology and conservation of large African mammals*. London: Macmillan. [49, 225–26, 235]

Emmons, L. H. 1980. Ecology and resource partitioning among nine species of African rainforest squirrels. *Ecol. Monogr.* 50: 31–54. [64, 149, 183, 244–45, 247]

Engels, W. L. 1947. Nest building and food storage of a captive chipmunk. *J. Mamm.* 28: 296–97. [237]

English, P. F. 1923. The dusky-footed wood rat (*Neotoma fuscipes*). *J. Mamm.* 4: 1–9. [261]

Enoksson, B. 1987. Local movements in the nuthatch *Sitta europaea*. *Acta Regiae Soc. Sci. Litt. Gothob. Zool.* 14: 36–47. [315, 364]

Erlinge, S. 1969. Food habits of the otter, *Lutra lutra* L., and the mink, *Mustela vision* Schreber, in a trout water in southern Sweden. *Oikos* 20: 1–7. [226, 232]

Erlinge, S., B. Bergsten, and H. Kristiansson. 1974. Hermelinen och dess byte—jaktbeteende och flyktreaktioner. *Fauna Flora* 69: 203–11. [226]

Estep, D. Q., D. L. Lanier, and D. H. Dewsbury. 1978. Variation of food hoarding with the estrous cycle of Syrian golden hamsters. *Horm. Behav.* 11: 259–63. [127–28, 134]

Etienne, A. S., E. Emmanuelli, and M. Zinder. 1982. Ontogeny of hoarding in the golden hamster: The development of motor patterns and their sequential coordination. *Develop. Psychobiol.* 15: 33–45. [114–15, 118–19]

Etienne, A. S., R. Matathia, E. Emmanuelli, M. Zinder, and D. C. de Caprona. 1983. The sequential organization of hoarding and its ontogeny in the golden hamster. *Behaviour* 83: 80–111. [33, 114, 118, 263]

Evans, A. C. 1948. The identity of earthworms stored by moles. *Proc. Zool. Soc. Lond.* 118: 356–59. [20, 222–23]

Evans, H. E. 1957a. Ethological studies on digger wasps of the genus *Astata*. *J. New York Entomol. Soc.* 65: 159–85. [340–42]

———. 1957b. *Studies on the comparative ethology of digger wasps of the genus Bembix.* New York: Comstock Publishing Associates. [87, 340]

———. 1958. The evolution of social life in wasps. *Proc. Tenth Int. Cong. Entomol.* 2: 449–57. [51, 53, 68]

———. 1959. Observations on the nesting behavior of digger wasps of the genus *Ammophila*. *Am. Midl. Nat.* 62: 449–73. [52, 340]

———. 1962a. A review of nesting behavior of digger wasps of the genus *Aphilanthops*, with special attention to the mechanics of prey carriage. *Behaviour* 19: 239–60. [13, 341–42]

———. 1962b. The evolution of prey-carrying mechanisms in wasps. *Evolution* 16: 468–83. [57–58, 340]

———. 1966a. The behaviour patterns of solitary wasps. *Ann. Rev. Entomol.* 11: 123–54. [52–53, 79, 87, 95–96, 337, 340, 342]

———. 1966b. *The comparative ethology and evolution of the sand wasps.* Cambridge, Mass.: Harvard University Press. [5, 6, 337]

———. 1971. Observations on the nesting behavior of wasps of the tribe Cercerini. *J. Kans. Entomol. Soc.* 44: 500–523. [38]

Evans, H. E., and M. J. W. Eberhard. 1970. *The wasps.* Ann Arbor: University of Michigan Press. [87, 104]

Evans, H. E., and A. W. Hook. 1982. Communal nesting in the digger wasp *Cerceris australis* (Hymenoptera: Sphecidae). *Aust. J. Zool.* 30: 557–68. [68, 340]

Evans, H. E., C. S. Lin, and C. M. Yoshimoto. 1953. A biological study of *Anoplius apiculatus autumnalis* (Banks) and its parasite, *Evagetes mohave* (Banks). *J.N.Y. Entomol. Soc.* 61: 61–78. [87, 338]

Evans, H. E., and C. M. Yoshimoto. 1962. The ecology and nesting behavior of the Pompilidae of the northeastern United States. *Misc. Publ. Entomol. Soc. Am.* 3: 65–119. [338]

Evans, R. A., J. A. Young, G. J. Cluff, and J. K. McAdoo. 1983. Dynamics of antelope bitterbrush seed caches. U.S. Department of Agriculture, Forest Service, Gen Tech. Rept. INT 152: 195–202. [185, 190, 204]

Everett, R. L., and A. W. Kulla. 1977. Rodent cache seedlings of shrub species in the Southwest. *Tree Planters' Notes* 27: 11–12. [202]

Ewer, R. F. 1965. Food burying in the African ground squirrel, *Xerus erythropus* (E. Geoff.). *Z. Tierpsychol.* 22: 321–27. [62, 94, 116, 118, 248, 277]

———. 1967. The behaviour of the African giant rat (*Cricetomys gambianus* Waterhouse). *Z. Tierpsychol.* 24: 6–79. [116, 119, 143, 264–65]

———. 1968. *Ethology of mammals.* New York: Plenum Press. [46, 48, 62, 100, 277]

———. 1973. *The carnivores.* London: Weidenfeld and Nicolson. [226]

Eyre, J. 1984. Willow tits storing food. *Br. Birds* 77:118–19. [310]

Fantino, M., and H. Brinnel. 1986. Body weight set-point changes during the ovarian cycle: Experimental study of rats using hoarding behavior. *Physiol. Behav.* 35:991–96. [123, 128–30]

Fantino, M., and M. Cabanac. 1980. Body weight regulation with a proportional hoarding response in the rat. *Physiol. Behav.* 24:939–42. [120, 123]

———. 1984. Effect of cold ambient temperature on the rat's food hoarding behavior. *Physiol. Behav.* 32:183–90. [135–37]

Fenner, M. 1985. *Seed ecology.* New York: Chapman and Hall. [187]

Ferguson, R. B. 1962. Growth of single bitterbrush plants vs. multiple groups established by direct seeding. Ogden, Utah: U.S. Department of Agriculture, Forest Service, Intermountain Forest and Range Experiment Station, *Res. Note* 90:1–2. [208]

Ferguson, R. B., and J. V. Basile. 1967. Effect of seedling numbers on bitterbrush survival. *J. Range Mgmt.* 20:380–82. [208]

Fincher, G. T., T. B. Stewart, and R. Davis. 1970. Attraction of coprophagous beetles to feces of various animals. *J. Parasitol.* 56:378–83. [323]

Finley, R. B., Jr. 1958. The wood rats of Colorado: Distribution and ecology. *Univ. Kans. Publ. Mus. Natl. Hist.* 10(6):213–552. [64, 260–62]

———. 1969. Cone caches and middens of *Tamasciurus* in the Rocky Mountain region. In *Contributions in mammalogy*, edited by J. K. Jones, Jr., 233–73. Lawrence: University Press of Kansas. [18, 84, 241–43]

Fisher, E. M. 1951. Notes on the red fox (*Vulpes vulpes*) in Missouri. *J. Mamm.* 32:296–99. [62, 227–29]

Fisher, H. J. 1945. Notes on voles in central Missouri. *J. Mamm.* 26:435–37. [270–71]

Fisher, H. I., and E. E. Dater. 1961. Esophageal diverticula in the redpoll, *Acanthis flammea. Auk* 78:528–31. [54]

Fisher, R. M., and M. T. Myers. 1979. A review of factors influencing extralimital occurrences of Clark's nutcracker in Canada. *Can. Field-Nat.* 94:43–51. [65]

Fitch, H. S., and D. G. Rainey. 1956. Ecological observations on the woodrat, *Neotoma floridana. Univ. Kans. Publ., Mus. Natl. Hist.* 8:499–533. [262]

Fitzpatrick, J. 1978. Alpine chough retrieving and re-hiding piece of orange. *Br. Birds* 71:134. [306]

Fleming, T. H., and G. J. Brown. 1975. An experimental analysis of seed hoarding and burrowing in two species of Costa Rican heteromyid rodents. *J. Mamm.* 56:301–15. [64, 126, 252]

Floyd, M. E. 1982. The interaction of piñon pine and Gambel oak in plant succession near Dolores, Colorado. *Southw. Nat.* 29:143–47. [185, 211]

Forbes, R. B. 1966. Fall accumulation of fat in chipmunks. *J. Mamm.* 47:715–16. [24]

Formanowicz, D. R., P. J. Bradley, and E. D. Brodie. 1989. Food hoarding by the least shrew (*Cryptotis parva*)—Intersexual and prey type effects. *Am. Midl. Nat.* 122:26–33. [126, 220]

Formozov, A. N. 1933. The crop of cedar nuts, invasion into Europe of the Siberian nutcracker (*Nucifraga caryocatactes macrorhynchus* Brehm) and fluctuations in numbers of the squirrel (*Sciurus vulgaris* L.). *J. Anim. Ecol.* 2:70–81. [65–66, 73, 194, 197, 367]

———. 1948. Small rodents and insectivores of the Sharinskii region of the Kostromskaya district during the 1930–1940 period. *Fauna and Ecology of Rodents. Materials on Rodents.* No. 3, MOIP Moscow, U.S.S.R. [272]

———. 1966. Adaptive modification of behaviour in mammals of the Eurasian steppes. *J. Mamm.* 47:208–23. [20, 26, 48, 67–68, 83, 266, 270–72]

Forster, G. H. 1967. Rook burying pine cones. *Br. Birds* 60:137–38. [305]

Foster, J. B. 1961. Life history of the *Phenacomys* vole. *J. Mamm.* 42:181–98. [272]

Fox, J. F. 1982. Adaptation of gray squirrel behavior to autumn germination by white oak acorns. *Evolution* 36:800–809. [75–76, 106, 191, 202]

Fox, N. 1979. Nest robbing and food storing by New Zealand falcons (*Falco novaeseelandiae*). *Raptor Res.* 13:51–56. [12–14, 284–85, 287–91]

Frank, C. L. 1988a. The effect of moldiness level on seed selection by *Dipodomys spectabilis. J. Mamm.* 69:358–62. [85]

————. 1988b. The influence of moisture content on seed selection by kangaroo rats. *J. Mamm.* 69: 353–57. [25, 85, 256]

Frank, L. G. 1969. Selective predation and seasonal variation in the diet of the fox (*Vulpes vulpes*) in N.E. Scotland. *J. Zool., London* 189: 526–32. [230]

Frazar, A. M. 1877. The mottled owl as a fisherman. *Bull. Nuttall Ornithol. Club* 2: 80. [290]

Free, J. B. 1955. The behavior of robber honeybees. *Behaviour* 7: 233–40. [89, 105]

————. 1965. The allocation of duties among worker honeybees. *Symp. Zool. Soc.* 14: 39–59. [116]

————. 1982. *Bees and mankind.* London: George Allen and Unwin. [359]

Free, J. B., and C. G. Butler. 1959. *Bumblebees.* London: Collins. [357–58]

Free, J. B., and J. H. Williams. 1972. Hoarding by honeybees (*Apis mellifera* L.). *Anim. Behav.* 20: 327–34. [119, 131, 135, 142, 144]

Freeman, B. E. 1981a. Parental investment and its ecological consequences on the solitary wasp *Sceliphron assimile* (Dahlbom) (Sphecidae). *Behav. Ecol. Sociobiol.* 9: 261–68. [37–38, 40–41]

————. 1981b. Parental investment, maternal size and population dynamics of a solitary wasp. *Am. Nat.* 117: 357–62. [37, 40]

Freeman, B. E., and C. A. Taffe. 1974. Population dynamics and nesting behaviour of *Eumenes colona* (Hymenoptera) in Jamaica. *Oikos* 25: 388–94. [343–44]

French, A. R. 1976. Selection of high temperature for hibernation by the pocket mouse, *Perognathus longimembris. Ecology* 57: 185–91. [26]

————. 1988. The patterns of mammalian hibernation. *Am. Sci.* 76: 569–75. [26, 63]

French, N. R. 1955. Foraging behavior and predation by Clark nutcracker. *Condor* 57: 61–62. [302]

Fridzen, K. E. 1984. Apparent short-term storage of food by hawk owl, *Surnia ulula. Vår Fågelvärld* 43: 495. [289]

Frisvad, J. C., O. Filtenborg, and D. T. Wicklow. 1987. Terverticillate penicillia isolated from underground seed caches and cheek pouches of banner-tailed kangaroo rats (*Dipodomys spectabilis*). *Can. J. Bot.* 65: 765–73. [86]

Frohlich, D. R. 1983. On the nesting biology of *Osmia (Chenosmia) bruneri* (Hymenoptera: Mega-

chilidae). *J. Kans. Entomol. Soc.* 56: 123–30. [351]

Frohlich, D. R., and F. D. Parker. 1983. Nest building behavior and development of the sunflower leafcutter bee: *Eumegachile (Sayapis) pugnata* (Say) (Hymenoptera: Megachilidae). *Psyche* 90: 193–209. [351]

————. 1985. Observations on the nest-building and reproductive behavior of a resin-gathering bee: *Dianthidium ulkei* (Hymenoptera: Megachilidae). *Ann. Entomol. Soc. Am.* 78: 804–10. [351]

Frohlich, D. R., and V. J. Tepedino. 1986. Sex ratio, parental investment, and interparent variability in nesting success in a solitary bee. *Evolution* 40: 142–51. [40]

Fulk, G. W. 1977. Food hoarding of *Bandicota bengalensis* in a rice field. *Mammalia* 41: 539–41. [274]

Funmilayo, O. 1979. Food consumption, preference and storage in the mole. *Acta Theriol.* 24: 379–89. [20, 222–24]

Furnier, G. R., P. Knowles, M. A. Clyde, and B. P. Danick. 1987. Effects of avian seed dispersal on the genetic structure of whitebark pine populations. *Evolution* 41: 607–12. [208]

Fye, R. E. 1965. The biology of the Vespidae, Pompilidae, and Sphecidae (Hymenoptera) from trap nests in northwestern Ontario. *Can. Entomol.* 97: 716–44. [338, 341–42]

Galil, J. 1967. On the dispersal of the bulbs of *Oxalis cernua* Thumb. by mole-rats (*Spalax ehrenbergi,* Nehring). *J. Ecol.* 55: 787–92. [107, 179, 205–6, 266]

Gashwiler, J. S. 1979. Deer mouse reproduction and its relationship to the tree seed crop. *Am. Midl. Nat.* 102: 95–104. [28, 30, 260]

Gates, J. E., and D. M. Gates. 1980. A winter food cache of *Microtus pennsylvanicus. Am. Midl. Nat.* 103: 407–8. [20, 270, 272]

Gauthier-Pitters, H. 1962. Beobachtungen an Feneks (*Fennecus zerda* Zimm.). *Z. Tierpsychol.* 19: 440–64. [226–27]

Genelly, R. E. 1965. Ecology of the common mole-rat (*Cryptomys hottentotus*) in Rhodesia. *J. Mamm.* 46: 647–65. [20, 108, 267–69, 290]

George, W. G., and T. Kimmel. 1977. A slaughter of mice by common crows. *Auk* 94: 782–83. [119, 305, 307]

George, W. G., and R. Sulski. 1984. Thawing of frozen prey by a great horned owl. *Can. J. Zool.* 62:314–15. [88]

Gerber, H. S., and E. C. Klostermeyer. 1972. Factors affecting the sex ratio and nesting behavior of the alfalfa leafcutter bee. *Wash. Agric. Exp. Stn., Tech. Bull.* 73:1–11. [39, 351]

Gerling, D., P. D. Hurd., Jr., and A. Hefetz. 1983. Comparative behavioral biology of two Middle East species of carpenter bees (*Xylocopa* Latreille) (Hymenoptera: Apoidea). *Smithson. Contr. Zool.* 369:1–33. [353]

Geyer, L. A., C. A. Kornet, and J. G. Robers, Jr. 1984. Factors affecting caching in the pine vole, *Microtus pinetorum. Mammalia* 48:165–72. [131, 140–41]

Gibb, J. A. 1954. Feeding ecology of tits, with notes on the treecreeper and goldcrest. *Ibis* 96:513–43. [179]

———. 1960. Populations of tits and goldcrests and their food supply in pine plantations. *Ibis* 102:163–208. [89, 310–14]

Gibson, L. 1922. Bird notes from North Greenland. *Auk* 39:350–63. [62, 228, 230]

Gignoux, C. 1921. The storage of almonds by the California woodpecker. *Condor* 23:118–21. [293–94]

Gill, F. B., and L. L. Wolf. 1977. Nonrandom foraging by sunbirds in a patchy environment. *Ecology* 58:1284–96. [169]

Gilliam, M. 1979. Microbiology of pollen and bee bread: The genus *Bacillus. Apidologie* 10:269–74. [88]

Gilliam, M., S. L. Buchmann, B. J. Lorenz, and D. W. Roubik. 1985. Microbiology of the larval provisions of the stingless bee, *Trigona hypogea,* an obligate necrophage. *Biotropica* 17:28–31. [87]

Giuntoli, M., and L. R. Mewaldt. 1978. Stomach contents of Clark's nutcracker collected in western Montana. *Auk* 95:595–98. [18, 28, 79, 305]

Gladkina, T. S., and N. Y. Chentsova. 1971. Food storage by *Microtus arvalis* Pall. *Vestn. Zool.* 5:17–22. [270]

Glanz, W. E., R. W. Thorington, Jr., J. Giacalone-Madden, and L. R. Heaney. 1982. Seasonal food use and demographic trends in *Sciurus granatensis.* In *The ecology of a tropical forest: Seasonal rhythms and long-term changes,* edited by E. G. Leigh, Jr., A. S. Rand, and D. M. Windsor, 239–52. Washington, D.C.: Smithsonian Institution. [244, 247]

Goertz, J. W., R. M. Dawson, and E. E. Mowbray. 1975. Response to nest boxes and reproduction by *Glaucomys volans* in northern Louisiana. *J. Mamm.* 56:933–39. [64, 249]

Goidanich, A., and C. E. Malan. 1964. Sulla nidificazionae pedotrofica di alcume specie di *Onthophagus europei* e microflora aerobica dell'apparato digerente della larva di *Onthophagus taurus* Schreber (Coleoptera: Scarabaeidae). *Ann. Fac. Sci. Agr. Univ. Studi Torino* 2:213–378. [324]

Gojmerac, W. L. 1980. *Bees, beekeeping, honey and pollination.* Westport, Conn.: AVI Publishing. [359]

Goldman, E. A. 1951. Biological investigations in Mexico. *Smithson. Misc. Coll.* 115:1–476. [295]

Goodwin, D. 1955. Jays and carrion crows recovering hidden food. *Br. Birds* 48:181–83. [305–6]

———. 1956. Further observations on the behaviour of the jay. *Ibis* 98:186–219. [89, 94, 141, 148–49, 299]

———. 1976. *Crows of the world.* Ithaca: Cornell University Press. [64, 298]

Gordon, R. C., and O. L. Cartwright. 1974. Survey of food preferences of some North American Canthonini (Coleoptera: Scarabaeidae). *Entomol. News* 85:181–85. [323]

Greaves, J. W. 1968. Food concealment by merlins. *Br. Birds* 61:310–11. [284–86]

Greenquist, E. A., and J. S. Rovner. 1976. Lycosid spiders on artificial foliage: Stratum choice, orientation preferences, and prey-wrapping. *Psyche* 83:196–209. [321–22]

Grey, G. W., and G. G. Naughton. 1971. Ecological observations on the abundance of black walnut in Kansas. *J. Forest.* 69:741–43. [198, 210]

Griffin, J. R. 1971. Oak regeneration in the upper Carmel Valley, California. *Ecology* 52:862–68. [179, 186–87, 248]

Grinnell, J. 1936. Up-hill planters. *Condor* 38:80–82. [89, 200, 211]

Grinnell, J., and J. Dixon. 1918. Natural history of the ground squirrels of California. *Calif. State Comm., Hort. Mon. Bull.* 7:597–708. [247]

Grinnell, J., J. S. Dixon, and J. M. Linsdale. 1930.

Vertebrate natural history of a section of northern California through the Lassen Peak region. *Univ. Calif. Publ. Zool.* 35:1–594. [279, 281–82]

Grinnell, J., and T. I. Storer. 1924. *Animal life in the Yosemite.* Berkeley: University of California Press. [238]

Grönwall, O., and A. Pehrson. 1984. Nutrient content in fungi as a primary food of the red squirrel *Sciurus vulgaris* L. *Oecologia* 64:230–31. [246]

Gross, N. B., and V. H. Cohn, Jr., 1954. The effect of vitamin-B deficiency on the hoarding behavior of rats. *Am. J. Psychol.* 67:124–28. [122, 139]

Gross, N. B., A. H. Fisher, and V. H. Cohn. 1955. The effect of a rachitogenic diet on the hoarding behavior of rats. *J. Comp. Physiol. Psychol.* 48:451–55. [122, 139]

Grubb, T. C., Jr., and T. A. Waite. 1987. Caching by red-breasted nuthatches. *Wilson Bull.* 99:696–99. [315]

Grulich, I. 1981. Die Baue des hamsters (*Cricetus cricetus* Rodentia, Mammalia). *Folia Zool.* 30:99–116. [263]

Guggisberg, C. A. W. 1975. *Wild cats of the world.* New York: Taplinger Publishing. [235]

Gurnell, J. 1984. Home range, territoriality, caching behavior and food supply of the red squirrel (*Tamiasciurus hudsonicus fremonti*) in a subalpine lodgepole pine forest. *Anim. Behav.* 32:1119–31. [18–9, 112, 241–42]

Gutierrez, R. J., and W. D. Koenig. 1978. Characteristics of storage trees used by acorn woodpeckers in two California woodlands. *J. Forest.* 76:162–64. [293, 295]

Guze, H. 1958. The effects of pre-weaning nursing deprivation on the later maternal hoarding and sexual behavior in the rat. Diss. Abst. 18:2227–29. [177]

Gwynne, D. T. 1979. Nesting biology of the spider wasps (Hymenoptera: Pompilidae) which prey on burrowing wolf spiders (Araneae: Lycosidae, *Geolycosa*). *J. Nat. Hist.* 13:681–92. [2, 338]

Gwynne, D. T., and G. N. Dodson. 1983. Nonrandon provisioning by the digger wasp, *Palmodes laeviventris* (Hymenoptera: Sphecidae). *Ann. Entomol. Soc. Am.* 76:434–36. [89, 96]

Gysel, L. W. 1971. A 10-year analysis of beechnut production and use in Michigan. *J. Wildl. Mgmt.* 35:516–19. [189, 367]

Habeck, J. R. 1960. Tree-caching behavior of the gray squirrel. *J. Mamm.* 41:125–26. [245]

Hadow, H. H. 1973. Winter ecology of migrant and resident Lewis' woodpeckers in southeastern Colorado. *Condor* 75:210–24. [296]

Haemig, P. D. 1989. Island jay (*Aphelocoma coerulescens insularis*) stores lizards in the relict tree *Lyonothamnus asplenifolius. Southw. Nat.* 34:146–47. [298]

Haftorn, S. 1954. Contributions to the food biology of tits especially about storing of surplus food. I. The crested tit (*Parus c. cristatus* L.). *Kgl. Norske Vidensk. Selsk. Skr.* 1953(4):1–122. [17, 67, 87, 94, 105, 107–8, 148, 156, 170, 310–13, 363]

———. 1956a. Contribution to the food biology of tits especially about storing of surplus food. II. The coal-tit (*Parus ater* L.). *Kgl. Norske Vidensk. Selsk. Skr.* 1956(2):1–52. [94, 107, 156, 310–13]

———. 1956b. Contribution to the food biology of tits especially about storing of surplus food. III. The willow-tit (*Parus atricapillus* L.). *Kgl. Norske Vidensk. Selsk. Skr.* 1956(3):1–78. [17, 67, 148, 156, 310–11, 313–14]

———. 1956c. Contribution to the food biology of tits especially about storing of surplus food. Part IV. A comparative analysis of *Parus atricapillus* L., *P. cristatus* L., and *P. ater* L. *Kgl. Norske Vidensk, Selsk. Skr.* 1956(4):1–54. [65, 105, 162, 170, 184, 203, 310–13]

———. 1959. The proportion of spruce seeds removed by the tits in a Norwegian spruce forest in 1954–55. *Det. Kgl. Norske Vidensk, Selsk, Forh.* 32:121–25. [17, 65, 313]

———. 1974. Storage of surplus food by the boreal chickadee *Parus hudsonicus* in Alaska, with some records on the mountain chickadee *Parus gambeli* in Colorado. *Ornis Scand.* 5:145–61. [65, 89, 107, 310–13]

Haga, R. 1960. Observations on the ecology of the Japanese pika. *J. Mamm.* 41:200–212. [280–82]

Haglund, B. 1966. Winter habits of the lynx (*Lynx lynx* L.) and wolverine (*Gulo gulo* L.) as revealed by tracking in the snow. *Viltrevy* 4:81–310. [226]

Halffter, G. 1977. Evolution of nidification in the Scarabaeinae (Coleoptera: Scarabaeidae). *Quaest. Entomol.* 13:231–53. [323]

———. 1982. Evolved relations between reproductive and subsocial behaviors in Coleoptera. In *The biology of social insects,* edited by M. D. Breed, C. D. Michener, and H. E. Evans, 164–70. Boulder, Colo.: Westview Press. [328]

Halffter, G., S. Anduaga, and C. Huerta. 1983. Nidification des *Nicrophorus. Bull. Soc. Entomol. France* 88:648–66. [58]

Halffter, G., and W. D. Edmonds. 1982. *The nesting behavior of dung beetles (Scarabaeinae): An ecological and evolutionary approach.* Mexico City: Instituto de Ecologia, District Federal. [7, 9, 49, 56, 126, 180, 322–23]

Hall, J. G. 1960. Willow and aspen in the ecology of beaver on Sagehen Creek, California. *Ecology* 41:484–94. [258]

———. 1981. A field study of the Kaibab squirrel in Grand Canyon National Park. *Wildl. Monogr.* 75:1–54. [244, 246]

Hallwachs, W. 1986. Agoutis (*Dasyprocta punctata*): The inheritors of guapinol (*Hymenaea courbaril:* Leguminosae). In *Frugivores and seed dispersal,* edited by A. Estrada and T. H. Fleming, 285–304. Dordrecht, The Netherlands: Dr. W. Junk Publishers. [75, 89, 182–84, 188–89, 198, 278]

Hamilton, W. J., Jr. 1930. The food of the Soricidae. *J. Mamm.* 11:26–39. [220]

———. 1933. The weasels in New York. *Am. Midl. Nat.* 14:289–344. [266]

———. 1939. *American mammals: Their lives, habits and economic relations.* New York: McGraw-Hill. [250, 259, 270, 272]

———. 1941. The food of small forest mammals in eastern United States. *J. Mamm.* 22:250–63. [277]

———. 1942. The buccal pouch of *Peromyscus. J. Mamm.* 23:449–50. [55, 258–59]

———. 1943. *The mammals of eastern United States.* Ithaca, N.Y.: Comstock. [141]

Hammen, T. van der, T. A. Wijmstra, and W. H. Zagwijn. 1971. The floral record of the late Cenozoic of Europe. In *The late Cenozoic glacial ages,* edited by K. K. Turekian, 391–424. New Haven, Conn.: Yale University Press. [212]

Hammer, L. R. 1972. Further hoarding preferences in hamsters. *Psychonom. Sci.* 26:139–40. [140]

Hammond, P. M. 1976. Kleptoparasitic behavior of *Onthophagus suturalis* Peringuey (Coleoptera:

Scarabaeidae) and other dung-beetles. *Coleopterist's Bull.* 30:245–49. [96]

Hampton, M. 1983. Blackbird catching and briefly hoarding worms. *Br. Birds* 76:88. [316]

Hannon, S. J., R. L. Mumme, W. D. Koenig, S. Spon, and P. A. Pitelka. 1987. Acorn crop failure, dominance, and a decline in numbers in the cooperatively breeding acorn woodpecker. *J. Anim. Ecol.* 56:197–207. [21]

Hansson, L. 1986. Geographic differences in the sociability of voles in relation to cyclicity. *Anim. Behav.* 34:1215–21. [91, 142]

Hardy, G. A. 1949. Squirrels cache of fungi. *Can. Field-Nat.* 63:86–87. [243]

Hardy, M. H., O. E. Vrablic, H. A. Covant, and S. V. Kandarkar. 1986. The development of the Syrian hamster cheek pouch. *Anat. Rec.* 214: 273–82. [54]

Hardy, R. 1945. The influence of types of soil upon the local distribution of some mammals in southwestern Utah. *Ecol. Monogr.* 15:71–108. [253]

———. 1949. Notes on mammals from Arizona, Nevada, and Utah. *J. Mamm.* 30:434–35. [183, 202]

Harlow, W. M., E. S. Harrar, and F. M. White. 1979. *Textbook of Dendrology.* 6th ed. New York: McGraw-Hill. [200]

Harper, R. J., and D. Jenkins. 1982. Food caching in European otters (*Lutra lutra*). *J. Zool., London* 197:297–98. [226, 232]

Harrington, F. H. 1981. Urine-marking and caching behavior in the wolf. *Behaviour* 76:280–88. [172, 225–26, 228–29]

———. 1982. Urine marking at food and caches in captive coyotes. *Can. J. Zool.* 60:776–82. [225–29]

Harrison, C. J. O. 1954. Jays recovering buried acorns. *Br. Birds* 47:406. [300]

Harrison, J. S., and P. A. Werner. 1984. Colonization by oak seedlings into a heterogeneous successional habitat. *Can J. Bot.* 62:559–63. [91, 184, 198, 210]

Hart, D. 1958a. Hoarding of food by coal tit. *Br. Birds* 51:122–23. [310]

———. 1958b. Hoarding of food by willow tit. *Br. Birds* 51:122. [310, 365]

Hart, E. B. 1971. Food preferences of the cliff chipmunk, *Eutamias dorsalis,* in northern Utah. *Great Basin Nat.* 31:182–88. [238–40]

Hastings, J. 1986. Provisioning by female western cicada killer wasps, *Sphecius grandis* (Hymenoptera: Sphecidae): Influence of body size and emergence time on individual provisioning success. *J. Kans. Entomol. Soc.* 59:262–68. [38, 40]

Hatt, R. T. 1929. The red squirrel: Its life history and habits, with special reference to the Adirondacks of New York and the Harvard Forest. *Roosevelt Wild Life Ann.* 2:10–146. [84, 200, 243]

———. 1930. The biology of the voles of New York. *Roosevelt Wild Life Bull.* 5:512–623. [270, 273]

———. 1931. Habits of a young flying squirrel (*Glaucomys volans*). *J. Mamm.* 12:233–38. [116, 249]

———. 1943. The pine squirrel in Colorado. *J. Mamm.* 24:311–45. [241–43]

Hawbecker, A. C. 1940. The burrowing and feeding habits of *Dipodomys venustus. J. Mamm.* 21:388–96. [62, 179, 202, 252–55]

———. 1944. The giant kangaroo rat and sheep forage. *J. Wildl. Mgmt.* 8:161–65. [252, 255]

———. 1945. Food habits of the barn owl. *Condor* 47:161–66. [289]

Hay, K. G. 1958. Beaver census methods in the Rocky Mountain region. *J. Wildl. Mgmt.* 22:395–401. [258]

Hay, O. P. 1887. The red-headed woodpecker a hoarder. *Auk* 4:193–96. [293]

Haydak, M. H. 1935. Brood rearing by honeybees confined to a pure carbohydrate diet. *J. Econ. Entomol.* 28:657–60. [362]

Hayman, R. 1958. Magpie burying and recovering food. *Br. Birds* 51:275. [306–7]

Hayward, C. L. 1952. Alpine biotic communities of the Uinta Mountains, Utah. *Ecol. Monogr.* 22:93–102. [83, 282]

Headlee, T. J., and G. A. Dean. 1908. The mound-building prairie ant (*Pogonomyrmex occidentalis* Cresson). *Kans. Agric. Exp. Stn., Bull.* 154:165–80. [329–30, 332]

Heaney, L. R., and R. W. Thorington, Jr. 1978. Ecology of neotropical red-tailed squirrels, *Sciurus granatensis,* in the Panama Canal Zone. *J.*

Mamm. 59:846–51. [64, 183, 192, 198, 244–45, 247, 278]

Hefetz, A., H. M. Fales, and S. W. T. Batra. 1979. Natural polyesters: Dufour's gland macrocyclic lactones from brood cell laminesters in *Colletes* bees. *Science* 204:415–17. [88, 346]

Heinrich, B. 1988. Food sharing in the raven, *Corvus corax.* In *The Ecology of social behavior,* edited by C. N. Slobodchikoff, 285–311. New York: Academic Press. [305]

Heinrich, B., and G. A. Bartholomew. 1979. The ecology of the African dung beetles. *Sci. Am.* 241(11):146–56. [49, 56, 366]

Hellmich, R. L., and W. C. Rothenbuhler. 1986a. Pollen hoarding and use by high and low pollen-hoarding honeybees during the course of brood rearing. *J. Apic. Res.* 25:30–34. [112]

———. 1986b. Relationship between different amounts of brood and the collection and use of pollen by the honey bee (*Apis mellifera*). *Apidologie* 17:13–20. [142]

Hemberg, E. 1918. Bokens (*Fagus silvatica* L.) invandring till Skandinavien och dess spridningsbiologi. *Svenska Skogvårdsföreningens Tidskrift* 16:157–81. [214]

Henderson, J. 1920. Migrations of the pinyon jay in Colorado. *Condor* 22:36. [66]

Henry, J. D. 1977. The use of urine marking in the scavenging behavior of the red fox (*Vulpes vulpes*). *Behaviour* 61:82–105. [172, 225, 228–29]

———. 1986. *Red fox: The catlike canine.* Washington, D.C.: Smithsonian Institution Press. [13, 95, 100, 149, 175, 225–27]

Henshaw, H. W. 1921. The storage of acorns by the California woodpecker. *Condor* 23:109–18. [295]

Henty, C. J. 1975. Feeding and food-hiding responses of jackdaws and magpies. *Br. Birds* 68:463–66. [139, 306]

Herberg, L. J., and J. E. Blundell. 1967. Lateral hypothalamus: Hoarding behavior elicited by electrical stimulation. *Science* 155:349–50. [118, 124]

———. 1970. Non-interaction of ventromedial and lateral hypothalamic mechanisms in the regulation of feeding and hoarding behavior in the rat. *Q. J. Exp. Psychol.* 22:133–41. [118, 120, 124–25]

Herberg, L. J., J. G. Pye, and J. E. Blundell. 1972.

Sex differences in the hypothalamic regulation of food hoarding: Hormones versus calories. *Anim. Behav.* 20:186–91. [112, 120, 125, 127]

Herberg, L. J., and D. N. Stephens. 1977. Interaction of hunger and thirst in the motivational arousal underlying hoarding behavior in the rat. *J. Comp. Physiol. Psychol.* 91:359–64. [122]

Herberg, L. J., and P. Winn. 1982. Body-weight regulatory mechanisms and food hoarding in hereditarily obese (fa/fa) and lean (Fa/Fa) Zucker rats. *Physiol. Behav.* 29:631–35. [123]

Hess, E. H. 1953. Shyness as a factor influencing hoarding in rats. *J. Comp. Physiol. Psychol.* 46:46–48. [138]

Heth, G., E. M. Golenberg, and E. Nevo. 1989. Foraging strategy in a subterranean rodent, *Spalax ehrenbergi:* A test case for optimal foraging theory. *Oecologia* 79:496–505. [266]

Hewson, R. 1981. Hoarding of carrion by carrion crows. *Br. Birds* 74:509–12. [305–6]

Heydecker, W. 1956. Establishment of seedlings in a field. I. Influence of sowing depth on seedling emergence. *J. Hort. Sci.* 31:76–88. [187]

Hibbard, E. A., and J. R. Beer. 1960. The plains pocket mouse in Minnesota. *Flicker* 32:89–94. [253–54]

Hibben, F. C. 1937. A preliminary study of the mountain lion (*Felis oregonensis* sp.). Univ. New Mexico Bull. Biol. Ser. 5(3):1–59. [233]

Hicks, C. H. 1927. *Pseudomasaris vespoides* (Cresson), a pollen provisioning wasp. *Can. Entomol.* 59:75–79. [345]

Higuchi, H. 1977. Stored nuts *Castanopsis cuspidata* as a food resource of nestling varied tits *Parus varius. Tori* 26:9–12. [17, 32, 310–11, 314]

Hill, E. P. 1982. Beaver. In *Wild mammals of North America: Biology, management, economics,* edited by J. A. Chapman and G. A. Feldhammer, 256–81. Baltimore: Johns Hopkins University Press. [257]

Hinde, R. A. 1952. Behaviour of the great tit (*Parus major*) and some other related species. *Behav. Suppl.* 2:1–201. [314]

Hoffmeister, D. F. 1971. *Mammals of Grand Canyon.* Chicago: University of Illinois Press. [244]

Höglund, N. H., and E. Lansgren. 1968. The great grey owl and its prey in Sweden. *Viltrevy* 5:364–421. [289]

Holland, J. G. 1954. The influence of previous experience and residual effects of deprivation on hoarding in the rat. *J. Comp. Physiol. Psychol.* 47:244–47. [117]

Hölldobler, B. 1982. Communication, raiding behavior and prey storage in *Cerapachys* (Hymenoptera; Formicidae). *Psyche* 89:3–23. [87, 336–37]

Hollyer, J. N. 1970. The invasion of nutcrackers in autumn 1968. *Br. Birds* 63:353–73. [65]

Holzworth, J. M. 1930. *Wild grizzlies of Alaska.* New York: The Knickerbocker Press. [226, 231]

Honacki, J. A., K. E. Kinman, and J. W. Koeppl. 1982. *Mammal species of the world: A taxonomic and geographic reference.* Lawrence, Kans.: Allen Press and Association of Systematic Collections. [217, 219]

Hormay, A. L. 1943. Bitterbrush in California. U. S. Department of Agriculture, Forest Service, *For. Res. Note* 34:1–13. [183, 192, 194, 202, 204, 206–7]

Horn, E. E., and H. S. Fitch. 1942. Interrelations of rodents and other wildlife on the range. *Calif. Dept. Agr., Exp. Stn. Bull.* 663:96–129. [250, 256]

Hornocker, M. 1970. An analysis of mountain lion predation upon mule deer and elk in the Idaho Primitive Area. *Wildl. Monogr.* 21:1–39. [226, 233]

Horton, J. S., and J. T. Wright. 1944. The wood rat as an ecological factor in southern California watersheds. *Ecology* 25:341–51. [20, 64, 83, 260, 262]

Houston, T. F. 1984. Bionomics of a pollen-collecting wasp, *Paragia tricolor* (Hymenoptera: Vespidae: Masarinae), in western Australia. *Rec. West. Aust. Mus.* 11:141–51. [345]

Howard, T. E. 1980. Sisyphean behavior in a red-headed woodpecker. *Chat* 44:17–18. [293]

Howard, W. E., and H. E. Childs, Jr. 1959. Ecology of pocket gophers with emphasis on *Thomomys bottae menia. Hilgardia* 29:277–354. [250]

Howard, W. E., and R. E. Cole. 1967. Olfaction in seed detection by deer mice. *J. Mamm.* 48:147–50. [171]

Howard, W. E., and F. C. Evans. 1961. Seeds

stored by prairie deer mice. *J. Mamm.* 42:260–63. [67, 258–59]

Howard, W. E., R. E. Marsh, and R. E. Cole. 1968. Food detection by deer mice using olfactory rather than visual cues. *Anim. Behav.* 16:13–17. [171–72]

Howe, H. F. 1989. Scatter- and clump-dispersal and seedling demography: Hypothesis and implications. *Oecologia* 79:417–26. [202, 206, 208]

Howe, H. F., and J. Smallwood. 1982. Ecology of seed dispersal. *Ann. Rev. Ecol. Syst.* 13:201–28. [178, 184, 188]

Howell, A. B. 1926. Voles of the genus *Phenacomys*. II. Life history of the red tree mouse (*Phenacomys longicaudus*). *North Am. Fauna* 48:39–64. [11, 48, 272]

Howell, A. H. 1924. Revision of the American pikas (genus *Ochotona*). *North Am. Fauna* 47:1–57. [280]

———. 1929. Revision of the American chipmunks. *North Am. Fauna* 52:1–157. [238]

———. 1938. Revision of the North American ground squirrels. *North Am. Fauna* 56:1–256. [25, 247–48]

Hubbard, C. A. 1922. Some data upon the rodent *Aplodontia*. *Murrelet* 3:14–18. [236]

Huerta, C., S. Anduaga, and G. Halffter. 1981. Relaciones entre nidificacion y ovario en *Copris* (Coleoptera, Scarabaeidae, Scarabaeinae). *Folia Entomol. Mexicana* 47:139–70. [325]

Hunt, J. McV. 1941. The effect of infant feeding-frustration upon adult hoarding behavior. *J. Abnorm. Psychol.* 36:338–60. [117]

Hunt, J. McV., H. Schlosberg, R. Solomon, and E. Stellar. 1947. Studies of the effects of infantile experience on adult behavior in rats. I. Effects of infantile feeding frustration on adult hoarding. *J. Comp. Physiol. Psychol.* 40:291–304. [117]

Hunt, J. McV., and R. R. Willoughby. 1939. The effect of frustration on hoarding in rats. *Psychosom. Med.* 1:309–10. [111, 120]

Huntly, N. J., A. T. Smith, and B. L. Ivins. 1986. Foraging behavior of the pika (*Ochotona princeps*), with comparisons of grazing versus haying. *J. Mamm.* 67:139–48. [280–81]

Hurd, P. D., Jr., and E. G. Linsley. 1976. The bee family Oxaeidae with a revision of the North American species (Hymenoptera: Apoidea). *Smithson. Contr. Zool.* 220:1–75. [347]

Hurly, T. A., and R. J. Robertson. 1987. Scatter-hoarding by territorial red squirrels: A test of the optimal density model. *Can. J. Zool.* 65:1247–52. [62, 101–2, 241–43]

Hutchins, H. E., and R. M. Lanner. 1982. The central role of Clark's nutcracker in the dispersal and establishment of whitebark pine. *Oecologia* 55:192–201. [63, 72, 90, 93, 108, 179, 181, 184–85, 189, 194, 208–9, 241, 301–5, 366–67]

Hutchins, R. E. 1966. *Insects*. Englewood Cliffs, N. J.: Prentice-Hall. [83, 89, 330]

———. 1967. *The ant realm*. New York: Dodd, Mead. [334]

Hutto, R. L. 1978. A mechanism for resource allocation among sympatric heteromyid rodent species. *Oecologia* 33:115–26. [252, 366]

Iersel, J. J. A. van. 1975. The extension of the orientation system of *Bembix rostrata* as used in the vicinity of its nest. In *Function and evolution of behaviour*, edited by G. P. Baerends, C. Beer, and A. Manning, 142–68. Oxford: Clarendon Press. [154]

Iersel, J. J. A., van, and J. van den Assem. 1964. Aspects of orientation in the digger wasp *Bembix rostrata*. *Anim. Behav. Suppl.* 1:145–62. [153–54, 175]

Ihering, H., von. 1903. Biologie der stachellosen Honigbienen Brasiliens. *Zool. Jahrb., Abt. Syst.* 19:179–287. [355]

Imaizumi, Y. 1979. Seed storing behavior of *Apodemus speciosus* and *Apodemus argentatus*. *Zool. Mag., Tokyo.* 88:43–49. [20, 62, 183, 274–76]

Immelmann, K. 1971. Ecological aspects of periodic reproduction. In *Avian biology*, Vol. 1, edited by D. S. Farner and J. R. King, 341–89. New York: Academic Press. [27]

Ingles, L. G. 1952. The ecology of the mountain pocket gopher (*Thomomys monticola*). *Ecology* 33:87–95. [251]

Ingram, W. M. 1942. Snail associates of *Blarina brevicauda talpoides* (Say). *J. Mamm.* 23:255–58. [220]

———. 1944. Snails hoarded by *Blarina* at Ithaca, New York. *Nautilus* 57:135–37. [220]

Iwata, K. 1976. Evolution of instinct: Comparative

ethology of *Hymenoptera.* Washington, D.C.: Smithsonian Institution. [7, 53, 337]

Jacobs, L. 1989. Cache economy of the gray squirrel. *Nat. Hist.* 98(10): 40–47. [174, 176]

Jaeger, M. M. 1982. Feeding pattern in *Peromyscus maniculatus:* The response to periodic food deprivation. *Physiol Behav.* 28: 83–88. [15]

Jaeger, R. 1969. Hibernation in Siberian chipmunks *Tamias (Eutamias) sibiricus* Laxmann, 1769. *Z. Saugetierkd,* 34: 361–70. [24, 240]

James, P. C., and N. A. M. Verbeek. 1983. The food storage behavior of the northwestern crow. *Behaviour* 85: 276–91. [86, 89, 94–95, 155–56, 305–6, 309]

———. 1984. Temporal and energetic aspects of food storage in northwestern crows. *Ardea* 72: 207–16. [12–15, 31, 42, 126, 305, 307, 309]

———. 1985. Clam storage in a northwestern crow (*Corvus caurinus*): Dispersion and sequencing. *Can. J. Zool.* 63: 857–60. [101, 156, 167–68, 305–6]

Jameson, E. W., Jr. 1947. Natural history of the prairie vole. *Univ. Kans. Publ., Mus. Nat. Hist.* 1: 125–51. [270]

———. 1964. Patterns of hibernation of captive *Citellus lateralis* and *Eutamias speciosus. J. Mamm.* 45: 455–60. [24, 240]

Janick, J., R. W. Schery, F. W. Woods, and V. W. Ruttan. 1974. *Plant science: An introduction to world crops,* 2d ed. San Francisco: W. H. Freeman. [201]

Jansson, C., J. Ekman, and A. von Brömssen. 1981. Winter mortality and food supply in tits *Parus* spp. *Oikos* 37: 313–22. [17]

Janzen, D. H. 1971. Seed predation by animals. *Ann. Rev. Ecol. Syst.* 2: 465–92. [182, 184, 198]

———. 1977. Why fruits rot, seeds mold, and meat spoils. *Am. Nat.* 111: 691–713. [82–83, 88]

———. 1980. When is it coevolution? *Evolution* 34: 611–12. [70]

———. 1984. Dispersal of small seeds by big herbivores: Foliage is the fruit. *Am. Nat.* 123: 338–53. [180]

Jarvis, J. U. M., and J. B. Sale. 1971. Burrowing and burrow patterns of East African mole-rats *Tachyoryctes, Heliophobius,* and *Heterocephalus. J. Zool., London* 163: 451–79. [267–68]

Jarvis, P. G. 1963. The effects of acorn size and provenance on the growth of seedlings of sessile oak. *Q. J. For.* 57: 11–19. [201]

Jayasingh, D. B. 1980. A new hypothesis on cell provisioning in solitary wasps. *Biol. J. Linn. Soc.* 13: 167–70. [40]

Jayasingh, D. B., and C. A. Taffe. 1982. The biology of the eumenid mud-wasp *Pachodynerus nasidens* in trapnests. *Ecol. Entomol.* 7: 283–89. [38–39]

Jehl, J. R., Jr. 1979. Pine cones as granaries for acorn woodpeckers. *West. Birds* 10: 219–20. [295]

Jennings, T. J. 1975. Notes on the burrow systems of woodmice (*Apodemus sylvaticus*). *J. Zool., London* 177: 500–504. [274–75]

Jenny, J. P., and T. J. Cade. 1986. Observations on the biology of the orange-breasted falcon *Falco deiroleucus. Birds Prey Bull.* 3: 119–24. [285]

Jensen, T. S. 1982. Seed production and outbreaks of non-cyclic rodent populations in deciduous forests. *Oecologia* 54: 184–92. [28–30, 33]

———. 1985. Seed-seed predator interactions of European beech, *Fagus silvatica* and forest rodents, *Clethrionomys glareolus* and *Apodemus flavicollis. Oikos* 44: 149–56. [100, 171, 183, 199, 202, 275–76]

Jensen, T. S., and O. F. Nielsen. 1986. Rodents as seed dispersers in a heath-oak wood succession. *Oecologia* 70: 214–21. [191, 200, 210, 275]

Jeselnik, D. L., and I. L. Brisbin, Jr. 1980. Food-caching behavior of captive-reared red foxes. *Appl. Anim. Ethol.* 6: 363–67. [229–30]

Johnsen, S. 1969. Røyskatten. In *Norges dyr. I pattedyr,* edited by R. Frislid and A. Semb-Johansson, 128–37. Oslo: Cappelen. [226, 232]

Johnson, D. R., and M. H. Maxell. 1966. Energy dynamics of Colorado pikas. *Ecology* 47: 1059–61. [22, 280–82]

Johnson, H. C. 1900. In the breeding home of Clarke's nutcracker. *Condor* 2: 49–52. [305]

Johnson, L. S., J. M. Marzluff, and R. P. Balda. 1987. Handling of pinyon pine seed by the Clark's nutcracker. *Condor* 89: 117–25. [182, 301]

Johnson, M. D. 1981. Observations on the biology of *Andrena (Melandrena) dunningi* Cockerell (Hymenoptera: Andrenidae). *J. Kans. Entomol. Soc.* 54: 32–40. [347–48]

———. 1988. The relationship of provision weight

to adult weight and sex ratio in the solitary bee, *Ceratina calcarata. Ecol. Entomol.* 13:165–70. [37, 39, 353]

Johnson, T. K., and C. D. Jorgensen. 1981. Ability of desert rodents to find buried seeds. *J. Range Mgmt.* 34:312–14. [171–72, 185]

Johnson, W. C., and C. S. Adkisson. 1985. Dispersal of beech nuts by blue jays in fragmented landscapes. *Am. Midl. Nat.* 113:319–24. [91, 179, 181–84, 199, 211, 298–300]

———. 1986. Airlifting the oaks. *Nat. Hist.* 95(10): 40–47. [211]

Johnson, W. C., and T. Webb, III. 1989. The role of blue jays (*Cyanocitta cristata* L.) in the postglacial dispersal of fagaceous trees in eastern North America. *J. Biogeo.* 16:561–71. [214]

Jolly, A. 1972. *The evolution of primate behavior.* New York: Macmillan. [225]

Jones, E. W. 1959. Biological flora of the British Isles, *Quercus* L. *J. Ecol.* 47:169–222. [186]

Källander, H. 1978. Hoarding in the rook *Corvus frugilegus. Anser Suppl.* 3:124–28. [60, 75, 89, 94, 141, 148, 199, 305–7]

Källander, H., and H. G. Smith. 1990. Food storing in birds: An evolutionary perspective. In *Current Ornithology,* vol. 7, edited by D. M. Power, 147–207. New York: Plenum Press. [1]

Kamil, A. C. 1978. Systematic foraging by a nectar-feeding bird, the amakihi (*Loxops virens*). *J. Comp. Physiol. Psychol.* 92:388–96. [196]

Kamil, A. C., and R. P. Balda. 1985. Cache recovery and spatial memory in Clark's nutcracker (*Nucifraga columbiana*). *J. Exper. Psychol.: Anim. Behav. Proc.* 11:95–111. [102, 159, 162, 168–69, 175, 186, 190]

Karasawa, K. 1976. Observations on the impalements made by shrikes. *Tori* 25:94–100. [12, 317–18]

Kato, J. 1985. Food and hoarding behavior of Japanese squirrels. *Jap. J. Ecol.* 35:13–20. [183, 199, 244–46]

Kattan, G. 1988. Food habits and social organization of acorn woodpeckers in Columbia. *Condor* 90:100–106. [64, 293, 295]

Kaufman, D. W. 1973. Captive barn owls stockpile prey. *Bird-Banding* 44:225. [288–89]

Kawamichi, T. 1976. Hay territory and dominance rank of pikas (*Ochotona princeps*). *J. Mamm.* 57:133–48. [280]

Kawamichi, M. 1980. Food, food hoarding and seasonal changes of Siberian chipmunks. *Jap. J. Ecol.* 30:211–20. [24, 33, 94, 171, 181, 183, 237–40]

Kay, F. R., and W. Whitford. 1978. The burrow environment of the bannertailed kangaroo rat, *Dipodomys spectabilis,* in south-central New Mexico. *Am. Midl. Nat.* 99:270–79. [86]

Kemp, G. A., and L. B. Keith. 1970. Dynamics and regulation of red squirrel *Tamiasciurus hudsonicus*) populations. *Ecology* 51:763–79. [18, 29, 241]

Kenagy, G. J. 1973. Adaptations for leaf eating in the Great Basin kangaroo rat, *Dipodomys microps. Oecologia* 12:383–412. [25, 254]

Kendall, K. C. 1980. Bear-squirrel-pine nut interaction. In *Yellowstone grizzly bear investigations, Annual Report 1978–79,* edited by R. R. Knight, B. M. Blanchards, K. C. Kendall, and L. E. Oldenburg, 51–60. U.S. Department of Interior, National Park Service. [3, 242–43]

———. 1983. Use of pine nuts by grizzly and black bears in the Yellowstone area. In *International Conference on Bear Research Management,* Vol. 5, edited by E. C. Meslow, 166–73. International Association for Bear Research and Management. [89]

Kerle, J. A. 1984. The behavior of *Burramys parvus* Broom (Marsupialia) in captivity. *Mammalia* 48:317–25. [217]

Kerr, W. E., and H. H. Laidlaw. 1956. General genetics of bees. *Adv. Genet.* 8:109–53. [33, 356–57]

Kessler, K. J., Jr. 1979. Premature loss of developing black walnut fruit. In *Walnut insects and diseases,* edited by J. K. Kessler, Jr., and B. C. Weber, 1–4, U.S. Department of Agriculture, Forest Service, Gen Tech. Rept. NC-52. [189]

Kikuzawa, K. 1988. Dispersal of *Quercus mongolica* acorns in a broadleaved deciduous forest. I. Disappearance. *For. Ecol. Mgmt.* 25:1–8. [185]

Kilham, L. 1958a. Sealed-in winter stores of red-headed woodpeckers. *Wilson Bull.* 70:107–13. [90, 105, 109–10, 293, 296]

———. 1958b. Territorial behavior in pikas. *J. Mamm.* 39:307. [279–80]

———. 1958c. Territorial behavior of wintering red-headed woodpeckers. *Wilson Bull.* 70:347–58. [47, 62, 83, 104, 293, 296]

———. 1963. Food storing in red-bellied woodpeckers. *Wilson Bull.* 75:227–34. [47, 62, 90, 94, 105, 293, 296–97]

———. 1974a. Covering of stores by white-breasted and red-breasted nuthatches. *Condor* 76:108–9. [314–15]

———. 1974b. Play in hairy, downy, and other woodpeckers. *Wilson Bull.* 86:35–42. [293]

———. 1975. Association of red-breasted nuthatches with chickadees in a hemlock cone year. *Auk* 92:160–62. [315]

———. 1984a. American crow feeding on and storing river otter dung. *Fla. Field Nat.* 12:103–4. [305–7]

———. 1984b. Foraging and food-storing of American crows in Florida. *Fla. Field Nat.* 12:25–31. [30, 126–127, 305–6]

King, B. 1988. Carrion crow hiding food when attacked, and later recovering it. *Br. Birds* 81:184. [305]

King, C. M. 1983. The relationship between beech (*Nothofagus* sp.) seedfall and populations of mice (*Mus musculus*), and the demographic and dietary response of stoats (Mustela erminea), in three New Zealand forests. *J. Anim. Ecol.* 52:141–66. [28]

Kinsinger, F. E. 1962. The relationship between depth of planting and maximum foliage height of seedlings of Indian ricegrass. *J. Range Mgmt.* 15:10–13. [187–88, 204]

Kirikov, S. V. 1936. On the ecological relationship between the nutcracker (*Nucifraga caryocatactes* L.) and the spruces (*Picea*). *Izvestiia Akademii Nauk USSR, Otdelenie Mat. I Estest, Nauk, Seriia Biologicheskaia* 6:1235–50. [301]

Kiris, I. D. 1958. Squirrel migration in the USSR. I. in *Translation of Russian Game Reports, Vol. 5 (Sable and Squirrel), 1951–1955),* 91–145. Canadian Wildlife Service, Ottawa. [66]

Kirk, V. M. 1972. Seed-caching by larvae of two ground beetles, *Harpalus pennsylvanicus* and *H. erraticus. Ann. Entomol. Soc. Amer.* 65:1426–28. [179, 326]

Kishchinskii, A. A. 1968. Birds of the Kolyma High-lands. Translated by Leon Kelso, 100–109. [32, 148, 302–3, 305]

Kivett, V. K., J. O. Murie, and A. L. Steiner. 1976. A comparative study of scent-gland location and related behavior in some northwestern Nearctic ground squirrel species (Sciuridae): An evolutionary approach. *Can. J. Zool.* 54:1294–306. [172]

Klapste, J. 1981. Caching of food by the white-winged chough. *Aust. Bird Watcher* 9:25–26. [308]

Klemperer, H. G., and R. Boulton. 1976. Brood burrow construction and brood care by *Heliocopris japetus* (Klug) and *Heliocopris hamadryas* (Fabricius) (Coleoptera, Scarabaeidae). *Ecol. Entomol.* 1:19–29. [36, 324, 326]

Klostermeyer, E. E., S. J. Mech, Jr., and W. B. Rasmussen. 1973. Sex and weight of *Megachile rotundata* (Hymenoptera: Megachilidae) progeny associated with provision weights. *J. Kans. Entomol. Soc.* 46:536–48. [37–39, 352]

Klugh, A. B. 1927. Ecology of the red squirrel. *J. Mamm.* 8:1–32. [243]

Knee, W. J., and J. T. Medler. 1965. Sugar concentration of bumblebee honey. *Am. Bee J.* 105:174–75. [358]

Kobayashi, T., and M. Watanabe. 1980. Seed hoarding behavior of the Asiatic chipmunk *Tamias sibirica asiaticus. Zool. Mag. Toyko* 89:235–43. [238]

Koenig, L. 1960. Das Aktionsystem des Siebenschlafers. *Z. Tierpsychol.* 17:427–505. [20, 276]

Koenig, W. D. 1980. Acorn storage by acorn woodpeckers in an oak woodland: An energetic analysis. In *Proceedings of the Symposium on the Ecology, Management, and Utilization of California Oaks,* edited by T. R. Plumb, 265–69. Pacific Southwest Forest and Range Experiment Station, *Gen. Tech. Rpt.* PSW-44. [21, 296–97]

Koenig, W. D., and R. L. Mumme. 1987. *Population ecology of the cooperatively breeding acorn woodpecker.* Princeton, N. J.: Princeton University Press. [21–22, 28, 30, 32–33, 68–69, 89, 93, 295, 297]

Koenig, W. D., and P. L. Williams. 1979. Notes on the status of the acorn woodpecker in central Mexico. *Condor* 81:317–18. [295]

Kondratov, A. V. 1953. On the restoration of the Siberian cedar pine in the wild by cluster sewing. *Agrobiologiia* 3:161–64. [206, 303]

Korelov, M. N. 1948. On the ecology of the kedrovka, *Nucifraga caryocatactes rothschildi* Hart. Translated by Leon Kelso. *Vestnik Akad. Nauk Kazakhkoi USSR* 5:72–75. [148, 301]

Korpimäki, E. 1987. Prey caching of breeding Tengmalm's owls *Aegolius funereus* as a buffer against temporary food shortage. *Ibis* 129:499–510. [12, 28, 49, 63–64, 288–90, 365]

———. 1989. Breeding performance of Tengmalm's owl *Aegolius funereus:* Effects of supplemental feeding in a peak vole year. *Ibis* 131:51–56. [28, 30]

Korstian, C. F. 1927. Factors controlling germination and early survival in oaks. *Yale Univ. Sch. For. Bull.* 19:1–115. [187, 189–90, 192–93, 201]

Korstian, C. F., and F. S. Baker. 1925. Forest planting in the Intermountain Region. U. S. Department of Agriculture, *Bull.* 1264:1–56. [241–42]

Korytin, S. A., and N. N. Solomin. 1969. Information on the ethology of the Canidae. *Sb. Trud. vses. nauchno-issled, Inst. Zhivotnogo Syrya Pushniny* 22:235–70. [172]

Koski, J. 1963. The bimodal distribution of hoarding scores in the golden hamsters. *Proc. Penn. Acad. Sci.* 37:291–94. [112]

Kramer, D. L., and W. Nowell. 1980. Central place foraging in the eastern chipmunk, *Tamias striatus. Anim. Behav.* 28:772–78. [53]

Krauch, H. 1945. Influence of rodents on natural regeneration of Douglas fir in the southwest. *J. Forest.* 43:585–89. [194]

Kraus, B. 1983. A test of the optimal-density model for seed scatterhoarding. *Ecology* 64:608–10. [100–101, 171, 190]

Kraus, K. E., and C. C. Smith. 1987. Fox squirrel use of prairie habitats in relation to winter food supply and vegetation density. *Prairie Nat.* 19:115–20. [209]

Krebs, J. R., D. F. Sherry, S. D. Healy, V. H. Perry, and A. L. Vaccarino. 1989. Hippocampal specialization of food-storing birds. *Proc. Natl. Acad. Sci.* 86:1388–92. [170]

Krog, J. 1954. Storing of food items in the winter nest of the Alaska ground squirrel, *Citellus undulatus. J. Mamm.* 35:586–87. [24, 247–48]

Krombein, K. V. 1967. *Trap-nesting wasps and bees: Life histories, nests, and associates.* Washington, D.C.: Smithsonian Institution Press. [6, 9, 38–39, 337, 341]

———. 1970. Behavioral and life-history notes on three Floridian solitary wasps (Hymenoptera: Sphecidae). *Smithson. Contr. Zool.* 46:1–26. [341]

Krombein, K. V., P. D. Hurd, Jr., D. R. Smith, and B. D. Burks. 1979. *Catalog of Hymenoptera in America North of Mexico,* Vol. 2. Washington, D.C.: Smithsonian Institution Press. [337]

Krott, P. 1960. Ways of the wolverine. *Nat. Hist.* 69:16–29. [225–26, 233]

Krushinskaya, N. L. 1966. Some complex forms of feeding behaviour of nutcracker *Nucifraga caryocatactes,* after removal of old cortex. *Zh. Evol. Biochim. Fisiol.* 11:563–68. [162–63]

———. 1970. On the problem of memory. *Priroda* 9:75–78. [162]

Kruuk, H. 1964. Predation and anti-predator behaviour of the black-headed gull (*Larus ridibundis* L.). *Behav. Suppl.* 11:1–130. [2, 149, 225–30]

———. 1972. *The spotted hyena: A study of predation and social behavior.* Chicago: University of Chicago Press. [48, 91, 119, 226, 230, 233]

Kulincevic, J. M., and W. C. Rothenbuhler. 1973. Laboratory and field measurements of hoarding behavior in the honeybee (*Apis mellifera* L.). *J. Apic. Rec.* 12:179–82. [112]

Kulincevic, J. M., V. C. Thompson, and W. C. Rothenbuhler. 1974. Relationship between laboratory tests of hoarding behavior and weight gained by honey bee colonies in the field. *Am. Bee J.* 114:93–94. [112]

Kulwich, R., L. Struglia, and P. B. Pearson. 1953. The effect of coprophagy on the excretion of B vitamins by the rabbit. *J. Nutr.* 49:639–45. [281]

Kumari, P. V., and J. A. Khan. 1979. Food hoarding by Indian gerbil, *Tatera indica indica* (Hardwicke). *Proc. Indian Acad. Sci.* 88:131–35. [121, 126, 264]

Kurczewski, F. E. 1968. Nesting behavior of *Plenoculus davisi* (Hymenoptera: Sphecidae, Larrinae). *J. Kans. Entomol. Soc.* 41:179–207. [36, 97, 340, 342]

————. 1987. A review of nesting behavior in the *Tachyshex pompiliformis* group with observations on five species (Hymenoptera: Sphecidae). *J. Kans. Entomol. Soc.* 60:118–26. [340]

Kurczewski, F. E., and E. J. Kurczewski. 1972. Host records for some North American Pompilidae (Hymenoptera), second supplement. Tribe Pepsini. *J. Kans. Entomol. Soc.* 45:181–93. [337–38]

————. 1973. Host records for some North American Pompilidae, third supplement. Tribe Pompilini. *J. Kans. Entomol. Soc.* 46:65–81. [338]

————. 1987. Nesting behavior and ecology of *Tachyshex antennatus* (Hymenoptera: Sphecidae). *J. Kans. Entomol. Soc.,* 60:408–20. [340]

Kurczewski, F. E., E. J. Kurczewski, and R. A. Norton. 1987. New prey records for species of Nearctic Pompilidae (Hymenoptera). *J. Kans. Entomol. Soc.* 60:467–75. [36]

Kurczewski, F. E., and R. C. Miller. 1984. Observations on the nesting of three species of *Cerceris* (Hymenoptera: Sphecidae). *Fla. Entomol.* 67:146–55. [36, 340, 342]

Kurczewski, F. E., and N. F. R. Snyder. 1964. Observations on the nesting of *Pompilus* (*Ammosphex*) *michiganensis* (Dreisbach) (Hymenoptera; Pompilidae). *Proc. Biol. Soc. Wash.* 77:215–22. [338]

Kurczewski, F. E., and M. G. Spofford. 1985. Observations on the nesting and unique cachement behavior of *Calicurgus hyalinatus* (Hymenoptera: Pompilidae). *Great Lakes Entomol.* 18:41–44. [338–39]

————. 1986. Observations on the behaviors of some Scoliidae and Pompilidae (Hymenoptera) in Florida. *Fla. Entomol.* 69:636–44. [57, 338]

————. 1987. Further observations on the nesting behavior of *Liris argentatus* (Hymenoptera: Sphecidae). *Great Lakes Entomol.* 20:121–25. [87]

Kurt, F., and A. G. Jayasuriya. 1968. Notes on a dead bear. *Loris* 11:182–83. [12, 226, 235]

Kuznetsov, N. I. 1959. On the ecology of the nutcracker in the mid-Urals. Translated by Leon Kelso. *Byulleten Moskovskovo Obshchestva Ispytatelei Prirody. Otdel Biologicheskii* 46(2):132–33. [93, 302–5]

Lack, D. 1968. *Ecological adaptations for breeding in birds.* London: Methuen. [32]

Lal, H., M. C. Nautiyal, and R. M. Sharma. 1984. Walnut (*Juglans regia*) seed germination. I. Effect of planting depth and seed position in soil. *Prog. Hort.* 16:6–8. [187]

Lane, P. A., and J. R. Duncan. 1987. Observations on northern hawk-owls nesting in Roseau County. *Loon* 59:165–74. [289–90]

Lanier, D. L., D. Q. Estep, and D. A. Dewsbury. 1974. Food hoarding in muroid rodents. *Behav. Biol.* 11:177–87. [138, 140–41]

Lanner, R. M. 1981. *The piñon pine: A natural and cultural history.* Reno: University of Nevada Press. [72]

————. 1982a. Adaptations of whitebark pine seed dispersal by Clark's nutcracker. *Can. J. Forest Res.* 12:391–402. [71–72, 182, 195–96]

————. 1982b. Avian seed dispersal as a factor in the ecology and evolution of limber and whitebark pines. In *Proc. 6th North Am. For. Biol. Workshop,* 15–48. Edmonton, Alberta, Canada. [71–73, 195–96, 206–8]

————. 1983. The expansion of singleleaf pinon in the Great Basin. In *The archaeology of Monitor Valley, 2, Gatecliff Shelter,* Anthropological Papers, Vol. 59, Part 1, edited by D. H. Thomas, 167–71. New York: American Museum of Natural History. [215]

————. 1988. Dependence of Great Basin bristlecone pine on Clark's nutcracker for regeneration at high elevations. *Arctic Alpine Res.* 20:358–62. [206–8, 211]

Lanner, R. M., H. E. Hutchins, and H. A. Lanner. 1984. Bristlecone pine and Clark's nutcracker: Probable interaction in the White Mountains, California. *Great Basin Nat.* 44:357–60. [192, 208, 211, 301]

Lanner, R. M., and S. B. Vander Wall. 1980. Dispersal of limber pine seed by Clark's nutcracker. *J. Forest.* 78:637–39. [181, 192, 197, 207, 210, 302–4]

Lantz, D. E. 1907. An economic study of field mice (genus *Microtus*). *USDA Bull. Biol. Surv.* 31:1–64. [270]

Laskey, A. R. 1942. Blue jays burying food. *Migrant* 13:72–73. [298]

———. 1943. The seeds buried by blue jays. *Migrant* 14:58. [203]

La Tourrette, J. E., J. A. Young, and R. A. Evans. 1971. Seed dispersal in relation to rodent activities in seral big sagebrush communities. *J. Range Mgmt.* 24:118–20. [181, 202, 209, 255]

Lavender, D. P., and W. H. Engstrom. 1956. Viability of seeds from squirrel-cut Douglas fir cones. *Oregon State Board Forest. Res., Note* 27:1–19. [241]

Lavigne, R. J. 1969. Bionomics and nest structure of *Pogonomyrmex occidentalis* (Hymenoptera: Formicidae). *Ann. Entomol. Soc. Am.* 62:1166–75. [3, 21, 330–32]

Law, J. E. 1929. Another Lewis' woodpecker stores acorns. *Condor* 31:233–38. [296–97]

Lawhon, D. K., and M. S. Hafner. 1981. Tactile discriminatory ability and foraging strategies in kangaroo rats and pocket mice. *Oecologia* 50:303–9. [25, 252, 254, 256]

van Lawick-Goodall, J., and H. van Lawick-Goodall. 1970. *Innocent killers.* London: Collins. [62, 226, 228]

Lawrence, L. de K. 1968. Notes on hoarding nesting material, display, and flycatching in the gray jay (*Perisoreus canadensis*). *Auk* 85:139. [298]

Layne, J. N. 1954. The biology of the red squirrel, *Tamiasciurus hudsonicus loquax* (Bangs), in central New York. *Ecol. Monogr.* 24:227–67. [199, 243]

Lea, S. E. G., and R. M. Tarpy. 1986. Hamsters' demand for food to eat and hoard as a function of deprivation and cost. *Anim. Behav.* 34:1759–68. [121]

Leech, H. B. 1935. The family history of *Nicrophorus conversator* Walker. *Proc. Entomol. Soc. Br. Col.* 31:36–40. [327]

Le Louarn, H. 1974. Etude par marquage et recaptures du campagnol des champs et du campagnol de Fatio et montague. *Mammalia* 38:54–63. [20, 270, 273]

Leonard, H. 1978. Forest raven caching food. *Aust. Bird Watcher* 7:212. [305]

Levin, D. A., and H. W. Kerster. 1974. Gene flow in seed plants. In *Evolutionary biology*, vol. 7, edited by T. Dobzhansky, M. K. Hecht, and W. C. Steere, 139–220. New York: Plenum Press. [209]

Lewis, A. R. 1982. Selection of nuts by gray squirrels and optimal foraging theory. *Am. Midl. Nat.* 107:250–57. [76]

Lewis, C. F. 1978. Little raven caching food. *Aust. Bird Watcher* 7:272. [305]

Lewis, I. M. 1911. The seedlings of *Quercus virginiana*. *Plant World* 14:119–23. [76, 202]

Lewis, T. 1923. Coal tit hiding beech-nuts. *Br. Birds* 16:216–17. [311]

Licklider, L. C., and J. C. R. Licklider. 1950. Observations on the hoarding behavior of rats. *J. Comp. Physiol. Psychol.* 43:129–34. [120, 123, 140]

Ligon, J. D. 1968. The biology of the elf owl, *Micrathene whitneyi. Misc. Publ. Mus. Zool. Univ. Mich.* 136:1–70. [49, 288–90]

———. 1974. Green cones of the piñon pine stimulate late summer breeding in the piñon jay. *Nature* 250:80–82. [28]

———. 1978. Reproductive interdependence of piñon jays and piñon pines. *Ecol. Monogr.* 48:111–26. [18, 27–28, 32, 72–73, 181, 194, 197–98, 202, 209–10, 216, 304–5, 364]

Ligon, J. D., and D. J. Martin. 1974. Piñon seed assessment by the piñon jay, *Gymnorhinus cyanocephalus. Anim. Behav.* 22:421–29. [79, 83, 181, 196, 301]

Lima, S. 1986. Predation risk and unpredictable feeding conditions: Determinants of body mass in birds. *Ecology* 67:377–85. [14, 26, 86]

Lindsey, G. D. 1977. Evaluation of control agents for conifer seed protection. In *Test Methods for Vertebrate Pest Control and Management Materials,* edited by W. B. Jackson and R. E. Marsh, 5–13. Philadelphia: American Society for Testing and Materials, Spec. Tech. Publ. 625. [194]

Lindzey, G., and M. Manosevitz. 1964. Hoarding in the mouse. *Psychonom. Sci.* 1:35–36. [112]

Linhart, Y. B., and D. F. Tomback. 1985. Seed dispersal by nutcrackers causes multi-trunk growth form in pines. *Oecologia* 67:107–10. [208]

Linsdale, J. M. 1937. The natural history of the magpies. *Pac. Coast. Avifauna* 25:1–234. [306]

———. 1946. *The California ground squirrel.* Berkeley: University of California Press. [247–48]

Linsdale, J. M., and L. P. Tevis, Jr. 1951. *The dusky-*

footed wood rat. Berkeley: University of California Press. [261]

Linsley, E. G., and C. D. Michener. 1962. Brief notes on the habits of *Protoxaea* (Hymenoptera: Andrenidae). *J. Kans. Entomol. Soc.* 35:385–89. [347]

Lisk, R. D., L. A. Ciaccio, and C. Catanzaro. 1983. Mating behaviour of the golden hamster under seminatural conditions. *Anim. Behav.* 31:659–66. [33]

Litvinov, N. I., and G. I. Vasil'ev. 1973. Storage of food by the reddish-gray (large-toothed red-backed) vole. *Soviet J. Ecol.* 4:74–75. [20, 269–72]

Lloyd, H. G. 1968. Observations on nut selection by a hand-reared grey squirrel (*Sciurus carolinensis*). *J. Zool., London* 155:240–44. [182]

Lockard, R. B. 1968. The albino rat: A defensible choice or a bad habit. *Am. Psychol.* 23:734–42. [144]

Lockard, R. B., and J. S. Lockard. 1971. Seed preference and buried seed retrieval of *Dipodomys deserti. J. Mamm.* 52:219–21. [171, 185–86]

Lockley, R. M. 1938. The little owl inquiry and the Stockholm storm petrels. *Br. Birds* 31:278–79. [289]

Lockner, F. R. 1972. Experimental study of food hoarding in the red-tailed chipmunk (*Eutamias ruficaudus*). *Z. Tierpsychol.* 31:410–18. [120, 144, 238]

Löhrl, H. 1958. Das Verhalten des Kleibers (*Sitta europaea caesia* Wolf). *Z. Tierpsychol.* 15:191–252. [315]

Long, C. A. 1976. Evolution of mammalian cheek pouches and a possible discontinuous origin of a higher taxon (*Geomyoidea*). *Am. Nat.* 110:1093–97. [54, 250]

Long, D. A. C. 1950. Concealment of food by coal-tit. *Br. Birds* 43:335–36. [89]

Lore, R., and R. Moyer. 1973. Early feeding experience and food hoarding in rats. *Devel. Psychol.* 8:313. [117]

Lorenz, K. Z. 1970. *Studies in animal and human behaviour,* vol. 1. London: Methuen. [306]

Lorenz, K. Z., and U. von Saint Paul. 1968. Die Entwicklung des Spiessens und Klemmens bei den drei Wurgerarten *Lanius collurio, L. senator* und *L. excubitor. J. Ornithol.* 109:137–56. [115, 317, 319]

Loukashkin, A. S. 1940. On the pikas of north Manchuria. *J. Mamm.* 21:402–5. [83, 279–82]

Lovegrove, B., and J. U. M. Jarvis. 1986. Coevolution between mole-rats (Bathyergidae) and a geophyte, *Micranthus* (Iridaceae). *Cimbebasia* 8:79–85. [205, 268]

Ludescher, F. B. 1980. Feeding and caching of seeds in willow tits *Parus montanus* in the course of the year under constant feeding conditions. *Ökol. Vögel.* 2:135–44. [132–33, 313]

Lyall-Watson, M. 1964. The ethology of food-hoarding in mammals. Ph.D. diss., University of London. [277]

Lyman, C. P. 1954. Activity, food consumption and hoarding in hibernators. *J. Mamm.* 35:545–52. [26, 263]

Lynch, G. R., C. B. Lynch, and H. Dingle. 1973. Photoperiodism and behavior in a small mammal. *Nature* 244:46–47. [132]

Lyons, D., and J. A. Mosher. 1982. Food caching by raptors and caching of a nestling by the broad-winged hawk. *Ardea* 70:217–19. [31, 87, 285–86, 288–89]

Maccarone, A. D., and W. A. Montevecchi. 1981. Predation and caching of sea birds by red foxes (*Vulpes vulpes*) on Baccalieu Island, Newfoundland. *Can. Field-Nat.* 95:352–53. [62, 88, 228, 230]

McAdoo, J. K., C. C. Evans, B. A. Roundy, J. A. Young, and R. A. Evans. 1983. Influence of heteromyid rodents on *Oryzopis hymenoides* germination. *J. Range Mgmt.* 36:61–64. [181, 183, 186, 191–92, 204, 207–8, 254–55]

McCain, G., B. L. Garret, C. Reed, G. Mead, and R. Kuenstler. 1964. Effect of deprivation on hoarding of objects other than the deprived material. *Anim. Behav.* 12:409–15. [120]

McCarty, R., and C. H. Southwick. 1975. Food hoarding by the southern grasshopper mouse (*Onychomys torridus*) in laboratory enclosures. *J. Mamm.* 56:708–12. [126–27, 259]

McCleary, R. A., and C. T. Morgan. 1946. Food hoarding in rats as a function of environmental temperature. *J. Comp. Psychol.* 39:371–78. [135–36]

MacClintock, D. 1970. *Squirrels of North America.* New York: Van Nostrand Reinhold. [244, 248]

McComb, A. L. 1934. The relation between acorn weight and the development of one year chestnut oak seedlings. *J. Forest* 32: 479–84. [201]

McCook, H. C. 1879. *The natural history of the agricultural ant of Texas.* Philadelphia: Academy of Natural Sciences of Philadelphia. [4, 329]

———. 1882. *The honey ants of the Garden of the Gods and the occident ants of the American plains.* Philadelphia: J. B. Lippincott. [330, 332, 334–36]

McCord, C. M., and J. E. Cardoza. 1982. Bobcat and lynx. In *Wild mammals of North America: Biology, management, economics,* edited by J. A. Chapman and G. A. Feldhamer, 728-66. Baltimore: Johns Hopkins University Press. [226, 233]

McCord, F. 1941. The effect of frustration on hoarding in rats. *J. Comp. Physiol. Psychol.* 32: 531–41. [117, 120]

McCorquodale, D. B. 1986. Digger wasp (Hymenoptera: Sphecidae) provisioning flights as a defence against a nest parasite, *Senotainia trilineata* (Diptera: Sarcophagidae). *Can. J. Zool.* 64: 1620–27. [96]

Macdonald, D. W. 1976. Food caching by red foxes and some other carnivores. *Z. Tierpsychol.* 42: 170–85. [62, 87, 104, 119, 173–75, 225–30]

———. 1978. The hungry fox. *Wildlife, London* 20: 320–24. [227]

McDonald, P. M. 1969. Silvical characteristics of California black oak (*Quercus kelloggii* Newb.). U.S. Department of Agriculture, Forest Service, *Res. Paper PSW* 53: 1–20. [189, 200]

MacDonald, W. H. 1965. Bears and people. *Land For. Wildl.* 8: 16–32. [226]

MacDougall, R. S. 1942. The mole. Its life-history, habits, and economic importance. *Trans. Highl. Agric. Soc. Scot.* 54: 80–107. [20, 222, 223]

MacKay, W. P. 1981. A comparison of the nest phenologies of three species of *Pogonomyrmex* harvester ants (Hymenoptera: Formicidae). *Psyche* 88: 25–74. [179, 331–32]

MacKay, W. P., and E. E. MacKay. 1984. Why do harvester ants store seeds in their nests? *Sociobiology* 9: 31–47. [10, 21, 67, 179, 331–33]

McKeever, S. 1964. Food habits of the pine squirrel in northeastern California. *J. Wildl. Mgmt.* 28: 402–4. [243]

McKelvey, R. K., and M. H. Marx. 1951. Effects of infantile food and water deprivation on adult hoarding in the rat. *J. Comp. Physiol. Psychol.* 44: 423–30. [117]

McLean, I. G., and A. J. Towns. 1981. Differences in weight changes and the annual cycle of male and female arctic ground squirrels. *Arctic* 34: 249–54. [24, 127, 248, 363]

MacMillen, R. E. 1983. Adaptive physiology of heteromyid rodents. *Great Basin Nat. Memoirs* 7: 65–76. [25]

McNair, D. B. 1982. Tufted titmice store acorns. *Oriole* 47: 12–13. [310]

———. 1983. Brown-headed nuthatches store pine seeds. *Chat* 47: 47–48. [94, 105, 315]

———. 1985. An auxiliary with a mated pair and food-caching behavior in the fish crow. *Wilson Bull.* 97: 123–25. [30, 126, 306–7]

McQuade, D. B., E. H. Williams, and H. B. Eichenbaum. 1986. Cues used for localizing food by the gray squirrel (*Sciurus carolinensis*). *Ethology* 72: 22–30. [173]

MacRoberts, M. H. 1970. Notes on the food habits and food defense of the acorn woodpecker. *Condor.* 72: 196–204. [97, 104, 297]

———. 1974. *Acorns, woodpeckers, grubs, and scientists.* Pac. Disc. 27(5): 9–15. [297]

———. 1975. Food storage and winter territory in red-headed woodpeckers in northwestern Louisiana. *Auk.* 92: 382–85. [90, 293]

MacRoberts, M. H., and B. R. MacRoberts. 1976. Social organization and behavior of the acorn woodpecker in central coastal California. *Ornithol. Monogr.* 21: 1–115. [32, 62, 64, 67, 69, 79, 93, 105, 179, 189, 293–96]

———. 1985. Gila woodpecker stores acorns. *Wilson Bull.* 97: 571. [179, 292]

Madison, D. M. 1984. Group nesting and its ecological and evolutionary significance in overwintering microtine rodents. In *Winter ecology of small mammals,* edited by J. F. Merritt, 267–74. Pittsburgh: Carnegie Museum of Natural History, Spec. Publ. 10 [26]

Madison, D. M., R. FitzGerald, and W. McShea. 1984. Dynamics of social nesting in overwintering meadow voles: Possible consequences for population cycling. *Behav. Ecol. Sociobiol.* 15: 9–17. [26]

Madson, J. 1964. *Gray and fox squirrels.* East Alton, Ill.: Olin Mathieson Chemical Corp. [185]

Magoun, A. J. 1979. Summer scavenging activity in northeastern Alaska. U.S. Department of the Interior, National Park Service, *Trans. Proc. Ser.* 5:335–40. [62, 67, 226–28, 231, 305–6]

Mailliard, J. 1927. The birds and mammals of Modoc County, California. *Proc. Calif. Acad. Sci. Ser. 4,* 16:261–359. [66]

———. 1931. Redwood chickaree testing and storing hazel nuts. *J. Mamm.* 12:68–70. [83, 182, 199, 243]

Malcolm, J. R. 1980. Food caching by African wild dogs (*Lycaon pictus*). *J. Mamm.* 61:743–44. [45, 62, 67, 226–27, 230]

Malhi, C. S. 1986. Burrowing and hoarding behaviour of *Bandicota bengalensis* (Wardi) in wheat crop at Srinager (Garhwal)-Himalayas. *J. Environ. Biol.* 7:197–200. [273]

———. 1987. Hoarding behaviour in common myna (*Acridotheres tristis*). *Z. Angew Zool.* 74:247–48. [319]

Malyshev, S. I. 1968. *Genesis of the Hymenoptera and the phases of their evolution.* London: Methuen. [51–53, 337]

Manley, G. V. 1971. A seed-caching carabid (Coleoptera). *Ann. Entomol. Soc. Am.* 64:1474–75. [326]

Manosevitz, M. 1965. Genotype, fear, and hoarding. *J. Comp. Physiol. Psychol.* 60:412–16. [112–14, 138]

———. 1967. Hoarding and inbred strains of mice. *J. Comp. Physiol. Psychol.* 63:148–50. [113]

———. 1970. Hoarding: An exercise in behavioral genetics. *Psychol. Today* 4(8):56–58, 76. [113, 117]

Manosevitz, M., R. B. Campenot, and C. F. Swencionis. 1968. Effects of enriched environment upon hoarding. *J. Comp. Physiol. Psychol.* 66:319–24. [113–14, 117]

Manosevitz, M., and G. Lindzey. 1967. Genetics of hoarding: A biometrical analysis. *J. Comp. Physiol. Psychol.* 63:142–44. [113, 145]

———. 1970. Genetic variation and hoarding. In *Contributions to Behavior-Genetic Analysis: The Mouse as a Prototype,* edited by G. Lindzey and D. D. Thiessen, 91–113. New York: Appleton-Century-Crofts. [112]

Marriott, B. M., and E. A. Salzen. 1979. Food-storing behavior in captive squirrel monkeys (*Saimiri sciureus*). *Primates* 20:307–11. [104, 224]

Martin, I. G. 1981. Venom of the short-tailed shew (*Blarina brevicauda*) as an insect immobilizing agent. *J. Mamm.* 62:189–92. [86, 219]

———. 1984. Factors affecting food hoarding in the short-tailed shew *Blarina brevicauda. Mammalia* 48:65–72. [220–22]

Martin, J. W., and J. C. Kroll. 1975. Hoarding of corn by golden-fronted woodpeckers. *Wilson Bull.* 87:553. [292]

Martin, P. 1971. Movements and activities of the mountain beaver (*Aplodontia rufa*). *J. Mamm.* 52:717–23. [48, 236]

Martin, R. F. 1971. The canyon wren *Catherpes mexicanus* raiding food stores of a trypoxylid wasp. *Auk* 88:677. [89, 94]

Marx, M. H. 1950a. A stimulus-response analysis, of the hoarding habit in the rat. *Psychol. Rev.* 57:80–93. [117, 125]

———. 1950b. Experimental analysis of the hoarding habit in the rat. I. Preliminary observations. *J. Comp. Physiol. Psychol.* 43:295–308. [112, 117]

———. 1951. Experimental analysis of the hoarding habit of the rat. II. Terminal reinforcement. *J. Comp. Physiol. Psychol.* 44:168–77. [117, 176]

———. 1952. Infantile deprivation and adult behavior in the rat: Retention of increased rate of eating. *J. Comp. Physiol. Psychol.* 45:43–49. [117]

———. 1957. Experimental analysis of the hoarding habit in the rat. III. Terminal reinforcement under low drive. *J. Comp. Physiol. Psychol.* 50:168–71. [176]

Marx, M. H. and A. J. Brownstein. 1957. Experimental analysis of the hoarding habit in the rat. IV. Terminal reinforcement followed by high drive at test. *J. Comp. Physiol. Psychol.* 50:617–20. [176]

Marx, M. H., S. Iwahara, and A. J. Brownstein. 1957. Hoarding behavior in the hooded rat as a function of varied alley illumination. *J. Genet. Psychol.* 90:213–18. [131]

Maschwitz, U., M. Hahn, and P. Schönegge. 1979. Paralysis of prey in ponerine ants. *Naturwissenschaften* 66:213–14. [87, 336–37]

Masuda, A. and T. Oishi. 1988. Effects of photo-period and temperature on body weight, food intake, food storage, and pelage color in the Djungarian hamster. *J. Exp. Zool.* 248:133–39. [124, 132]

Matson, J. O., and D. P. Christian. 1977. A laboratory study of seed caching in two species of *Liomys* (Heteromyidae). *J. Mamm.* 58:670–71. [135]

Mattes, H. 1982. Die Lebensgemeinschaft von Tannenhäher und Arve. *Swiss Fed. Inst. For. Res. Rept.* 241:1–74. [63, 71–73, 91, 166, 179, 181, 183, 194–95, 207, 211, 215, 301–5]

Matthews, E. G. 1972. A revision of the Scarabaeine dung beetles of Australia, I. Tribe Onthophagini. *Aust. J. Zool.* (Suppl. Ser.) 9:3–330. [323]

———. 1974. A revision of the Scarabaeine dung beetles of Australia, II. Tribe Scarabaeini. *Aust. J. Zool.* (Suppl. Ser.) 24:1–211. [322–23, 325]

May, J. T., and H. G. Posey. 1958. The effect of radiation by Cobalt-60 gamma rays on germination of slash pine seed. *J. For.* 56:854–55. [191]

Mayer, W. V. 1953. A preliminary study of the Barrow ground squirrel, *Citellus parryii barrowensis*. *J. Mamm.* 34:334–44. [247]

Mayer, W. V., and E. T. Roche. 1954. Developmental patterns in the Barrow ground squirrel *Spermophilus undulatus barrowensis. Growth* 18:53–69. [24, 248]

Mayr, E. 1970. *Populations, species, and evolution.* Cambridge, Mass.: Belknap Press of Harvard University Press. [113]

Mech, L. D. 1967. Telemetry as a technique in the study of predation. *J. Wildl. Mgmt.* 31:492–96. [228]

Mech, L. D. 1970. *The wolf.* New York: Natural History Press. [12, 226–28]

Melchior, E. 1975. 28 Feldmäuse als Vorrat in Mäusebussardhorst (*Buteo buteo*). *Regulus* 10:402. [285]

Mellanby, K. 1968. The effects of some mammals and birds on regeneration of oak. *J. Appl. Ecol.* 5:359–66. [209]

Mendelson, J., and G. Maul. 1974. Carrying behavior induced by shuttle-box self-stimulation in rats: Effect of food deprivation on object preference. *Behav. Biol.* 10:199–209. [120]

Merriam, C. H. 1910. The California ground squirrel. U.S. Bureau of Biological Survey, *Circular* 76:1–15. [247]

Merritt, J. F. 1986. Winter survival adaptations of the short-tailed shrew (*Blarina brevicauda*) in an Appalachian montane forest. *J. Mamm.* 67:450–64. [220–22]

Mewaldt, L. R. 1956. Nesting behavior of the Clark nutcracker. *Condor* 58:3–23. [28, 32, 79, 303, 305]

Meylan, A. 1977. Fossorial forms of the water vole *Arvicola terrestris* (L.) in Europe. *Eppo Bull.* 7:209–21. [270]

Mezhennyi, A. A. 1964. Biology of the nutcracker (*Nucifraga caryocatactes macrorhynchus*) in southern Yakutiya. *Zool. Zh.* 43(11):1679–87. [32, 303–5]

Miceli, M. O., and C. W. Malsbury. 1982. Availability of a food hoard facilitates maternal behaviour in virgin female hamsters. *Physiol. Behav.* 28:855–56. [33, 127]

Michael, C. W. 1926. Acorn storing methods of the California and Lewis' woodpeckers. *Condor* 28:68–69. [296]

Michener, C. D. 1961. Observations on the nests and behavior of *Trigona* in Australia and New Guinea (Hymenoptera, Apidae). *Am. Mus. Novit.* 2026:1–46. [348, 355]

———. 1964. Evolution of the nests of bees. *Am. Zool.* 4:227–39. [50, 52]

———. 1974. *The social behavior of the bees: A comparative study.* Cambridge, Mass.: Belknap Press of Harvard University Press. [7, 33, 52–53, 345, 354–57, 359]

Michener, C. D., M. L. Winston, and R. Jander. 1978. Pollen manipulation and related activities and structures in bees of the family Apidae. *Univ. Kans. Sci. Bull.* 51:575–601. [59, 79, 360]

Michener, H., and J. R. Michener. 1945. California jays, their storage and recovery of food, and observations at one nest. *Condor* 47:206–10. [203, 298]

Millar, J. S. 1979. Energetics of lactation in *Peromyscus maniculatus. Can. J. Zool.* 57:1015–19. [33, 127]

Millar, J. S., and F. C. Zwickel. 1972a. Characteristics and ecological significance of hay piles of pikas. *Mammalia* 36:657–67. [2, 22, 60–61, 83, 111, 279–82]

———. 1972b. Determination of age, age structure, and mortality of the pika, *Ochotona princeps* (Richardson). *Can. J. Zool.* 50:229–32. [22]

Miller, A. H. 1931. Systematic revision and natural history of the American shrikes (*Lanius*). *Univ. Calif. Publ. Zool.* 38:11–242. [87, 317–18]

———. 1937. A comparison of behavior or certain North American and European shrikes. *Condor* 39:119–22. [317]

———. 1963. Seasonal activity and ecology of the avifauna of an American equatorial cloud forest. *Univ. Calif. Publ. Zool.* 66:1–78. [295]

Miller, G. A. 1945. Concerning the goal of hoarding behavior in the rat. *J. Comp. Psychol.* 38:209–12. [142]

Miller, G. A., and L. Postman. 1946. Individual and group hoarding in rats. *Am. J. Psychol.* 59:652–68. [141–42]

Miller, G. A., and P. Viek. 1944. An analysis of the rat's response to unfamiliar aspects of the hoarding situation. *J. Comp. Psychol.* 37:221–31. [50, 137–38]

Miller, R. C. 1950. Oldest bird nest. *Pac. Disc.* 3(4):29–30. [61]

Milne, C. P. 1985. A heritability estimate of honey bee hoarding behavior. *Apidologie* 16:413–20. [112–13, 145]

Milne, C. P., Jr. and K. J. Pries. 1986. Honeybees with larger corbiculae carry larger pollen pellets. *J. Apic. Res.* 25:53–54. [58]

Milne, L. J., and M. J. Milne. 1944. Notes on the behavior of burying beetles (*Nicrophorus* spp.). *J. New York Entomol. Soc.* 52:311–27. [57, 327]

———. 1976. The social behavior of burying beetles. *Sci. Amer.* 235(2):84–89. [57, 327–28]

Mitchell, P. S., and J. H. Hunt. 1984. Nutrient and energy assays of larval provisions and feces in the black and yellow mud dauber, *Sceliphron caementarium* (Drury)(Hymenoptera: Sphecidae). *J. Kans. Entomol. Soc.* 57:700–704. [37–38]

Miyaki, M. 1987. Seed dispersal of the Korean pine, *Pinus koraiensis,* by the red squirrel, *Sciurus vulgaris. Ecol. Res.* 2:147–57. [183, 194]

Miyaki, M., and K. Kikuzawa. 1988. Dispersal of *Quercus mongolica* acorns in a broadleaved decid-

uous forest. 2. Scatterhoarding by mice. *For. Ecol. Mgmt.* 25:9–16. [183, 191, 200]

Moggridge, J. T. 1873. *Harvesting ants and trapdoor spiders: Notes and observations on their habits and dwellings.* London: L. Reeve. [329]

Mohana Rao, A. M. K. 1980. Demography and hoarding among the lesser bandicoot rat, *Bandicota bengalensis* (Gray and Hardwicke, 1835) in rice fields. *Saugetierkd. Mitt.* 28:312–14. [29, 274–76]

Mohr, C. O. 1943. Cattle droppings as ecological units. *Ecol. Monogr.* 13:276–98. [323]

Moller, H. 1982. Red squirrel ecology and behaviour. *Scott. Wildl.* 18:18–23 [244, 246]

———. 1983. Foods and foraging behaviour of red (*Sciurus vulgaris*) and grey (*Sciurus carolinensis*) squirrels. *Mamm. Rev.* 13:81–98. [84, 245–46]

Monson, G. 1943. Food habits of the banner-tailed kangaroo rat in Arizona. *J. Wildl. Mgmt.* 7:98–102. [253, 256]

Monson, G., and W. Kessler. 1940. Life history notes on the banner-tailed kangaroo rat, Merriam's kangaroo rat, and the white-throated wood rat in Arizona and New Mexico. *J. Wildl. Mgmt.* 4:37–43. [254, 256]

Monteith, G. H., and R. I. Storey. 1981. The biology of *Cephalodesmius,* a genus of dung beetles which synthesizes "dung" from plant material (Coleoptera: Scarabaeidae: Scarabaeinae). *Memoirs Queensl. Mus.* 20:253–71. [323, 325]

Moore, J. C. 1957. The natural history of the fox squirrel, *Sciurus niger shermani. Bull. Am. Mus. Nat. Hist.* 113:1–71. [171, 244]

Moreno, J., A. Lundberg, and A. Carlson. 1981. Hoarding of individual nuthatches *Sitta europaea* and marsh tits *Parus palustris. Holarctic Ecol.* 4:263–69. [44, 105, 126, 157, 179, 310–12, 314]

Morgan, C. T. 1945. The statistical treatment of hoarding data. *J. Comp. Psychol.* 38:247–56. [112]

———. 1947. The hoarding instinct. *Psychol. Rev.* 54:335–41. [117]

Morgan, C. T., E. Stellar, and O. Johnson. 1943. Food-deprivation and hoarding in rats. *J. Comp. Psychol.* 35:275–95. [112, 119–21, 123, 125]

Morris, B. 1962. A denizen of the evergreen forest. *Afr. Wild Life* 16:117–21. [89, 264]

———. 1963. Notes on the giant rat (*Cricetomys gambianus*) in Nyasaland. *Afr. Wild Life* 17:102–7. [89, 264]

Morris, D. 1962. The behaviour of the green acouchi (*Myoprocta pratti*) with special reference to scatter hoarding. *Proc. Zool. Soc. London* 139:701–31. [3, 119, 140, 277–79]

Morrison, F. B. 1946. *Feeds and feeding*, 20th ed. Ithaca, N.Y.: Morrison. [236]

Morrison, P. 1960. Some interrelations between weight and hibernation function. *Bull. Mus. Comp. Zool.* 124:75–91. [26, 63]

Morton, E. S. 1973. On the evolutionary advantage and disadvantages of fruit eating in tropical birds. *Am. Nat.* 107:8–22. [32]

Morton, M. L., C. S. Maxwell, and C. E. Wade. 1974. Body size, body composition, and behavior of juvenile Belding ground squirrels. *Great Basin Nat.* 34:121–34. [247]

Morton, M. L., and P. W. Sherman. 1978. Effects of a spring snowstorm on behavior, reproduction, and survival of Belding's ground squirrels. *Can. J. Zool.* 56:2578–90. [25, 248]

Morton, S. R., R. Hinds, and R. E. MacMillen. 1980. Cheek pouch capacity in heteromyid rodents. *Oecologia* 46:143–46. [54–55, 252]

Morton, S. R., and R. E. MacMillen. 1982. Seeds as sources of preformed water for desert-dwelling granivores. *J. Arid Environ.* 5:61–67. [25, 85, 256]

Moskovits, D. 1978. Winter territorial and foraging behavior of red-headed woodpeckers in Florida. *Wilson Bull.* 90:521–35. [62, 83, 293, 296–97]

Mueller, H. C. 1974. Food caching behavior in the American kestrel (*Falco sparverius*). *Z. Tierpsychol.* 34:105–14. [13, 94, 120, 149, 284–85, 287]

Mumford, R. E., and R. L. Zusi. 1958. Notes on movements, territory, and habitat of wintering saw-whet owls. *Wilson Bull.* 70:188–91. [47, 289–90]

Mumme, R. L., and A. de Queiroz. 1985. Individual contributions to cooperative behaviour in the acorn woodpecker: Effects of reproductive status, sex, and group size. *Behaviour* 95:290–313. [68, 294–96]

Munger, T. T. 1917. Western yellow pine in Ore-gon. U.S. Department of Agriculture, *Bulletin* 418:1–48. [193, 206–8, 211]

Murie, A. 1936. Following fox trails. University of Michigan, Museum of Zoology, *Misc. Publ.* 32:1–45. [12, 89, 149, 225, 229]

———. 1944. *The wolves of Mount McKinley*. U.S. Department of the Interior, National Park Service, *Fauna Ser.* 5:1–238. [226–27]

———. 1961. *A naturalist in Alaska*. New York: Devin-Adair. [20, 83, 269–72]

Murie, J. O. 1977. Cues used for cache-finding by agoutis (*Dasyprocta punctata*). *J. Mamm.* 58:95–96. [171, 277]

Murie, O. J. 1927. The Alaska red squirrel providing for winter. *J. Mamm.* 8:37–40. [243]

Murphy, M. R. 1971. Natural history of the Syrian golden hamster—A reconnaissance expedition. *Am. Zool.* 11:632. [26, 263]

Muul, I. 1968. Behavioral and psychological influences on the distribution of the flying squirrel, *Glaucomys volans*. University of Michigan, Museum of Zoology, *Misc. Publ.* 134:1–66. [172, 248–49]

———. 1970. Day length and food caches. In *Field studies in natural history*, 78–86. New York: Van Nostrand Reinhold. [26, 62, 64, 132, 136, 148, 172, 199, 248–49]

Myllek, G. 1986. Auf Grimbarts. *Fahrte. Kosmos Stuttg.* 82:11–17. [226]

Mysterud, I. 1973. Behaviour of the brown bear (*Ursus arctos*) at moose kills. *Norw. J. Zool.* 21:267–72. [88, 226, 231]

———. 1975. Sheep killing and feeding behaviour of the brown bear (*Ursus arctos*) in Trysil, south Norway 1973. *Norw. J. Zool.* 23:243–60. [87, 226, 231]

Nagorsen, D. W. 1987. Summer and winter food caches of the heather vole, *Phenacomys intermedius*, in Quetico Provincial Park, Ontario. *Can. Field-Nat.* 101:82–85. [48, 272]

Nakamura, H., and Y. Wako. 1988. Food storing behaviour of willow tit *Parus montanus*. *J. Kamashina Inst. Ornithol.* 20:21–36. [17, 170, 311, 314]

Naumov, N. P., and V. S. Lobachev. 1975. Ecology of desert rodents of the USSR (Jerboas and Gerbils). In *Rodents in desert environments* edited by I. Prakash and P. K. Ghosh, 465–598. The Hague: Dr. W. Junk Publishers. [20, 83, 141, 264–65, 277]

Neff, J. L., and B. B. Simpson. 1981. Oil-collecting structures in the Anthophoridae (Hymenoptera): morphology, function, and use in systematics. *J. Kans. Entomol. Soc.* 54:95–123. [59, 354]

Nel, J. A. J. 1967. Burrow systems of *Desmodillus auricularis* in the Kalahari Gemsbok National Park. *Koedoe* 10:118–21. [264]

Nellis, C. H., and L. B. Keith. 1968. Hunting activities and success of lynxes in Alberta. *J. Wildl. Mgmt.* 32:718–22. [12, 225–26, 233, 235]

Nelson, E. W. 1893. Description of a new species of *Arvicola,* of the Mynomes group, from Alaska. *Proc. Biol. Soc. Wash.* 8:139–41. [270]

Nevo, E. 1961. Observations on Israeli populations of the mole rat *Spalax e. ehrenbergi* Nehring 1898. *Mammalia* 25:127–44. [20, 107–8, 179, 191, 205, 266]

Nichols, J. T. 1927. How do squirrels find buried nuts? *J. Mamm.* 8:55–57. [171]

Nielsen, B. O. 1977. Beech seed as an ecosystem component. *Oikos* 29:268–74. [189]

Nielsen, R. R. 1973. Dehusking black walnuts controls rodent pilferage. *Tree Planter's Notes* 24(3): 33. [172]

Nikolskiy, A. A., N. P. Guricheva, and P. P. Dmitriyev. 1984. Winter reserves of the dauric haymaker in steppe pastures. *Byull. Mosk. Obshch. Ispyt. Prir. (Otd. Biol.)* 89(6):9–22. [281]

Nilsson, S. G. 1985. Ecological and evolutionary interaction between reproduction of beech *Fagus silvatica* and seed eating animals. *Oikos* 44:157–64. [179, 184, 189, 202, 211, 214]

Nixon, C. M., M. W. McClain, and R. W. Donohoe. 1975. Effects of hunting and mast crops on a squirrel population. *J. Wildl. Mgmt.* 39:1–25. [20]

Nixon, C. M., and M. W. McClain, and L. P. Hansen. 1980. Six years of hickory seeds yields in southeastern Ohio. *J. Wildl. Mgmt.* 44:534–39. [189, 367]

Nogueira-Neto, P. 1954. Notas bionomicas sobre meliponineos, Ill-Sobre a enxameagem. *Arq. Mus. Nacional, Rio de Janeiro* 42:219–452. [33]

Norden, B. B. 1984. Nesting biology of *Anthophora abrupta* (Hymenoptera: Anthophoridae). *J. Kans. Entomol. Soc.* 57:243–62. [353–54]

Norden, B. B., S. W. T. Batra, H. M. Fales, A. Hefetz, and G. T. Shaw. 1980. *Anthophora* bees: Unusual glycerides from maternal Dufour's glands serve as larval food and cell lining. *Science* 207:1095–97. [354]

Norden, B. B., and A. G. Scarbrough. 1979. Nesting biology of *Andrena (Larandrena) miserabilis* Cresson and description of the prepupa (Hymenoptera: Andrenidae). *Brimleyana* 2:141–46. [347]

Norris, R. A. 1958. Comparative biosystematics and life history of the nuthatches *Sitta pygmaea* and *Sitta pusilla. Univ. Calif. Publ. Zool.* 56: 119–300. [315]

Northcott, T. 1971. Feeding habits of beaver in Newfoundland. *Oikos* 22:407–10. [258]

Novakowski, N. S. 1967. The winter bioenergetics of a beaver population in northern latitudes. *Can. J. Zool.* 45:1107–18. [23, 26, 42, 67, 257–58, 363]

Novikov, G. A. 1948. On the dispersal of oaks by jays. *Priroda* 37:69–70. [55, 299–300, 307]

Nowak, R. M., and J. L. Paradiso. 1983. *Walker's mammals of the world,* 4th ed. Baltimore: Johns Hopkins University Press. [26, 55, 106, 141, 247, 259, 263–64, 266–67, 274, 276–77, 279]

Nunn, G. L., D. Klem Jr., T. Kimmel, and T. Merriman. 1976. Surplus killing and caching by American kestrels (*Falco sparverius*). *Anim. Behav.* 24:759–63. [14, 119, 284]

Nyby, J., and D. D. Thiessen. 1980. Food hoarding in the Mongolian gerbil (*Meriones unguiculatus*): Effects of food deprivation. *Behav. Neural Biol.* 30:39–48. [111, 119, 121, 125, 134, 144]

Nyby, J., P. Wallace, K. Owen, and D. D. Thiessen. 1973. An influence of hormones on hoarding behavior in the Mongolian gerbil (*Meriones unguiculatus*). *Horm. Behav.* 4:283–88. [125–28, 134]

Nyholm, E. S. 1978. Lichens stored by the bank vole (*Clethrionomys glareolus* Schreber) in Kuusamo (Ks) in 1977. *Luonnon Tutkija* 82:71. [272]

O'Brien, M. F., and F. E. Kurczewski. 1982. Nesting and overwintering behavior of *Liris argentata* (Hymenoptera: Larridae). *J. Ga. Entomol. Soc.* 17:60–68. [340]

Ognev, S. I. 1935. *Mammals of the USSR and adjacent countries. Vol. 3. Carnivora (Fissipedia and Pinnepedia).* Translated and published by Israel Program for Scientific Translations, Jerusalem, 1962. [226, 233]

———. 1964. *Mammals of the USSR and adjacent countries. Vol. 7 Rodents.* Translated and published by Israel Program for Scientific Translations, Jerusalem, 1964. [272]

Oksanen, T. 1983. Prey caching in the hunting strategy of small mustelids. *Acta Zool. Fenn.* 174:197–99. [49, 226, 231–32]

Okunev, L. P., and G. B. Zonov. 1980. Ecological adaptations of Mongolian pika (*Ochotona pallasi* Gray) to life in mountain-steppe terrain. *Soviet J. Ecol.* 11:374–78. [22, 91, 279–80, 282]

Olberg, G. 1959. *Das Verhalten der Solitären Wespen Mitteleuropas (Vespidae, Pompilidae, Sphecidae).* Berlin: Veb Deutscher Verlag der Wissenschaften. [338]

Oliphant, L. W., and W. J. P. Thompson. 1976. Food caching behavior in Richardson's merlin. *Can. Field-Nat.* 90:364–65. [12, 13, 31, 47, 285–88]

Olmsted, C. E. 1937. Vegetation of certain sand plains of Connecticut. *Bot. Gaz.* 99:209–300. [179, 206, 208, 210–11]

Olsson, V. 1985. The winter habits of the great grey shrike, *Lanius excubitor.* IV. Handling of prey. *Vår Fågelvärld* 4:269–83. [89, 317–18]

Ordway, E. 1984. Aspects of the nesting behavior and nest structure of *Diadasia opuntiae* Ckll. (Hymenoptera: Anthophoridae). *J. Kans. Entomol. Soc.* 57:216–30. [353]

Orians, G. H., and N. E. Pearson. 1979. On the theory of central place foraging. In *Analysis of ecological systems,* edited by D. J. Horn, G. R. Stairs, and R. Mitchell, 155–77. Columbus: Ohio State University Press. [53]

Osgood, W. H., E. A. Preble, and G. H. Parker. 1915. The fur seals and other life of the Pribilof Islands, Alaska in 1914. *Senate Documents,* Vol. 6, No. 980. Washington, D.C. [228–29]

Oswald, H. 1956. Beobachtungen über die Samen-verbreitung bei der Zirbe (*Pinus cembra*). *Allgem. Forstztg. Wien* 67(15/16):200–202. [211, 276, 302]

Ovington, J. D., and C. MacRae. 1960. The growth of seedlings of *Quercus petraea. J. Ecol.* 48:549–55. [192]

Ovington, J. D., and G. Murray. 1964. Determination of acorn fall. *Q. J. Forest.* 58:152–59. [189, 191, 246]

Owen, H., and P. Owen. 1956. Tufted titmice plant sunflower seeds. *Kentucky Warbler* 32:62. [179]

Owen, J. H. 1929. Food of the red-backed shrike. *Br. Birds* 23:95–96. [317]

———. 1931. The feeding-habits of the sparrow-hawk. *Br. Birds* 25:151–55. [285]

———. 1948. The larder of the red-backed shrike. *Br. Birds* 41:200–203. [12–13, 317–18]

Owens, C. D. 1971. The thermology of wintering honey bee colonies. U.S. Department of Agriculture, *Tech. Bull.* 1429:1–32. [16]

Panuska, J. A. 1959. Weight patterns and hibernation in *Tamias striatus. J. Mamm.* 40:554–66. [24]

Panuska, J. A., and N. J. Wade. 1956. The burrow of *Tamias striatus. J. Mamm.* 37:23–31. [237, 240]

le Pape, G., and J. M. Lassalle. 1986. Behavioral development in mice: Effects of maternal environment and the albino locus. *Behav. Genet.* 16:531–41. [117]

Park, W. 1922. Time and labor factors in honey and pollen gathering. *Am. Bee J.* 62:254–55. [58]

Parker, A. 1977. Kestrel hiding food. *Br. Birds* 70:339–40. [285]

Parker, F. D. 1966. A review of the North American species in the genus *Leptochilus* (Hymenoptera: Eumenidae). *Misc. Publ. Entomol. Soc. Am.* 5:151–229. [343]

———. 1967. Notes on the nests of three species of *Pseudomasaris* Ashmead (Hymenoptera: Masaridae). *Pan-Pac. Entomol.* 43:213–16. [345]

———. 1978. Biology of the bee genus *Proteriades* Titus (Hymenoptera: Megachilidae). *J. Kans. Entomol. Soc.* 51:145–73. [352]

———. 1980. Nests of *Osmia marginipennis* Cresson with a description of the female (Hymenoptera: Megachilidae). *Pan-Pac. Entomol.* 56:38–42. [89]

————. 1984. The nesting biology of *Osmia (Trichinosmia) latisculcata* Michener. *J. Kans. Entomol. Soc.* 57:430–36. [351–52]

————. 1985. Nesting habits of *Osmia grinnelli* Cockerell (Hymenoptera: Megachilidae). *Pan-Pac. Entomol.* 61:155–59. [351]

Parker, F. D., and G. E. Bohart. 1982. Notes on the biology of *Andrena (Callandrena) helianthi* Robertson (Hymenoptera: Andrenidae). *Pan-Pac. Entomol.* 58:111–16. [347]

Parker, F. D., and H. W. Potter. 1973. Biological notes on *Lithurgus apicalis* Cresson (Hymenoptera: Megachilidae). *Pan-Pac. Entomol.* 49:294–99. [351]

Parker, F. D., and V. J. Tepedino. 1982. Behavior of *Osmia (Nothosmia) marginata* Michener in the nest (Hymenoptera: Megachilidae). *Pan-Pac. Entomol.* 58:231–35. [351]

Parker, F. D., V. J. Tepedino, and D. L. Vincent. 1980. Observations on the provisioning behavior of *Ammophila aberti* Haldeman (Hymenoptera: Sphecidae). *Psyche* 87:249–58. [5, 52]

Parrack, D. W. 1969. A note on the loss of food to the lesser bandicoot rat, *Bandicota bengalensis. Current Sci.* 38(4):93–94. [274, 276]

Parrack, D. W., and J. Thomas. 1970. The behavior of the lesser bandicoot rat, *Bandicota bengalensis.* (Gray and Hardwicke). *J. Bombay Nat. Hist. Soc.* 67:67–80. [276]

Parrott, J. 1980. Frugivory by great grey shrikes *Lanius excubitor. Ibis* 122:532–33. [309, 317–18]

Patton, D. R., and J. R. Vahle. 1986. Cache and nest characteristics of the red squirrel in an Arizona mixed-coniferous forest. *West. J. Appl. For.* 1:48–51. [241]

Paulsen, H. A., Jr. 1950. Mortality of velvet mesquite seedlings. *J. Range Mgmt.* 3:281–86. [204]

Payne, N. F. 1981. Accuracy of aerial censusing for beaver colonies in Newfoundland. *J. Wildl. Mgmt.* 45:1014–16. [258]

Pearson, O. P. 1959. Biology of the subterranean rodents, *Ctenomys* in Peru. *Memorias Museo de Hist. Nat. "Javier Prado"* 9:1–56. [279]

Peck, M. E. 1921. On the acorn-storing habit of certain woodpeckers. *Condor* 23:31. [295]

Peck, S. B., and A. Forsyth. 1982. Composition, structure, and competitive behaviour in a guild of Ecuadorian rainforest dung beetles (Coleoptera; Scarabaeidae). *Can. J. Zool.* 60:1624–34. [89, 96]

Peckham, D. J., and F. E. Kurczewski. 1978. Nesting behavior of *Chlorion aerarium. Ann. Entomol. Soc. Am.* 71:758–61. [340–42]

Pedersen, A. 1966. *Polar animals.* New York: Taplinger Publishing. [62, 226, 228, 230]

Perkins, L. A., G. K. Bienek, and L. G. Klikoff. 1976. The diet of *Dipodomys merriami vulcani. Am. Midl. Nat.* 95:507–12. [185]

Perrins, C. M. 1970. The timing of birds' breeding seasons. *Ibis* 112:242–55. [27]

Petit, D. R., L. J. Petit, and K. E. Petit. 1989. Winter caching ecology of deciduous woodland birds and adaptations for protection of stored food. *Condor* 91:766–76. [311, 315]

Pettifer, H. L., and J. A. J. Nel. 1977. Hoarding in four southern African rodent species. *Zool. Afr.* 12:409–18. [264–65]

Phares, R. E., D. T. Funk, and C. M. Nixon. 1974. Removing black walnut hulls before direct seeding not always protection against rodent pilferage. *Tree Planter's Notes* 25(4):23–24. [172]

Phelan, F. J. S. 1977. Food caching in the screech owl. *Condor* 79:127. [88, 288–89]

Phillips, A. R., J. Marshall, and G. Monson. 1964. *The birds of Arizona.* Tucson: University of Arizona Press. [65]

Phillips, J. K., and E. C. Klostermeyer. 1978. Nesting behavior of *Osmia lignaria propinqua* Cresson (Hymenoptera: Megachilidae). *J. Kans. Entomol. Soc.* 51:91–108. [37, 39, 59, 351–52]

Phillips, R. A. 1978. Common crow observed caching living fish. *Migrant* 49:85–86. [305–6]

Piek, T., and R. T. Simon Thomas. 1969. Paralysing venoms of solitary wasps. *Comp. Biochem. Physiol.* 30:13–31. [87]

Pinkowski, B. C. 1977. Food storage and re-storage in the red-headed woodpecker. *Bird-Banding* 48:74–75. [89, 94]

Pirozynski, K. A., and D. W. Malloch. 1988. Seeds, spores and stomachs: Coevolution in seed dispersal mutualisms. In *Coevolution of fungi with plants and animals,* edited by K. A. Pirozynski and D. L. Hawksworth, 227–46. San Diego: Academic Press. [70, 194]

Pitelka, F. A., P. Q. Tomich, and G. W. Treichel. 1955. Ecological relations of jaegers and owls as lemming predators near Barrow, Alaska. *Ecol. Monogr.* 25: 85–117. [10, 12, 14, 31–32, 49, 97, 288–90]

Pivnik, S. A. 1960. Restoration of cedar-pine stlannik (*Pinus pumila* Rgl.) in plant communities of the Cislenan uplands (Yakutia). In *Problemy kedra. Trudy po Lesnomy Khozyaistvo Sibiri,* No. 6: 129. Publ. Sibirsk. Otdel. Akad. Nauk USSR. [184, 190, 303]

Pizzey, G. 1980. *A field guide to the birds of Australia.* Princeton, N. J.: Princeton University Press. [307–8]

Plateaux-Quénu, C. 1972. *La biologie des abeilles primitives.* Paris: Masson. [345]

Platt, W. J. 1976. The social organization and territoriality of short-tailed shrew populations in old field habitats. *Anim. Behav.* 24: 305–18. [220]

Poché, R. M., Y. Mian, E. Haque, and P. Sultana. 1982. Rodent damage and burrowing characteristics in Bangladesh wheat fields. *J. Wildl. Mgmt.* 46: 139–47. [273–74]

Ponugaeva, A. G. 1953. *Homeostatic regulation of gas exchange in crowded groups of rodents.* Moscow, Leningrad: Trudy Instituta Fiziologii AN SSSR, tom 2. [26]

Portenko, L. A. 1948. Neck pouches in birds. *Priroda* 10: 50–54. [54]

Porter, J. H., F. A. Webster, and J. C. R. Licklider. 1951. The influence of age and food deprivation upon the hoarding behavior of rats. *J. Comp. Physiol. Psychol.* 44: 300–309. [117, 120]

Powell, J. A. 1964. Additions to the knowledge of the nesting behavior of North American *Ammophila* (Hymenoptera: Sphecidae). *J. Kans. Entomol. Soc.* 37: 240–58. [52]

Powell, R. A. 1978. A comparison of fisher and weasel hunting behavior. *Carnivore* 1: 28–34. [226]

Powlesland, R. G. 1980. Food-storing behaviour of the South Island robin. *Mauri Ora* 8: 11–20. [12–14, 47, 89, 94–95, 126, 316]

———. 1981. The foraging behaviour of the South Island robin. *Notornis* 28: 89–102. [316]

Pravosudov, V. V. 1984. The storage of food by the Siberian jay *Perisoreus infaustus* (Passeriformes, Corvidae) in spring. *Zool. Zh.* 63: 950–53. [299–300]

———. 1985. Search for and storage of food by *Parus cinctus lapponicus* and *P. montanus borealis* (Paridae). *Zool. Zh.* 64: 1036–43. [313]

———. 1986. Individual differences in the behavior of the Siberian tit (*Parus cinctus* Bodd.) and the willow tit (*Parus montanus* Bald.) in foraging and storing food. *Ekologiya* 4: 60–64. [164, 311–13]

Price, M. V., and S. H. Jenkins. 1986. Rodents as seed consumers and dispersers. In *Seed dispersal,* edited by D. R. Murray, 191–235. Sydney, Australia: Academic Press. [178]

Pruett-Jones, M. A., and S. G. Pruett-Jones. 1982. Spacing and distribution of bowers in MacGregor's bowerbird (*Amblyornis macgregoriae*). *Behav. Ecol. Sociobiol.* 11: 25–32. [28]

———. 1985. Food caching in the tropical frugivore, MacGregor's bowerbird (*Amblyornis macgregoriae*). *Auk* 102: 334–41. [2, 11–12, 28, 51, 83, 95, 126, 179, 308, 310]

Pulliainen, E., and V. Hyypia. 1975. Winter food and feeding habits of lynxes (*Lynx lynx*) in southeastern Finland. *Suomen Riista* 26: 60–63. [226]

Pulliainen, E., and J. Keränen. 1979. Composition and function of beard lichen stores accumulated by bank voles, *Clethrionomys glareolus* Schreb. *Aquilo Ser. Zool.* 19: 73–76. [272]

Purchas, T. P. G. 1975. Rooks (*Corvus frugilegus frugilegus* L.) hiding nuts in Hawke's Bay. *Proc. N. Z. Ecol. Soc.* 22: 111–12. [305]

———. 1980. Feeding ecology of rooks (*Corvus frugilegus*) on the Heretaunga Plains, Hawke's Bay, New Zealand. *N. Z. J. Zool.* 7: 557–78. [199, 305–7]

Quay, W. B. 1965. Comparative survey of the sebaceous and sudoriferous glands of the oral lips and angle in rodents. *J. Mamm.* 46: 23–37. [172]

Quick, E., and A. W. Butler. 1885. The habits of some Arvicolinae. *Am. Nat.* 19: 113–18. [270]

Quink, T. F., H. G. Abbott, and W. J. Mellen. 1970. Locating tree seed caches of small mammals with a radioisotope. *For. Sci.* 16: 147–48. [105]

Räber, H. 1944. Versuche zur Ermittlung des Beuteschemus an einem Hausmarder (*Martes foina*) und einem Iltis (*Putorius putorius*). *Rev. Suisse Zool.* 51:293–332. [226]

———. 1950. Das Verhalten gefangener Waldohreulen (*Asio otus otus*) und Waldkäuze (*Strix aluco aluco*) zur Beute. *Behaviour* 2:1–95. [289]

Radvanyi, A. 1966. Destruction of radio-tagged seeds of white spruce by small mammals during summer months. *For. Sci.* 12:307–15. [194]

Rainey, D. G. 1956. Eastern woodrat, *Neotoma floridana:* Life history and ecology. *Univ. Kans. Publ. Mus. Nat. Hist.* 8:535–646. [262]

Raspopov, M. P., and J. A. Isakov. 1935. On the biology of the squirrel. In *The biology of the hare and the squirrel, and their diseases,* edited by P. A. Mantejfel, 33–79. Moscow. [245]

Rathmayer, W. 1962. Paralysis caused by the digger wasp *Philanthus. Nature* 196:1148–51. [87]

Raw, F. 1966. The soil fauna as a food source for moles. *J. Zool., London* 149:50–54. [222, 224]

Rebar, C., and O. J. Reichman. 1983. Ingestion of moldy seeds by heteromyid rodents. *J. Mamm.* 64:713–15. [84]

Reed, K. M. 1987. Caloric content of an excavated food cache of *Perognathus flavescens. Texas J. Sci.* 39:191–92. [25]

Reese, J. G. 1972. A Chesapeake barn owl population. *Auk* 89:106–14. [97, 288–89]

Regnier, R., and R. Pussard. 1926. La Constitution des Mogasins de Réserve du *Microtus arvalis* Pallas (campagnol des champs) et son Importante pour la Pullulation de ce Ronquer. *Compt. Rend. Acad. Sci.* 183:92–94. [273]

Reichman, O. J. 1981. Factors influencing foraging in desert rodents. In *Foraging behavior; Ecological, ethological, and psychological approaches,* edited by A. C. Kamil and T. D. Sargent, 195–213. New York: Garland STPM Press. [173, 186–87]

———. 1988. Caching behavior by eastern woodrats, *Neotoma floridana,* in relation to food perishability. *Anim. Behav.* 36:1525–32. [139, 260]

Reichman, O. J., A. Fattaey, and K. Fattaey. 1986. Management of sterile and mouldy seeds by a desert rodent. *Anim. Behav.* 34:221–25. [84–86]

Reichman, O. J., and P. Fay. 1983. Comparison of the diets of a caching and non-caching rodent. *Am. Nat.* 122:576–81. [41]

Reichman, O. J., and D. Oberstein. 1977. Selection of seed distribution types by *Dipodomys merriami* and *Perognathus amplus. Ecology* 58:636–43. [171–72, 185–86, 366]

Reichman, O. J., and C. Rebar. 1985. Seed preferences by desert rodents based on levels of mouldiness. *Anim. Behav.* 33:726–29. [84]

Reichman, O. J., D. T. Wicklow, and C. Rebar. 1985. Ecological and mycological characteristics of caches in the mounds of *Dipodomys spectabilis. J. Mamm.* 66:643–51. [105, 167, 252–54]

Reid, W. H., and H. J. Fulbright. 1981. Impaled prey of the loggerhead shrike in the northern Chihuahuan Desert. *Southw. Nat.* 26:204–5. [317]

Reig, O. A. 1970. Ecological notes on the fossorial octodont rodent *Spalacopus cyanus* (Molina). *J. Mamm.* 51:592–601. [279]

Reimers, N. F. 1953. The food of the nutcracker and its role in the dispersal of the cedar-pine in the mountains of Khamar-Daban. *Lesn. Khoz.* 1:36–37. [72, 194, 210–11, 303]

———. 1958. The reforestation of burns and forest tracts devastated by silkworms in the mountain cedar-pine taiga of south Cisbaikal and the role of vertebrate animals in this process. *Byulleten Moskooskogo Obshchestva Ispytatelei Prirody. Otdel Biologicheskii* 63:49–56. [210, 304]

———. 1959a. The nesting of the long-billed nutcracker in central Siberia. *Zool. Zh.* 38:907–15. [32, 305]

———. 1959b. The nutcracker (*Nucifraga caryocatactes macrorhynchus* Brehm). *Trudy Biol. Inst. (Zoologicheskii) Sibirsk Otdel. Akad. Nauk. USSR* 5:121–66. [303]

Rettenmeyer, C. W. 1963. Behavioral studies of army ants. *Kans. Univ. Sci. Bull.* 44:281–465. [336]

Reynolds, E. G. 1969. Feeding behaviour of the pied currawong. *Emu* 69:186. [308]

Reynolds, H. G. 1954. Some interrelationships of the Merriam's kangaroo rat to velvet mesquite. *J. Range Mgmt.* 7:176–80. [204]

———. 1958. The ecology of the Merriam kangaroo rat (*Dipodomys merriami* Mearns) on the grazing lands of southern Arizona. *Ecol. Monogr.* 28:111–27. [171, 183, 185, 191, 202, 204, 255]

Reynolds, H. G., and G. E. Glendening. 1949. Merriam kangaroo rat a factor in mesquite propagation on southern Arizona range lands. *J. Range Mgmt.* 2:193–97. [186, 202, 204, 206–7, 209–10, 253, 255]

Ribbands, C. R. 1953. *The behaviour and social life of honeybees.* London: Bee Research Association. [60]

———. 1955. Community defence against robber bees. *Am. Bee J.* 95:313, 320. [89, 105]

Rich, T., and B. Trentlage. 1983. Caching of long-horned beetles (Cerambycidae: *Prionus integer*) by the burrowing owl. *Murrelet* 64:25–26. [49, 86, 107, 289]

Richard, P. B. 1964. Les materiaux de construction du castor (*Castor fiber*), leur signification pour ce rongeur. *Z. Tierpsychol.* 21:592–601. [51, 257]

Richards, T. J. 1949. Concealment of food by nuthatch, coat-tit and marsh tit. *Br. Birds* 42:360–61. [311, 314]

———. 1958. Concealment and recovery of food by birds, with some relevant observations on squirrels. *Br. Birds* 51:497–508. [46, 171, 179, 190, 199, 244, 306, 314–15]

Rijnsdorp, A., S. Daan, and C. Dijkstra. 1981. Hunting in the kestrel, *Falco tinnunculus,* and the adaptive significance of daily habits. *Oecologia* 50:391–406. [13–14, 88, 285, 287–88]

Rinderer, T. E. 1981. Volatiles from empty comb increase hoarding by the honey bee. *Anim. Behav.* 29:1275–76. [144]

———. 1982. Maximal stimulation by comb of honey bee (*Apis mellifera*) hoarding behavior. *Ann. Entomol. Soc. Am.* 75:311–12. [144]

Rinderer, T. E., and J. R. Baxter. 1978a. Effects of empty comb on hoarding behavior and honey production of the honey bee. *J. Econ. Entomol.* 71:757–59. [144]

———. 1978b. Honey bees: The effect of group size on longevity and hoarding in laboratory cages. *Ann. Entomol. Soc. Am.* 71:732. [142]

———. 1979. Honey bee hoarding behaviour: Effects of previous stimulation by empty comb. *Anim. Behav.* 27:426–28. [144]

———. 1980a. Hoarding behavior of the honey bee: Effects of empty comb, comb color, and genotype. *Environ. Entomol.* 9:104–5. [144]

———. 1980b. Honey bee hoarding of high fructose corn syrup and cane sugar syrup. *Am. Bee J.* 120:817–18. [139]

Rinderer, T. E., A. B. Bolten, J. R. Harbo, and A. M. Collins. 1982. Hoarding behavior of European and Africanized honey bees (Hymenoptera: Apidae). *J. Econ. Entomol.* 75:714–15. [64, 112]

Rinderer, T. E., A. M. Collins, R. L. Hellmich II, and R. G. Danka. 1986. Regulation of the hoarding efficiency of Africanized and European honey bees. *Apidologie* 17:227–32. [16, 64, 112]

Rinderer, T. E., A. M. Collins, and K. W. Tucker. 1985. Honey production and underlying nectar harvesting activities of Africanized and European honeybees. *J. Apic. Res.* 24:161–67. [16, 64, 365]

Rinderer, T. E., and H. A. Sylvester. 1978. Variation in response to *Nosema apis,* longevity and hoarding behavior in a free-mating population of the honey bee. *Ann. Entomol. Soc. Am.* 71:372–74. [113]

Ristich, S. S., 1953. A study of the prey, enemies, and habits of the great-golden digger wasp *Chlorion ichneumoneum* (L.). *Can. Entomol.* 85:374–86. [89, 94, 339]

Ritchie, R. J. 1980. Food caching behaviour of nestling wild hawk-owls. *Raptor Res.* 14:59–60. [49, 288–90]

Ritter, W. E. 1921. Acorn-storing by the California woodpecker. *Condor* 23:3–14. [293, 295]

———. 1922. Further observations on the activities of the California woodpecker. *Condor* 24:109–22. [293]

———. 1938. *The California woodpecker and I.* Berkeley: University of California Press. [295, 297]

Roberts, A. 1951. *The mammals of South Africa.* Johannesburg, South Africa: Central News Agency. [268]

Roberts, R. B., and C. H. Dodson. 1967. Nesting biology of two communal bees, *Euglossa imperialis* and *Euglossa ignita* (Hymenoptera: Apidae), including description of larvae. *Ann. Entomol. Soc. Am.* 60:1007–14. [359]

Roberts, R. B., and S. R. Vallespir. 1978. Specialization of hairs bearing pollen and oil on the legs of bees (Apoidea: Hymenoptera). *Ann. Entomol. Soc. Am.* 71:619–27. [59, 354]

Roberts, R. C. 1979. The evolution of avian food-storing behavior. *Am. Nat.* 114:418–38. [1, 8, 67, 79, 364]

Roberts, T. S. 1937. How two captive young beavers constructed a food pile. *Proc. Minn. Acad. Sci.* 5:24–27. [257]

Robinette, W. L., J. S. Gashwiler, and O. W. Morris. 1959. Food habits of the cougar in Utah and Nevada. *J. Wildl. Mgmt.* 23:261–73. [233]

Robinson, D. E., and E. D. Brodie. 1982. Food hoarding behavior in the short-tailed shrew *Blarina brevicauda. Am. Midl. Nat.* 108:369–75. [95, 220–21]

Robinson, M. H. 1969. Predatory behavior of *Argiope argentata* (Fabricus). *Am. Zool.* 9:161–73. [320]

Robinson, M. H., H. Mirick, and O. Turner. 1969. The predatory behavior of some araneid spiders and the origin of immobilization wrapping. *Psyche* 76:487–501. [2, 12, 87, 89, 320, 322]

Robson, R. W. 1974. Food-burying and recovery by rook. *Br. Birds* 67:214–15. [305]

Rogers, C. M. 1987. Predation risk and fasting capacity: Do wintering birds maintain optimal body mass? *Ecology* 1051–61. [26]

Ross, S., and W. I. Smith. 1953. The hoarding behavior of the mouse. II. The role of deprivation, satiation, and stress. *J. Genet. Psychol.* 82:279–97. [120, 136]

Ross, S., W. I. Smith, and V. Denenberg. 1950. A preliminary study of individual and group hoarding in the white rat. *J. Genet. Psychol.* 77:123–27. [141]

Ross, S., W. I. Smith, and B. L. Woessner. 1955. Hoarding: An analysis of experiments and trends. *J. Gen. Psychol.* 52:307–26. [117]

Rothenbuhler, W. C., J. M. Kulincevic, and V. C. Thompson. 1979. Successful selection of honeybees for fast and slow hoarding of sugar syrup in the laboratory. *J. Apic. Res.* 18:272–78. [112]

Roubik, D. W. 1982a. Obligate necrophagy in a social bee. *Science* 217:1059–60. [356]

———. 1982b. Seasonality in colony storage, brood production and adult survivorship: Studies of *Melipona* in tropical forest (Hymenoptera: Apidae). *J. Kans. Entomol. Soc.* 55:789–800. [16–17, 356]

———. 1983. Nest and colony characteristics of stingless bees from Panama (Hymenoptera: Apidae). *J. Kans. Entomol. Soc.* 56:327–55. [88, 355–56]

Roubik, D. W., and C. D. Michener. 1984. Nesting biology of *Crawfordapis* in Panama (Hymenoptera, Colletidae). *J. Kans. Entomol. Soc.* 57:662–71. [347]

Rovner, J. S., and S. J. Knost. 1974. Post-immobilization wrapping of prey by lycosid spiders of the herbaceous stratum. *Psyche* 81:398–415. [321–22]

Rowley, I. 1973. The comparative ecology of Australian corvids, II. Social organization and behaviour. Melbourne, Australia: Commonwealth Scientific and Industrial Research Organization, *Wildl. Res.* 18:25–65. [306]

Roy, D. F. 1957. A record of tanoak acorn and seedling production in northwestern California. U.S. Department of Agriculture, Forest Service, California Forest and Range Experiment Station, *For. Res. Notes* 124:1–6. [189]

Roy, S. 1974. Pre-harvest loss of rice due to field rodents. *Econ. Polit. Week.* 9(6):A66–67. [273–74, 276]

Rozen, J. G., Jr. 1965. The biology and immature stages of *Melitturga clavicornis* (Latreille) and of *Sphecodes albilabris* (Kirby) and the recognition of the Oxaeidae at the family level (Hymenoptera, Apoidea). *Am. Mus. Novit.* 2224:1–18. [347–48]

———. 1967. Review of the biology of panurgine bees, with observations on North American forms (Hymenoptera, Andrenidae). *Am. Mus. Novit.* 2297:1–44. [347]

———. 1970. Biology, immature stages, and phylogenetic relationships of fideliine bees, with the description of a new species of *Neofidelia* (Hymenoptera, Apoidea). *Am. Mus. Novit.* 2427:1–25. [353]

———. 1973. Biological notes on the bee *Andrena accepta* Viereck. *J. N. Y. Entomol. Soc.* 81:54–61. [347]

———. 1977a. Biology and immature stages of the bee genus *Meganomia* (Hymenoptera, Melittidae). *Am. Mus. Novit.* 2630:1–14. [350]

———. 1977b. The ethology and systematic relationships of fideliine bees, including a description of the mature larva of *Parafidelia* (Hymenoptera, Apoidea). *Am. Mus. Novit.* 2637: 1–15. [353]

———. 1984a. Comparative nesting biology of the bee tribe Exomalopsini (Apoidea, Anthophoridae). *Am. Mus. Novit.* 2798: 1–37. [353]

———. 1984b. Nesting biology of diphaglossine bees (Hymenoptera, Colletidae). *Am. Mus. Novit.* 2786: 1–33. [347]

———. 1986. The natural history of the Old World nomadine parasitic bee *Pasites maculatus* (Anthophoridae: Nomadinae) and its host *Pseudapis diversipes* (Halictidae: Nomiinae). *Am. Mus. Novit.* 2861: 1–8. [349, 353]

———. 1987. Nesting biology of the bee *Ashmeadiella holtii* and its cleptoparasite, a new species of *Stelis* (Apoidea: Megachilidae). *Am Mus. Novit.* 2900: 1–10. [351]

Rozen, J. G., Jr., and N. R. Jacobson. 1980. Biology and immature stages of *Macropis nuda,* including comparisons to related bees (Apoidea, Melittidae). *Am. Mus. Novit.* 2702: 1–11. [348, 350]

Rozen, J. G., Jr., and R. J McGinley. 1976. Biology of the bee genus *Conauthalictus* (Halictidae, Dufoureinae). *Am. Mus. Novit.* 2602: 1–6. [349]

Rue, L. L., III. 1969. *The world of the red fox.* Philadelphia: J. B. Lippincott. [89, 227]

Ruffer, D. G. 1965. Burrows and burrowing behavior of *Onychomys leucogaster. J. Mamm.* 46: 241–47. [259]

Rusch, D. A., and W. G. Reeder. 1978. Population ecology of Alberta red squirrels. *Ecology* 59: 400–420. [18, 241]

Rust, R. W. 1980. Nesting biology of *Hoplitis biscutellae* (Cockerell) (Hymenoptera: Megachilidae). *Entomol. News* 91: 105–9. [351–52]

Rutter, R. J. 1969. A contribution to the biology of the gray jay (*Perisoreus canadensis*). *Can. Field-Nat.* 83: 300–316. [94, 148, 300, 363]

———. 1972. The gray jay: A bird for all seasons. *Nature, Canada* 1(4): 28–32. [18, 298, 300]

Ryan, J. M. 1986. Comparative morphology and evolution of cheek pouches in rodents. *J. Morph.* 190: 27–41. [54]

Sabilaev, A. S., and S. G. Borovskii. 1970. Aboveground food storages of the great gerbil and the probable causes of their occurrence (exemplified at Ustyurt). *Ekologiya* 1: 81–82. [264–65]

Saigo, B. W. 1969. The relationship of non-recovered rodent caches to the natural regeneration of ponderosa pine. Masters thesis, Oregon State University, Corvallis. [248]

Saito, S. 1983a. Caching of Japanese stone pine seeds by nutcrackers at the Shiretoko Peninsula, Hokkaido. *Tori* 32: 13–20. [72, 275]

———. 1983b. On relations of the caching by animals on the seed germination of Japanese stone pine, *Pinus pumila* Regel. *Bull. Shiretoko Mus.* 5: 23–40. [71–72, 187–88, 194, 207–8]

Sakagami, S. F., and C. D. Michener. 1962. *The nest architecture of the sweat bees.* Lawrence: University Press of Kansas. [348–49]

Salbert, P., and N. Elliott. 1979. Observations on the nesting behavior of *Cerceris watlingensis* (Hymenoptera: Sphecidae, Philanthinae). *Ann. Entomol. Soc. Am.* 72: 591–95. [68, 342]

Salvin, O. 1876. Exhibition of a piece of trunk of a pine tree from Guatemala perforated by *Melanerpes formicivorus. Proc. Zool. Soc. London.* 1876: 414. [295]

Sanchez, J. C., and O. J. Reichman. 1987. The effects of conspecifics on caching behavior of *Peromyscus leucopus. J. Mamm.* 68: 695–97. [142]

Sanderson, H. R. 1962. Survival of rodent cached bitterbrush seed. U. S. Department of Agriculture, Pacific Southwest Forest and Range Experiment Station, *Res. Note* 211: 1–3. [204, 210]

Scelfo, L. M., and L. R. Hammer. 1969. Stimulus preference in hoarding. *Psychonom. Sci.* 17: 155–56. [140]

Schaller, G. B. 1967. *The deer and the tiger.* Chicago: University of Chicago Press. [12, 48, 87, 104, 226, 233]

———. 1972. *The Serengeti lion: A study of predator-prey relations.* Chicago: University of Chicago Press. [233]

Scheffer, T. H. 1929. Mountain beavers in the Pacific Northwest: Their habits, economic status, and control. U.S. Department of Agriculture, *Farmer's Bull.* 1598: 1–18. [236]

———. 1931. Habits and economic status of the pocket gophers. U.S. Department of Agriculture, *Tech. Bull.* 224: 1–26. [250]

————. 1938. The pocket mice of Washington and Oregon in relation to agriculture. U.S. Department of Agriculture, *Tech. Bull.* 608:1–15. [62, 202, 252]

————. 1940. Excavation of the runway of the pocket gopher (*Geomys bursarius*). *Trans. Kans. Acad. Sci.* 43:473–78. [250]

Schenk, F. 1979. Comportements alimentaires du mulot sylvestre en actographe: relations avec l'activité nocturne dans le tombour, la photopériode, le sexe et la nouveauté. *Mammalia* 43:453–64. [33, 275–76]

Schmidt, F. 1934. Über die Fortptlanzungsbiologie von sibirischen Zobel (*Martes zibellina*) und europäischem Baummarder (*Martes martes* L.). *Z. Saugetierkd.* 9:392–403. [226]

Schmidt, U. 1979. Die Lokalisation vergrabenen Futters bei der Hausspitzmaus, *Crocidura russula* Hermann. *Z. Saugetierkd.* 44:59–60. [172]

Schnell, J. H. 1958. Nesting behavior and food habits of goshawks in the Sierra Nevada of California. *Condor* 60:377–403. [12, 31, 49, 285–88]

Schopmeyer, C. S. 1974. *Seeds of woody plants in the United States.* U.S. Department of Agriculture, Forest Service, *Agric. Handb. 450,* Washington, D.C. [186, 196–97, 200]

Schorger, A. W. 1949. Squirrels in early Wisconsin. *Trans. Wis. Acad. Sci., Arts, Letters* 39:195–247. [65–66]

Schrader, M. N., and W. E. LaBerge. 1978. The nest biology of the bees *Andrena* (*Melandrena*) *regularis* Malloch and *Andrena* (*Melandrena*) *carlini* Cockerell (Hymenoptera: Andrenidae). *Ill. Nat. Hist. Surv., Biol. Notes* 108:1–24. [347–48]

Schroder, G. D. 1979. Foraging behavior and home range utilization of the bannertail kangaroo rat (*Dipodomys spectabilis*). *Ecology* 60:657–65. [252]

Schuster, L. 1950. Über den Sammeltrieb des Eichelhähers (*Garrulus glandarius*). *Vogelwelt* 71:9–17. [181, 183, 200]

Schwarz, H. F. 1929. Honey wasps. *Nat. Hist.* 29:421–26. [345]

————. 1948. Stingless bees (Meliponidae) of the Western Hemisphere. *Bull. Am. Mus. Nat. Hist.* 90:1–546. [355]

Scott, M. P., and J. F. A. Traniello. 1989. Guardians of the underworld. *Nat. Hist.* 98(6):32–37. [327]

Scott, T. G. 1943. Some food coactions of the northern plains red fox. *Ecol. Monogr.* 13:427–79. [149, 227–29]

Sealy, S. G. 1984. Capture and caching of flying carpenter ants by pygmy nuthatches. *Murrelet* 65:49–51. [315]

Seastedt, T. R., O. J. Reichman, and T. C. Todd. 1986. Microarthropods and nematodes in kangaroo rat burrows. *Southw. Nat.* 31:114–16. [89]

Seeley, T. D. 1982. Adaptive significance of the age polyethism schedule in honeybee colonies. *Behav. Ecol. Sociobiol.* 11:287–93. [116]

————. 1983. The ecology of temperate and tropical honeybee societies. *Am. Sci.* 71:264–72. [359]

————. 1985. *Honeybee ecology: A study of adaptation in social life.* Princeton, N. J.: Princeton University Press. [16, 64, 79, 105, 119, 142, 359, 361–62]

Seeley, T. D., and R. A. Morse. 1976. The nest of the honey bee (*Apis mellifera*). *Insect. Soc.* 23:495–512. [3, 359–61]

Seeley, T. D., R. H. Seeley, and P. Akratanakul. 1982. Colony defense strategies of the honeybees in Thailand. *Ecol. Monogr.* 52:43–63. [3, 67, 89, 104–5, 110, 359–60]

Seeley, T. D., and P. K. Visscher. 1985. Survival of honeybees (*Apis mellifera*) in cold climates: The critical timing of colony growth and reproduction. *Ecol. Entomol.* 10:81–88. [16, 28, 34, 361–63]

Seitz, P. 1954. The effect of infantile experiences upon adult behavior in animal subjects. *Am. J. Psychiatry* 110:916–27. [117]

Semenov-Tian-Shanskii, O. 1972. The brown bear in the Lapland Reserve, USSR. *Aquilo Ser. Zool.* 13:98–102. [231]

Serventy, D. L., and H. M. Whittell. 1951. *A handbook of the birds of Western Australia,* 2d. ed. Perth, Australia: Paterson Brokensha Pty. [307]

Seton, E. T. 1928. *Lives of game animals.* Boston: Charles T. Branford. [250, 270]

————. 1953. *Lives of game animals,* Vol. IV, Part II. Boston: Charles T. Branford. [257]

Shaffer, L. 1980. Use of scatter hoards by eastern

chipmunks to replace stolen food. *J. Mamm.* 61:733–34. [105, 238–40]

Sharp, W. M. 1959. A commentary on the behavior of free-running gray squirrels. In *Proceedings of the 13th Annual Conference of the Southeastern Association of the Game and Fish Commission,* edited by J. W. Webb, 382–87. Columbia, S. C. [66, 89, 245]

Shaw, E. W. 1954. Direct seeding in the Pacific Northwest. *J. Forest.* 52:827–28. [194]

Shaw, M. W. 1968a. Factors affecting the natural regeneration of sessile oak (*Quercus petraea*) in north Wales: I. A preliminary study of acorn production, viability and losses. *J. Ecol.* 56:565–83. [191]

———. 1968b. Factors affecting the natural regeneration of sessile oak (*Quercus petraea*) in north Wales: II. Acorn losses and germination under field conditions. *J. Ecol.* 56:647–60. [189]

———. 1974. The reproductive characteristics of oak. In *The British oak: Its history and natural history,* edited by M. G. Morris and F. H. Perring, 162–81. Farington, England: E. W. Classey. [192]

Shaw, W. T. 1925. Notes on the ecology of the Columbian ground squirrel (*Citellus columbianus columbianus*) at Pullman, Washington. *Murrelet* 6:46–54. [24, 247–48]

———. 1934. The ability of the giant kangaroo rat as a harvester and storer of seeds. *J. Mamm.* 15:275–86. [62, 83, 93, 104–5, 181, 240, 252, 254–56]

———. 1936. Moisture and its relation to the cone-storing habit of the western pine squirrel. *J. Mamm.* 17:337–49. [84, 108, 139, 241–42]

Sheikher, C., and C. S. Malhi. 1983. Territorial and hoarding behavior in *Bandicota* spp. and *Mus* spp. of Garhwal Himalayas. *Proc. Indian Natl. Sci. Acad.* 49B:332–35. [273–74]

Shellhammer, H. S. 1966. Cone-cutting activities of Douglas squirrels in sequoia groves. *J. Mamm.* 47:525–26. [242]

Sherman, R. J., and W. W. Chilcote. 1972. Spatial and chronological patterns of *Purshia tridentata* as influenced by *Pinus ponderosa. Ecology* 53:294–98. [193, 204, 206, 208–9, 211]

Sherry, D. F. 1982. Food storage, memory, and marsh tits. *Anim. Behav.* 30:631–33. [102, 167]

———. 1984a. Food storage by black-capped chickadees: Memory for the location and contents of caches. *Anim. Behav.* 32:451–64. [18, 65, 67, 160–62, 167, 170, 175]

———. 1984b. What food-storing birds remember. *Can. J. Psychol.* 38:304–21. [1, 148]

———. 1985. Food storage by birds and mammals. *Adv. Study Behav.* 15:153–88. [1, 148]

———. 1987. Foraging for stored food. In *Quantitative analyses of behavior: Foraging* vol. 6, edited by M. L. Commons, A. Kacelnik, and S. J. Shettleworth, 209–27. London: Erlbaum Associates. [1, 92, 148]

———. 1989. Food storing in the Pardiae. *Wilson Bull.* 101:289–304. [170, 309]

Sherry, D. F., M. Avery, and A. Stevens. 1982. The spacing of stored food by marsh tits. *Z. Tierpsychol.* 58:153–62. [67, 100–102, 105, 110, 167–68, 184, 310]

Sherry, D. F., J. R. Krebs, and R. J. Cowie. 1981. Memory for the location of stored food in marsh tits. *Anim. Behav.* 29:1260–66. [102, 155–57, 160–62, 164, 170, 175]

Sherry, D. F., and A. L. Vaccarino. 1989. Hippocampus and memory for food caches in black-capped chickadees. *Behav. Neurosci.* 103:308–18. [163]

Shettleworth. S. J. 1983. Memory in food-hoarding birds. *Sci. Am.* 248:102–10. [1, 148, 169]

———. 1985. Food storing by birds: Implications for comparative studies of memory. In *Memory Systems of the Brain: Animal and Human Cognition Processes,* edited by N. M. Weinberger, J. L. McGaugh, and G. Lynch, 231–50. New York: Guilford Publishers. [1, 148, 169–70]

Shettleworth, S. J., and J. R. Krebs. 1982. How marsh tits find their hoards: The role of site preference and spatial memory. *J. Exp. Psychol.: Anim. Behav. Proc.* 8:354–75. [67, 102, 155–57, 159–62, 167–68, 175]

———. 1986. Stored and encountered seeds: A comparison of two spatial memory tasks in marsh tits and chickadees. *J. Exp. Psychol.: Anim. Behav. Proc.* 12:248–57. [160–62, 164–65, 170, 175]

Shettleworth, S. J., J. R. Krebs, S. D. Healy, and

C. M. Thomas. 1990. Spatial memory of food-storing tits (*Parus ater* and *P. atricapillus*): Comparison of storing and nonstoring tasks. *J. Comp. Psychol.* 104:71–81. [169]

Short, L. L. 1982. *Woodpeckers of the world.* Greenville, Del.: Delaware Museum of Natural History, Monogr. Ser. No. 4. [292]

Shortridge, G. C. 1934. *The mammals of South West Africa.* London: Heinemann. [106, 191, 205, 267]

Shubin, I. G. 1959a. Ecology of Streltsov voles in Kazakhstan uplands. *Trudy Inst. Zool. Akad. Nauk Kazakhskoy SSR,* Vol. 10. [271]

———. 1959b. Ecology of the Mongolian pika in the Kazakh Upland. *Trudy Inst. Zool. Akad. Nauk Kazakhskoy SSR. Alma-Ata* 10:114–43. [280]

Shull, A. F. 1907. Habits of the short-tailed shrew, *Blarina brevicauda. Am. Nat.* 41:495–522. [107, 219–20]

Silverman, H. J., and I. Zucker. 1976. Absence of post-fast food compensation in the golden hamster (*Mesocricetus auratus*). *Physiol. Behav.* 17:271–86. [121, 263]

Silvertown, J. W. 1980. The evolutionary ecology of mast seeding in trees. *Biol. J. Linn. Soc.* 14:235–50. [189, 202]

Simmons, K. E. L. 1968. Food-hiding by rooks and other crows. *Br. Birds* 61:228–29. [306]

———. 1970. Food-hiding by rooks. *Br. Birds* 63:175–77. [306]

Simpson, B. B., J. L. Neff, and D. Seigler. 1977. *Krameria,* free fatty acids and oil-collecting bees. *Nature* 267:150–51. [354]

Skoczen, S. 1961. On food storage of the mole, *Talpa europaea* Linnaeus 1758. *Acta Theriol.* 5:23–43. [20, 86, 107, 222–23]

———. 1970. Food storage of some insectivorous mammals (Insectivora). *Przegl. Zool.* 14:243–48. [222]

Skutch, A. F. 1969. *Life histories of Central American birds.* Vol. 3, Pacific Coast Avifauna 35, Berkeley, Calif.: Cooper Ornithological Society. [62, 64, 295]

Sladen, F. W. L. 1912. *The humble-bee, its life history and how to domesticate it.* London: Macmillan. [358]

Slough, B. G. 1978. Beaver food cache structure and utilization. *J. Wildl. Mgmt.* 42:644–46. [108, 257–58]

Slough, B. G., and R. M. F. S. Sadleir. 1977. A land capability classification system for beaver (*Castor canadensis* Kuhl). *Can. J. Zool.* 55:1324–35. [258]

Smallwood, P. D., and W. D. Peters. 1986. Grey squirrel food preferences: The effects of tannin and fat concentration. *Ecology* 67:168–74. [76]

Smith, A. T. 1974. The distribution and dispersal of pikas: Influences of behavior and climate. *Ecology* 55:368–76. [281]

Smith, C. C. 1968. The adaptive nature of social organization in the genus of tree squirrel *Tamiasciurus. Ecol. Monogr.* 38:31–63. [18, 29, 71, 83, 97–98, 104, 197, 240–42, 245]

———. 1970. The coevolution of pine squirrels (*Tamiasciurus*) and conifers. Ecol. Monogr. 40:349–71. [70, 71, 79, 195]

———. 1975. The coevolution of plants and seed predators. In *The coevolution of animals and plants,* edited by L. E. Gilbert and P. H. Raven, 53–77. Austin: University of Texas Press. [198, 202]

———. 1981. The indivisible niche of *Tamiasciurus:* An example of nonpartitioning of resources. *Ecol. Monogr.* 51:343–63. [29, 240, 242]

Smith, C. C., and R. P. Balda. 1979. Competition among insects, birds, and mammals for conifer seeds. *Am. Zool.* 19:1065–83. [65, 89, 189, 197, 367]

Smith, C. C., and D. Follmer. 1972. Food preferences of squirrels. *Ecology* 53:82–91. [76, 79, 199, 202]

Smith, C. C., and O. J. Reichman. 1984. The evolution of food caching by birds and mammals. *Ann. Rev. Ecol. Syst.* 15:329–51. [1–2, 49, 63, 83, 176, 178]

Smith, C. F. 1948. A burrow of the pocket gopher (*Geomys bursarius*) in eastern Kansas. *Trans. Kans. Acad. Sci.* 51:313–15. [205, 250]

Smith, C. F., and S. E. Aldous. 1947. The influence of mammals and birds in retarding artificial and natural reseeding of coniferous forests in the United States. *J. Forest.* 45:361–69. [194]

Smith, J. P., J. S. Maybee, and F. M. Maybee. 1979. The effects of increasing distance to food and deprivation level on food-hoarding behavior in *Rattus norvegicus. Behav. Neural. Biol.* 27:302–18. [53, 120]

Smith, K. G., and T. Scarlett. 1987. Mast production and winter populations of red-headed woodpeckers and blue jays. *J. Wildl. Mgmt.* 51:459–67. [202]

Smith, M. C. 1968. Red squirrel responses to spruce cone failure in interior Alaska. *J. Wildl. Mgmt.* 32:305–17. [1, 18–19, 110, 240, 242, 363]

Smith, S. M. 1972. The ontogeny of impaling behavior in the loggerhead shrike, *Lanius ludovicianus* L. *Behaviour* 42:232–47. [2, 47, 115, 118, 317, 319]

Smith, W. I., and E. K. Powell. 1955. The role of emotionality in hoarding. *Behaviour* 8:57–62. [138]

Smith, W. I., and S. Ross. 1950a. Hoarding behavior in the golden hamster. *J. Genet. Psychol.* 77:211–15. [33, 121, 125]

———. 1950b. The effect of continued food deprivation on hoarding in the albino rat. *J. Genet. Psychol.* 77:117–21. [120]

———. 1953a. The hoarding behavior of the mouse. I. The role of previous feeding experience. *J. Genet. Psychol.* 82:279–97. [117, 120]

———. 1953b. The hoarding behavior of the mouse. III. The storing of "non-relevant" material. *J. Genet. Psychol.* 82:309–16. [121]

Smyth, M. 1966. Winter breeding in woodland mice, *Apodemus sylvaticus,* and voles, *Clethrionomys glareolus* and *Microtus agrestis,* near Oxford. *J. Anim. Ecol.* 35:471–85. [28]

Smythe, N. D. E. 1970. Relationships between fruiting seasons and seed dispersal methods in a Neotropical forest. *Am. Nat.* 104:25–35. [75, 188, 198, 202, 277–78]

———. 1978. The natural history of the Central American agouti (*Dasyprocta punctata*). *Smithson. Contr. Zool.* 257:1–52. [75, 148, 172, 188, 198, 201, 277–78]

Smythe, N. D. E., W. F. Glanz, and E. G. Leigh, Jr. 1982. Population regulation in some terrestrial frugivores. In *The ecology of a tropical forest: Seasonal rhythms and long-term changes,* edited by E. G. Leigh, Jr., A. S. Rand, and D. M. Windsor, 227–38. Washington, D.C.: Smithsonian Institution. [198, 278]

Snead, E. W., and G. O. Hendrickson. 1942. The food habits of the badger in Iowa. *J. Mamm.* 23:380–91. [226, 233]

Snelling, R. R. 1976. A revision of the honey ants, genus *Myrmecocystus. Nat. Hist. Mus. Los Angeles Co. Sci. Bull.* 24:1–163. [334–35]

Solheim, R. 1984. Caching behavior, prey choice and surplus killing by pygmy owls *Glaucidium passerinum* during winter, a functional response of a generalist predator. *Ann. Zool. Fenn.* 21:301–8. [14, 87–88, 90, 288–91]

Sollberger, D. E. 1940. Notes on the life history of the small eastern flying squirrel (*Glaucomys volans volans*). *J. Mamm.* 21:282–93. [249]

Sonerud, G. A., and P. E. Fjeld. 1985. Searching and caching behaviour in hooded crows: An experiment with artificial nests. *Fauna Norv. Ser. C. Cinclus* 8:18–23. [101, 306–7]

Sonnenschein, E., and H.-U. Reyer. 1984. Biology of the slatecoloured boubou and other bush shrikes. *Ostrich* 55:86–96. [317]

Sooter, C. A. 1946. Habits of coyotes in destroying nests and eggs of waterfowl. *J. Wildl. Mgmt.* 10:33–38. [226, 228]

Soper, M. F. 1976. *New Zealand birds.* Christchurch, New Zealand: Whitcoulls. [316]

Sorenson, M. W. 1962. Some aspects of water shrew's behavior. *Am. Midl. Nat.* 68:445–62. [220–21]

Sork, V. L. 1983a. Mammalian seed dispersal of pignut hickory during three fruiting seasons. *Ecology* 64:1049–56. [182]

———. 1983b. Mast-fruiting in hickories and availability of nuts. *Am. Midl. Nat.* 109:81–88. [189, 202]

———. 1984. Examination of seed dispersal and survival in red oak, *Quercus rubra* (Fagaceae), using metal-tagged acorns. *Ecology* 65:1020–22. [183]

Sork, V. L., P. Stacey, and J. E. Averett. 1983. Utilization of red oak acorns in non-bumper crop year. *Oecologia* 59:49–53. [180]

Southern, J. B. 1946. Unusual feeding behaviour of tits. *Br. Birds* 39:214. [314]

Spencer, D. A., and A. L. Spencer. 1941. Food habits of the white-throated wood rat in Arizona. *J. Mamm.* 22:280–84. [262]

Sperber, I., and C. Sperber. 1963. Notes on the food consumption of merlins. *Zool. Bidr. Uppsala* 35:263–68. [288]

Spradbery, J. P. 1973. *Wasps: An account of biology*

and natural history of solitary and social wasps. Seattle: University of Washington Press. [7, 337, 342–44]

Sridhara, S., and K. Srihari. 1980. Food hoarding behavior of *Tatera indica cuvieri* (Waterhouse) in captivity. *Mysore J. Agric. Sci.* 14:78–81. [126, 264–65]

Stacey, P. B., and C. E. Bock. 1978. Social plasticity in the acorn woodpecker. *Science* 202:1298–300. [62, 65, 68–69]

Stacey, P. B., and R. Jansma. 1977. Storage of piñon nuts by the acorn woodpecker in New Mexico. *Wilson Bull.* 89:150–51. [293]

Stacey, P. B., and J. D. Ligon. 1987. Territory quality and dispersal options in the acorn woodpecker, and a challenge to the habitat-saturation model of cooperative breeding. *Am. Nat.* 130:654–76. [21–22, 33–34, 68–69, 297]

Stallcup, P. L. 1968. Spatio-temporal relationships of nuthatches and woodpeckers in ponderosa pine forests in Colorado. *Ecology* 49:831–43. [315]

Stamatopoulos, C. V., and J. C. Ondrias. 1987. Notes on the ingestive behavior of the vole *Pitymys atticus* (Mammalia, Rodentia). *Mammalia* 51:39–41. [50, 269–70]

Stamm, J. S. 1954. Genetics of hoarding. I. Hoarding differences between homozygous strains of rats. *J. Comp. Physiol. Psychol.* 47:157–61. [112, 145]

———. 1956. Genetics of hoarding II. Hoarding behavior of hybrid and backcrossed strains of rats. *J. Comp. Physiol. Psychol.* 49:349–52. [113]

Stapanian, M. A. 1986. Seed dispersal by birds and squirrels in the deciduous forests of the United States. In *Frugivores and seed dispersal,* edited by A. Estrada and T. H. Fleming, 225–36. Dordrecht, The Netherlands: Dr. W. Junk Publishers. [184]

Stapanian, M. A., and C. C. Smith. 1978. A model for seed scatterhoarding: Coevolution of fox squirrels and black walnuts. *Ecology* 59:884–96. [3, 62, 98–99, 101–4, 110, 171, 174, 181, 183–84, 190, 199, 244–45]

———. 1984. Density-dependent survival of scatterhoarded nuts: An experimental approach. *Ecology* 65:1387–96. [100–101, 110, 171, 190, 209–10]

———. 1986. How fox squirrels influence the invasion of prairies by nut-bearing trees. *J. Mamm.* 67:326–32. [91, 100, 184, 198, 209–10]

States, J. B. 1976. Local adaptions in chipmunk (*Eutamias amoenus*) populations and evolutionary potential of species borders. *Ecol. Monogr.* 46:221–56. [24, 238, 240]

Stearns, R. E. C. 1882. The acorn-storing habit of the California woodpecker. *Am. Nat.* 16:353–57. [203]

Steiner, A. L. 1975. "Greeting" behavior in some Sciuridae, from an ontogenetic, evolutionary and socio-behavioral perspective. *Nature, Canada* 102:737–51. [172]

———. 1978. Evolution of prey-carrying mechanisms in digger wasps: Possible role of a functional link between prey-paralyzing and carrying studied in *Oxybelus uniglumis* (Hymenoptera, Sphecidae, Crabroninae). *Quaest. Entomol.* 14:393–409. [58]

Steiniger, F. 1949. Biologische Beobachtungen an freilebenden Wanderratten auf der Hallig Norderoog. *Verhandl. Deutsch. Zool.* pp. 152–56. [274]

Steinmann, E. 1976. Über die Nahorientierung solitäer Hymenopteren: Individwelle Markierung der Nesteingänge. *Mitt. Schweiz. Ent. Ges.* 49:253–58. [150]

———. 1985. Die Wand-Pelzbiene *Anthophora plagiata* (Illiger) (Hymenoptera: Apoidea). *Jber. Natf. Ges. Graubünden* 102:137–42. [152]

Stellar, E. 1943. The effect of epinephrine, insulin, and glucose upon hoarding in rats. *J. Comp. Psychol.* 36:21–32. [122]

———. 1951. The effect of experimental alteration of metabolism on the hoarding behavior of the rat. *J. Comp. Physiol. Psychol.* 44:290–99. [122]

Stellar, E., J. McV. Hunt, H. Schlosberg, and R. L. Solomon. 1952. The effect of illumination on hoarding behavior. *J. Comp. Physiol. Psychol.* 45:504–7. [131]

Stellar, E., and C. T. Morgan. 1943. The roles of experience and deprivation in the onset of hoarding behavior in the rat. *J. Comp. Psychol.* 36:47–55. [120]

Stendell, R., and L. Waian. 1968. Observations on food caching by an adult female sparrow hawk

(*Falco sparverius*). *Condor* 70:187. [12, 47, 285, 287]

Stephen, W. P. 1966. *Andrena (Cryptandrena) viburnella*. I. Bionomics. *J. Kans. Entomol. Soc.* 39:42–51. [347]

Stephen, W. P., G. E. Bohart, and P. F. Torchio. 1969. *The biology and external morphology of bees.* Corvallis, Ore.: Oregon State University, Agricultural Experiment Station. [5–6, 345]

Stephens, D. N. 1982. Hoarding behavior and the defence of body weight in adult rats following early development. *Q. J. Exp. Psychol. Comp. Physiol. Psychol.* 34B:183–94. [117]

Stephens, D. W., and J. R. Krebs. 1986. *Foraging theory.* Princeton, N.J.: Princeton University Press. [101]

Stevens, T. A., and J. R. Krebs. 1986. Retrieval of stored seeds by marsh tits *Parus palustris* in the field. *Ibis* 128:513–25. [12, 18, 312, 314]

Stiles, E. W., and E. T. Dobi. 1987. Scatterhoarding of horsechestnuts by eastern gray squirrels. *Bull. N. J. Acad. Sci.* 32:1–3. [182–83, 185, 244–45]

Stilmark, F. R. 1963. On the ecology of the Siberian chipmunk (*Eutamias sibiricus* Laxm.) in the stonepine forests of western Sayan. *Zool. Zh.* 42:92–102. [238]

Stone, E. R., and M. C. Baker. 1989. The effects of conspecifics on food caching by black-capped chickadees. *Condor* 91:886–90. [94, 141]

Stone, R. C., and C. L. Hayward. 1968. Natural history of the desert woodrat, *Neotoma lepida. Am. Midl. Nat.* 80:458–76. [260–61]

Strassman, J. E. 1979. Honey caches help female paper wasps (*Polistes annularis*) survive Texas winters. *Science* 204:207–9. [23, 34, 42, 67, 344–45]

Stuart, C. T. 1986. The incidence of surplus killing by *Panthera pardus* and *Felis caracal* in Cape Province, South Africa. *Mammalia* 50:556–58. [226, 233, 235]

Stuebe, M. M., and D. C. Andersen. 1985. Nutritional ecology of a fossorial herbivore: protein N and energy value of winter caches made by the northern pocket gopher, *Thomomys talpoides. Can. J. Zool.* 63:1101–5. [20, 41, 139, 251]

Stumper, R. 1961. Radiobiologische Untersuchungen über den sozialen Nahrungshaushalt der

Honigameise *Proformis nasuta* (Nyl). *Naturwissenschaften* 48:735–36. [336]

Sudd, J. H. 1967. *An introduction to the behavior of ants.* New York: Saint Martin's Press. [329]

Sulkava, S., and E. S. Nyholm. 1987. Mushroom stores as winter food of the red squirrel, *Sciurus vulgaris,* in northern Finland. *Aquilo Ser. Zool.* 25:1–8. [83, 246–47]

Sullivan, T. P. 1978. Lack of caching of direct-seeded Douglas fir seeds by deer mice. *Can. J. Zool.* 56:1214–16. [194]

Suthers, R. A. 1978. Sensory ecology of birds. In *Sensory ecology: Review and perspectives,* edited by M. A. Ali, 217–51. New York: Plenum Publishing. [156]

Svärdson, G. 1957. The "invasion" type of bird migration. *Br. Birds* 50:314–43. [65]

Svendsen, G. E. 1982. Weasels. In *Wild mammals of North America: Biology, management, economics,* edited by J. A. Chapman and G. A. Feldhamer, 613–28. Baltimore: Johns Hopkins University Press. [226, 232]

Sviridenko, P. A. 1971. Role of *Sciurus vulgaris* L. in distribution of walnut. *Vest. Zool.* 5:87–88. [185, 199, 244]

Swanberg, P. O. 1951. Food storage, territory and song in the thick-billed nutcracker. *Proc. Xth Intern. Ornithol. Cong.* 10:545–54. [63, 73, 147, 149, 166, 199, 301–5]

———. 1956. Territory in the thick-billed nutcracker *Nucifraga caryocatactes. Ibis* 98:412–19. [199, 301, 303]

———. 1969. Jays recovering buried food from under snow. *Br. Birds* 62:239–40. [299–300]

———. 1981. Clutch size in the thick-billed nutcracker *Nucifraga c. caryocatactes* in Scandinavia in relation to the supply of hazel nuts in the individual winter stores. *Vår Fågelvärld* 40:399–408. [30]

Swarth, H. S. 1919. Some Sierran chipmunks. *Sierra Club Bull.* 10:401–13. [238]

Swenson, J. E., and S. J. Knapp. 1980. Composition of beaver caches on the Tongue River in Montana. *Prairie Nat.* 12:33–36. [107, 258]

Swenson, J. E., S. J. Knapp, P. R. Martin, and T. C. Hinz. 1983. Reliability of aerial cache surveys to monitor beaver (*Castor canadensis*)

population trends on prairie rivers in Montana. *J. Wildl. Mgmt.* 47:697–703. [258]

Sylvester, H. A. 1978. Response of honey bees to different concentrations of sucrose in a hoarding test. *Am. Bee J.* 118:746–47. [139]

———. 1979. Honey bees: Response to galactose and lactose incorporated into sucrose syrup. *J. Econ. Entomol.* 72:81–82. [139]

Tadlock, C. C., and H. G. Klein. 1979. Nesting and food-storage behavior of *Peromyscus maniculatus gracilis* and *P. leucopus noveboracensis. Can. Field-Nat.* 93:239–42. [259]

Taffe, C. A. 1983. The biology of the mud-wasp *Zeta abdominale* (Drury) (Hymenoptera: Eumenidae). *Zool. J. Linn. Soc.* 77:385–93. [344]

Takahashi, L. K., and R. K. Lore. 1980. Foraging and food hoarding of wild *Rattus norvegicus* in an urban environment. *Behav. Neural Biol.* 29:527–31. [111, 144, 274]

Talbot, M. 1943. Population studies of the ant, *Prenolepis imparis* Say. *Ecology* 24:31–44. [335–36]

Tannenbaum, M. G., and E. B. Pivorum. 1984. Differences in daily torpor patterns among three species of *Peromyscus. J. Comp. Physiol. B* 154:233–36. [26]

———. 1987. Variation in hoarding behaviour in southeastern *Peromyscus. Anim. Behav.* 35:297–99. [26, 64, 138, 365]

Tappe, D. T. 1941. Natural history of the Tulare kangaroo rat. *J. Mamm.* 22:117–48. [252–54, 256]

Taylor, W. P., and W. T. Shaw. 1927. *Mammals and birds of Mount Rainier National Park.* Washington, D.C.: U.S. Department of the Interior, National Park Service, Government Printing Office. [236, 281]

Tengö, J., and G. Bergström. 1977. Cleptoparasitism and odor mimetism in bees: Do *Nomada* males imitate the odor of *Andrena* females? *Science* 196:1117–19. [89]

Tepedino, V. J., J. M. Loar, and N. L. Stanton. 1979. Experimental trapnesting: Notes on nest recognition in three species of megachilid bees (Hymenoptera: Megachilidae). *Pan-Pac. Entomol.* 55:195–98. [96, 152]

Tepedino, V. J., and P. F. Torchio. 1982. Phenotypic variability in nesting success among *Osmia lignaria propinqua* females in a glasshouse environment (Hymenoptera: Megachilidae). *Ecol. Entomol.* 7:453–62. [40]

Tevis, L., Jr. 1953. Stomach contents of chipmunks and mantled squirrels in northeastern California. *J. Mamm.* 34:316–24. [194, 202]

———. 1958. Interrelations between the harvester ant *Veromessor pergandei* (Mayr) and some desert ephemerals. *Ecology* 39:695–704. [330, 332]

Thiessen, D., and P. Yahr. 1977. *The gerbil in behavioral investigations.* Austin: University of Texas Press. [126, 264]

Thomas, K. R. 1974. Burrow systems of the eastern chipmunk (*Tamias striatus pipilans* Lowery) in Louisiana. *J. Mamm.* 55:454–59. [237–38]

Thompson, D. C., and P. S. Thompson. 1980. Food habits and caching behavior of urban grey squirrels. *Can. J. Zool.* 58:701–10. [10, 19–20, 89–90, 171, 176, 181, 185, 190, 199–200, 244, 246]

Tigner, J. C., and R. J. Wallace. 1972. Hoarding of food and non-food items in blind, anosmic and intact albino rats. *Physiol. Behav.* 8:943–48. [140]

Tinbergen, N. 1935. Über die Orientierung des Bienenwolfes (*Philanthus triangulum* Fabr.). II. Die Bienenjagd. *Z. Vergl. Physiol.* 21:699–716. [149–50]

———. 1965. Von den Vorratskammern des Rotfuchses (*Vulpes vulpes* L.). *Z. Tierpsychol.* 22:119–49. [100, 171, 174–75, 225–28, 230]

———. 1972. *The animal in its world.* Vol. I Field studies. London: George Allen and Unwin. [149–50, 154, 175, 225–28, 230]

Tinbergen, N., M. Impekoven, and D. Franck. 1967. An experiment on spacing-out as a defence against predation. *Behaviour* 28:307–21. [98]

Toates, F. M. 1978. A circadian rhythm of hoarding in the hamster. *Anim. Behav.* 26:631. [15, 263]

Toland, B. 1984. Unusual predatory and caching behavior of American kestrels in central Missouri. *Raptor Res.* 18:107–10. [284–87]

Tomasi, T. E. 1978. Function of venom in the short-tailed shrew, *Blarina brevicauda. J. Mamm.* 59:852–54. [86, 219]

Tomback, D. F. 1978. Foraging strategies of Clark's

nutcrackers. *Living Bird* 16:123–61. [18, 94, 109, 179, 183–84, 194–96, 301–4]

———. 1980. How nutcrackers find their seed stores. *Condor* 82:10–19. [89, 160, 190]

———. 1982. Dispersal of whitebark pine seeds by Clark's nutcracker: A mutualism hypothesis. *J. Anim. Ecol.* 51:451–67. [18, 179, 181, 301, 303–4]

———. 1983. Nutcrackers and pines: Coevolution or coadaptation? In *Coevolution,* ed. M. H. Nitecki, pp. 179–223. Chicago: University of Chicago Press. [72]

———. 1986. Post-fire regeneration of krummholz whitebark pine: A consequence of nutcracker seed caching. *Madroño* 33:100–110. [210]

Tonkin, J. M. 1983. Activity patterns of the red squirrels (*Sciurus vulgaris*). *Mamm. Rev.* 13:99–111. [199, 244, 246]

Topachevskii, V. A. 1976. *Fauna of the USSR, Mammals: Mole rats, Spalacidae.* New Delhi: Amerind Publishing. [266]

Torchio, P. F. 1965. Observations on the biology of *Colletes ciliatoides* (Hymenoptera: Apoidea, Colletidae). *J. Kans. Entomol. Soc.* 38:182–87. [347]

———. 1970. The ethology of the wasp *Pseudomasaris edwardsii* (Cresson), and a description of its immature forms (Hymenoptera: Vespoidea, Masaridae). *Los Angeles Co. Mus. Contr. Sci.* 202:1–32. [345]

———. 1971. The biology of *Anthophora* (*Micranthophora*) *peritomae* Cockerell (Hymenoptera: Apoidea, Anthophoridae). *Los Angeles Co. Mus. Contr. Sci.* 206:1–14. [353]

———. 1975. The biology of *Perdita nuda* and descriptions of its immature forms and those of its *Sphecodes* parasite (Hymenoptera: Apoidea). *J. Kans. Entomol. Soc.* 48:257–79. [347]

———. 1984. The nesting biology of *Hylaeus bisinuatus* Forster and development of its immature forms (Hymenoptera: Colletidae). *J. Kans. Entomol. Soc.* 57:276–97. [347]

Torchio, P. F., G. E. Trostle, and D. J. Burdick. 1988. The nesting biology of *Colletes kincaidii* Cockerell (Hymenoptera: Colletidae) and development of its immature forms. *Ann. Entomol. Soc. Am.* 81:605–25. [346]

Torchio, P. F., and N. N. Youssef. 1968. The biology of *Anthophora* (*Micranthophora*) *flexipes* and it cleptoparasite, *Zacosmia maculata,* including a description of the immature stages of the parasite (Hymenoptera: Apoidea, Anthophoridae). *J. Kans. Entomol. Soc.* 41:289–302. [89]

Tordoff, H. B. 1955. Food-storing in the sparrow hawk. *Wilson Bull.* 67:139–40. [13, 149, 288]

Townsend, J. E. 1953. Beaver ecology in western Montana with special reference to movements. *J. Mamm.* 34:459–79. [257–58]

Tranquillini, W. 1979. *Physiological ecology of the alpine timberline: Tree existence at high altitudes with special reference to the European Alps.* Berlin: Springer-Verlag. [211]

Trealeaven, R. B. 1980. Observations on the peregrine: One session's hunting and food caching. *Hawk Trust Ann. Rep.* 10:30–31. [285]

Trivers, R. L., and H. Hare. 1976. Haplodiploidy and the evolution of the social insects. *Science* 191:249–63. [40]

Tuck, L. M. 1960. *The murres: Their distribution, populations and biology.* Ottawa: Canadian Wildlife Service. [228]

Tulloch, R. J. 1968. Snowy owls breeding in Shetland in 1967. *Br. Birds* 61:119–32. [289–90]

Turcek, F. J. 1951. On the relation of the jay (*Garrulus glandarius* L.) to oak (*Quercus* sp.). *Lesnick A. Prace* 29:385–96. [200–201, 211]

Turcek, F. J., and L. Kelso. 1968. Ecological aspects of food transportation and storage in the Corvidae. *Comm. Behav. Biol., Part A* 1:277–97. [46, 183, 195]

Turner, B. C. 1959. Feeding behaviour of choughs. *Br. Birds* 52:388–90. [306]

Turner, G. T., R. M. Hanson, V. H. Reid, H. P. Tietjen, and A. L. Ward. 1973. Pocket gophers and Colorado mountain rangeland. *Colo. State Univ. Exp. Stn. Bull.* 5545:1–90. [250]

Vahle, J. R., and D. R. Patton. 1983. Red squirrel cover requirements in Arizona mixed conifer forests. *J. Forest.* 81:14–15, 22. [241]

Van Camp, L. F., and C. J. Henny. 1975. The screech owl: Its life history and population ecology in northern Ohio. *North Am. Fauna* 71:1–65. [31, 49, 288–90]

Vandermeer, J. H. 1979. Hoarding of captive *Heteromys desmarestianus* (Rodentia) on the fruits of *Welfia georgii*, a rainforest dominant palm in Costa Rica. *Brenesia* 16:107–16. [75, 143]

Vander Wall, S. B. 1982. An experimental analysis of cache recovery in Clark's nutcracker. *Anim. Behav.* 30:84–94. [95–96, 102, 106, 119, 148, 155–59, 162–63, 165, 169, 173, 175–76, 186]

———. 1988. Foraging of Clark's nutcrackers on rapidly changing pine seed resources. *Condor* 90:621–31. [18, 94, 142, 181, 183, 189–90, 301, 304]

Vander Wall, S. B., and R. P. Balda. 1977. Coadaptations of the Clark's nutcracker and the piñon pine for efficient seed harvest and dispersal. *Ecol. Monogr.* 47:89–111. [3, 18, 59, 67, 71–73, 83, 94, 110, 179–83, 192, 194–96, 209, 301–4, 364]

———. 1981. Ecology and evolution of food-storage behavior in conifer-seed-caching corvids. *Z. Tierpsychol.* 56:217–42. [2, 4, 18, 28, 55, 57, 59–60, 63, 66, 72, 77, 79, 81, 108–9, 170, 181, 183–84, 198, 216, 299, 302–4]

———. 1983. Remembrances of seeds stashed. *Nat. Hist.* 92:60–65. [304]

Vander Wall, S. B., S. W. Hoffman, and W. K. Potts. 1981. Emigration behavior of Clark's nutcracker. *Condor* 83:162–70. [65–66, 301, 363]

Vander Wall, S. B., and H. E. Hutchins. 1983. Dependence of Clark's nutcracker, *Nucifraga columbiana*, on conifer seeds during the postfledging period. *Can. Field-Nat.* 97:208–14. [18, 32, 79, 91, 106, 155, 158, 162, 164, 192, 216, 303, 305, 367]

Vander Wall, S. B., and K. G. Smith. 1987. Cache-protecting behavior of food-hoarding animals. In *Foraging behavior*, ed. A. C. Kamil, J. R. Krebs, and R. H. Pulliam, pp. 611–44. New York: Plenum Press. [1, 6, 82, 209]

Van Devender, T. R., and W. G. Spaulding. 1979. Development of vegetation and climate in the southwestern United States. *Science* 204:701–10. [212, 215]

van Tyne, J., and A. J. Berger. 1959. *Fundamentals of ornithology.* New York: Dover Publishers. [307]

Vaughn, R. 1961. *Falco eleonorae. Ibis* 103:114–28. [285, 287]

Vaughan, T. A. 1967. Food habits of the northern pocket gopher on short-grass prairie. *Am. Midl. Nat.* 77:176–89. [250]

Vestal, E. H. 1938. Biotic relations of the wood rat (*Neotoma fuscipes*) in the Berkeley Hills. *J. Mamm.* 19:1–36. [260, 262]

Victor, R., and J. C. Wigwe. 1989. Hoarding—a predatory behaviour of *Sphaerodema nepoides* Fabricius (Heteroptera: Belostomatidae). *Arch. Hydrobiol.* 116:107–11. [320]

Viek, P., and G. A. Miller. 1944. The cage as a factor in hoarding. *J. Comp. Psychol.* 37:203–10. [137]

Vinson, S. B., G. W. Frankie, and R. E. Coville. 1987. Nesting habits of *Centris flavofasciata* Friese (Hymenoptera: Apoidea: Anthophoridae) in Costa Rica. *J. Kans. Entomol. Soc.* 60:249–63. [89, 96, 353–54]

Visscher, P. K., and T. D. Seeley. 1982. Foraging strategy of honeybee colonies in temperate deciduous forest. *Ecology* 63:1790–801. [360]

Volker, V., and W. Rudat. 1978. On the postfledging behaviour of nutcrackers (*Nucifraga caryocatactes* L.). *Zool. Jb. Syst. Bd.* 105:386–88. [32, 199, 301, 305]

Vollrath, F. 1979. Behaviour of the kleptoparasitic spider *Argyrodes elevatus* (Araneae, Theridiidae). *Anim. Behav.* 27:515–21. [2, 320, 322]

Voorhies, M. R. 1974. Fossil pocket mice burrows in Nebraska. *Am. Midl. Nat.* 91:492–98. [61, 86]

Vorhies, C. T., and W. P. Taylor. 1922. Life history of the kangaroo rat, *Dipodomys spectabilis spectabilis* Merriam. *USDA Bull.* 1091:1–40. [3, 25, 252–54, 256]

———. 1940. Life history and ecology of the white-throated wood rat, *Neotoma albigula albigula* Hartley, in relation to grazing in Arizona. *Univ. Ariz. Agric. Exp. Stn. Tech. Bull.* 86:455–529. [262]

Voth, E. H. 1968. Food habits of the Pacific mountain beaver, *Aplodontia rufa pacifica* Merriam. Ph. D. diss., Oregon State University, Corvallis, p. 263. [236]

Waddell, D. 1951. Hoarding behavior in the golden hamster. *J. Comp. Physiol. Psychol.* 44:383–88. [131–32]

Waite, R. K. 1985. Food caching and recovery by farmland corvids. *Bird Study* 32:45–49. [149, 156, 307]

———. 1986. Carrion crow recovering bread from beneath snow. *Br. Birds* 79:659–60. [166, 307]

Waite, T. A. 1988. A field test of density dependent survival of simulated gray jay caches. *Condor* 90:247–49. [100, 298]

Waite, T. A., and T. C. Grubb, Jr. 1988. Diurnal caching rhythm in captive white-breasted nuthatches (*Sitta carolinensis*). *Ornis Scand.* 19:68–70. [13–14, 26, 315–16]

Walker, A. 1923. A note on the winter habits of *Eutamias townsendi*. *J. Mamm.* 4:257. [237]

Wallace, G. J. 1948. The barn owl in Michigan: Its distribution, natural history, and habits. *Mich. State Coll. Agric. Exp. Tech. Bull.* 208:1–61. [12, 49, 289–90]

———. 1963. *An introduction to ornithology.* New York: Macmillan. [316]

Wallace, R. J. 1976. Tail-hoarding in the albino rat. *Anim. Behav.* 24:176–80. [140]

———. 1978. Hoarding of inedible objects by albino rats. *Behav. Biol.* 23:409–14. [140]

———. 1979. Novelty and partibility as determinants of hoarding in the albino rat. *Anim. Learn. Behav.* 7:549–54. [140]

———. 1982. Studies of object retrieval by Australian bush rats (*Rattus fuscipes* Waterhouse). *Z. Tierpsychol.* 59:141–56. [274]

———. 1983. Saccharin hoarding by albino rats: Further evidence on incentive and object retrieval. *J. Genet. Psychol.* 108:211–24. [122, 140]

———. 1984. A sequential analysis of saccharin and water hoarding by albino rats. *J. Genet. Psychol.* 111:241–52. [122]

———. 1988. The classification of burrows of Norway rats (*Rattus norvegicus* Berkenhout) and its relation to food hoarding. *Mammalia* 52:45–49. [111, 274]

Waller, D. M. 1979. Models of mast fruiting in trees. *J. Theor. Biol.* 80:223–32. [202]

Walter, H. 1979. *Eleonora's falcon: Adaptations to prey and habitat in a social raptor.* Chicago: University of Chicago Press. [49, 88, 97, 285]

Walters, I. N. 1979. Torresian crow burying and hiding food. *Sunbird* 10:23–24. [305–6]

———. 1980. Caching of food by grey butcherbirds. *Sunbird* 11:47–48. [307]

Walton, M. A. 1903. *A hermit's wild friends.* Boston: C. H. Simonds. [192]

Warkentin, I. G., and L. W. Oliphant. 1985. Observations on winter food caching by the Richardson's merlin. *Raptor Res.* 19:100–101. [13, 285–86, 288]

Wästljung, U. 1989. Effects of crop size and stand size on seed removal by vertebrates in hazel *Corylus avellana. Oikos* 54:178–84. [199, 202]

Watt, A. S. 1919. On the causes of failure of natural regeneration in British oakwoods. *J. Ecol.* 7:173–203. [186]

———. 1923. On the ecology of British beech-woods with special reference to their regeneration. *J. Ecol.* 7:1–48. [184–85, 187, 192]

Wcislo, W. T. 1984. Gregarious nesting of a digger wasp as a "selfish herd" response to a parasitic fly (Hymenoptera: Sphecidae; Diptera: Sarcophagidae). *Behav. Ecol. Sociobiol.* 15:157–60. [97]

Webb, S. L. 1986. Potential role of passenger pigeons and other vertebrates in rapid Holocene migrations of nut trees. *Quaternary Res.* 26:367–75. [214]

Webb, S. L. 1987. Beech range extension and vegetation history: Pollen stratigraphy of two Wisconsin lakes. *Ecology* 68:1993–2005. [212–13]

Wehner, R. 1981. Spatial vision in arthropods. In *Handbook in sensory physiology.* VII/6C, edited by H. Autrum, 287–616. Berlin: Springer-Verlag. [150, 152–54, 175]

Weininger, O. 1953. The performance of white rats as a function of dominance and accumulating activity. *J. Comp. Physiol. Psychol.* 46:200–203. [141]

Wells, P. H., and J. Giacchino, Jr. 1968. Relationship between the volume and the sugar concentration of loads carried by honeybees. *J. Apic Res.* 7:77–82. [58]

Wells, P. V. 1983. Paleobiogeography of montane islands in the Great Basin since the last glacio-pluvial. *Ecol. Monogr.* 53:341–82. [212, 215]

Wemmer, C. 1969. Impaling behavior of the log-gerhead shrike, *Lanius ludovicianus* Linnaeus. *Z. Tierpsychol.* 26:208–24. [115–16, 118–20, 317, 319]

West, N. E. 1968. Rodent-influenced establishment of ponderosa pine and bitterbrush seedlings in central Oregon. *Ecology* 49:1009–11. [106, 158, 192–94, 202, 204, 206–7]

Westcott, P. W. 1964. Invasion of Clark nutcrackers and piñon jays into southeastern Arizona. *Condor* 66:441. [66]

———. 1969. Relationships among three species of jays wintering in southeastern Arizona. *Condor* 71:353–59. [66]

Wetherill, C. M. 1852. Chemical investigation of the Mexican honey art. *Proc. Natl. Acad. Sci. Phil.* 6:111–13. [335]

Wheeler, W. M. 1908. Honey ants, with a revision of the American Myrmecocysti. *Am. Mus. Nat. Hist. Bull.* 24:345–97. [50, 333–35]

———. 1910. *Ants: Their structure, development, and behavior.* New York: Columbia University Press. [33, 179, 329–30, 333–35]

———. 1930. The ant *Prenolepis imparis* Say. *Ann. Entomol. Soc. Am.* 23:1–26. [335]

———. 1933. *Colony-founding among ants: With an account of some primitive Australian species.* Cambridge, Mass.: Harvard University Press. [87, 336]

Whisaw, I. Q., and J. A. Tomie. 1989. Food-pellet size modifies the hoarding behavior of foraging rats. *Psychobiology* 17:93–101. [139]

White, E. 1962. Nest building and provisioning in relation to sex in *Sceliphron spirifex* L. (Sphecidae). *J. Anim. Ecol.* 31:317–29. [38–39]

White, J. W., Jr. 1966. Inhibine and glucose oxidase in honey—A review. *Am. Bee J.* 106:214–16. [88]

———. 1975. Honey. In *The hive and the honey bee,* edited by Dadant and Sons, 491–530. Hamilton, Ill.: Dadant and Sons. [88, 361]

Wicklow, D. T., R. Kumar, and J. E. Lloyd. 1984. Germination of blue grama seeds buried by dung beetles (Coleoptera: Scarabaeidae). *Envir. Entomol.* 13:878–81. [180]

Wicklow, D. T., and J. C. Zak. 1983. Viable grass seeds in herbivore dung from a semi-arid grassland. *Grass Forage Sci.* 38:25–26. [180]

Wightman, J. A., and V. M. Rogers. 1978. Growth, energy and nitrogen budgets and efficiencies of the growing larvae of *Megachile pacifica* (Panzer) (Hymenoptera: Megachilidae). *Oecologia* 36:245–57. [37–38]

Willard, J. R., and H. H. Crowell. 1965. Biological activities of the harvester ant, *Pogonomyrmex owyheei,* in central Oregon. *J. Econ. Entomol.* 58:484–89. [10, 21]

Willie, A. 1983. Biology of the stingless bees. *Ann. Rev. Entomol.* 28:41–64. [89, 355–56]

Willgohs, J. F. 1974. The eagle owl *Bubo bubo* (L.) in Norway. I. Food ecology. *Sterna* 13:129–77. [289]

Williams, F. X. 1919. *Epyris extraneus* Bridwell (Bethylidae), a fossorial wasp that preys on the larva of the tenebrionid beetle *Gonocephalum seriatum* (Boisduval). *Proc. Hawaii Entomol. Soc.* 4:55–63. [51]

Williams, H. J., M. R. Strand, G. W. Elzen, S. B. Vinson, and S. J. Merritt. 1986. Nesting behavior, nest architecture, and use of Dufour's gland lipids in nest provisioning by *Megachile integra* and *M. mendica mendica* (Hymenoptera: Megachilidae). *J. Kans. Entomol. Soc.* 59:588–97. [351]

Willmer, P. G. 1985. Thermal ecology, size effects, and the origins of communal behaviour in *Cerceris* wasps. *Behav. Ecol. Sociobiol.* 17–151–60. [40]

Wilson, D. S. 1985. Adaptive indirect effects. In *Community ecology,* edited by J. Diamond and T. Case, 437–44. New York: Harper and Row. [96, 327–28]

Wilson, D. S., and J. Fudge. 1984. Burying beetles: Intraspecific interactions and reproductive success in the field. *Ecol. Entomol.* 9:195–203. [126, 327–28, 366]

Wilson, E. O. 1971. *The insect societies.* Cambridge, Mass.: Belknap Press of Harvard University Press. [52–53, 359]

Wilsson, L. 1971. Observations and experiments on the ethology of the European beaver (*Castor fiber* L.). *Viltrevy* 8:115–260. [258]

Wingfield, J. C., G. F. Ball, A. M. Dufty, Jr., R. E. Hegner, and M. Ramenofsky. 1987. Testosterone and aggression in birds. *Am. Sci.* 75:602–8. [27]

Wolfe, J. B. 1939. An exploratory study of food-storing in rats. *J. Comp. Physiol. Psychol.* 28: 97–108. [111, 120, 125]

Wolff, J. O. 1984. Overwintering behavioral strategies in taiga voles (*Microtus xanthognathus*). In *Winter ecology of small mammals,* edited by J. F. Merritt, 315–18. Pittsburgh: Carnegie Museum of Natural History. [20]

Wolff, J. O., and G. C. Bateman. 1978. Effects of food availability and ambient temperature on torpor cycles of *Perognathus flavus* (Heteromyidae). *J. Mamm.* 59: 707–16. [15, 25, 42, 256]

Wolff, J. O., and W. Z. Lidicker, Jr. 1981. Communal winter nesting and food sharing in taiga voles. *Behav. Ecol. Sociobiol.* 9: 237–40. [26, 67–68, 79, 270–71]

Wong, M. 1989. The implications of germinating acorns in the granaries of acorn woodpeckers in Panama. *Condor* 91: 724–26. [105, 179, 295]

Wong, R. 1984. Hoarding versus the immediate consumption of food among hamsters and gerbils. *Behav. Proc.* 9: 3–11. [121]

Wong, R., and C. H. Jones. 1985. A comparative analysis of feeding and hoarding in hamsters and gerbils. *Behav. Proc.* 11: 301–8. [125–26]

Wood, O. M. 1938. Seedling reproduction of oak in southern New Jersey. *Ecology* 19: 276–93. [76, 106, 187, 191]

Woodmansee, R. G. 1977. Clusters of limber pine trees: A hypothesis of plant-animal coaction. *Southw. Nat.* 21: 511–17. [193, 206–9, 211]

Woodroof, J. G. 1979. *Tree nuts: Production, processing, products,* 2d ed. Westport, Conn.: AVI Publishing. [201]

Woods, C. A., and D. K. Boraker. 1975. *Octodon degus. Mamm. Species* 67: 1–5. [279]

Woods, K. D., and M. B. Davis. 1989. Paleoecology of range limits: Beech in the upper peninsula of Michigan. *Ecology* 70: 681–96. [213]

Wray, D. L. 1938. Notes on the southern harvester ant (*Pogonomyrmex badius* Latr.) in North Carolina. *Ann. Entomol. Soc. Am.* 31: 196–201. [330–31]

Wrazen, J. A. 1980. Late summer activity changes in populations of eastern chipmunks (*Tamias striatus*). *Can. Field-Nat.* 94: 305–10. [240]

Wrazen, J. A., and L. A. Wrazen. 1982. Hoarding,

body mass dynamics, and torpor as components of the survival strategy of the eastern chipmunk. *J. Mamm.* 63: 63–72. [24, 42, 89, 145, 240]

Wright, G. M. 1934. Cougar surprised at well-stocked larder. *J. Mamm.* 15: 321. [48, 226, 233–34]

Wyatt, T. D. 1986. How a subsocial intertidal beetle, *Bledius spectabilis,* prevents flooding and anoxia in its burrow. *Behav. Ecol. Sociobiol.* 19: 323–31. [328]

Wyman, J. 1967. The jackals of the Serengeti. *Animals* 10: 79–83. [226–28, 230]

Yabe, T. 1981. Hoarding of peanuts in fields by the Norway rat (*Rattus norvegicus*). *J. Mamm. Soc. Japan* 8: 201–2. [111, 144, 274]

Yahner, R. H. 1975. The adaptive significance of scatter hoarding in the eastern chipmunk. *Ohio J. Sci.* 75: 176–77. [62, 238–40]

Yeager, L. E. 1937. Cone-piling by Michigan red squirrels. *J. Mamm.* 18: 191–94. [19, 241–42]
———. 1943. The storing of muskrat and other foods by minks. *J. Mamm.* 24: 100–101. [226, 232–33]

Yom-Tov, Y., and R. Hilborn. 1981. Energetic constraints on clutch size and time of breeding in temperate zone birds. *Oecologia* 48: 234–43. [27]

Yosef, R., and B. Pinshow. 1989. Cache size in shrikes influences female mate choice and reproductive success. *Auk* 106: 418–21. [29–30, 317–18]

Young, S., and E. Goldman. 1946. *The puma, mysterious American cat.* New York: Dover Publications. [233–34]

Youngman, P. M. 1975. *Mammals of the Yukon Territory.* Ottawa: National Museum of Natural Sciences. [270]

Yudin, B. S. 1972. Storing of earthworms by Siberian mole is one of the adaptation to the life under Siberian climatic conditions. *Izv. Sib. Otdel Akad. Nauk SSSR (Ser. Biol. Med. Nauk)* 1972: 133–37. [223–24]

Zucchi, R., S. F. Sakagami, J. M. F. de Camargo. 1969. Biological observations on a Neotropical parasocial bee, *Eulaema nigrita,* with a review on the biology of Euglossinae (Hymenoptera, Apidae). A comparative study. *J. Fac. Sci. Hokkaido Univ. Ser. 6 Zool.* 17:271–379. [359]

Zucker, I., and M. Boshes. 1982. Cirannual body weight rhythms of ground squirrels: Role of gonadal hormones. *Am. J. Physiol.* 243:546–51. [130]

Zusi, R. L. 1987. A feeding adaptation of the jaw articulation in New World jays (Corvidae). *Auk* 104:665–80. [79]